Analog and Switching
Circuit Design

Analog and Switching Circuit Design

Using Integrated and Discrete Devices

Second Edition

J. WATSON
University of Wales
Swansea, Wales, U.K.

WILEY

A Wiley-Interscience Publication

JOHN WILEY & SONS

New York • Chichester • Brisbane • Toronto • Singapore

Library of Congress Cataloging in Publication Data:

Watson, J. (Joseph)
    Analog and switching circuit design: using integrated and
discrete devices / J. Watson. -- 2nd ed.
        p.   cm.
    "A Wiley-Interscience publication."
    Bibliography: p.
    Includes index.

    1. Switching circuits. 2. Analog electronic systems. 3. Solid
state electronics. I. Title.
TK7868.S9W33   1989
621.3815'37--dc19   89-5426
                                                       CIP

Printed in the Republic of Singapore

10 9 8 7 6 5 4 3 2 1

Preface

Electronics is a subject of enormous diversity and depth, for which reason attempts have been made to compartmentalize the various fields of interest, whilst nevertheless recognizing the major overlaps which must occur. For example, three such divisions have been microelectronic fabrication, computer technology and communications, which have now overlapped to result in the modern major area of information technology (IT). Another form of division has been between analog and digital electronic circuit design, and these areas have now overlapped in integrated circuits which support both technologies, such as analog–digital and digital–analog converters, multiplexers, switched-capacitor filters and certain microprocessors.

At the same time, mathematical techniques have been developed[1] to describe and analyze the various functions which have been made possible by the application of modern electronic technology, particularly in the area of signals. Here, Fourier transform methods for signal analysis have been augmented by Z-transform techniques[2], which are relevant to sampled-data systems; and the widespread availability of cheap computing power has made possible such departures as the use of the fast Fourier transform (FFT) for the implementation of signal processing[3].

The interdependence of all these developments in both technology and analytical techniques can be observed when complete systems are considered. For example, figure P.1 shows a block diagram for an instrumentation system involving signal acquisition, processing and conversion, data manipulation and finally utilization. A case in point might be a chemical production line where a number of measurements must be made of such quantities as temperature, pressure, humidity and pH. Various forms of transducer (including semiconductor types) are used for this purpose and are represented by the blocks on the left of figure P.1. The variables are normally continuous and the transducers respond to them in various known ways which result in output signals of diverse amplitude and impedance levels. So, it is usually necessary to employ analog techniques of signal condition-

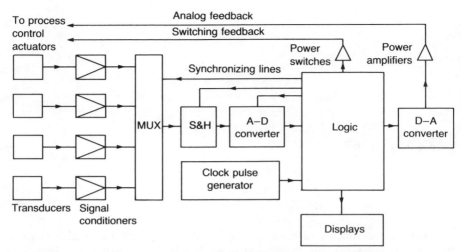

Figure P.1. A process-control instrumentation system block diagram.

ing, many of which have descended from the analog computer developments of earlier decades. The signal conditioning blocks are shown next in figure P. 1.

It is now possible to sample sequentially the outputs of the bank of signal conditioners, and this is done with a multiplexer (MUX) which is essentially a solid-state switch which applies each signal to a single line in turn. In figure P.1, the MUX is shown feeding a sample-and-hold module which, as its name suggests, samples the signal appearing at the MUX output (after the inevitable switching transient is over) and holds it long enough for further processing.

In this system, analog-to-digital (A–D) conversion follows[4], after which the digitized data can be processed in many ways. For example, though the analog signal may be filtered (if necessary) using analog techniques within the signal-conditioning module, such filtering may also be accomplished digitally after the A–D converter.

The information acquired by the transducers is now in the form of digital data which may be stored and combined according to the programming of the logic unit also shown in figure P.1. This logic may take several forms, from hard-wired combinatorial and sequential logic circuits[5] (often involving custom-interconnected uncommitted logic arrays, or ULAs in monolithic microcircuit form), to micro- or mainframe computers[6].

Outputs to actuators may be in the form of ON–OFF signals which can be interfaced by power transistors to operate solenoids; or as sequences of pulses to drive stepper motors; or they may appear again in analog form via D–A converters[4].

Finally, a whole family of solid-state and other displays may be operated via appropriate microcircuit interfaces and drivers.

This system is but one example of the complexity and breadth of modern electronics, and serves to show that to gain an understanding of it implies that a number of texts must be studied in parallel, for no one book of reasonable size can treat more than a small part of the subject in other than a trivial manner. The aim of the present volume is to introduce a selection of contemporary semiconductor devices and integrated circuit structures and, through an understanding of their operation and characteristics, establish the basic tenets of analog and low-power switching design which will lead to the realization of useful and economical circuits.

Parallel reading should therefore include material on digital logic[5], in which there is already a plethora (perhaps even a surfeit) of excellent works, and also on high-frequency topics such as are necessary in transmitter-receiver design[7]. A natural follow-up would then be the immensely important field of computer-aided-design, or CAD. Computer techniques, programs (such as SPICE) and models[8] are invaluable for optimizing a design, determining sensitivity to parameter changes, taking account of tolerances and laying out interconnection schemes for complex networks from printed boards to very large scale integrated (VLSI)[9] circuits. But before such powerful tools can be used, the basic circuit configuration must have been created by a designer, who must therefore know what devices and integrated structures are appropriate and how they may be combined to perform the tasks in hand. It is hoped that the present book will form an introduction to this latter body of knowledge insofar as analog and low-power switching circuits are concerned, and will be useful in the later years of university courses as a practical complement to analytical material[†], and so that the necessarily heavy emphasis on digital methods does not lead to the obscuration of semiconductor devices and structures other than gates and memory cells. It is for these reasons that the mathematical level is limited to presuppose only a knowledge of elementary calculus; and given a basic prerequisite of circuit theory, the book will be found to be complete within itself.

Chapter 1 offers an introduction to some of the most common semiconductor devices from a circuit-oriented viewpoint and chapter 2 introduces the solid-state physics which explains their electrical characteristics. This has been done at a very elementary level, and largely qualitatively; other texts should be consulted when a deeper understanding of semiconductor physics and technology becomes necessary[10, 11]. Chapter 2 concludes with further notes on transistor characterization and introduces the concept of the *equivalent circuit* and *modeling*.

Chapter 3 shows how incremental changes in transistor operating conditions can lead to parameters which describe the *small-signal performance* of amplifier stages and chapter 4 establishes methods of *biasing* which place

---

[†]As Horowitz and Hill point out in their preface to *The Art of Electronics*, 'there is a tendency among textbook writers to represent the theory, rather than the art, of electronics'.

and maintain a transistor into an operating condition that makes these small-signal parameters valid.

Chapter 5 extends the work on amplifier stages to include frequency response and feedback, and then goes on to introduce the *difference stage*, initially as a DC amplifier. The major advantages of monolithic integrated circuit realization of such stages are presented, and new basic circuit elements are introduced which depend upon the close matching made possible by IC techniques, such as the *current mirror*. The chapter concludes by combining the difference stage and current mirror to demonstrate how the monolithic transconductance multiplier becomes possible.

Chapter 6 (which may be omitted on a first reading) is concerned solely with the question of electrical noise, and presents concepts that are used in the design of low-noise amplifiers, which are also met in later chapters.

Chapter 7 is entirely devoted to integrated circuits, mainly in their analog manifestations. It has been rightly pointed out[†] that "To make the most of a chip, you have to understand what goes on inside, particularly with circuits which are very flexible—they're not just black boxes," and this sentiment is followed throughout the chapter insofar as bipolar structures are concerned.

Chapter 8 introduces frequency-dependent circuits from a hardware-oriented viewpoint, and it is here that further parallel reading is necessary, for topics relating to filters, oscillators, and waveform generators require not only specialized treatment, but an understanding of mathematical techniques outside the scope of this volume[1].

The design of power amplifiers is treated in chapter 9 and switching topics (other than logic) are presented in chapter 10 for bipolar structures.

Discrete and IC field-effect devices are described in chapter 11, and comparisons between bipolar, BIFET and BIMOS operational amplifiers are highlighted. The chapter concludes with introductions to modern insulated-gate developments including charge-coupled-devices (CCDs) and power-FETS.

The family of switching devices from unijunction transistors to triacs is introduced in chapter 12 and some of the increasingly important range of semiconductor transducers are described in chapter 13.

Chapter 14 is devoted to mains-driven power supplies, both linear and switching; and also includes a section on battery technology, for a great deal of modern electronic equipment is now fully portable.

Finally, chapter 15 introduces the concept of computer-aided design (CAD), and shows how application-specific integrated circuits (ASICs) may now be designed and produced, even if only a low level of computing power is available. It continues with a brief treatment of the relationship between analog and digital electronics, which affords a natural return to the type of system exemplified by figure P.1; and concludes with a brief introduction to the analog-to-digital and digital-to-analog converters which make such interfacing possible.

[†] By Paul Brokaw, quoted in *EDN*, May 20th, 1979.

A series of questions will be found at the end of each chapter, along with a set of relevant numerical problems. The book concludes with appendices which amplify certain points in the text.

REFERENCES

1. Bird, G. J. A., 1980, *Design of Continuous and Digital Electronic Systems* (New York: McGraw-Hill).
2. Lynn, P. A., 1982, *Electronic Signals and Systems* (London: Macmillan).
3. Brigham, E., 1974, *The Fast Fourier Transform* (Englewood Cliffs, NJ: Prentice Hall).
4. Seitzer, D., Pretzl, G. and Hamdy, N. A., 1983, *Electronic Analog-to-Digital Converters* (New York: John Wiley).
5. Lee, S. C., 1976, *Digital Circuits and Logic Design* (Englewood Cliffs, NJ: Prentice Hall).
6. Sloan, M., 1980, *Introduction to Minicomputers and Microcomputers* (New York: Addison-Wesley).
7. Roddy, D. and Coolen, J., 1981, *Electronic Communications* (Reston, VA: Reston).
8. Getreu, I., 1976, *Modelling the Bipolar Transistor* (Tektronix).
9. Mead, C. and Conway, L., 1980, *An Introduction to VLSI Systems* (New York: Addison-Wesley).
10. Bar-Lev, A., 1984, *Semiconductors and Electronic Devices*, Ed. II (Englewood Cliffs, NJ: Prentice Hall).
11. Morgan, D. V. and Board, K., 1983, *An Introduction to Semiconductor Microtechnology* (New York: John Wiley).

J. WATSON

Swansea, U.K.
July 1989

Contents

Symbols

A	Voltage gain of a "perfect" amplifier	$C_{b'c}$	Base–collector capacitance ⎫ hybrid-π
A_i	Current gain	$C_{b'e}$	Base–emitter capacitance ⎬ para-meters
$A_{i(FB)}$	Current gain with feedback	C_{ds}	Drain–source capacitance ⎭
A_{ov}	Voltage gain measured from a Thévenin input generator	C_{gd} (or C_{rss})	Gate–drain capacitance
$A_{ov(FB)}$	Voltage gain, with feedback, measured from a Thévenin input generator	C_{gs}	Gate–source capacitance
		C_{iss} (or C_{gss})	Input capacitance of FET with output short-circuited to AC (CS mode)
A_v	Terminal voltage gain	C_{oss} (or C_{dss})	Output capacitance of FET with input short-circuited to AC (CS mode)
$A_{v(CM)}$	Common-mode voltage gain		
$A_{v(FB)}$	Terminal voltage gain with feedback	CMR	Common mode rejection ratio
B	Feedback fraction	D^*	Detectivity
B	Flux density		
B_{sat}	Saturation flux density	E_e	Irradiance
		E_{os}	Direct component of offset voltage
b	Transport factor, device constant	E_v	Illuminance
		e	Voltage generated by a "perfect" voltage source
C_C	Coupling capacitor		
C_E	Emitter bypass capacitor	e_{os}	Total offset voltage

F_T	Gain–bandwidth product	I_{CBO}	Collector–base leakage current with emitter open-circuited
f	Frequency		
f_H	High-frequency cut-off point	I_{CEO}	Collector–emitter leakage current with base open-circuited
f_L	Low-frequency cut-off point		
f_m	A frequency near the middle of a pass-band	I_D	Direct drain current
		I_{DSS}	Drain current when $V_{GS} = 0$
f_1	Frequency at which the gain is unity	I_E	Direct emitter current
Δf	Equivalent noise bandwidth	I_G	Direct gate current
		I_P	Photo-current
g_{fs}	Transconductance of a FET	I_Q	Quiescent collector (or drain) current
g_{fso}	g_{fs} when $V_{GS} = 0$	I_Z	Current through Zener diode
g_m	Transconductance of a bipolar transistor	I_o	Reverse leakage current of a diode
H	Magnetizing force	i_b	Incremental base current
h_{FB}	Common-base direct current gain	i_b^-, i_b^+	Bias currents at inverting and noninverting inputs, respectively
h_{FC}	Common-collector direct current gain		
h_{FE}	Common-emitter direct current gain	i_c	Incremental collector current
$h_{FE(O)}$	CE current gain immediately available	i_d	Incremental drain current
		i_d	Offset current ($i_b^- - i_b^+$)
h_{fb}	Common-base incremental current gain	i_e	Incremental emitter current
h_{fc}	Common-collector incremental current gain	i_f	Incremental feedback current
		i_{in}	Incremental input current
h_{fe}	Common-emitter incremental current gain	i_s	Incremental source current
$\left.\begin{array}{l}h_f\\h_i\\h_o\\h_r\end{array}\right\}$	h-parameters		
		K	Bias stability factor $\partial I_C/\partial I_{CEO}$
		k	Boltzmann's constant
I_B	Direct base current		
I_C	Direct collector current	L_v	Luminance

M_v	Luminous exitance	R_g	Generator internal resistance
NEP	Noise equivalent power	R_{in}	Input resistance of active device
NF	Noise factor	$R_{in(CM)}^-$ ⎤	Common-mode resistances at inverting and noninverting
P_C	Power dissipated at collector junction		
P_L	Full load power	$R_{in(CM)}^+$ ⎦	inputs, respectively
P_S	Power extracted from supply at full load	$R_{in(FB)}$	Input resistance with feedback
P_{diss}	Power dissipated within a transistor	R_{ni}	Noise current resistance
$P_{diss(FL)}$	Power dissipated under conditions of maximum sinusoidal signal	R_{nv}	Noise voltage resistance
		R_{out}	Output resistance of active device
$P_{diss(m)}$	Maximum power dissipation possible	$R_{out(FB)}$	Output resistance with feedback
$P_{diss(Q)}$	Power dissipation under quiescent (no signal) conditions	r_{DS}	DC or chord channel resistance of a FET
		$r_{DS(ON)}$	r_{DS} when FET or IGFET is fully ON.
P_{tot}	Maximum allowed power dissipation within a transistor	r_Z	Incremental resistance of a Zener diode
Q	Selectivity	$r_{bb'}$ ⎤	
q	Charge on an electron	$r_{b'c}$	
		$r_{b'e}$ ⎬ Hybrid-π parameters	
R_B	Base resistor	r_{ce} ⎦	
R_{BP}	Parallel combination of R_{in} and bias resistors	r_{ds}	Incremental channel resistance of a FET
R_C	Collector resistor	S	Bias stability factor $\partial I_C / \partial I_{CBO}$
R_{CS} (or R_{CES})	Collector saturation resistance	S_L	Load stability factor
R_E	Emitter resistor	S_T	Temperature stability factor
R_F	Feedback resistor		
R_G	Total resistance external to an input point	S_V	Transfer stabilization factor
R_L	Load resistance	T	Temperature
R_D	Drain resistor for a FET	T_{amb}	Ambient temperature
		T_j	Junction temperature
R_S	Source resistor for a FET	t_{OFF}	OFF-time
		t_{ON}	ON-time
R_X	Long-tail resistor	t_d	Delay time

S_L, S_T, S_V — for power supplies

t_f	Fall time	V_Z	Zener voltage
t_r	Rise time	V_c	Control voltage for chopper or analog gate
t_s	Storage time		
V_B	Bias voltage at base or gate w.r.t. common line	v_{be}	Incremental base–emitter voltage
V_{BB}	Voltage across the "bases" of a unijunction transistor	v_{cb}	Incremental collector–base voltage
V_{BE}	Direct base–emitter voltage	v_{ce}	Incremental collector–emitter voltage
V_C	Voltage at collector w.r.t. common line	v_{ds}	Incremental drain–source voltage
V_{CB}	Direct collector–base voltage	v_{eb}	Incremental emitter–base voltage
V_{CC}, V_{DD} or V_{EE}	Direct voltage supply	v_{gd}	Incremental gate–drain voltage
V_{CE}	Direct collector–emitter voltage	v_{gs}	Incremental gate–source voltage
$V_{CE(m)}$	Maximum designed collector–emitter voltage	v_{in}	Incremental input voltage
V_{CEM}	Maximum allowed collector–emitter voltage	$v_{in(CM)}$	Common-mode input voltage
V_{CES}	Collector–emitter saturation voltage	$v_{in(ECM)}$	Common-mode error voltage
V_{DS}	Direct drain–source voltage	v_{no}	Incremental noise output voltage
V_E	Voltage at emitter w.r.t. common line	v_{ni}	Incremental noise input voltage
V_{EB}	Direct emitter–base voltage	v_{out}	Incremental output voltage
V_{GD}	Direct gate–drain voltage		
V_{GS}	Direct gate–source voltage	W_f	Work function
$V_{GS(th)}$	Threshold voltage of IGFET	W_g	Energy gap
V_{GSQ}	Quiescent gate–source voltage	Z_F	Feedback impedance
V_P	Pinch-off voltage	Z_{in}	Input impedance
V_Q	Quiescent collector–emitter (or drain–source) voltage	$Z_{in(FB)}$	Input impedance with feedback
		Z_{out}	Output impedance
		$Z_{out(FB)}$	Output impedance with feedback
V_T	kT/q (taken as 26 mV at 25°C)	$\bar{e}_n$	Spot noise voltage referred to input
		$\bar{i}_n$	Spot noise current referred to input

$\bar{\alpha}, \bar{\alpha}', \bar{\alpha}''$	Direct current gain for CB, CE, and CC modes	η	Intrinsic stand-off ratio for a unijunction transistor
$\alpha, \alpha', \alpha''$	Incremental current gain for CB, CE and CC modes		
		θ	Thermal resistance
β	Incremental current gain for CE mode	Φ	Total flux
		ϕ	Phase angle
$\bar{\beta}$	Direct current gain for CE mode		
		τ_c	Collector time factor
γ	Emitter efficiency		
γ	Correlation coefficient	τ_{co}	Collector time factor measured at $V_{CB} = 0$
λ	Channel length modulation parameter	τ_s	Saturation time factor

Charge control parameters: τ_c, τ_{co}, τ_s

1

Semiconductor Devices and Integrated Circuits

In electronic design, components are divided into two classes, *active* and *passive*, and for both analysis and design purposes these are represented by *equivalent circuits* or *models*. If such an equivalent circuit contains one or more voltage and/or current generators, an active device is being modeled. Otherwise, a passive device is involved. This means that diodes and transistors count as active devices, whereas resistors, capacitors and inductors do not. The derivation and use of such models will be treated extensively in this volume.

Like individual resistors and capacitors, diodes and transistors are available as *discrete* (i.e., individually packaged) devices; or they may appear within *microcircuits* as multiple structures. The term "microcircuit" is used to encompass both *monolithic* and *hybrid integrated circuits* (ICs), and these terms will be further explained in later chapters. For the present, it will suffice to note that many thousands of active structures can be diffused into a single-crystal chip of silicon, and that this technology has made possible such developments as the microprocessor, the various forms of semiconductor memory and the uncommitted logic array (ULA). Such digital integrated structures are paralleled in the analog world by complete operational amplifiers and transconductance multipliers, for example; and there are also ICs that contain both analog and digital circuitry, including analog-to-digital converters (ADCs), digital-to-analog converters (DACs), and switched-capacitor filters. Furthermore, complete subunits, both digital and analog (e.g., "logic cells" and "gain blocks") may be designed into complete application-specific integrated circuits (ASICs), using sophisticated computer work stations, for subsequent custom fabrication.

In order to understand how electronic circuits may be designed to fulfil specific functions it is necessary to have a grasp of the properties of the individual IC and other components. For purely digital design this implies an

1

understanding of both the hardware and software of the relevant logic, and of the heirarchy of programming where microprocessors and their associated memories are involved. Conversely, in real-world situations it is found that phenomena which it is desired to measure, and so convert to data, are almost invariably analog in nature; and very often the utilization of such data for control purposes necessitates an analog approach, so that a thorough understanding of basic circuit design becomes necessary. This may be most easily acquired by first listing the family of semiconductor devices available (several of which appear as multiple structures within integrated circuits) and proceeding to consider both their properties and the basic (qualitative) physics of their operation. Their use in circuits can then be treated in detail.

1.1 SEMICONDUCTOR DEVICES

(i) The diode. Available in discrete form, as multiple arrays and within microcircuits. The basic attribute is that of rectification but it also has applications as a voltage reference (Zener diode) and in the detection and measurement of radiation, including light.

(ii) The bipolar transistor. Also available in discrete form and within microcircuits. Can perform almost all electronic functions including amplification and switching, and can act as a constant-current source and a transducer.

(iii) The field-effect transistor. Again available individually or within microcircuits and can provide amplification and switching. Can also be used as a voltage-controlled resistor (VCR), constant-current source (or current limiter (CL)) and as a transducer.

(iv) Four-layer devices. The family of Shockley diodes, thyristors and triacs which are essentially controllable switches. Most, but not all, can only be turned ON (but not OFF) using a third electrode and so are normally found as discrete power control devices involving AC supplies.

Less important structures are also available and will be treated in the course of the text. Meanwhile, the basic properties of the foregoing list are presented.

1.1.1 Diodes and Zener Diodes

Figure 1.1(a) shows the symbol for a rectifier, or diode, while figure 1.1(b) illustrates an idealized electrical characteristic for the device.

When a *forward* voltage is applied, the resistance of the diode would

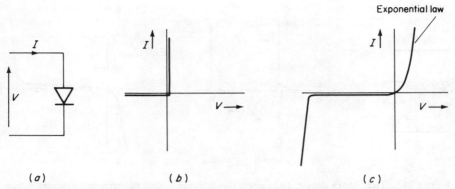

Figure 1.1. The diode: (*a*) diode symbol and polarities; (*b*) ideal diode characteristics; (*c*) practical diode characteristics.

ideally be zero and a forward current would flow (in the direction of the arrow in the diode symbol) which would be limited only by the external circuitry. On the other hand, the resistance of the diode would be infinite for a reverse voltage and no current would flow.

This ideal situation cannot be realized, however, and the characteristic of a practical diode is as shown in figure 1.1(*c*). Here, the application of a forward voltage produces a current which is an exponential function of that voltage. Conversely, a reverse voltage results in a small reverse leakage current which is largely a function of temperature.

These nonideal characteristics imply that power is dissipated in a diode. For example, a diode passing 1 A of forward current would, typically, drop about 0.6 V, so that the device would have to be capable of dissipating a power of 0.6 W in the form of heat. If it were not capable of doing this at a temperature below its rated maximum then it would catastrophically fail. For this reason, the ambient temperature is clearly a deciding factor in the power rating of a diode.

When a reverse voltage is applied, there appears, in addition to the small leakage current, a capacitance which is an inverse function of this voltage. Consequently, the diode can be used as a voltage-variable capacitor or *varactor*. Such diodes are used in high-frequency work for tuning purposes and, less commonly, in parametric amplifiers down to zero frequency.

Also shown in figure 1.1(*c*) is a sharp rise in reverse current when the reverse voltage reaches the *breakdown* level. This breakdown is due to both tunneling and avalanche effects (both discussed later in the text) and may result in irreversible damage unless the current is limited by the external circuitry.

Nondestructive breakdown is also possible and is utilized in the Zener diode. The symbol used in this book is illustrated in figure 1.2(*a*), while

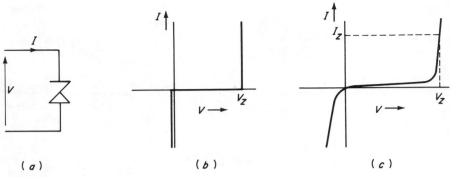

Figure 1.2. The Zener diode: (*a*) Zener diode symbol and polarities; (*b*) ideal Zener diode characteristic; (*c*) practical Zener diode characteristic.

diagrams (*b*) and (*c*) give idealized and attainable characteristics, respectively. From these characteristics it will be seen that the Zener diode is, from a circuitry point of view, essentially a normal diode operated beyond breakdown, that is, in the reverse-current quadrant. (Note that in the relevant symbol, the arrow still points in the direction of forward current, for reasons of consistency.)

Ideally, the Zener diode would present zero resistance after breakdown so defining a precise voltage irrespective of current level, as is implied by figure 1.2(*b*). The attainable characteristic, however, shows that a small but finite breakdown resistance appears, so that it becomes necessary for the Zener voltage V_Z to be defined at a specified Zener current I_Z. Deviations from this current will then result in voltage variations, as described in chapter 14. If the incremental resistance is small, however, the power dissipated will approximate to $V_Z I$. Hence, for a family of Zener diodes of a given physical size, the maximum permissible current will be inversely proportional to V_Z.

Whereas the normal diode is used as a rectifier, the basic application of the Zener diode is as a voltage reference device. Figure 1.3 shows how a

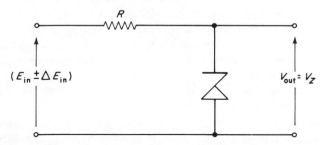

Figure 1.3. Zener diode voltage reference.

direct but variable voltage may be applied to a resistor in series with a Zener diode so that V_Z appears as a well-regulated reference voltage. However, in both cases, the current must be regulated by external means, which is not necessarily so for the transistor, which has a third, or control electrode.

1.1.2   The Bipolar Transistor

The two electrodes which carry the main current flow in a transistor are termed the *emitter* and the *collector*. Figures 1.4(*a*) and (*d*) illustrate the symbols for the two versions of the transistor, the npn and the pnp types. From a circuit point of view these differ only in that the voltage polarities and current directions in the two types are the converse of each other. The direction of current flow is delineated by the arrow on the emitter, which does in fact refer to the forward current flow in a diode structure existing between the emitter and the *base*, which is the control electrode for the transistor. In this sense the symbol is consistent with that of the diode, and

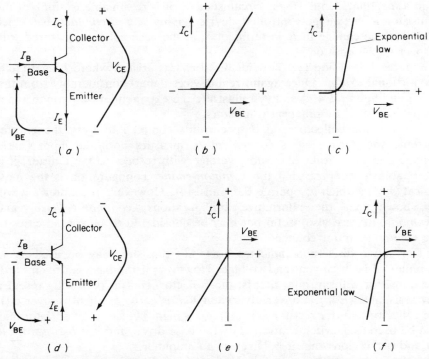

Figure 1.4. The transistor: (*a*) npn transistor symbol and polarities; (*b*) ideal transconductance characteristic (npn); (*c*) practical transconductance characteristic (npn); (*d*) pnp transistor symbol and polarities; (*e*) ideal transconductance characteristic (pnp); (*f*) practical transconductance characteristic (pnp).

from it all the other polarities may be determined. For example, in the pnp transistor, current enters the emitter and leaves from the collector and the base. Hence, both the collector and the base must be negative with respect to the emitter.

The collector current, I_C, is a function of the much smaller base current, I_B, under normal conditions of 'linear' operation, so that the device can be considered to be a current amplifier having a *current gain* given by:

$$\frac{I_C}{I_B} = h_{FE}.$$

This is a highly useful parameter for many circuit design calculations, though its value is not constant, but depends on the actual magnitude of I_C (see figure 2.10). As $|I_C|$ falls, so does h_{FE}, but at the same time the incremental resistance at the base rises. For a modern planar transistor, h_{FE} will still be of the order of 100 when $|I_C|$ has fallen to a few microamps (that is, the fall-off will be much less rapid than for the older type to which figure 2.10 refers) and the base resistance will have risen from a few kilohms to more than a megohm. Under these circumstances, it is reasonable to think of the transistor as a voltage-controlled device having a *transconductance* which describes the change in I_C in terms of V_{BE}, the base voltage measured with respect to the emitter.

Figures 1.4(*b*) and (*e*) show idealized characteristics where I_C is directly proportional to V_{BE}. Once again, real devices depart considerably from this ideal, and figures 1.4(*c*) and (*f*) illustrate the exponential transconductance curves exhibited by practical transistors.

Because the collector and base currents combine to form the emitter current, and because it is convenient to measure both the input (base) voltage and the output (collector) voltage with respect to the emitter, it is reasonable to suppose that the *common-emitter* configuration is the most logical way in which to operate the transistor. However, in chapter 4 it will be shown that the alternative configurations, *common-collector* and *common-base*, are also useful but may be considered to be special cases of the common-emitter connection.

The above discussion indicates how the transistor may be used as an amplifier of both current and voltage. However, if the base is driven hard, the transistor will saturate; that is, the collector–emitter path will present a low resistance, R_{CES}. Conversely, when V_{BE} is zero (or slightly reversed), the collector–emitter resistance is extremely high. Hence, the transistor may also be used as a switch controlled by the base drive; and it is *between* these ON and OFF regions that it behaves as an amplifier.

Although the transistor may be operated at collector currents low enough to make the resistance presented by the base rise to a few megohms, it is often necessary to exceed such values by several orders of magnitude. Leaving aside for the present the various feedback circuits which increase the apparent input resistance, a useful solution is to employ a *unipolar* or *field-effect* transistor (FET), which has precisely this characteristic.

1.1.3 The Field-Effect Transistor

Field-effect transistors may take several forms, each having different circuit properties. For the present, however, only the junction FET is briefly considered, and a comprehensive treatment is postponed until chapter 11.

Figures 1.5(*a*) and (*d*) show the symbols for both n-type and p-type FETs. Notice that the main current flows via the *source* and the *drain* electrodes, which correspond to the emitter and collector electrodes of the bipolar transistor: and that the n-channel drain-source voltage polarity is similar to the collector–emitter voltage polarity for the npn transistor. Also, V_{DS} for the p-channel FET corresponds to V_{CE} for the pnp transistor.

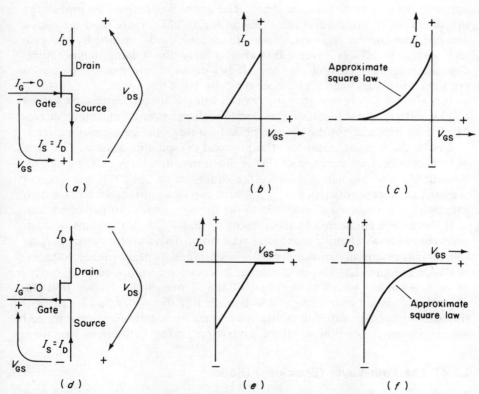

Figure 1.5. The FET: (*a*) n-channel FET symbol; (*b*) ideal transconductance characteristic (n-channel); (*c*) practical transconductance characteristic (n-channel); (*d*) p-channel FET symbol; (*e*) ideal transconductance characteristic (p-channel); (*f*) practical transconductance characteristic (p-channel).

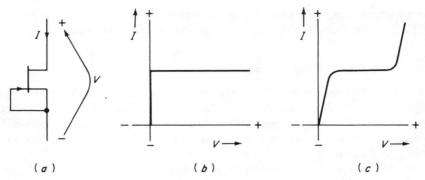

Figure 1.6. The field-effect current limiter: (*a*) the field-effect current limiter symbol; (*b*) idealized characteristic; (*c*) practical characteristic.

The FET control electrode is called the *gate*, and this requires a voltage of opposite polarity to that of the drain. The arrow on the gate symbol shows the direction of forward current *but a FET is always operated so that a forward gate current can never flow*. Hence, the diode formed by the gate and source is always reverse-biased and only the minute temperature-dependent leakage current can flow. It is for this reason that the gate of a FET presents a very high resistance, typically 10^{10} Ω or more.

It is also for this reason that the FET is a voltage-driven device, and figures 1.5(*b*) and (*e*) show idealized, or constant-slope, transconductance curves which take into account the reverse-biased nature of the gate-emitter circuit.

Unlike the bipolar transistor, the practical FET exhibits a square-law type of transconductance curve, and this is illustrated in figures 1.5(*c*) and (*f*). Also unlike the bipolar transistor, the drain-source path through the FET presents its lowest resistance, $r_{DS(on)}$, with zero gate drive, whereas a high gate voltage is required to switch it to the high-resistance, or OFF condition.

If the gate is connected to the source, so that $V_{GS} = 0$, the transconductance curve shows that I_D reaches a maximum. This current, termed I_{DSS}, will remain essentially constant over a wide range of drain-source voltages, as shown in figure 1.6. Hence the device becomes a constant-current source, or *current-limiter*, which is the dual of the Zener diode. Notice that the attainable constant-current region is bounded at the upper end by a rapid rise due to breakdown and at the lower end by a fall-off as the applied voltage becomes too low to allow saturation of the field-effect structure.

1.1.4 The Four-Layer (Shockley) Diode

This device exhibits what is essentially a two-state characteristic, as is shown in both idealized and attainable forms in figure 1.7. Here an increasing voltage will result in very little current rise (ideally none) until the *break-over* voltage is reached, when the resistance suddenly falls to a low value

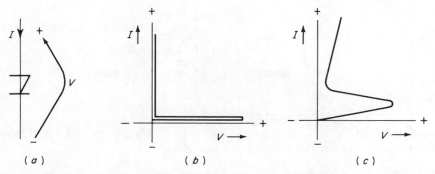

Figure 1.7. The four-layer or Shockley diode: (*a*) four-layer diode symbol; (*b*) idealized characteristic; (*c*) practical characteristic.

(ideally zero). For a real four-layer diode a sustaining current exists below which the device will revert to its high-resistance state.

The four-layer diode is clearly of use in certain switching circuits: for example, it may be connected in parallel with a capacitor to form a timing element as shown in figure 12.5.

1.1.5   The Thyristor

A modified version of the four-layer diode involves a third electrode which is able to switch the device from its high-resistance state to its low-resistance state when the applied voltage is *below* the break-over value. This device, the *thyristor* or silicon controlled rectifier, has characteristics sketched in figure 12.10. Note that the actual effect of a current applied to the control electrode, or *gate*, is to make the thyristor behave as a four-layer diode having a break-over voltage which is a function of this gate current (or, more correctly, the *charge* which is injected into the gate).

1.2   VOLTAGE AND CURRENT SOURCES AND MODELING

Later in the book, it will be shown how some of the aforementioned devices may be represented by *equivalent circuits* or *models*. These consist of networks of passive elements—resistances, capacitances, and inductances—plus ideal voltage and current sources. The symbols for the latter, as used in this book, are shown in figure 1.8. Here, diagram (*a*) represents an ideal independent voltage source, the value of which is not a function of any other variable, whereas diagram (*b*) represents a voltage source having a value dependent upon some defined variable. Both have zero internal impedances.

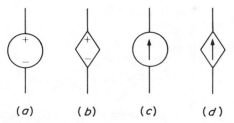

Figure 1.8. Symbols for voltage sources, (*a*) independent and (*b*) dependent; and current sources, (*c*) independent and (*d*) dependent.

Diagram (*c*) represents an ideal independent current source, the value of which is not dependent upon any other variable, whereas diagram (*d*) represents a current source having a value which is a function of some defined variable. Both have infinitely high internal impedances.

Note that if a voltage source is *suppressed*, it simply remains as a short circuit; whereas if a current source is suppressed, it becomes an open circuit.

As an example of the use of an independent voltage source, a battery having an open circuit voltage *E* and an internal resistance *R*, may be modeled as in figure 1.9(*b*). This is a very simple and approximate equivalent circuit for the battery, and may be alternatively represented by converting to a current source (using the Thevenin–Norton transformation) to result in diagram (*c*).

Whereas this is inherently a DC model, waveform generators and other primary sources of signals (or noise!) may also be represented in the same way. However, sources which depend upon other variables will utilize the "lozenge" symbol. For example, a dependent current source which appears frequently in the text has a value given by:

$$i = g_m v \; . \tag{1.1}$$

This means that the current *i* is given by the (variable) voltage *v* multiplied

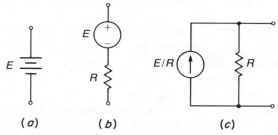

Figure 1.9. A battery (*a*) modeled using an independent voltage source (*b*) or an independent current source (*c*).

by the parameter g_m. Clearly, for the units to be correct, g_m must be a conductance (or more generally, an admittance).

As the text progresses, it will become apparent that most of the models used describe *small-signal* or *incremental* operation of the device in question. To illustrate this concept, consider the diode of figure 1.1. To a close approximation, the current-voltage relationship is given by:

$$I = I_0 \left[\exp\left(\frac{qV}{kT}\right) - 1 \right] \qquad (1.2)$$

where I_0 is the very small reverse leakage current which flows on application of a reverse voltage, as in the third quadrant of figure 1.1(c). In the first quadrant, which represents forward operation, a *working point Q* may be defined by any current, say I_1 and its associated voltage drop, V_1, as shown in figure 1.10(b). At this point, the tangent to the curve represents an *incremental* or *slope resistance r* (which has been labelled $1/r$ in the diagram because current is on the ordinate and voltage on the abscissa, as is usual in electronics).

The value of r may be found quite simply as follows from equation (1.2),

$$\frac{1}{r} = \frac{dI}{dV} = \frac{qI_0}{kT} \exp\left(\frac{qV}{kT}\right) = \frac{q}{kT} (I + I_0)$$

which reduces to qI/kT when $I \gg I_0$. Hence,

$$r \simeq \frac{kT}{qI} \qquad (1.3)$$

and at 25°C

$$r \simeq \frac{26}{I} \ \Omega \ \text{if } I \text{ is in mA} . \qquad (1.4)$$

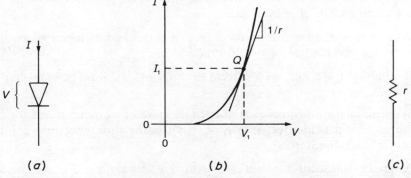

Figure 1.10. The diode forward characteristic showing a working point Q, and an incremental model.

For small or incremental excursions of I and V very near to Q, the diode may be modeled as in figure 1.10(c), where the value of r is defined entirely by the position of Q; that is, by the current I_1 (and of course the temperature T).

Note that the incremental resistance has been given a lower-case letter symbol. This uses the convention that small-signals and incremental parameters normally follow. Another example in the present context would be that small excursions of I and V around Q would be represented by $\delta I = i$ and $\delta V = v$.

As this model refers to small-signal excursions only, there is no need to include either a voltage source or an "ideal" diode in the equivalent circuit. Had a model been developed for large signals, however, both would have been needed. Also, a more sophisticated model would have taken account of the fact that the current-voltage characteristic defined by equation (1.2) is not strictly accurate, but only an approximation; and that the diode presents capacitance as well as incremental resistance. The present very crude model does, however, serve to illustrate the concept of incremental behavior.

1.3   SUMMARY

The foregoing introduction to the electrical characteristics exhibited by a selection of common semiconductor devices indicates that it should be possible to realize a vast selection of circuit functions by suitably combining such devices. A brief and basic introduction to modeling was also given, and it is the combination of the two concepts which, coupled with a considerable degree of flair and originality, can be said to define the design process.

QUESTIONS

1. Explain two separate and distinct uses for a semiconductor diode having regard to its I/V characteristic.

2. What base variables control the collector current of a bipolar transistor? How can this control be put to practical use?

3. If a bipolar transistor has a current gain h_{FE} of 200, what percentage is I_C of I_E?

4. Semiconductor junctions exist at both the base of a bipolar transistor and the gate of a field-effect transistor: what is the difference between the two in operating terms?

5. A bipolar transistor is sometimes termed a "normally–OFF" device and the FET a "normally–ON" device. Why?

6. Why can a junction FET be used as a two-terminal current-limiting device?

PROBLEMS

1.1. In the circuit of figure 1.3, a 12 V Zener diode is used which should not be operated at a current below 4 mA. If $R = 100\,\Omega$ and a load resistor of $333\,\Omega$ is connected across the output, what is the minimum acceptable value of the input voltage $(E_{in} - \Delta E_{in})$?

1.2. The collector current I_C in a transistor is 20 mA and its emitter current I_E is 20.5 mA. What is its current gain h_{FE} (deduced from figure 1.4)?

1.3. The base–emitter voltage V_{BE} is measured for the working transistor of problem 1.2 and found to be +0.7 V. Is the transistor pnp or npn? If the same voltage polarity were correctly applied to the gate–source electrodes of a junction field-effect transistor, would the FET be a p-channel or n-channel type?

1.4. Which of the circuits in figure 1.11 are impossible and why?

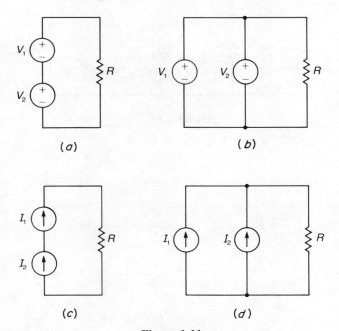

Figure 1.11.

1.5. What is the ratio of v_{out} to v_{in} (in magnitude and sense) for the circuit of figure 1.12 if $v_{in} = 22.3$ mV, $g_m = 40$ mA/V and $R_L = 5.6$ kΩ? What is the value of v_{out}?

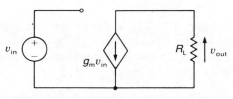

Figure 1.12.

1.6. For the circuit of figure 1.13, determine the magnitude and sign of the current labelled i_L and hence the voltage v_{out}, given that:-

$$e = 12.5 \text{ mV} \qquad g_m = 20 \text{ mA/V} \qquad R_g = 2 \text{ k}\Omega$$

$$R_{in} = 8 \text{ k}\Omega \qquad R_{out} = 47 \text{ k}\Omega \qquad R_L = 5.6 \text{ k}\Omega$$

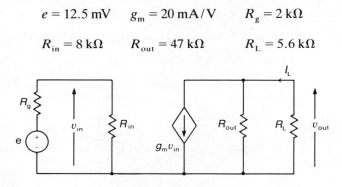

Figure 1.13.

2

Physical Principles and DC Characteristics of Diodes and Transistors

There are already a large number of explanations of how the properties of a semiconducting material can be used to achieve diode and transistor action. Nevertheless, in order that semiconductor circuitry should be understood, the basic physics of the components must be kept in mind at all times and the following account of the relevant phenomena is therefore included without apology. The treatment has been made as brief as possible but without omitting any of the topics necessary for an insight into the workings of the diode and transistor.

The intrinsically semiconducting element silicon is predominant in the manufacture of all forms of diode and transistor, but germanium remains useful for some power transistors and diodes where low device voltage drops are needed; and certain intermetallic compounds such as gallium arsenide are becoming important in high-frequency amplification and fast logic. Silicon and germanium lie in the fourth group of the periodic table (see table 2.1), the atoms of both having four valence electrons. In the germanium or silicon crystal the energies of the valence electrons lie within certain limits, as shown in the energy-level diagram of figure 2.1(b). A forbidden region, or energy gap, separates the valence band from the conduction band. This energy gap is a measure of the energy that an electron at the top of the valence band must acquire before it can jump to the conduction band and thus become free to move within the crystal.

For a valence electron to cross the forbidden region, it must acquire a minimum energy of about 0.7 eV for Ge or 1.15 eV for Si at room temperature. These facts help to explain certain phenomena connected with germanium and silicon devices. For example, an injection of energy in the form of radiation or heat should assist the valence electrons to cross the gap to

TABLE 2.1. Part of the Periodic Table.

Group III		Group IV		Group V	
Boron	B	Carbon	C	Nitrogen	N
Aluminum	Al	Silicon	Si	Phosphorus	P
Gallium	Ga	Germanium	Ge	Arsenic	As
Indium	In	Tin	Sn	Antimony	Sb
Thallium	Tl	Lead	Pb	Bismuth	Bi

(Group number gives the number of valence electrons in an atom of each element)

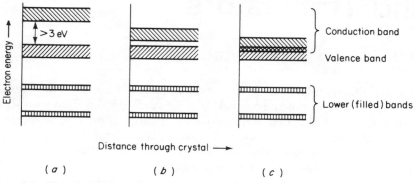

Figure 2.1. Energy-level diagrams for: (a) insulator; (b) semiconductor: and (c) conductor.

the conduction band, which suggests that the conductivity of the crystal should be a function of incident light and ambient temperature. This is in fact so, and at normal temperatures the characteristics of germanium devices vary more pronouncedly than those of silicon, since germanium has the smaller energy gap.

When an electron leaves the valence band, obviously one atom of the semiconductor is left with only three valence electrons. However, it is possible for a valence electron from a neighboring atom to fill this vacancy and this process can carry on, the result being that the "hole" left by the missing electron can drift through the crystal. This constitutes a net movement of negative charge in the direction opposite to the "motion" of the hole, which is precisely the same thing as considering the hole to have a positive charge.

This phenomenon, known as "hole conduction," is restricted to the valence band, whereas electron conduction takes place in the conduction band. Although hole conduction is really a net movement of electrons, mathematically it is possible to derive an effective mass and charge for a hole and to consider it as a positively charged entity. This effective mass is different from that of an electron as is the effective charge, though the latter difference is small.

When electron and hole conduction occur in a material by the mechanism described above, the material is known as an *intrinsic* semiconductor.

Suppose that a small amount of a group-five element has been added to an intrinsic semiconductor in such a manner that the foreign atoms replace some of the semiconductor atoms in the crystal lattice. This means that for each impurity atom there exists a very lightly bound electron, and if the valence band of the impurity is very near to the conduction band of the semiconductor, then, at normal temperatures, almost all these free electrons will be able to jump to the conduction band. This situation is true for silicon containing phosphorus, and the energy levels for such a crystal are sketched in figure 2.2(a). The addition of such an impurity to a semiconductor is known as *doping*, and depending upon the materials and their application, the impurity concentration ranges from about 2 atoms in 10^9 to 2 in 10^3 atoms of semiconductor. The energy gap between the phosphorus donor band and the silicon conduction band is only about 0.04 eV.

Owing to all the donated electrons in the conduction band, the crystal is now much more conductive than it was originally; it has become an *extrinsic* semiconductor. Naturally, the intrinsic semiconductor properties remain, and some electrons still succeed in passing from the valence to the conduction band, leaving holes. These, however, are greatly outnumbered by the donated electrons, hence the crystal is known as an n-type semiconductor (the "n" meaning that the majority of the charge carriers are negative). The

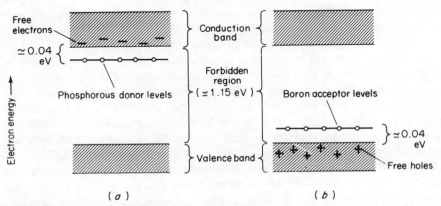

Figure 2.2. Energy-level diagrams for: (a) silicon doped with phosphorus (n-type); and (b) silicon doped with boron (p-type).

donor valence level does not contain mobile holes because for each impurity atom four valence electrons remain to form the four crystal-lattice bonds. However, because of the absence of one valence electron, the impurity atom has a net positive charge, so becoming an immobile or bound ion.

The reverse case—a surfeit of holes over electrons—can be attained by doping an intrinsic semiconductor with an impurity element from group III in the periodic table. Such an impurity, having only three valence electrons and being incorporated into the crystal lattice, should be able to accept an extra valence electron from the valence band of the semiconductor if the energy levels are not too far apart. This requirement is met by doping silicon with boron, and a sketch of the relevant energy-level diagram is given in figure 2.2(b). Once again the forbidden band is only about 0.04 eV wide and thus at room temperature nearly all the vacancies in the boron atom valence orbits will be filled by electrons from the silicon valence band, leaving a large number of holes which are then free to move through that band. Again, the intrinsic semiconductor effect remains, the few electrons in the silicon conduction band being negligible compared with the holes in the valence band. The crystal is now a p-type extrinsic semiconductor in which the acceptor atoms have become bound negative ions, whilst the free carriers are nearly all (positive) holes.

2.1   THE pn JUNCTION

It is possible to produce a single crystal of semiconductor material that is part p-type and part n-type. At the junction of the two regions, certain phenomena occur which make the diode and transistor possible. Figure 2.3(a) represents such a junction, and indicates the predominance of free holes in the p-type material (the majority carriers) and the few free electrons (the minority carriers) present because of intrinsic semiconductor effects. Similarly, the majority carriers in the n-type material are free electrons and the minority carriers are holes.

From figure 2.3(a) it is apparent that the concentration of free holes in the p-material should give rise to a diffusion of some of these holes across the junction to the right, and similarly a diffusion of electrons from the n-material to the left. This tendency does in fact exist, but since such a flow would constitute an electric current thus violating the law of conservation of energy, the result is that an electric field is set up which prevents the continuation of this flow. This space charge field results from the diffusion of some holes to the n-material and of some electrons to the p-material, as shown in figure 2.3(a). The resulting potential difference between the two sides is indicated in figure 2.3(b), which shows the potential gradient existing across the *depletion layer*. This depletion layer is so called because any electron diffusing into it will quickly drift to the right and any hole to the left, thus leaving very few charge carriers within its boundaries at any

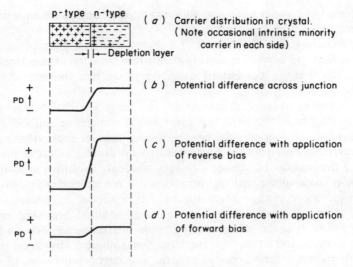

Figure 2.3. The pn junction.

time. The presence of an electric field across the crystal cannot, of course, be detected merely by connecting a voltmeter across the ends, for the contact potentials formed when the circuit is completed oppose this field.

Having seen how the existence of the electric field maintains charge equilibrium in the crystal as far as the majority carriers in each side are concerned, it becomes apparent that the *minority* carriers are *assisted* across the depletion layer. Thus, a free electron from the p-material can easily cross to the n-material, and similarly a hole may easily cross from the n- to the p-material. However, such an occurrence is equivalent to reducing very slightly the potential difference between the two sides, and so a few of the most energetic majority carriers are able to surmount the potential barrier and restore equilibrium.

2.1.1 Application of a Reverse Voltage

Suppose that a battery is placed across the crystal with the negative terminal connected to the p-material and the positive to the n-material. The potential at each side of the depletion layer will now be augmented as shown in figure 2.3(c). Consequently, the minority carriers will be able to cross the junction even more easily, whereas the majority carriers will be prevented more effectively from doing so. When the voltage of the battery is only of the order of a few millivolts, practically all the minority carriers generated by intrinsic semiconductor action at the prevailing temperature will cross the junction, thereby constituting a saturation current which will then be almost independent of any increase in the applied reverse voltage. It will, of course, be a function of temperature and will also depend on the amount of light, if any, reaching the crystal.

A semiconductor diode consists of a crystal containing a pn junction, and its reverse leakage current is merely the saturation current described above. At 20°C the leakage current of a silicon diode can be as low as a few nano-amps.

If the reverse voltage is increased, the drift velocity of the minority carriers as they cross the depletion layer will also increase and will eventually be great enough to ionize the neutral atoms therein, thus releasing further charge carriers. This process could lead to eventual avalanche breakdown and cause the leakage current to rise very suddenly as shown in figure 2.4. If the diode is so designed that this breakdown is not necessarily destructive, it can be used as a voltage regulator or Zener diode, as discussed in the previous chapter. Strictly speaking, a form of field emission occurs at voltages lower than those associated with avalanche diodes and diodes designed to make use of this are the true Zener diodes. However, the name "Zener diode" has now become generic and covers all forms of voltage-reference device.

A further effect which is put to use is that the depletion layer increases in width with the reverse voltage applied. This means that the capacitance which exists between the n and p regions is decreased with an increase in reverse voltage, and this effect makes possible the solid-state variable capacitors which are currently available. The capacity ranges available involve a few tens or hundreds of picofarads and so these *varactors* can be used in high frequency applications and are commonly found in the tuning circuits of many television receivers. They are also useful in parametric amplifiers down to zero frequency.

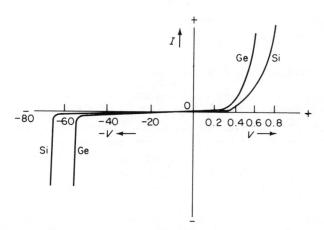

Figure 2.4. Form of diode characteristic.

2.1.2 Application of a Forward Voltage

When a small forward voltage is applied, the potential difference between the p- and n-materials is reduced, as shown in figure 2.3(d). This enables the majority carriers to cross the junction very easily and when the forward voltage has increased to a few hundred millivolts a very low resistance is presented, which, taken in conjunction with the very low reverse leakage current described above, indicates how the device becomes of use as a rectifier.

2.2 THE SEMICONDUCTOR DIODE

A quantitative physical treatment of junction phenomena[1] shows that the diode current in both directions is given by:

$$I = I_0(e^{qV/kT} - 1)$$

where I_0 is the reverse leakage current, V the applied voltage, q the electronic charge, k Boltzmann's constant, and T absolute temperature (in units of degrees kelvin, K).

For forward conduction, $e^{qV/kT} \gg 1$, so that the equation describes the observed exponential current rise; whereas when V is negative, this term rapidly becomes small, so that $I = -I_0$, the reverse leakage current.

Figure 2.4 shows a typical characteristic for semiconductor diodes. The forward volt drop is about 0.8 V for silicon and about 0.5 V for germanium. The power dissipated by such a device will be simply the forward drop multiplied by the forward current. The small size of semiconductor diodes implies also a small thermal capacity, which means that if the power to be dissipated is too great the resulting temperature rise will destroy the device. Naturally, the heat generated must be lost at the surface, and the rate of loss of heat becomes less with an increase in ambient temperature. This means that the maximum permissible current is a function of ambient temperature, and a *de-rating* curve is normally included in manufacturers' data. Figure 2.5 shows a set of three characteristic curves for a typical silicon diode. If a diode is attached to a heat sink, the de-rating characteristic is improved. Many power rectifiers, that is those capable of passing a forward current of 2.0 A or more, are fitted with a threaded terminal that can be screwed through a heat sink. This terminal is connected electrically to one of the electrodes and care must therefore be taken not to short-circuit the diode accidently via its own heat sink. Mica washers are normally supplied with such diodes for use where it is necessary to insulate the diode fully from the heat sink and these are very thin so as not to hinder heat flow unduly.

The maximum reverse voltage of the diode must never be exceeded, even by transients, otherwise the diode may be permanently damaged. The use of

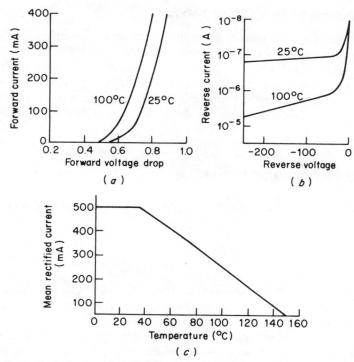

Figure 2.5. Typical diode characteristics (*a*) forward; (*b*) reverse; (*c*) temperature de-rating.

diodes as small power rectifiers is discussed in chapter 14, as are the characteristics of Zener diodes.

The reverse characteristic of figure 2.5(*b*) indicates that the reverse current does not actually approach a limit for a practical device, since a small amount of the total reverse current originates from the purely resistive, or ohmic, leakage between the leadout wires of the device. This component is, of course, also present as part of the forward current but is usually insignificant compared with the magnitude of the forward current considered. The constant part of the reverse current is due to the small number of holes and electrons produced in the intrinsic semiconductor, and, as stated earlier, is dependent on the ambient temperature.

2.3 THE JUNCTION TRANSISTOR

Having established the basic phenomena occurring at a pn junction it is now possible to examine the action of a junction transistor, which consists of three layers of extrinsic semiconductor material forming two junctions inside a single crystal. The crystal regions may be either npn as shown in figure 2.6(*a*), or pnp. If no external potential is applied to the crystal, depletion layers will exist at the junctions, and the charge and potential distributions will be as shown in figures 2.6(*b*) and (*c*).

Suppose that voltages are applied to the regions as indicated by figure 2.7(*a*). The two batteries are connected so that they bias the first junction in the forward direction so that the majority carriers (electrons) in the first region, or *emitter*, are able to cross easily into the middle region, or *base*. The lowering of the potential barrier in this manner also means that the majority carriers in the base region (holes) are able to cross easily to the emitter, and since this is not desirable for reasons which will become

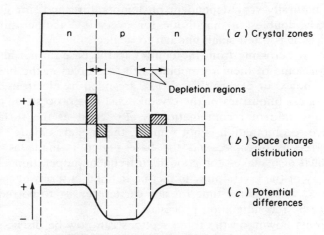

Figure 2.6. npn transistor regions.

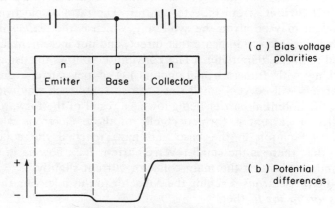

Figure 2.7. Biasing of npn transistor.

apparent later, the doping of the emitter is much heavier than that of the base. Hence the ratio of electrons to holes crossing the depletion layer is very high.

As the base is field-free, the electrons entering this region now pass, by diffusion processes only, to the second depletion layer. Here, owing to the high reverse field produced by the second battery, they are swept rapidly into the final region, the *collector*. This field also effectively prevents collector electrons from passing through to the base, though any minority carriers (holes) formed by intrinsic semiconductor phenomena will, of course, be assisted in their transit to the base. Similarly, any electrons formed in the base will be assisted across to the collector. The small current formed by these "intrinsic" carriers is precisely the same as the reverse current of a diode, and in the transistor is usually given the symbol I_{CO} or I_{CBO}. It is naturally very dependent on temperature and, for germanium, approximately doubles in magnitude for every 9°C temperature rise. In silicon devices it is often small enough to neglect.

The electrons crossing from the emitter to the base do not all reach the collector, for some of them recombine with free holes in the base; that is, some of the holes, or electron vacancies, are filled by electrons having the correct energies. Impurities in the crystal assist this process, as do dislocations, and the rate of recombination is also great at the surface of the crystal. Since recombination is not wanted, the crystal is made as pure and dislocation-free as possible, and the base region is made as narrow as possible which also makes for good high-frequency operation. The purity and regularity of the crystal may in fact be defined in terms of "recombination lifetime," which is the time for any carrier density to be reduced by a factor of $1/e$ by recombination processes.

The currents flowing in the three sections can now be discussed.

The emitter current is composed mainly of electron current from emitter to base, and a much smaller hole current from base to emitter. These currents are, of course, additive, since they both constitute a flow of "conventional current" from base to emitter. The ratio of electron current to total current may be given the symbol $\bar{\gamma}$, which is the *emitter efficiency*, and the bar is included to indicate that direct, and not incremental, currents are referred to in this parameter. (This convention follows the usual practice when working with transistor parameters.) The net current due to the intrinsic carriers will be very small since the currents each way will tend to cancel out, the potential barrier being low as a result of the forward biasing.

The collector current is formed chiefly of those electrons which have escaped recombination in the base and have reached the collector. In addition to this, there is the small reverse current I_{CBO} flowing in the same direction and augmenting the main collector current slightly.

If the ratio of electrons reaching the collector to those leaving the emitter is the *transport factor b*, then

$$\frac{I_C - I_{CBO}}{I_E} = \bar{\alpha} = \bar{\gamma}b \tag{2.1}$$

or, if $I_{CBO} \ll I_C$, then $I_C \simeq \bar{\alpha} I_E$. Because the derivation of this ratio implies that the base is common to both emitter and collector circuits (as shown in figure 2.8(b)), $\bar{\alpha}$ is called the *common-base direct current gain*. Its value, for most modern transistors, exceeds 0.95.

Using equation (2.1) the two other current ratios may be simply derived as follows.

Putting $I_E = I_B + I_C$ into equation (2.1) gives:

$$\frac{I_C - I_{CBO}}{I_B + I_C} = \bar{\alpha}$$

so that

$$I_C - \frac{I_{CBO}}{1 - \bar{\alpha}} = \frac{\bar{\alpha}}{1 - \bar{\alpha}} I_B \, .$$

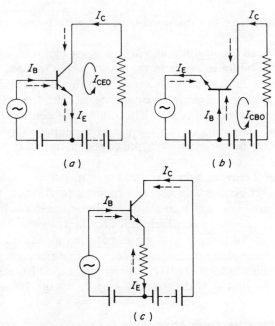

Figure 2.8. Transistor configurations for: (a) common emitter; (b) common base; and (c) common collector.

Defining

$$\bar{\alpha}' = \frac{\bar{\alpha}}{1 - \bar{\alpha}} \quad \text{then} \quad \frac{1}{1 - \bar{\alpha}} = (1 + \bar{\alpha}')$$

giving

$$I_C - (1 + \bar{\alpha}')I_{CBO} = \bar{\alpha}'I_B .$$

If $I_B = 0$, then $I_C = (1 + \alpha')I_{CBO}$, which is the collector–emitter leakage current for an open-circuited base, I_{CEO}. Hence,

$$\frac{I_C - I_{CEO}}{I_B} = \bar{\alpha}' \tag{2.2}$$

where $\bar{\alpha}'$ is the *common-emitter direct current gain*, and is relevant to the circuit of figure 2.8(*a*).

Note that I_{CEO} is greater then I_{CBO}, but if it is still small compared with I_C, then $I_C \simeq \bar{\alpha}'I_B$.

Putting $I_C = I_E - I_B$ into equation (2.2) gives

$$I_E - I_{CEO} = (1 + \bar{\alpha}')I_B = \bar{\alpha}''I_B . \tag{2.3}$$

Here $\bar{\alpha}''$ is defined as the common-collector direct current gain, and if $I_{CEO} \ll I_E$, then $I_E = \bar{\alpha}''I_B$.

It should be noted that although current gains or ratios discussed above are referred to as being relevant to the CB, CE or CC modes, this is only by analogy with the small-signal or incremental gains which will be discussed later. Clearly, if the transistor is operating at all, the three ratios are relevant to any mode of connection. This point has been illustrated in figure 2.8(*c*), where the collector is not common to the input loop. The base bias battery would have been returned to the collector to make this so, which is an unusual connection.

The most useful current gain is $\bar{\alpha}'$, and it is also given the symbols $\bar{\beta}$, $\bar{h}_{21}$ or h_{FE}. In this text both $\bar{\beta}$ and h_{FE} will be used in addition to $\bar{\alpha}'$ in order that the reader may become accustomed to the common conventions.

If $\bar{\alpha} = 0.95$, then $\bar{\alpha}' = 19.0$, which indicates that, in the common-emitter configuration, a useful current gain exists between input and output. Clearly, a high numerical value of $\bar{\alpha}$ is desirable, which implies a high value for $\bar{\gamma}$, which in turn implies a high emitter/base doping ratio. In fact $\bar{\gamma}$ can closely approach unity so that $\bar{\alpha}$ is determined largely by the transport factor b.

2.3.1 Transistor Characteristics

Any electrical device can be described by sets of graphs showing how the input and output voltages and currents vary with respect to each other and with respect to certain external parameters, of which temperature is usually the most important. Transistors are no exception to this rule and manufacturers' graphs are published which provide the data necessary for circuit design.

Transistors are three-terminal devices and can therefore be connected in three different ways. The common-emitter, common-base and common-collector modes are illustrated in figure 2.8 and the common-collector mode is more usually termed "emitter-follower." The terms "grounded" or "earthed" emitter, base, or collector are also used, but are rather confusing since they refer to the fact that the electrode mentioned is common to both input and output circuits, but do not mean that they are necessarily connected to ground.

It is therefore apparent that three sets of characteristics are necessary to describe the operation of a transistor connected in each of its three modes. However, since the common-emitter connection is by far the most widely used, it being the only one which produces both voltage and current amplification, the others are very often omitted.

2.3.2 The Output Characteristic

The most useful curve for the establishment of the DC operating point of a transistor is the output characteristic, which relates the collector current and voltage. Figure 2.9 illustrates common-emitter output characteristics for a typical npn planar transistor.

It will be seen that for a given base current, the collector current rises sharply as the collector depletion-layer field is set up on application of a

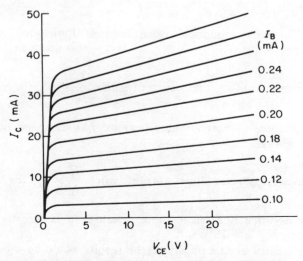

Figure 2.9. Output characteristics for a typical npn small-signal transistor.

reverse voltage, but soon reaches a limit determined by the base current, and levels off. The family of curves do not lie quite parallel or horizontal which is indicative of the fact that the output resistance is by no means constant.

The voltage at the collector V_{CE} is, of course, positive, this being an npn transistor. If the convention is that currents flowing into the transistor are considered positive and those flowing out are considered negative, then the collector current I_C, along the ordinate, takes a positive value.

From this family of curves may be obtained the DC current gain, h_{FE} or $\bar{\alpha}'$, for any combination of collector voltage and base current. For example, using $V_{CE} = 8\,V$ and $I_B = 0.2\,mA$, I_C is seen to be 20 mA. Hence, $h_{FE} = 20/0.2 = 100$.

At this point it is convenient to consider the relations between the $\bar{\alpha}'$ and the h_{FE} parameters. In the early days of transistors, it was the common-base configuration which was first investigated, and the DC current gain in this mode was assigned the symbol $\bar{\alpha}$. The DC current gains in the common-emitter and common-collector modes were designated $\bar{\alpha}'$ and $\bar{\alpha}''$, respectively. Again, the bar was included to indicate DC conditions only, and in the next chapter, it will be seen how the small-signal or incremental parameters omit the bar and become α, α', and α''.

In figure 2.8, the continuous arrows show the current flow for a real npn transistor, while the broken arrows show the positive direction of current flow for the positive-in, negative-out convention. So, when writing the current-gain parameter for the three modes, the signs of the currents must be taken into account as follows. Neglecting I_{CBO} and I_{CEO},

$$\text{common-base direct current gain} = \frac{I_C}{I_E} = h_{FB} \text{ (or } \bar{h}_{21})$$

which has a *negative* numerical value because I_C is positive and I_E is negative. Hence $h_{FB} = -\bar{\alpha}$.

$$\text{Common-emitter direct current gain} = \frac{I_C}{I_B} = h_{FE} \text{ (or } \bar{h}_{21}')$$

which has a *positive* numerical value because I_C and I_B are both positive. Hence $h_{FE} = \bar{\alpha}' = \bar{\beta}$.

$$\text{Common-collector direct current gain} = \frac{I_E}{I_B} = h_{FC} \text{ (or } \bar{h}_{21}'')$$

which has a *negative* numerical value because I_E is negative and I_B is positive. Hence $h_{FC} = -\bar{\alpha}''$.

Had the transistor been a pnp type, the results would have been identical since *all* the real current flows would have changed direction.

The parameters in parentheses are sometimes used in place of the *h*-parameters given. Occasionally the subscripts E, B or C are given immediately after the numbers in place of the superscript dashes. The numbers 1 and 2 themselves refer to input and output respectively, while the presence of a bar above the *h* means that the parameter refers to DC conditions. The capital subscripts in the h_{FE} convention render such a bar unnecessary. In the next chapter, which deals with incremental changes, lower case subscripts (e.g., h_{fe}) will be introduced.

The initial discussion of current ratios made the definition that

$$\bar{\alpha}' = \frac{\bar{\alpha}}{1 - \bar{\alpha}}. \tag{2.4}$$

Taking current directions into account, this becomes

$$h_{FE} = \frac{-h_{FB}}{1 + h_{FB}}. \tag{2.5}$$

Therefore, $\bar{\alpha} = -h_{FB}$. Thus, if

$$h_{FB} = -0.95 \quad \text{then} \quad h_{FE} = \frac{0.95}{1 - 0.95} = 19.$$

Looking again at figure 2.9, it is apparent that h_{FE} is not constant, otherwise the family of curves would be parallel and equidistant for equal increments of I_B. Instead, h_{FE} varies with I_C, and a curve showing this can be plotted from figure 2.9 by taking a specific value for V_{CE}. Such a curve is shown in figure 2.10, where the value of h_{FE} is plotted with respect to I_C. The value of V_{CE} is maintained throughout at 10 V.

The increasing slopes of the I_C/V_{CE} characteristics as V_{CE} rises are due largely to base-width modulation effects. Most of V_{CE} appears as the reverse voltage, V_{CB}, and as this rises, the width of the collector–base depletion

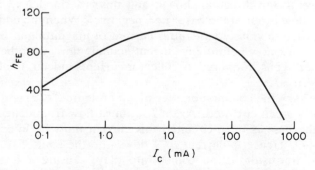

Figure 2.10. h_{FE} versus I_C for a typical npn small-signal transistor. $V_{CE} = 10$ V and $T = 25°$C.

layer increases. Hence, the width of the effective base region decreases so that the slope of the charge concentration in the base, from emitter to collector, becomes steeper. Hence, I_C increases with V_{CE} resulting in a nonhorizontal I_C/V_{CE} characteristic.

The reason for the fall in h_{FE} as I_C decreases is that there is always a small, fairly constant emitter–base leakage current, I_{RE}, which takes no part in the transistor action. This leakage is associated with the existence of recombination centers in the emitter–base junction, particularly near the surface. As I_E falls, I_{RE} becomes an increasingly significant fraction of it, so that the ratio I_E/I_B and hence I_C/I_B also falls. The reduction in h_{FE} at high values of I_C is due to "current crowding" in the emitter and to conductivity modulation in the base. The former is the term given to non-homogeneous current distribution in the emitter–base junction at high currents because of increasing voltage drops in the base material itself; whilst the latter refers to increased conductivity in the base region (due to high charge-carrier densities) leading to a reduction in the emitter efficiency.

In addition to its variation with collector current, h_{FE} also increases almost linearly with temperature. If the value at 25°C is given, it is reasonable to expect this to have increased by some 20–30% at 50°C. Curves of h_{FE} versus temperature are not usually included in manufacturers' data sheets and are in fact not often necessary.

2.3.3 Leakage Currents

If figure 2.9 were magnified such that a curve at $I_B = 0$ were shown, it would be seen that a small collector current would still flow. This is the collector–emitter leakage current I_{CEO} or I'_{CO}, and it is actually due to the reverse leakage current mentioned previously in connection with diodes.

In order to understand the nature of this leakage current, it is best to begin by considering the common-base mode of figure 2.8(b). In this case, if the emitter is disconnected, only the base–collector junction is left in circuit, and this constitutes a reverse-biased diode. Consequently, the saturation current flows in the direction shown, and this is termed I_{CBO} or I_{CO}. The direction of flow is out of the base (i.e., negative). When the emitter is again connected, then obviously an emitter current of magnitude at least equivalent to I_{CBO} must flow before any current actually flows into the base in the usual manner. Thus, for very low collector currents indeed, the base current actually reverses.

Consider now the common-emitter mode of figure 2.8(a) and let the base connection be open-circuited. Again I_{CBO} must flow from collector to base, but since the base is open-circuited this current must be injected into the emitter circuit. Hence, a current I_{CBO} flows into the emitter from the base, and normal transistor action comes into play, causing a further current $h_{FE} I_{CBO}$ to flow from collector to emitter.

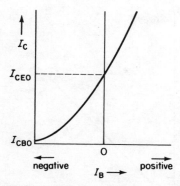

Figure 2.11. The I_C/I_B curve near its origin, for an npn transistor.

Hence,

$$I_{CEO} = I_{CBO} + h_{FE}I_{CBO} = I_{CBO}(1 + h_{FE}) \simeq I_{CBO}h_{FE}$$

if $h_{FE} \gg 1$. This expression is exactly the same as was used in the discussion of current ratios; that is, $I_{CEO} = (1 + \bar{\alpha}')I_{CBO}$, and its meaning is illustrated in figure 2.11, which is an expanded version of figure 3.2 near the origin. Here it will be seen that as I_B falls to zero, the total collector current reduces to I_{CEO} and that to produce a current in the collector lead below I_{CEO}, the base must be reverse biased. When this is done the collector current may fall to I_{CBO} but the reverse-biased emitter junction will also exhibit a leakage current, I_{EBO}.

For reasons explained earlier in this chapter, the leakage currents for silicon devices would be expected to be much lower than those for germanium devices and this is well illustrated by comparing two small transistors:

TI 2N404 (germanium pnp) $I_{CBO} = -5.0\ \mu A$ ⎫ maximum values
TI BC178 (silicon pnp) $I_{CBO} = -0.1\ \mu A$ ⎬ at 25°C and -10 V .

2.3.4   Variation of Leakage with Temperature

As leakage is essentially an intrinsic semiconductor phenomenon, it varies considerably with temperature. A general rule already quoted for germanium transistors is that it will double in value for every 9°C temperature rise. For silicon, it is rather less than this, and since the absolute value for silicon is so much lower, it is often possible to ignore it in circuit design, a procedure which is not advised for germanium devices. Figure 2.12 shows curves of I_{CBO} versus temperature for typical germanium and silicon transis-

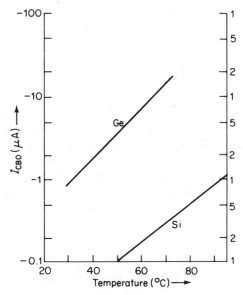

Figure 2.12. I_{CBO} versus temperature curves for typical silicon and germanium transistors.

tors. It is for reasons such as this that silicon transistors have become widespread, and at the time of writing all integrated circuits are based on silicon technology.

2.3.5 Transistor Fabrication

Most contemporary small-signal transistors, and all monolithic integrated circuits, are fabricated using *planar* or *planar epitaxial* technology. In the planar process, a lightly-doped n-type (for example) single-crystal silicon wafer is passivated; that is, a layer of silicon dioxide is grown on the surface by heating in oxygen. Portions of this layer are etched away using hydrofluoric acid (which does not attack pure silicon) and an acceptor impurity is diffused through the exposed silicon surface to form a p-type region as shown in figure 2.13(a). The surface is again passivated and selectively etched and a final dopant is diffused in to form a second n-type region. Gold ohmic contacts are deposited after further passivation and etching, and the final npn structure exhibits the (idealized) cross-section of figure 2.13(a).

Because this is essentially a double-diffusion process the initial wafer must be lightly doped; that is, its conductivity must be low, which leads to a high saturation resistance in the transistor, R_{CES}. To avoid this restriction, it is possible to grow a thin low conductivity layer on the surface of a high conductivity wafer by gaseous phase deposition. This is the process of

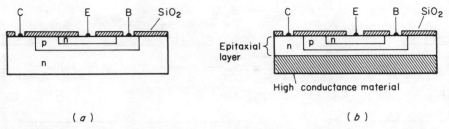

Figure 2.13. (a) The planar structure; (b) the planar epitaxial structure (idealized cross-sections, not to scale).

epitaxy, and it results in a collector region whose bulk is of the required high conductivity material. Hence, the planar epitaxial transistor is capable of exhibiting, concurrently, a high frequency response (>500 MHz), low saturation resistance, and high breakdown voltage.

It is also possible to fabricate both resistor and capacitor structures using these techniques and this has made possible complete self-contained monolithic microcircuits such as the operational amplifiers described in chapter 7.

Solid-state technology is now a vast subject in its own right, and it is currently possible to diffuse many thousands of structures into a single silicon chip[2]. It is this which has made possible both the microprocessor and the solid-state memory.

2.3.6 The Ebers–Moll Equations

A limited quantitative physical analysis of the transistor can lead to a pair of equations which describe its complete electrical characteristics (in an idealized form) at low injection levels for both forward and reverse biasing of either junction[3]. These are the Ebers–Moll equations[4], which have been used as the basis for several computer-aided design (CAD) programs. In their original form, with signs appropriate to a pnp transistor, they are:

$$I_{\mathrm{E}} = I_{\mathrm{ES}}(e^{qV_{\mathrm{EB}}/kT} - 1) - \bar{\alpha}_{\mathrm{R}} I_{\mathrm{CS}}(e^{qV_{\mathrm{CB}}/kT} - 1) = I_{\mathrm{F}} - \bar{\alpha}_{\mathrm{R}} I_{\mathrm{R}} \qquad (2.6)$$

$$I_{\mathrm{C}} = -\bar{\alpha} I_{\mathrm{ES}}(e^{qV_{\mathrm{EB}}/kT} - 1) + I_{\mathrm{CS}}(e^{qV_{\mathrm{CB}}/kT} - 1) = -\bar{\alpha} I_{\mathrm{F}} + I_{\mathrm{R}} . \qquad (2.7)$$

Figure 2.14(a) shows the conventional voltage polarities and current directions which lead to the signs in equations (2.6) and (2.7) (which would of course be reversed for an npn transistor). Figure 2.14(b) illustrates an equivalent circuit (originally suggested by Adler) which is described by these equations.

In equation (2.6), if the collector and base are short-circuited ($V_{\mathrm{CB}} = 0$), the second term disappears. If V_{EB} is negative—that is, the emitter junction is reverse biased—then the current flowing in the emitter lead is simply I_{ES},

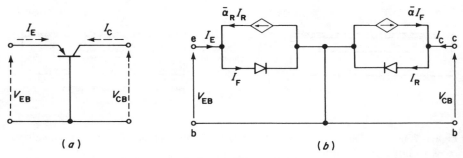

Figure 2.14. An equivalent circuit for the Ebers–Moll equations to a pnp transistor.

the emitter–base leakage or the saturation current for the short-circuited collector condition. Conversely, if V_{EB} goes positive, I_E rises exponentially in the same way as the diode current mentioned earlier in the chapter.

If the collector is now reverse biased by a few volts in the usual manner, V_{CB} becomes a negative quantity of a magnitude such that the second term tends to $-\bar{\alpha}_R I_{CS}$. (Here, I_{CS} is the collector–base leakage current with the emitter short-circuited and $\bar{\alpha}_R$ is the common-base current gain for the "inverted operation" condition. The geometry of most transistors is asymmetric in such a way that $\bar{\alpha} > \bar{\alpha}_R$.)

Hence, for normal operation, equation (2.6) may be written:

$$I_E = I_{ES}(e^{qV_{EB}/kT} - 1) + \bar{\alpha}_R I_{CS}.\tag{2.8}$$

The first term in this equation is by far the larger, and describes that component of the emitter current which consists of holes injected across the depletion layer to become minority carriers in the base region. Some few holes also manage to cross to the base from the collector and a proportion $\bar{\alpha}_R$ reach the emitter: these are represented by the second term.

The collector current has been seen to be given by $\bar{\alpha} I_E$ and this is reflected in the first term of equation (2.7). The second, leakage current, term is again small because V_{CB} is sufficiently negative to reduce it to $-I_{CS}$. Hence, equation (2.7) becomes

$$I_C = -\bar{\alpha} I_{ES}(e^{qV_{EB}/kT} - 1) - I_{CS}.\tag{2.9}$$

This equation shows that the collector current is an exponential function of the emitter–base voltage (as sketched in figure 1.4(f) along with a small leakage component). That the collector current is controlled by the emitter–base voltage is a very useful way of describing transistor action from a circuit point of view. It is, of course, equally valid to consider the collector current to be a function of emitter *current*, as has been qualitatively discussed earlier in the chapter. Equation (2.9) can easily be converted to this form by inserting equation (2.8):

$$I_C = -\bar{\alpha}(I_E - \bar{\alpha}_R I_{CS}) - I_{CS}$$
$$= -\bar{\alpha}I_E - (1 - \bar{\alpha}_R\bar{\alpha})I_{CS}. \tag{2.10}$$

This is of the same form as equation (2.1), which implies that $I_{CBO} = (1 - \bar{\alpha}_R\bar{\alpha})I_{CS}$. (Note that as the "positive-in, negative-out" sign convention is now being used, $\bar{\alpha}I_E$ must have a *negative* numerical value, because it flows *out* of the collector. This accounts for the presence of the negative sign in the equation itself.)

The collector current can, of course, be represented as a function of base current, which is convenient when the common-emitter configuration is to be considered. To do this in terms of the Ebers–Moll model it is simply necessary to redraw the equivalent circuit with the emitter as the common point, and extract an expression for I_C in terms of I_B, where $I_B = -I_C - I_E$.

Equation (2.9) already expresses I_C in terms of V_{EB}, and it is common practice to generalize and approximate this equation as follows:

(a) the leakage term, I_{CS}, is omitted as being negligibly small;
(b) $\bar{\alpha}$ is omitted on the assumption that it is close to unity;

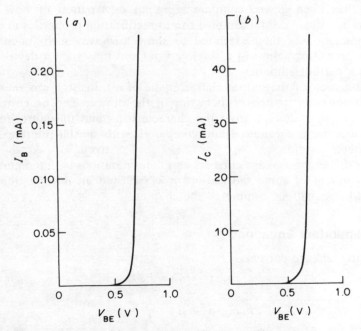

Figure 2.15. Typical input (a) and transconductance (b) curves for a small-signal silicon npn transistor (at room temperature and constant V_{CE}).

(*c*) it is assumed that the current directions for both npn and pnp transistors are well known, so that the equation is written without specifying them, and for this reason;

(*d*) V_{EB} is written V_{BE}.

The basic transconductance equation for the bipolar transistor now results:

$$I_C \simeq I_{ES}(e^{qV_{BE}/kT} - 1) \tag{2.11a}$$

and if the working point is well away from the origin of the I_C/V_{BE} characteristic, so that $e^{qV_{BE}/kT} \gg 1$, it becomes:

$$I_C \simeq I_{ES}\, e^{qV_{BE}/kT} . \tag{2.11b}$$

At room temperature, $kT/q \simeq 26$ mV, so that V_{BE} need only be greater than 120 mV to establish that the omission of the 1 leads to an error below 1%. Both an input and a transconductance characteristic are shown in figure 2.15 and are related by the current gain parameter h_{FE}.

2.4   SUMMARY

In this chapter a qualitative discussion of the relevant properties of semiconductors has been given, culminating in an explanation of how the pn junction in a single crystal can give rise to rectification properties in a diode. The argument was then extended to show how two such junctions can produce transistor action in a single crystal and how such a device can be used as a current amplifier.

A discussion of the output characteristic of a transistor was then given, and was restricted to the DC behavior of the device. The DC current gain parameters were derived from the characteristics and the more important conventions were considered and compared, with special reference to the *h*-parameter system.

After discussing leakage currents and their variations with temperature, a brief treatment of some fabrication processes and an introduction to the Ebers–Moll equations completed the chapter.

2.4.1   Important Equations

Neglecting leakage currents,

$$h_{FB}(=-\bar{\alpha}) = I_C/I_E$$

$$h_{FE}(=\bar{\alpha}' = \bar{\beta}) = \frac{I_C}{I_B} = \frac{-h_{FB}}{1 + h_{FB}}$$

$$h_{FC}(=-\bar{\alpha}'') = I_E/I_B = -(h_{FE} + 1)$$

$$I_C \simeq I_{ES}(e^{qV_{BE}/kT} - 1) .$$

REFERENCES

1. Gray, P. E., De Witt, D., Boothroyd, A. R. and Gibbons, J. F. 1964, *Physical Electronics and Circuit Models of Transistors* (New York: Wiley) p. 43.
2. Till, W. C. and Luxon, J. T., 1982, *Integrated Circuits: Materials, Devices and Fabrication* (Englewood Cliffs, NJ: Prentice Hall).
3. Gray, P. E., De Witt, D., Boothroyd, A. R. and Gibbons, J. F., 1964, *Physical Electronics and Circuit Models of Transistors* (New York: Wiley) chap. 9.
4. Ebers, J. J. and Moll, J. L., 1954, Large-signal behaviour of junction transistors *Proc. IRE* **42**, 1761–1772.

QUESTIONS

1. What is the difference between intrinsic and extrinsic semiconduction, and how is the latter achieved?

2. What is a depletion layer, and why does it exist?

3. The reverse current in a diode is temperature sensitive. Why?

4. Suggest two uses for a diode other than as a rectifier.

6. Why is I_{CEO} larger than I_{CBO}? By how much is it larger?

7. Why is h_{FB} the negative of $\bar{\alpha}$?

8. In figure 2.9, the lowest few curves in the family are nearly horizontal. This implies that the transistor might be used as an approximation to a theoretical circuit element. Which one, and why?

PROBLEMS

2.1. A diode has a reverse leakage current I_0 of 10^{-9} A. What is the forward voltage drop across this diode when it is conducting 3.6 A? What is the power dissipation in the diode? Assume that $kT/q \simeq 26$ mV and criticize this assumption.

2.2. What is the incremental (or slope) resistance of the diode in problem 2.1 when conducting (i) 1 A, (ii) 0.1 mA? Again assume that $kT/q \simeq 26$ mV.

2.3. The direct current entering the collector of a transistor is 5 mA, and that leaving the emitter is 5.05 mA. Neglecting leakage current, what is h_{FE} for the transistor working under the specified conditions?

2.4. By examining the characteristics of figure 2.9, decide whether the transistor collector terminal behaves more like a current source when a base current of 0.1 mA or 0.2 mA flows.

2.5. For the transistor of problem 2.3, the collector-to-base leakage current I_{CBO} is 0.1 μA. What is its collector-to-emitter leakage I_{CEO}?

2.6. A small power transistor is known to have an emitter–base leakage current for the short-circuit collector condition of $I_{ES} = 1$ nA at 25°C (where $kT/q \simeq 26$ mV). If a base–emitter voltage of 0.55 V is applied, what is the collector current under normal working conditions?

2.7. A transistor connected in the common-base mode is found to have a DC current gain α of 0.98. What are the values (and signs) of h_{FB}, h_{FE} and h_{FC}?

3

Small-signal Parameters and Equivalent Circuits

In order to make calculations which are meaningful in the context of circuit design, it is necessary to represent the transistor by an equivalent circuit which exhibits characteristics similar to those of the transistor itself. For example, the Ebers–Moll equivalent circuit discussed at the end of chapter 2 describes the transistor in terms of two back-to-back diodes and two current generators. This circuit, within its limits of accuracy, will predict the currents flowing at low base injection levels and low frequencies.

Contemporary computer aided design (CAD) methods model the transistor more comprehensively; for example, the full SPICE II equivalent contains some forty parameters, though fewer are needed for most computations. Commercially-available programs allow such models to be called up automatically, when parameter values are assigned, or "default" parameters accepted by the designer. The model is then inserted into its host circuit, and the program allows various bias and signal analyses to be carried out.

For the more basic calculations, however, hand methods and simple models are entirely adequate. Furthermore, they offer the insights needed into the fundamental electronic concepts which allow the aforementioned host circuit to be designed in the first place, and without which the application of CAD techniques would be impossible.

Insofar as small-signal amplifiers are concerned, it is not necessary to know the total currents flowing, but only the incremental changes in them. Hence, provided the transistor is biased into a suitable operating condition, it is valid to consider only *incremental* equivalent circuits. A treatment of biasing methods will appear in chapter 4 and the present discussion will concentrate on two of the commonest incremental equivalent circuits.

There are two approaches to modeling. First, the transistor may be thought of as a "black box" or system, and a circuit having similar input and output characteristics may be postulated. Second, a physical evaluation of the transistor may be made and circuit components chosen to represent the function of each physical mechanism. Clearly, the latter method is likely to produce the more complex equivalent, though this may then be reduced by approximation to yield a usable circuit.

The h-parameter equivalent circuit is the most common example of the "black-box" approach and the 'h' (meaning "hybrid") refers to the fact that the two equations describing the circuit give the input voltage and output current, respectively. This is one of the six possible ways of describing a two-port (or four-terminal) network and the others are listed in Appendix II.

The chief advantage of the h-parameter model is that all the variables are measurable, which makes it easy to determine the h-parameters themselves, as will be seen below. Conversely, the hybrid-π equivalent, which is the most common of the simplified physically-oriented models, has some parameters which cannot be determined directly but which are easily expressed in terms of the h-parameters. It is for this reason that the h-parameter model will be treated first, but thereafter the hybrid-π model will be used as the preferential instrument of analysis whenever possible.

3.1   THE h-PARAMETER SYSTEM

If an unknown circuit is placed in a two-port "black box" as shown in figure 3.1, one method of characterizing it is to write the following equations which relate the input and output voltages and currents:

$$V_{in} = f(I_{in}, V_{out}) \tag{3.1a}$$

$$I_{out} = f(I_{in}, V_{out}) . \tag{3.1b}$$

If incremental changes in V_{in} and I_{out} occur, the relevant total differentials are:

$$\delta V_{in} = \frac{\partial V_{in}}{\partial I_{in}} \, \delta I_{in} + \frac{\partial V_{in}}{\partial V_{out}} \, \delta V_{out} \tag{3.2a}$$

$$\delta I_{out} = \frac{\partial I_{out}}{\partial I_{in}} \, \delta I_{in} + \frac{\partial I_{out}}{\partial V_{out}} \, \delta V_{out} . \tag{3.2b}$$

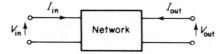

Figure 3.1. The two-port "black box."

Such incremental changes could refer to the small signals which are superimposed on the much larger, but direct, biasing currents and voltages of an operating transistor *in any of its three modes*. Equations (3.2) could therefore be written as follows:

$$v_{in} = h_i i_{in} + h_r v_{out} \qquad (3.3a)$$

$$i_{out} = h_f i_{in} + h_o v_{out} \qquad (3.3b)$$

and being simultaneous equations, their determinant would be:

$$\triangle = h_i h_o - h_r h_f . \qquad (3.3c)$$

To define the four parameters in terms of these small-signal currents and voltages, first short-circuit the output so that $v_{out} = 0$. Then,

$$h_i = v_{in}/i_{in} \quad (v_{out} = 0) \qquad (3.4a)$$

and

$$h_f = i_{out}/i_{in} \quad (v_{out} = 0) . \qquad (3.4b)$$

Now, open-circuiting the input so that $i_{in} = 0$ gives

$$h_r = v_{in}/v_{out} \quad (i_{in} = 0) \qquad (3.4c)$$

and

$$h_o = i_{out}/v_{out} \quad (i_{in} = 0) . \qquad (3.4d)$$

Note that h_i has the dimensions of an impedance and that h_o is an admittance whereas h_r and h_f are simply numbers, being a voltage ratio and a current ratio, respectively. The four parameters may now be defined in words:

h_i = incremental input impedance with output short-circuited to AC
h_f = incremental current gain with output short-circuited to AC
h_r = incremental reverse voltage feedback ratio with input open-circuited to AC
h_o = incremental output admittance with input open-circuited to AC

Suppose now that a transistor working in the common-emitter mode were inside the "black box." Equations (3.1) and (3.2) would become:

$$V_{BE} = f(I_B, V_{CE}) \qquad (3.5a)$$

$$I_C = f(I_B, V_{CE}) \qquad (3.5b)$$

and

$$\partial V_{BE} = \frac{\partial V_{BE}}{\partial I_B}\, \delta I_B + \frac{\partial V_{BE}}{\partial V_{CE}}\, \delta V_{CE} \qquad (3.6a)$$

$$\delta I_C = \frac{\partial I_C}{\partial I_B}\, \delta I_B + \frac{\partial I_C}{\partial V_{CE}}\, \delta V_{CE} \qquad (3.6b)$$

or

$$v_{be} = h_{ie} i_b + h_{re} v_{ce} \qquad (3.7a)$$

$$i_c = h_{fe} i_b + h_{oe} v_{ce} \qquad (3.7b)$$

from which

$$h_{ie} = v_{be}/i_b \quad (v_{ce} = 0) \qquad (3.8a)$$

$$h_{fe} = i_c/i_b \quad (v_{ce} = 0)\,. \qquad (3.8b)$$

Here, i_b is the base (input) current due to the input voltage v_{be}, both being small signals, or incremental changes in I_B and V_{BE}. Also, i_c is the small-signal component of the collector current, I_C.

The $v_{ce} = 0$ condition is met by connecting a large capacitor from the collector to the emitter, so short-circuiting the output to AC.

Similarly,

$$h_{re} = v_{be}/v_{ce} \quad (i_b = 0) \qquad (3.9a)$$

$$h_{oe} = i_c/v_{ce} \quad (i_b = 0) \qquad (3.9b)$$

where a large inductor has been placed in series with the base, so open-circuiting it to AC.

Clearly, a transistor working in the CB or CC (emitter-follower) modes could have been characterized in the same manner, when equations (3.3) would have been equally valid. However, the parameters themselves would have been defined in terms of the input and output incremental voltages and currents for those modes, and the numerical values eventually obtained would have been different from those relevant to the CE mode.

For example, in the CB mode, the current-gain parameter would relate the collector (output) current to the emitter (input) current with the collector and base short-circuited to AC:

$$h_{fb} = i_c/i_e \quad (v_{cb} = 0)\,.$$

It is left as an exercise for the reader to define the other h-parameters in these alternative terms.

The h-parameters are actually the slopes of the transistor characteristic curves, which is why they can be easily determined numerically. For example, h_{oe} is simply the slope of an output characteristic, such as appear in figure 2.9, at some given working point. Further, h_{fe} can also be obtained from figure 2.9 by taking a given value of V_{CE} and plotting a curve of I_C/I_B, as has been done in figure 3.2.

The fact that the characteristics of figure 2.9 are neither parallel nor equidistant shows that the numerical values of h_{oe} and h_{fe} are very dependent upon the working point chosen. This is why such values are always quoted at specified working points.

Owing to manufacturing tolerances, transistors of a given type exhibit ranges of parameter values for a given working point, and such spreads must also be defined. For the typical small-signal silicon planar transistor of table 3.2

$$h_{fe} = 50 \text{ to } 200 \quad \text{at} \quad V_{CE} = 10 \text{ V} \quad \text{and} \quad I_C = 1 \text{ mA} .$$

Spreads even greater than these are not uncommon.

The input impedance, h_{ie}, is the slope of the V_{BE}/I_B characteristic at some specified working point. This curve is usually drawn with I_B on the ordinate and V_{BE} on the abscissa, as in figure 2.15(a). Again, the parameter dependence on working point is very obvious, for near the origin, h_{ie} is clearly high, whereas when V_{BE} exceeds about 0.7 V (for silicon transistors) it becomes quite low.

Although it is not obvious from curves such as figure 2.15(a) V_{BE} is a weak function of V_{CE}, this being due to the *base-width modulation* effect, which refers to the narrowing of the base region as the collector depletion layer widens with V_{CB}. This results in a slight fall in V_{BE} as V_{CE} rises (for

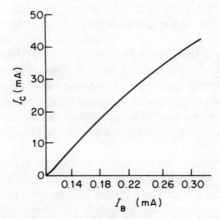

Figure 3.2. I_C/I_B curve derived from figure 2.9. $V_{CE} = 10$ V.

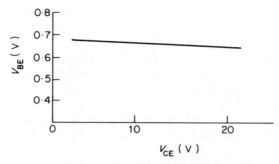

Figure 3.3. V_{BE}/V_{CE} characteristic showing small negative slope. $I_C = 5$ mA.

constant I_C) as is shown in figure 3.3. This slight slope accounts for the low value of h_{re}.

3.1.1   The *h*-Parameter Equivalent Circuit

In order that actual current and voltage gains or input and output impedances may be calculated for practical networks, an equivalent circuit must be constructed for the transistor which will involve only the fluctuating signal currents. This circuit is immediately suggested by the definitions of the *h*-parameters and will be the same for all the configurations; only the actual *h*-parameter numerical values will be dependent on the mode of connection.

Figure 3.4 depicts the equivalent circuit, which is built up as follows. Firstly, v_{in} is applied to the input of the transistor in the conventional sense. This input circuit consists of the input impedance h_i, in series with the voltage source $h_r v_{out}$. Thus the input current i_{in} (again drawn in the conventional direction) is defined by both v_{in} and v_{out}. Note that $h_r v_{out}$ opposes v_{in}. This is due to the negative slope of the V_{BE}/V_{CE} curve shown in figure 3.3.

The input current is then amplified by the current gain h_f, and appears as a current source $h_f i_{in}$ in the output circuit. This is shunted by the output admittance, across which is developed the output voltage v_{out}. Both v_{out} and i_{out} are shown in the conventional sense and the circuit is complete.

Later, it will be shown how this equivalent circuit can be inserted into a practical amplifier network in preparation for making performance calculations.

Since the *h*-parameters for the three modes of connection describe the same transistor it is apparent that there must be a relation between them. Consider for instance a transistor connected in the common-emitter mode and suppose that the value of h_{fb} is known. By inserting the CB *h*-parameter equivalent circuit into the CE configuration, an analysis may be made to

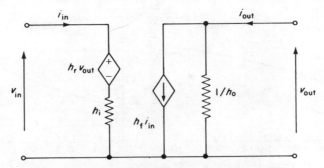

Figure 3.4. The h-parameter equivalent circuit.

extract an expression for h_{fe}:

$$h_{fe} = \frac{-h_{fb}}{1 + h_{fb}} \tag{3.10}$$

(compare the DC expression of equation (2.5)).

For example, if $h_{fb} = -0.98$, then $h_{fe} = 49$, which is positive since both δI_C and δI_B have the same sign.

Similarly,

$$h_{fc} = \frac{-1}{1 + h_{fb}} \tag{3.11}$$

and

$$h_{fb} = \frac{-h_{fe}}{1 + h_{fe}}. \tag{3.12}$$

A list of the relations between the h-parameters for the three modes is given in table 3.1 at the end of the chapter.

3.1.2 Variations of h-Parameters

All the h-parameter numerical values depend upon the working point at which they are measured. Figure 3.5 shows such variations and is relevant to a silicon planar small-signal npn transistor, the Motorola 2N3903. Figure 3.5(a) shows how h_{fe} increases with I_C, and this is simply due to the increasing slope of the h_{FE} versus I_C curve over the relevant range. (Figure 2.10 illustrates such a characteristic and is accompanied by a brief explanation.) The increase of h_{oe} with I_C, illustrated in figure 3.5(b) results from the increasing slopes of the family of output characteristics as the collector current rises (see figure 2.9). As will be shown in the next section, the input

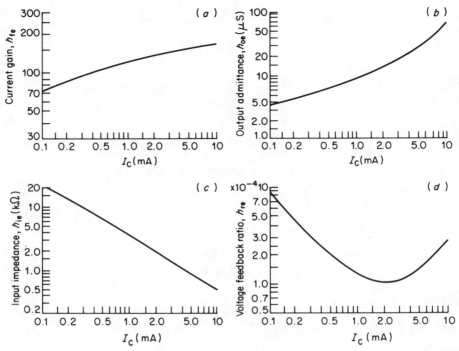

Figure 3.5. Variations in *h*-parameters with collector current for a 2N3903 npn silicon planar transistor. $V_{CE} = 10$ V, $f = 1$ kHz and $T = 25°C$. (Reproduced by permission of Motorola (Semiconductors) Ltd.)

impedance, h_{ie}, is an inverse function of I_C, and this is reflected in figure 3.5(*c*). The reverse voltage feedback ratio, h_{re}, is a complex function of I_C because of base-width modulation effects and variations in interelectrode capacitances. This complexity leads to the characteristic shape shown in figure 3.5(*d*). Detailed explanations of these effects will be found in most books on semiconductor device fabrication.

Finally, it should be noted that all the *h*-parameters increase with temperature largely as a result of charge mobility changes in the crystal lattice.

3.2 THE HYBRID-π EQUIVALENT CIRCUIT

It will have been noticed that although h_{ie} and h_{oe} have been termed an impedance and an admittance, respectively, they have nevertheless been represented as a resistance and a conductance. This reflects the generality that, although strictly speaking the *h*-parameters have complex values, they are normally used only in those frequency ranges where the real parts are

predominant. For high-frequency calculations they give way to the complex
y-parameters (see Appendix II) or to the more convenient hybrid-π formulation. The latter incorporates an equivalent circuit which is physically
related to the common-emitter mode and includes capacitances which take
account of the frequency response characteristics of the transistor. By this
means complex parameter values are avoided.

The CE hybrid-π equivalent circuit is shown in figure 3.6 and its components may be explained as follows.

The actual ohmic resistance of the base region is represented by $r_{bb'}$, and
a hypothetical active base location is assumed to exist at the point b'. This
point is, of course, not susceptible to practical measurement. Indeed it may
be assumed to shift with, for example, the applied collector–base voltage,
for the effective base width is a function of the width of the collector
depletion layer. Thus $r_{bb'}$, like all small-signal transistor parameters, is
variable.

The base–emitter resistance is represented by $r_{b'e}$ and the capacitance by
$C_{b'e}$. Here $C_{b'e}$ contains two components, the *emitter diffusion capacitance*
C_D and the much smaller *emitter transition capacitance* C_T. Although the
diffusion capacitance is, strictly speaking, due to a distributed effect in the
base region, it is nevertheless approximated by a lumped parameter in the
hybrid-π circuit for reasons of simplification. It comes about because when
minority carriers are injected from the emitter into the base and diffuse
across to the collector, those actually present in the base region constitute a
charge, the density of which is a maximum near the emitter junction and is
zero at the collector junction. Consequently, with variations in current
through the transistor, this charge must vary. Because I_C is a function of V_{BE}
the variation of the base charge with V_{BE} constitutes the emitter diffusion
capacitance, dQ_B/dV_{BE}. This ratio is of paramount importance when the
transistor is switched from a low to a high current state and has led to the
charge-control concept mentioned in chapter 10.

The *transition* or *barrier capacitance* of a junction has been discussed

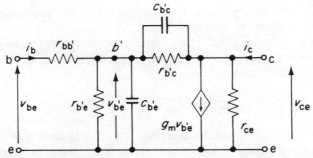

Figure 3.6. The hybrid-π equivalent circuit. (Note: in some texts, alternative
symbols for the hybrid-π or Giacoletto model appear as follows: $r_\pi = r_{b'e}$; $r_\mu = r_{b'c}$;
$r_x = r_{bb'}$; $C_\pi = C_{b'e}$; $C_\mu = C_{b'c}$.)

earlier and it has been seen that it is true capacitance; that is, it comes about because of charges existing at either side of a depletion layer. The value of this capacitance is an inverse function of the thickness of the depletion layer and hence of the potential difference across it. For this reason the transition capacitance of a reverse-biased junction is much smaller than that for a forward-biased junction; hence C_T for the emitter junction is much larger than $C_{b'c}$, which is the transition capacitance of the collector junction. Even so, $C_T \ll C_D$ so that $C_{b'e} \gg C_{b'c}$. The resistance $r_{b'c}$ is very large and can often be ignored.

The magnitude of the single, dependent, current generator in the hybrid-π equivalent, $g_m v_{b'e}$, being a function of the voltage across the capacitive b'–e junction, must fall with frequency. The rate of fall of $v_{b'e}$ is simply that of a single low-pass RC circuit (discussed in Appendix I) and accounts for the utility of the hybrid-π circuit in high-frequency calculations. This topic will be continued in chapter 5.

3.2.1 The Hybrid-π Parameters

Although the hybrid-π equivalent circuit (first proposed by Giacoletto[1]) is an extremely useful model, most of its parameters are not susceptible to direct measurement but must be inferred from physical quantities or other measurable parameters. In particular, g_m and $r_{b'e}$ are immediately obtainable from the equation relating I_C and $V_{B'E}$, the total collector current and the voltage across the active part of the base–emitter junction. This relationship may be obtained from the Ebers–Moll model[2] which represents the transistor in any operating condition and from which equations for I_E and I_C may be derived. Equation (2.11a) is an approximation derived from this model and for base current levels low enough to make the voltage drop in $r_{bb'}$ negligible ($V_{B'E} \simeq V_{BE}$), may be rewritten as follows:

$$I_C \simeq I_{ES}(e^{qV_{B'E}/kT} - 1). \tag{3.13}$$

Here, I_{ES}, the emitter–base leakage current with the collector short-circuited is best found by extrapolation.

For normal operation, with the emitter junction forward-biased to a few hundred millivolts, the exponential term is much greater than unity. Hence, equation (3.13) further approximates to

$$I_C \simeq I_{ES}\, e^{qV_{B'E}/kT}. \tag{3.14}$$

This exponential approximation describes the transconductance characteristic of a transistor (figure 2.15(b)) and its accuracy for a modern low-level planar transistor at low current levels is particularly good, as is demonstrated by figure 3.7. Here, the measured plot is on semi-logarithmic axes, the straight-line result confirming the exponential relationship.

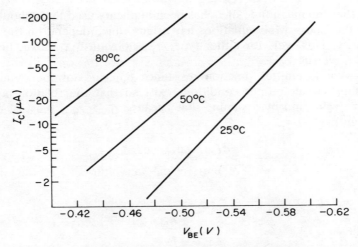

Figure 3.7. Transconductance characteristics for a BFW 20 silicon planar low-level pnp transistor.

Notice that $|I_C|$ is very temperature sensitive, not only because of the existence of T in the exponent, but also because I_{ES}, being a reverse leakage current, is itself highly temperature dependent. If $|I_C|$ is held constant, $|V_{BE}|$ will be found to fall by 2 to 3 mV/°C, whereas if $|V_{BE}|$ is held constant $|I_C|$ obviously increases exponentially with temperature. In the next chapter attention will be paid to biasing circuits whose function is to minimize the temperature dependence of $|I_C|$.

The transconductance, g_m, is easily obtained from equation (3.14):

$$g_m = \frac{dI_C}{dV_{B'E}} = \frac{q}{kT} I_C .$$ (3.15a)

This is an interesting result in that g_m is seen to be directly proportional to the collector current and inversely proportional to the temperature but is independent of the transistor itself.

Using the fact that $kT/q \simeq 26$ mV at 25°C,

$$g_m \simeq \frac{I_C}{0.026} \text{ A/V} \quad \text{at 25°C}$$

or in more practical units,

$$g_m \simeq 39 \, I_C \text{ mA/V} \quad \text{at 25°C}$$ (3.15b)

if I_C is in milliamps.

Note that the magnitude of I_C has been implicitly used throughout; this is because the Ebers–Moll equations have signs which depend on the transistor polarity. However, for either type, g_m is obviously positive, being the slope of I_C versus $V_{B'E}$.

The dynamic emitter junction resistance $r_{b'e}$ can now be obtained by letting short-circuit output conditions exist so that the relationship $h_{fe} = dI_C/dI_B$ is valid and by assuming that because $r_{b'c} \gg r_{b'e}$, most of δI_B will flow down $r_{b'e}$,

$$\frac{1}{r_{b'e}} = \frac{dI_B}{dV_{B'E}} = \frac{dI_B}{dI_C}\frac{dI_C}{dV_{B'E}} \simeq \frac{g_m}{h_{fe}}$$

or

$$r_{b'e} \simeq \frac{h_{fe}}{g_m}. \qquad (3.16a)$$

Again assuming $kT/q \simeq 26\,\text{mV}$ at 25°C,

$$r_{b'e} \simeq \frac{26 h_{fe}}{I_C}\,\Omega \quad \text{at 25°C} \qquad (3.16b)$$

if I_C is in milliamps.

Clearly, although $r_{b'e}$ varies inversely with the collector current, its value is modified by the fact that h_{fe} also varies with collector current, though in the opposite sense, as shown in figure 3.5.

If it is assumed that $r_{b'c}$, being associated with a reverse-biased junction, is much greater than $r_{b'e}$, then at low frequencies the input resistance must be:

$$R_{in} \simeq r_{bb'} + r_{b'e}$$

and if short-circuit output conditions prevail,

$$R_{in} = h_{ie} \simeq r_{bb'} + r_{b'e}. \qquad (3.17)$$

Because $r_{b'e}$ rises rapidly as I_C falls (equation (3.16b)) then h_{ie} must do the same, for $r_{bb'}$ is not very dependent on I_C, being a bulk resistance. This explains the rapid rise of h_{ie} with falling I_C in figure 3.5.

Equation (3.17) also suggests a method for determining $r_{bb'}$. If h_{ie} is measured and $r_{b'e}$ is calculated then

$$r_{bb'} \simeq h_{ie} - r_{b'e}. \qquad (3.18)$$

Earlier, it was suggested that $r_{b'c}$ must be large, being associated with a reverse-biased junction. This can be demonstrated by determining h_{re}, which involves applying a voltage source at the output and measuring the resultant voltage at the (open-circuited to AC) input. A glance at the hybrid-π equivalent circuit will suffice to show that, for low frequencies,

$$\frac{v_{be}}{v_{ce(i_b=0)}} = h_{re} = \frac{r_{b'e}}{r_{b'c} + r_{b'e}}$$

so that

$$r_{b'c} = \frac{r_{b'e}(1 - h_{re})}{h_{re}} \simeq \frac{r_{b'e}}{h_{re}} .$$ (3.19)

To justify numerically that $r_{b'c} \gg r_{b'e}$, consider the data in table 3.2. Because these data are relevant to a collector current of 1 mA,

$$r_{b'e(min)} \simeq \frac{26 \times 50}{1} = 1300\ \Omega \quad r_{b'e(max)} \simeq \frac{26 \times 200}{1} = 5200\ \Omega$$

giving

$$r_{b'c(max)} \simeq \frac{1300}{0.5 \times 10^{-4}} = 26\ \text{M}\Omega \quad r_{b'c(min)} \simeq \frac{5200}{5 \times 10^{-4}} = 10.4\ \text{M}\Omega .$$

If the input is now short-circuited to AC a simple analysis of the circuit will show that the measured output resistance is comprised largely of r_{ce}. That is at low frequencies, assuming that $r_{b'c} \gg r_{ce}$,

$$\frac{v_{ce}}{i_c} = R_{out} \simeq r_{ce} .$$ (3.20a)

This is normally a fairly large resistance and, like $r_{b'c}$, it varies as a result of base-width modulation (or the *Early effect*) described on p. 29. The slope of an output characteristic at the working point obviously gives r_{ce} and this is again a small-signal or dynamic resistance, the value of which is a function of that working point.

Figure 3.8 shows a family of output characteristics which have been produced backwards to coalesce at a point on the horizontal axis. This represents the Early voltage V_A, and if it is known then r_{ce} can be calculated at any point on the straight line part of any characteristic:

$$r_{ce} = \frac{v_{ce}}{i_c} = \frac{V_{CE} - V_A}{I_C} .$$ (3.20b)

Note that r_{ce} is inversely proportional to I_C and therefore to g_m also.

In the h-parameter system $1/h_{oe}$ corresponds to r_{ce} and if graph 3.5(b) is consulted it will be seen that for the transistor in question, working at $I_C = 1$ mA, where h_{oe} is about 8.6 μS,

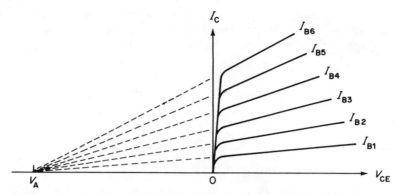

Figure 3.8. Output characteristics produced backwards to define the Early voltage V_A.

$$r_{ce} = \frac{10^3}{8.6} = \frac{10 - V_A}{1}$$

or

$$V_A = 10 - \frac{10^3}{8.6} \simeq -106 \text{ V}.$$

(Note that Early voltages are negative for npn, positive for pnp transistors.)

This is a typical value within the usual wide range of about 60 to 140 V.

To determine the value of $C_{b'c}$, it is necessary only to observe that if the transistor is connected in the CB mode with the input open-circuited to AC, then the output capacitance C_{obo} is simply $C_{b'c}$ (neglecting header capacitances):

$$C_{b'c} \simeq C_{obo}. \tag{3.21}$$

Finally, to determine the value of $C_{b'e}$, it is necessary to measure h_{fe} as a function of frequency. To do this the output is short-circuited (by definition); that is, in figure 5.4, the collector load $R_L = 0$. If now a sinusoidal input signal is applied between base and emitter, V_{be}, then an input current I_b will flow so that

$$I_b = V_{b'e}\left(\frac{1}{r_{b'e}} + j\omega(C_{b'e} + C_{b'c})\right).$$

Also, the output current is

$$I_c = g_m V_{b'e}$$

so that

$$h_{fe} = \frac{I_c}{I_b} = \frac{g_m}{(1/r_{b'e}) + j\omega(C_{b'e} + C_{b'c})}$$

or

$$|h_{fe}| = \frac{g_m}{[(1/r_{b'e}^2) + \omega^2(C_{b'e} + C_{b'c})^2]^{1/2}} \, .$$

At frequencies where $\omega^2(C_{b'e} + C_{b'c})^2 \gg 1/r_{b'e}^2$

$$|h_{fe}| \simeq \frac{g_m}{\omega(C_{b'e} + C_{b'c})}$$

or

$$|h_{fe}|f \simeq \frac{g_m}{2\pi(C_{b'e} + C_{b'c})} \, .$$

In chapter 5 it will be shown that, at the frequencies where $r_{b'e}$ ceases to be important, $|h_{fe}|f = F_T$, the *gain–bandwidth product*. This parameter is usually given in data sheets so that $C_{b'e}$ may be easily determined from

$$C_{b'e} = \frac{g_m}{2\pi F_T} - C_{b'c} \tag{3.22}$$

$$\simeq \frac{g_m}{2\pi F_T} \quad \text{if} \quad C_{b'c} \ll C_{b'e} \, .$$

All the hybrid-π parameters have now been defined; some in terms of physical quantities and some in terms of the (measurable) h-parameters. In chapter 5, it will be shown how the circuit is of use in high-frequency calculations and also how it may be simplified to facilitate the design of very low-level stages.

3.3 TYPICAL PARAMETER VALUES

It is useful in circuit design to have a knowledge of typical numerical values for the more important parameters. Taking both the h-parameters and the hybrid-π parameters some general conclusions can be drawn regarding their magnitudes, namely:

(1) The short-circuit CE current gain, h_{fe}, is usually in the range 50 to 300 for a modern small-signal silicon planar transistor, which implies that h_{fb} is not less than -0.98 (equation (3.12)). This means that those approximations which assume h_{fb} to to close to -1 do not introduce

much error. Also, h_{fc} is always $-(h_{fe} + 1)$, so that these two parameters can often be used interchangeably, *provided* that it is recalled that h_{fc} has a negative numerical value.

(2) The input resistance to a CE stage (with output short-circuited) is h_{ie} or $(r_{bb'} + r_{b'e})$. Since $r_{b'e} \simeq 26h_{fe}/I_C$, it is usually larger than $r_{bb'}$. Typically, h_{ie} is a few thousand ohms for a small-signal silicon planar transistor, while $r_{bb'}$ may be only a few hundred or even tens of ohms.

(3) The output resistance, r_{ce} or $1/h_{oe}$, is usually between $20\,k\Omega$ and $0.5\,M\Omega$ so that approximations which assume that it is much larger than a collector resistor of a few thousand ohms are justified.

(4) In the hybrid-π formulation, internal feedback occurs via $r_{b'c}$, whereas this function is carried out by the dependent voltage generator, $h_{re}v_{ce}$, in the h-parameter case. Since this effect is small, h_{re} must also be small and is typically about 10^{-4}. Conversely, $r_{b'c}$ (being the resistance across a reverse-biased junction) must be large and is commonly more than $10\,M\Omega$ (equation (3.19)).

In the case of the CC or emitter-follower stage, the load impedance appears in the emitter circuit and so is common to both the input and output loops. This means that the full output signal voltage is fed back to the input, so that the reverse voltage transfer ratio, h_{rc}, *is unity*.

Table 3.2 at the end of the chapter gives h-parameter values for a typical small-signal silicon planar transistor, and various approximations will be made in chapter 5 using such values.

3.4 SUMMARY

The two most common incremental, or small-signal equivalent circuits for the bipolar transistor have been presented, and some relationships between their parameters have been established. This has been done using only the "positive-in, negative-out" current convention and the node relationship that $i_c + i_b + i_e = 0$.

3.4.1 Important Equations

1. *h-parameters*

$$v_{in} = h_i i_{in} + h_r v_{out}$$

$$i_{out} = h_f i_{in} + h_o v_{out}$$

$$\Delta h = h_i h_o - h_r h_f$$

$$h_{fe} = \frac{-h_{fb}}{h_{fb} + 1} \qquad h_{fb} = \frac{-h_{fe}}{1 + h_{fe}} \qquad h_{fc} = -(h_{fe} + 1).$$

2. *Hybrid-π parameters*

$$g_m \simeq \frac{q}{kT} |I_C| \quad \text{or} \quad 39|I_C| \text{ mA/V} \quad \text{at } 25°\text{C} \, (I_C \text{ in mA})$$

$$r_{b'e} \simeq \frac{h_{fe}}{g_m} \quad \text{or} \quad \frac{26 h_{fe}}{|I_C|} \, \Omega \quad \text{at } 25°\text{C} \, (I_C \text{ in mA})$$

$$r_{b'c} \simeq \frac{r_{b'e}}{h_{re}}$$

$$r_{bb'} \simeq h_{ie} - r_{b'e}$$

$$C_{b'c} \simeq C_{obo}$$

$$C_{b'e} = C_T + C_D \simeq \frac{g_m}{2\pi F_{T'}} - C_{b'c}.$$

TABLE 3.1 CB and CC *h*-Parameters in Terms of CE *h*-Parameters.

$h_{ib} = \dfrac{h_{ie}}{1 + h_{fe}}$	$h_{ic} = h_{ie}$
$h_{ob} = \dfrac{h_{oe}}{1 + h_{fe}}$	$h_{oc} = h_{oe}$
$h_{rb} = \dfrac{h_{ie} h_{oe}}{1 + h_{fe}} - h_{re}$	$h_{rc} = \dfrac{1}{1 + h_{re}}$
$h_{fb} = \dfrac{-h_{fe}}{1 + h_{fe}}$	$h_{fc} = -(1 + h_{fe})$

TABLE 3.2 Parameters for a Typical Small-Signal npn Silicon Planar Transistor.

$V_{CEM} = 40$ V; $I_{CM} = 250$ mA; $P_{tot} = 250$ mW at 25°C

Direct current gain at $I_C = 1$ mA:

	minimum	maximum
h_{FE}	35	170

Small-signal *h*-parameters at $I_C = 1$ mA; $V_{CE} = 10$ V; $f = 1$ kHz

	minimum	maximum
h_{ie}	1800 Ω	5800 Ω
h_{oe}	2 μS	40 μS
h_{re}	0.5×10^{-4}	5.0×10^{-4}
h_{fe}	50	200
⚠$_e$	1.1×10^{-3}	132×10^{-3}

TABLE 3.2 (*Continued*)

Other small-signal parameters:

$$r_{bb'} < 50\,\Omega$$

$C_{b'e} \simeq 15\,\text{pF}; \; C_{b'c} \simeq 2\,\text{pF}$ at $V_{CE} = 10\,\text{V}$ and $f = 100\,\text{kHz}$

$$F_{T(min)} = 300\,\text{MHz}$$

REFERENCES

1. Giacoletto, L. J., 1954, Study of PNP alloy junction transistors from d.c. through medium frequencies *RCA Rev.* **15**, 506, December.
2. Searle, C. L., Boothroyd, A. R., Angelo, E. J., Gray, P. E. and Pederson, D. O., 1974, *Elementary Circuit Properties of Transistors* (New York: Wiley, SEEC Series) chap. 2.

QUESTIONS

1. What is the difference between a full model and an incremental model?
2. What advantage does the h-parameter model have over the hybrid-π model?
3. How are the three basic transistor configurations, CE, CB, and CC, taken into account in the h-parameter model?
4. Can the fact that h_{fe} is slightly different from h_{FE} be deduced from figure 3.2? If so, how?
5. If the quiescent collector current of a 2N3903 transistor were decreased from about 5 mA to about 0.1 mA, by how much would the incremental input resistance at the base increase? (see figure 3.5).
6. For a transistor working at a quiescent collector current of 1 mA, plot the approximate change in g_m over a temperature range of 0°C to 40°C.
7. If $C_{b'e}$ is assumed to remain constant, how will the gain–bandwidth product F_T change with the quiescent collector current I_C? (see equation (3.22)).
8. For the typical small-signal npn transistor of Table 3.2, what is the lowest value of h_{fb} to be expected for the working conditions given?

PROBLEMS

3.1. Using the characteristics of a typical 2N3903 small-signal npn transistor given in figure 3.5, determine the four CE h-parameters at $V_{CE} = 10\,\text{V}$;

$I_C = 5\,\text{mA}$ and $f = 1\,\text{kHz}$. Also find the determinant $\triangle_e$ using these parameters.

3.2. If the transistor of problem 3.1 were to be connected in the common-base mode, what would be the value of h_{fb}?

3.3. What is the transconductance g_m of the transistor of problem 3.1 if it is working at room temperature? Also, what is its incremental input resistance $r_{b'e}$?

3.4. A transistor has a collector current of $1\,\text{mA}$, and exhibits internal capacitances of $C_{b'e} = 15\,\text{pF}$ and $C_{b'c} = 2\,\text{pF}$. What is its approximate gain–bandwidth product F_T?

3.5. A transistor connected in the CE mode and working at room temperature, has a quiescent collector current of $1\,\text{mA}$. If the measured value of h_{ie} is $2850\,\Omega$ and the incremental current gain h_{fe} is 100, what is the value of $r_{bb'}$? Is this value likely to be accurate?

3.6. A circuit design calls for a transistor to have an incremental collector resistance of not less than $50\,\text{k}\Omega$ when $I_C = 2\,\text{mA}$ and $V_{CE} = 10\,\text{V}$. What must be the minimum magnitude of the Early voltage for the transistor type used?

4

The Transistor as an Amplifier

The DC and the AC characteristics of transistors described in chapters 2 and 3, respectively, will now be combined and used to establish the circuitry necessary for constructing practical transistor amplifiers. Initially, only small signal amplifiers will be considered since the large excursions of current involved in power amplifiers make necessary a rather different treatment.

Transistors require to be biased in such a manner that known *quiescent* currents flow when no alternating signal is applied. These currents must make the operating conditions of the device such that it does not dissipate too much power and that, when the signal is applied, a minimum of distortion occurs from input to output. It is usually the circuit components necessary for correct biasing which dictate the gain and input and output impedances for the amplifier. These can be calculated by combining an equivalent circuit for the transistor with the biasing components and solving the relevant circuit equations.

To illustrate the concept that a transistor must be properly biased into a working condition before it is able to accept small signals, it is convenient to consider the most common configuration, which is the common-emitter mode of figure 2.8(a). This is repeated in figure 4.1(a), which also shows the simplest (but most unsatisfactory) method of biasing; that is, the inclusion of a base bias resistor R_B taken directly to the supply V_{CC}.

Figure 4.1(b) shows the *load-line diagram* relevant to this circuit, which is assumed to be biased so that a no-signal or *quiescent operating point Q* is defined. The load-line itself is the locus of possible operating points on an output characteristic diagram like that of figure 2.9, given a fixed value of R_C. It is obvious from the circuit that:

$$I_C = \frac{V_{CC} - V_{CE}}{R_C}$$

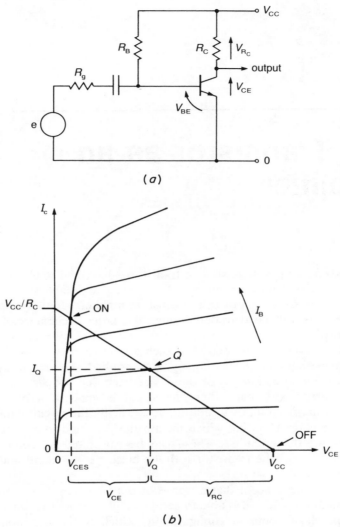

(a)

(b)

Figure 4.1. A simple but unsatisfactory biasing method (a), and the load-line diagram (b).

or

$$I_C = - \frac{V_{CE}}{R_C} + \frac{V_{CC}}{R_C}$$

which is a straight line of slope $-1/R_C$, and cutting the abscissa at V_{CC}. From a physical point of view, this means if no base current were injected, then the transistor would be cut off at $I_C \rightarrow 0$. In this OFF condition there can be no voltage drop across R_C so that $V_{CE} = V_{CC}$.

Conversely, if a large enough base current were injected to saturate the transistor, then $I_C \approx V_{CC}/R_C$ and only a small saturation voltage V_{CES} would appear across that transistor. This is the ON, or saturated, condition.

Transistors used as switches are required to move rapidly between these two conditions, and such applications will be treated in chapter 10. However, for small-signal amplifiers, a quiescent working point Q between these extremes must first be defined by a biasing circuit, the aim of which is not only to establish this working point at $I_C = I_Q$ and $V_{CE} = V_Q$, but to maintain its position on the load-line as closely as possible under all circumstances. Then, a small signal may be injected as is also shown in figure 4.1(a) as e and this will make the working point "wobble" along the load-line about its quiescent position Q. Since $\Delta I_C \gg \Delta I_B$, this will constitute both signal current and hence voltage amplification. (Further notes on this, including signal distortion, appear in chapter 9.)

This is the way in which the CE stage operates, and it can be seen that if I_B increases slightly, then I_C must increase considerably more, and so V_{CE} must decrease, because $V_{CE} = V_{CC} - I_C R_C$ (and *vice versa*). This accounts for the phase inversion property of the CE stage.

To define the quiescent position of Q, therefore, a base bias current I_B must be injected, and figure 4.1(a) shows this being accomplished by simply connecting R_B from the base to the supply. Unfortunately, this leads to disastrous consequences engendered by transistor *parameter tolerances*, which can be considerable, and make necessary better circuit designs which result in proper biasing irrespective of these tolerances. These problems will now be treated in detail.

4.1 BIASING OF TRANSISTORS

The serious problems posed by the wide variations in parameter values encountered in a series of transistors—even of one given type—can be exemplified by considering the effect of variations in the DC current gain, h_{FE}, on the bias condition of a simple circuit. Consider figure 4.1(a) wherein a quiescent collector current of 2 mA is chosen for an npn transistor type having maximum and minimum values of h_{FE} quoted as 100 and 40, respectively. For a "minimum" transistor $I_B = 2/40 = 0.05$ mA, which gives R_B as:

$$R_B = \frac{V_{CC} - V_{BE}}{I_B} = \frac{12 - 0.7}{0.05} = 226 \text{ k}\Omega$$

(where V_{BE} has been taken as 0.7 V and V_{CC} as a typical 12 V).

If the quiescent voltage drops across the transistor and its load are made equal at 6 V each (which is a reasonable condition, as will be seen later) then:

$$R_C = 6/2 = 3 \text{ k}\Omega \,.$$

Now suppose that a "maximum" transistor is substituted. The base current will remain much the same, but the collector current will try to rise to $0.05 \times 100 = 5$ mA. If this were possible the voltage drop across R_C would become $5 \times 3 = 15$ V! What actually happens is that the transistor *bottoms* or *saturates*, which means that the current is limited by R_C and some 0.3 V only is dropped across the transistor (for a silicon device). In this condition, the emitter gives rise to an electron current of $h_{FE}I_B$ and most of these electrons reach the collector. The excess electrons bias the collector in the forward direction and eventually recross to the base, for R_C limits the external current according to Ohm's law. The bottoming voltage is therefore the difference between the forward voltage drop of the emitter–base and collector–base diodes. It is given by the voltage at the "knee" of the I_C/V_{CE} curves (hence the alternative term "knee voltage") and is also usually quoted in the manufacturers' data sheets. Here it is given the symbol V_{CES}, where the final subscript means saturation.

When saturation has occurred, the transistor no longer works as an amplifier and in the present example the only way to bring it into operation again would be to increase R_B. This illustrates the futility of using the published I_C/V_{CE} characteristic as more than a general guide to the nature of the transistor. It also shows that the biasing of a transistor by means of a quiescent base current can be very bad practice.

A considerable improvement can be effected by connecting R_B between the base and the collector, as shown in figure 4.2. In this circuit, if the collector current rises above the chosen quiescent value, a greater voltage drop appears across R_C, and V_{CE} therefore decreases. This means that the current flowing into the base is reduced since $I_B \simeq V_{CE}/R_B$. Hence, the DC *feedback* so provided tends to keep the transistor in an operative state

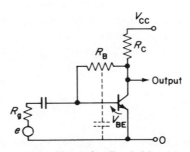

Figure 4.2. Shunt feedback bias.

irrespective of changes in h_{FE}. A circuit of this nature actually provides *shunt* DC feedback since the DC feedback path is in parallel with the AC signal path.

The prime disadvantage of this method is that since the base can never go negative with respect to the emitter then the collector current can never drop below I_{CEO} (see figure 2.11).

Another disadvantage is that feedback of the AC signal also occurs, which will act to reduce the gain of the circuit. This can be obviated by connecting a capacitor to ground from a point midway along R_B, which will shunt the AC component of the feedback current to ground before it contributes to the base current. It will, of course, also shunt the incoming signal by a resistance of $\frac{1}{2}R_B$ if the reactance of the capacitor is small compared with R_B at the frequencies of interest.

A biasing circuit that is in very common use is shown in figure 4.3. Here, the bias voltage V_B is kept fairly constant by the resistors R_1 and R_2. If I_C (and hence I_E) rises, the voltage drop across R_E increases and V_E becomes more positive. Now obviously $V_{BE} = V_B - I_E R_E$ and since V_B remains constant V_{BE} must fall, which reduces I_C. The control involved is very good, for a small change in V_{BE} leads to a large change in I_C, as can be seen from the curve of figure 2.15(b). However, the input signal is again shunted, this time by R_1 and R_2 in parallel, the power pack impedance being very small. In contrast to the circuit of figure 4.2, I_C can in this case fall below I_{CEO} and may approach I_{CBO}, for V_E can go positive with respect to the base if the incoming signal goes very negative.

It is apparent that the degree of control exercised by this circuit is dependent on the degree of invariance of V_B, which in turn is dependent on the values of R_1 and R_2. The best possible control would occur if R_2 were zero, in which case V_{BE} would be equal to $-V_E$. This, of course, would be

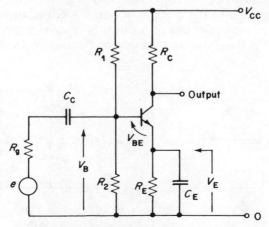

Figure 4.3. Series feedback biasing method.

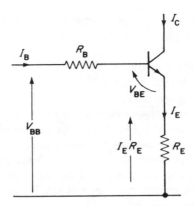

Figure 4.4. DC equivalent circuit for analyzing the series feedback biasing method of figure 4.3.

useless for a common-emitter amplifier since the signal would be short-circuited at the input, but if the signal were injected as shown in figure 4.5(a) the circuit would become a common-base amplifier. Such a circuit obviously uses an emitter current as its biasing parameter and the above argument shows that it provides good stability, which is in contrast to the poor stability of the base-current biased circuit of figure 4.1(a). As has also been shown, it is a special case of figure 4.3 and since by definition $I_C < I_E$ it can produce no current gain but only voltage (and power) gain.

Another special case is that of figure 4.5(b) which shows a common-collector (or emitter-follower) circuit. Here, $V_E = V_B - V_{BE}$, which shows that no voltage gain but only current (and power) gain is possible.

In the circuits shown in figures 4.3 and 4.5(a), the emitter resistor must be shunted by a capacitor, otherwise the fluctuations in I_E owing to the signal will also tend to be reduced and a loss in gain owing to *degeneration* will occur. In contrast to the method of figure 4.2, the circuit of figure 4.3 has a DC feedback path (through R_E) which is in series with the input signal path. Figure 4.3 is therefore an example of a *series DC feedback* circuit.

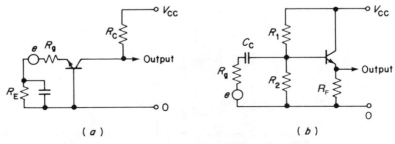

Figure 4.5. (a) Common-base stage; (b) emitter-follower stage.

4.1.1 Temperature Variations and Stability Factor

The consideration of biasing networks given above arose from an example showing how an increase in h_{FE} owing to transistor replacement could cause saturation to occur. This increase in h_{FE} could, however, have been due to an increase in temperature and the subsequent rise in I_C would then have been augmented by two further phenomena.

Firstly, I_{CEO} rises with temperature (this can be deduced from figure 2.12) and this contributes directly to I_C.

Secondly, V_{BE} falls with a rise in temperature, as shown in figure 4.6. In a typical circuit, such as that of figure 4.3, $V_{BE} = V_B - I_E R_E$, which means that as V_{BE} falls, I_E and hence I_C must rise, assuming that V_B remains constant.

If a transistor amplifier is not to become inoperative because of a rise in temperature the cumulative effect of the changes in h_{FE}, I_{CEO}, and V_{BE} must be compensated for in the circuit itself. If this is not done the transistor will saturate when a certain temperature is reached, or may even be destroyed owing to *thermal runaway*. This term refers to the case where, as a result of an increase in temperature, I_C increases and heats up the transistor, thus causing I_C to increase even more, until finally the maximum rate of heat dissipation is exceeded.

In order to calculate the effectiveness of a biasing circuit in counteracting the increase in I_C owing to temperature rise, partial differential coefficients giving the rate of change of I_C with respect to I_{CEO}, V_{BE} and h_{FE} must first be derived. Next, each of these derivatives must be multiplied by the rate of

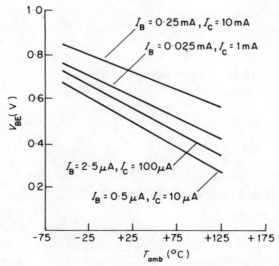

Figure 4.6. V_{BE}/T_{amb} curves for Texas 2N930 (low-level, low-noise silicon npn transistor). (Reproduced by permission of Texas Instruments Ltd.)

change of I_{CEO}, V_{BE}, and h_{FE}, respectively, with temperature and the sum of these multiples will then give dI_C/dT.

Actually, it is rarely necessary to go as far as this, for modern silicon transistors exhibit very low values of I_{CEO} which do not become sufficiently large to adversely affect biasing conditions even at high temperatures. (In point of fact, the temperature variations in V_{BE} have a more marked effect on I_C.) Furthermore, modern methods of biasing developed for monolithic microcircuits (such as the use of current mirrors) can be applied to discrete devices also, as will be shown, and these are very little affected by temperature variations.

However, as an example of the use of a DC equivalent circuit, a derivation of bias stability for the circuit of figure 4.3 now follows.

Firstly, resistors R_1 and R_2 are combined to give R_B, shown in figure 4.4:

$$R_B = \frac{R_1 R_2}{R_1 + R_2}.$$

A voltage V_{BB} is now assumed to drive I_B around the base-emitter circuit. Summing voltages round this input loop gives:

$$V_{BB} = I_E R_E + V_{BE} + I_B R_B. \tag{4.1}$$

To eliminate V_{BB} and I_E from equation (4.1) and hence arrive at an expression for I_C, two relationships can be employed:

$$I_E = I_C + I_B \tag{4.2}$$

$$I_B = \frac{I_C - I_{CEO}}{h_{FE}}. \tag{4.3}$$

Substituting for I_B from (4.3) into (4.2) gives

$$I_E = I_C + \frac{I_C - I_{CEO}}{h_{FE}}$$

$$= \frac{I_C(1 + h_{FE})}{h_{FE}} - \frac{I_{CEO}}{h_{FE}}. \tag{4.4}$$

Inserting (4.3) and (4.4) into (4.1) gives

$$V_{BB} = I_C R_E \left(\frac{1 + h_{FE}}{h_{FE}} \right) - \frac{I_{CEO} R_E}{h_{FE}} + V_{BE} + \frac{I_C R_B}{h_{FE}} - \frac{I_{CEO} R_B}{h_{FE}}$$

whence

$$I_C(R_B + R_E + R_E h_{FE}) = h_{FE}(V_{BB} - V_{BE}) + I_{CEO}(R_E + R_B)$$

and

$$I_C\left(1 + \frac{R_E h_{FE}}{R_E + R_B}\right) = h_{FE}\left(\frac{V_{BB} - V_{BE}}{R_E + R_B}\right) + I_{CEO} . \tag{4.5}$$

Differentiating (4.5) partially with respect to I_{CEO} gives

$$\frac{\partial I_C}{\partial I_{CEO}}\left(1 + \frac{R_E h_{FE}}{R_E + R_B}\right) = 1$$

or

$$\frac{\partial I_C}{\partial I_{CEO}} = K$$

where K, the stability factor†, is given by

$$K = \frac{1}{1 + [R_E h_{FE}/(R_E + R_B)]} . \tag{4.6}$$

As stated above, the variation of I_C with I_{CEO} is of paramount importance for germanium transistors but the discussion of the ramifications of the stability factor K will be postposed until all the derivations have been completed. The reader may, however, wish to pass on to this discussion immediately (see p. 68) and can do so without impairing the continuity of the argument.

Differentiating (4.5) partially again, this time with respect to V_{BE}, gives

$$\frac{\partial I_C}{\partial V_{BE}}\left(1 + \frac{R_E h_{FE}}{R_E + R_B}\right) = \frac{-h_{FE}}{R_E + R_B}$$

or

$$\frac{\partial I_C}{\partial V_{BE}} = \frac{-K h_{FE}}{R_E + R_B} . \tag{4.7}$$

This equation shows how I_C varies with V_{BE} and the negative sign confirms that I_C increases as V_{BE} decreases.

Finally, differentiating (4.5) partially with respect to h_{FE} gives

---

†Note that K is one form of stability factor. Another form, usually denoted by S, defines the change of I_C with I_{CBO}, not I_{CEO}:

$$S = \frac{\partial I_C}{\partial I_{CBO}} = K(1 + h_{FE}) .$$

$$\frac{\partial I_C}{\partial I_{FE}}\left(1 + \frac{R_E h_{FE}}{R_E + R_B}\right) + \frac{R_E I_C}{R_E + R_B} = \frac{V_{BB} - V_{BE}}{R_E + R_B}$$

or

$$\frac{\partial I_C}{\partial h_{FE}} = K\left(\frac{V_{BB} - V_{BE}}{R_E + R_B} - \frac{R_E I_C}{R_E + R_B}\right).$$

Using (4.1) and (4.2) this becomes

$$\frac{\partial I_C}{\partial h_{FE}} = K\left(\frac{I_E R_E + I_B R_B}{R_E + R_B} - \frac{I_E R_E - I_B R_E}{R_E + R_B}\right)$$

or

$$\frac{\partial I_C}{\partial h_{FE}} = K I_B . \tag{4.8}$$

If it is now assumed that the manner in which I_{CEO}, V_{BE}, and h_{FE} vary with temperature is known, it becomes possible to collect together the partial derivatives (4.6), (4.7), and (4.8) to produce a total derivative dI_C/dT:

$$\frac{dI_C}{dT} = \frac{\partial I_C}{\partial I_{CEO}}\frac{dI_{CEO}}{dT} + \frac{\partial I_C}{\partial V_{BE}}\frac{dV_{BE}}{dT} + \frac{\partial I_C}{\partial h_{FE}}\frac{dh_{FE}}{dT} \tag{4.9}$$

$$= K\frac{dI_{CEO}}{dT} - \frac{K h_{FE}}{R_E + R_B}\frac{dV_{BE}}{dT} + K I_B \frac{dh_{FE}}{dT} . \tag{4.10}$$

Consider the formula for K, the stability factor:

$$K = \frac{\partial I_C}{\partial I_{CEO}} = \frac{1}{1 + [h_{FE} R_E/(R_E + R_B)]} . \tag{4.6}$$

This confirms the earlier argument that the stability improves if R_B is small, for this makes K small and hence I_C changes but little with variations in I_{CEO}. It is also apparent that a high value of h_{FE} makes for good stability and it is for this reason that, when K is calculated, the *lowest* possible value for h_{FE} should be used. By this means the worst stability factor likely to occur for a given transistor type can be predicted.

In general K must be less than unity but a useful rule of thumb is to make it less than about 0.2†.

As an example, consider a small power germanium transistor in a circuit

†The stability factor $S(= K(1 + h_{FE}))$, is always greater than unity and a comparable rule of thumb is that it should be less than 5.

such that $K = 0.2$ at the lowest quoted value of h_{FE}, which is 33. At 25°C, I_{CBO} is given as 1.2 μA, rising to 62 μA at 75°C.

It would be simple to convert I_{CBO} to I_{CEO} for each case considered but it would be more useful to quote equation (4.6) directly in terms of I_{CBO} thus:

$$\frac{\partial I_C}{\partial I_{CBO}} = \frac{\partial I_C}{\partial I_{CEO}} \frac{\partial I_{CEO}}{\partial I_{CBO}} \ (= S)$$

$$= K(1 + h_{FE}) \tag{4.11}$$

(since $I_{CBO} = I_{CEO}/(1 + h_{FE})$).

Over the temperature range 25–75°C, I_{CBO} changes by 60.8 μA. This would result in a change in I_C given by

$$\Delta I_C = 0.2 \times 34 \times 60.8 = 414 \ \mu\text{A}.$$

If for this transistor a typical value of I_C is 20 mA the change of 0.414 mA is only some 2% of this.

If spot values or curves of I_{CBO}/T are not given, it is safe to assume that I_{CBO} will double for every 9–10°C temperature rise in germanium transistors. For silicon, I_{CBO} is several orders of magnitude lower and at normal temperatures the ohmic component of this leakage current forms a significant part of the total. This means that the rate of change of I_{CBO} with temperature is rather indeterminate until about 80–90°C is reached, when, as a general rule, it begins to double every 13°C approximately. For normal or room temperature operation it is usually safe to neglect the effect of changes in I_{CBO} for silicon transistors.

The change in V_{BE} with respect to temperature usually becomes of importance for silicon transistors at high temperatures and for both silicon and germanium this change is approximately −0.002 V/°C. The change in I_C due to this can be calculated from equation (4.7).

Equation (4.8) is rarely used, since calculations such as that performed at the beginning of this chapter generally suffice.

The foregoing equations relating to stability were derived for the circuit of figure 4.3. Those for figure 4.2 could be derived in a similar manner and would result in the following expressions:

$$\frac{\partial I_C}{\partial I_{CEO}} = K \tag{4.12}$$

$$\frac{\partial I_C}{\partial V_{BE}} = \frac{-K h_{FE}}{R_C + R_B} \tag{4.13}$$

and

$$\frac{\partial I_C}{\partial h_{FE}} = K I_B \quad (\text{if } R_B \gg R_C) \tag{4.14}$$

where

$$K = \frac{1}{1 + [h_{FE}R_C/(R_C + R_B)]} .$$ (4.15)

4.1.2 Thermal Runaway

If a transistor undergoes a temperature rise dT_j, the subsequent increase in I_C will cause more power to be dissipated, mainly at the collector junction where most of the internal voltage drop occurs. This will cause a further rise in temperature at this junction, $r\,dT_j$. This series of events will be repetitive and, ignoring the variations in the rate of heat dissipation, the final temperature will be

$$T_j = T_{amb} + dT_j + r\,dT_j + r^2\,dT_j + \cdots .$$ (4.16)

This is T_{amb} plus a geometric series having a common ratio r, and if $r < 1$, the sum of the series is given by

$$dT_j\left(\frac{1}{1-r}\right)$$ (4.17)

or the final temperature is given by $T_{amb} + dT_j[1/(1 - r)]$.

If the final temperature is sufficiently high to destroy the device (or if $r > 1$ so that the temperature continues to rise *ad infinitum*, theoretically), thermal runaway is said to have taken place.

If r is very small so that only the first few terms of (4.16) are significant and the final temperature is not excessive the circuit is said to be thermally stable. In this case r can be expressed in terms of the circuit parameters and a condition for thermal stability can be derived.

To find r, let the power increase due to the temperature rise dT_j be dP_C. The further temperature rise $r\,dT_j$ due to dP_C will be $\theta\,dP_C$, where θ is the collector-junction temperature rise in degrees per watt.

Therefore

$$r = \frac{r\,dT_j}{dT_j} = \frac{\theta\,dP_C}{dT_j}$$

$$= \theta\,\frac{dP_C}{dI_C}\frac{dI_C}{dT_j} .$$ (4.18)

Now $P_C = I_C V_{CE}$, which for the simple circuit of figure 4.1(a) becomes

$$P_C = I_C(V_{CC} - I_C R_C) .$$

Therefore

$$\frac{dP_C}{dI_C} = V_{CC} - 2I_C R_C = 2V_{CE} - V_{CC} .$$

Putting this in equation (4.18) gives

$$r = \theta(2V_{CE} - V_{CC})\frac{dI_C}{dT_j} . \tag{4.19}$$

Here again the derivative dI_C/dT of equation (4.9) appears (the transistor has merely been assumed to have worked in a steady state long enough for the average temperature to be little different from the collector junction temperature in equation (4.19)), but once again an examination of the equation shows that it is rarely necessary to calculate dI_C/dT. Obviously, if $2V_{CE} = V_{CC}$ then $r = 0$. This simply means that if the voltage drop across the transistor is the same as the voltage drop across the load, then $r = 0$. It is usually possible to achieve this in small signal amplifiers and circuits utilizing this *half-voltage principle* can be made very stable indeed.

However, the relevant derivation has been performed for the basic, unstabilized circuit of figure 4.1(*a*) and although thermal runaway has been shown to be impossible, changes in operating point—perhaps leading to saturation—can still occur owing to changes in ambient temperature. It is therefore necessary to show that the half-voltage principle is equally valid for the stabilized circuits of figures 4.2 and 4.3.

Taking figure 4.2 and returning to equation (4.18),

$$r = \theta \frac{dP_C}{dI_C} \frac{dI_C}{dT_j}$$

$$P_C = I_C V_{CE} = I_C[V_{CC} - R_C(I_C + I_B)]$$

$$= \begin{cases} I_C\left[V_{CC} - R_C I_C\left(1 + \dfrac{1}{h_{FE}}\right)\right], & \text{if } I_{CEO} \text{ is small} \\ I_C(V_{CC} - R_C I_C), & \text{if } 1/h_{FE} \ll 1 \end{cases}$$

giving

$$\frac{dP_C}{dI_C} = 2V_{CE} - V_{CC} .$$

This leads again to equation (4.19) thus confirming the half-voltage principle for this circuit providing the foregoing assumptions are true.

Repeating the process for figure 4.3 gives

$$P_C = I_C V_{CE} = I_C(V_{CC} - I_C R_C - I_E R_E)$$

$$= \begin{cases} I_C\left[V_{CC} - I_C R_C - I_C R_E\left(1 + \dfrac{1}{h_{FE}}\right)\right], & \text{if } I_{CEO} \text{ is small} \\ I_C[V_{CC} - I_C(R_C + R_E)], & \text{if } 1/h_{FE} \ll 1 \end{cases}$$

leading to

$$\frac{dP_C}{dI_C} = V_{CC} - 2I_C(R_C + R_E)$$

$$= [V_{CC} - I_C(R_C + R_E)] - [I_C(R_C + R_E)]$$

$$= V_{CE} - [V_{CC} - V_{CE}]$$

$$= 2V_{CE} - V_{CC}.$$

Again the result confirms equation (4.19) providing that the assumptions are true. In this case, however, care must be taken to include R_E with R_C when designing for the correct voltage division.

In either case, a more exact analysis must be made if I_C is to be so small that it compares with I_{CEO}.

The biasing techniques described above are relevant to discrete, or single-unit transistors. When transistor structures are fabricated concurrently on the same silicon chip their characteristics become accurately matched and it is possible to use this fact in the design of both simpler and better biasing circuits. This approach will be employed in chapters 5 and 7. Meanwhile, the small-signal performance of the discrete transistor amplifier stage is considered. In the first instance, the hybrid-π equivalent circuit will be used since it leads to simple expressions, though the h-parameter system does give more information about the small-signal performance, as will be seen.

4.2 GAIN AND RESISTANCE CALCULATIONS (HYBRID-π PARAMETER METHOD)

Having seen how biasing components determine the degree of operating point stability, it is now possible to assume that a properly biased transistor can be represented by a small-signal equivalent circuit, and its dynamic performance determined. To do this, the biasing components must be combined with the source and load components, before calculations of current and voltage gain and input and output impedance for the circuit can be made.

In the context of the present work it is convenient to assume that only resistive sources and loads are involved because later it will be shown how these combine with the circuit capacitances to define a frequency response. However, the same equations would result if source and load impedances were involved, except that complex quantities would have to be included.

Consider figure 4.3, in which an input, or source, generator having an internal resistance R_g is shown. The total effective resistance, R_G, at the input of the amplifier stage is obviously the parallel combination of R_g and the biasing components R_1 and R_2:

$$\frac{1}{R_G} = \frac{1}{R_g} + \frac{1}{R_1} + \frac{1}{R_2}.$$

Again in figure 4.3, the only load shown is R_C, but had a further load R_L' been present (such as the input of a second stage) then the total effective load R_L would have been:

$$\frac{1}{R_L} = \frac{1}{R_C} + \frac{1}{R_L'}.$$

(The effect of R_E is removed at the frequencies of interest by the shunting capacitor C_E.)

At frequencies below which the internal capacitances of the transistor are important (which can be a megahertz or more for modern planar transistors), the hybrid-π equivalent circuit of figure 3.6 can be much simplified by omitting these capacitances, as shown in figure 4.7. Here $r_{b'c}$ has also been omitted because it is very large. The result is that the circuit has become a *unilateral approximation*, which implies that the signal is transmitted forwards by the dependent current generator $g_m v_{b'e}$ but not backwards, $r_{b'c}$ and $C_{b'c}$ having been omitted. (In the *h*-parameter case this means that h_{re} has been taken as negligibly small.) A Thévenin equivalent involving the effective source resistance R_G has been included at the input and a load R_L at the output. Also, the frequency of interest has been assumed to be such that the reactance of C_c is negligible.

The input resistance (to the right of the b and e terminals) is now:

$$R_{in} = r_{bb'} + r_{b'e} \tag{4.20}$$

while the output resistance (to the left of the c and e terminals) is:

$$R_{out} = r_{ce}. \tag{4.21}$$

The current gain A_i is,

$$A_i = \frac{i_{out}}{i_{in}} = \frac{i_c}{i_b} = \frac{g_m v_{b'e} r_{ce}}{r_{ce} + R_L} \frac{r_{b'e}}{v_{b'e}}.$$

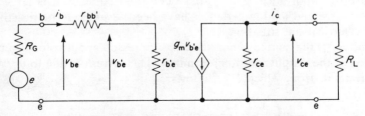

Figure 4.7. Simplified hybrid-π equivalent circuit for low frequencies.

Usually $r_{ce} \gg R_L$ so that this becomes:

$$A_i \simeq g_m r_{b'e}$$
$$= h_{fe} \tag{4.22}$$

by equation (3.16a).

The voltage gain is:

$$A_v = \frac{v_{out}}{v_{in}} = \frac{v_{ce}}{v_{be}}$$

$$= \frac{-g_m v_{b'e} r_{ce} R_L}{r_{ce} + R_L} \frac{r_{b'e}}{v_{b'e}(r_{bb'} + r_{b'e})}$$

or

$$A_v \simeq \frac{-g_m R_L r_{b'e}}{r_{bb'} + r_{b'e}} \tag{4.23a}$$

if $r_{ce} \gg R_L$.

If the transistor is run at a low collector current, then (according to equation (3.16b)) $r_{b'e}$ can become much larger than $r_{bb'}$, when

$$A_v \simeq -g_m R_L . \tag{4.23b}$$

In these equations, the negative sign means that a complete signal inversion has occurred between the input and output in addition to amplification. This is characteristic of the CE stage, which is a simple example of an *inverting amplifier*.

Although the hybrid-π equivalent circuit is basically relevant to the CE configuration it can nevertheless be used to calculate the performance of the CC and CB modes if the input and output are properly oriented. For example, the CC, or emitter-follower circuit of figure 4.8(a) can be represented as in figure 4.8(b). In these figures, R_L is the parallel combination of the unbypassed emitter resistor R_F and any other load placed across it.

Here again *mid-frequency operation* has been assumed, where the reactances of neither the external coupling capacitors nor the internal capacitances are of importance and $r_{b'c}$ has been omitted since it is large. Also, both the source and load impedances have been assumed to be resistive, this being the most usual situation.

To find R_{out}, the source generator e is suppressed and a voltage generator v_o applied at the output (emitter) terminal. The current i_o due to v_o can now be determined, from which R_{out} follows:

$$i_o = v_o \left(\frac{1}{R_L} + \frac{1}{r_{ce}} + \frac{1}{r_{b'e} + r_{bb'} + R_G} \right) - g_m v_{b'e}$$

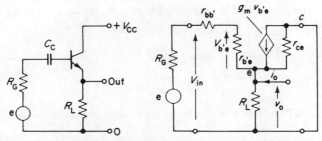

Figure 4.8. (*a*) An emitter-follower; and (*b*) its equivalent circuit. (Biasing resistors are not shown.)

but

$$v_{b'e} = - \frac{v_o r_{b'e}}{r_{b'e} + r_{bb'} + R_G}$$

whence

$$i_o = v_o \left(\frac{1}{R_L} + \frac{1}{r_{ce}} + \frac{g_m r_{b'e} + 1}{r_{b'e} + r_{bb'} + R_G} \right).$$

From this,

$$R_{out} = \frac{v_o}{i_o}$$

or

$$G_{out} = \frac{1}{R_{out}} = \frac{1}{R_L} + \frac{1}{r_{ce}} + \frac{g_m r_{b'e} + 1}{r_{b'e} + r_{bb'} + R_G}.$$

Knowing that $1/r_{ce} \ll 1/R_L$ and that $(g_m r_{b'e} + 1) = (h_{fe} + 1) = -h_{fc}$, this reduces to

$$G_{out} \simeq \frac{1}{R_L} - \frac{h_{fc}}{r_{b'e} + r_{bb'} + R_G}. \tag{4.24}$$

This means that R_{out} is given by the parallel combination of R_L and $(r_{bb'} + r_{b'e} + R_G)/(-h_{fc})$ and since R_L is usually much the larger of the two components,

$$R_{out} \simeq - \frac{r_{bb'} + r_{b'e} + R_G}{h_{fc}}. \tag{4.25a}$$

This is a very important equation since it shows that the emitter-follower is a *resistance transfer stage* and for all normal values of R_G will present a low output impedance. (The minus sign results from h_{fc} having a negative numerical value.)

If $R_G \gg (r_{bb'} + r_{b'e})$ then

$$R_{out} \simeq - \frac{R_G}{h_{fc}} . \qquad (4.25b)$$

Since $-h_{fc}$ differs from h_{fe} only by one, this is usually written:

$$R_{out} \simeq \frac{R_G}{h_{fe}} . \qquad (4.25c)$$

The input resistance may be determined as follows:

$$R_{in} = \frac{v_{in}}{i_{in}}$$

where v_o has been removed and e replaced, so that

$$i_{in} = v_{out}\left(\frac{1}{R_L} + \frac{1}{r_{ce}}\right) - g_m v_{b'e} .$$

But $v_{b'e} = i_{in} r_{b'e}$, so that,

$$i_{in}(1 + g_m r_{b'e}) = v_{out}\left(\frac{1}{R_L} + \frac{1}{r_{ce}}\right) .$$

Now $v_{out} = v_{in} - i_{in}(r_{bb'} + r_{b'e})$ so that,

$$\frac{i_{in}(1 + g_m r_{b'e})}{(1/R_L) + (1/r_{ce})} = v_{in} - i_{in}(r_{bb'} + r_{b'e}) .$$

Letting $(1 + g_m r_{b'e}) = -h_{fc}$ and dividing through by i_{in} gives:

$$\frac{-h_{fc}}{(1/R_L) + (1/r_{ce})} = R_{in} - (r_{bb'} + r_{b'e}) .$$

Since $1/r_{ce} \ll 1/R_L$, this can be written:

$$R_{in} \simeq (r_{bb'} + r_{b'e}) - h_{fc}R_L \qquad (4.26a)$$

or

$$R_{in} \simeq r_{bb'} + r_{b'e} + h_{fe}R_L \qquad (4.26b)$$

and if $h_{fe}R_L \gg (r_{bb'} + r_{b'e})$, this becomes

$$R_{in} \simeq h_{fe}R_L . \qquad (4.26c)$$

This equation shows that the resistance transfer property of the emitter-follower is symmetrical; that is, the load resistance is multiplied by the current gain and is presented at the input whereas the source resistance is divided by the current gain and is presented at the output. Hence, the emitter-follower can be used as a *buffer stage* to present a high input and a low output resistance.

The voltage gain must obviously be nearly unity, since $v_{in} = v_{out} + v_{be}$ and v_{be} is very small.

Formally, summing currents at the output:

$$i_{in} + g_m v_{b'e} = v_{out}\left(\frac{1}{r_{ce}} + \frac{1}{R_L}\right)$$

or

$$i_{in}(1 + g_m r_{b'e}) = v_{out}\left(\frac{1}{r_{ce}} + \frac{1}{R_L}\right).$$

Now

$$i_{in} = \frac{v_{in} - v_{out}}{r_{bb'} + r_{b'e}}$$

so

$$\frac{v_{in}(1 + g_m r_{b'e})}{r_{bb'} + r_{b'e}} = v_{out}\left(\frac{1}{r_{ce}} + \frac{1}{R_L} + \frac{1 + g_m v_{b'e}}{r_{bb'} + r_{b'e}}\right).$$

Now $(1 + g_m r_{b'e}) = -h_{fc}$ and $1/r_{ce} \ll 1/R_L$, so

$$\frac{v_{out}}{v_{in}} = \frac{-h_{fc}}{(r_{bb'} + r_{b'e})} \frac{1}{(1/R_L) - [h_{fc}/(r_{bb'} + r_{b'e})]}$$

$$= \frac{-h_{fc} R_L}{r_{bb'} + r_{b'e} - h_{fc} R_L}. \qquad (4.27)$$

This voltage gain approaches unity if $(r_{bb'} + r_{b'e}) \ll -h_{fc} R_L$, a condition which is *not* necessarily true.

4.3 GAIN AND RESISTANCE CALCULATIONS (h-PARAMETER METHOD)

R_G and R_L can now be included in the h-parameter equivalent circuit as shown in figure 4.9.

Consider first the current gain A_i:

$$A_i = \frac{i_{out}}{i_{in}}.$$

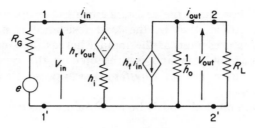

Figure 4.9. h-parameter equivalent circuit with external resistances.

i_{out} can be obtained by calculating the fraction of $h_f i_{\text{in}}$ which passes through R_L thus:

$$i_{\text{out}} = \frac{h_f i_{\text{in}}}{(1/h_\text{o}) + R_L} \frac{1}{h_\text{o}} .$$

Therefore

$$A_i = \frac{h_f}{1 + h_\text{o} R_L} . \qquad (4.28)$$

If the circuit relates to a common-emitter amplifier, the relevant h-parameters should be h_{fe} and h_{oe}, but as precisely the same expression would result for the common-base or emitter-follower modes, the second subscripts have been omitted, thereby making equation (4.28) quite general. This statement will apply to each of the h-parameter derivations given below and serves to illustrate the basic advantage of this form of equivalent circuit.

The voltage gain A_v is now easily obtained:

$$A_v = \frac{v_{\text{out}}}{v_{\text{in}}}$$

where

$$v_{\text{out}} = -i_{\text{out}} R_L \qquad (4.29)$$

and

$$v_{\text{in}} = i_{\text{in}} h_i + h_r v_{\text{out}} . \qquad (4.30)$$

Putting (4.29) into (4.30) gives

$$v_{\text{in}} = i_{\text{in}} h_i - h_r i_{\text{out}} R_L .$$

Therefore

$$A_v = \frac{-i_{out}R_L}{i_{in}h_i - h_r i_{out}R_L}$$

$$= \frac{-1}{(i_{in}/i_{out})(h_i/R_L) - h_r} .$$

But

$$\frac{i_{in}}{i_{out}} = \frac{1}{A_i}$$

and using equation (4.28)

$$A_v = \frac{-1}{[(1 + h_o R_L)/h_f](h_i/R_L) - h_r} \quad \text{or} \quad \frac{-h_f}{h_i(G_L + h_o) - h_r h_f}$$

$$(4.31a)$$

where $G_L = 1/R_L$, the load conductance. This may also be written

$$A_v = \frac{-h_f}{h_i G_L + \triangle} . \qquad (4.31b)$$

This equation shows that if h_f is negative, as for the common-base and common-collector modes, then A_v is positive. In other words, the output voltage is in the same phase as the input voltage. This could have been predicted by an examination of figures 2.8(b) and (c). In (b), when the emitter goes negative, I_C increases, so the voltage drop across the load resistor increases and the collector therefore also goes negative. In (c), if the base goes negative, I_E decreases, the voltage drop across the load decreases, so the emitter also goes negative.

For the common-emitter mode of figure 2.8(a), if the base goes positive, I_C increases, the voltage drop across the load increases and the collector goes negative, indicating a *phase reversal* from input to output. Equation (4.31a) also shows this since h_{fe} is a positive quantity and $h_{re}h_{fe}$ is always smaller than $h_{ie}(G_L + h_{oe})$.

The input and output resistances of transistor amplifier stages are dependent on the load and source resistances, respectively, and this may be shown as follows. Consider first the input resistance of the circuit of figure 4.9; that is, the resistance at points 1, 1′ looking only to the right.

Input resistance

$$R_{in} = \frac{v_{in}}{i_{in}} = \frac{i_{in}h_i + h_r v_{out}}{i_{in}}$$

and since

$$v_{out} = -i_{out}R_L = \frac{-h_f i_{in}}{(1/h_o) + R_L} \frac{1}{h_o} R_L$$

$$R_{in} = h_i - \frac{h_r h_f R_L}{1 + h_o R_L} . \tag{4.32}$$

If preferred, equation (4.32) can be quoted in terms of the load conductance G_L by dividing the top and bottom of the final term by R_L,

$$R_{in} = h_i - \frac{h_r h_f}{G_L + h_o} . \tag{4.33}$$

It is more convenient to derive an output conductance rather than an output resistance and this is done as follows.

Output conductance

$$G_{out} = \frac{i_{out}}{v_{out}} = \frac{v_{out} h_o + h_f i_{in}}{v_{out}}$$

$$= h_o + \frac{h_f i_{in}}{v_{out}} . \tag{4.34}$$

If the source voltage is suppressed then

$$i_{in} = \frac{-h_r v_{out}}{R_G + h_i} .$$

Putting this in (4.34) gives

$$G_{out} = h_o - \frac{h_r h_f}{R_G + h_i} . \tag{4.35}$$

The relationship between R_{in} and R_L and that between R_{out} and R_G can be displayed graphically. To do this equations (4.32) and (4.35) will be rewritten so that their solutions become obvious when R_G and R_L each tend to zero and infinity.

Take first equation (4.32):

$$R_{in} = h_i - \frac{h_r h_f R_L}{1 + h_o R_L} .$$

When

$$R_L = 0 \quad \text{then} \quad R_{in} = h_i .$$

To find R_{in} when $R_L \to \infty$, the equation may be again rewritten

$$R_{in} = \frac{h_i + h_i h_o R_L - h_r h_f R_L}{1 + h_o R_L}$$

$$= \frac{(h_i/R_L) + h_i h_o - h_r h_f}{(1/R_L) + h_o}$$

so when $R_L \to \infty$,

$$R_{in} = \frac{h_i h_o - h_r h_f}{h_o} = \frac{\triangle}{h_o}.$$

Here, $\triangle$ is the determinant of the h-parameter equations.

If the h-parameters are known for a given transistor then R_{in} can be plotted as a function of R_L for all three configurations. Table 4.1 lists the "typical" h-parameters for the BC237 silicon planar epitaxial npn transistor and these have been used to produce the computer-generated plots of figure 4.10(a).

Secondly, take equation (4.35) and rewrite:

$$R_{out} = \frac{1}{h_o - h_r h_f/(R_G + h_i)}$$

$$= \frac{R_G + h_i}{R_G h_o + h_o h_i - h_r h_f}.$$

When $R_G = 0$,

$$R_{out} = \frac{h_i}{\triangle}$$

When $R_G \to \infty$, it is best to rewrite the equation as follows:

$$R_{out} = \frac{1 + h_i/R_G}{h_o + \triangle/R_G}$$

TABLE 4.1. Typical h-parameters for the BC237 silicon planar epitaxial npn general purpose transistor (at $I_C = 1.6$ mA, $V_{CE} = 5$ V and $f = 1$ kHz). If only CE h-parameters are available the CB and CC values may be calculated using table 3.1.

	CE	CB	CC
h_i	3600 Ω	15.93 Ω	3600 Ω
h_o	24 μS	0.106 μS	24 μS
h_r	1.7×10^{-4}	2.123×10^{-4}	1
h_f	225	-0.996	-226
$\triangle$	48×10^{-3}	2.13×10^{-4}	226

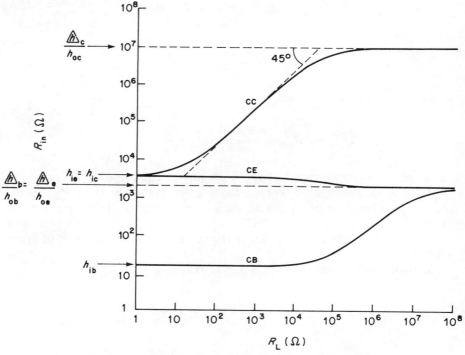

Figure 4.10. (*a*) Input resistance as a function of load.

so that in the limit,

$$R_{out} = 1/h_o \,.$$

Plotting equation (4.35) results in figure 4.10(*b*).

An important observation relating to the two graphs of figure 4.10 is that the plot referring to the cc or emitter-follower mode is at 45° to the (logarithmic) axes over much of its length. This implies a linear relationship and equations (4.33) and (4.35) should reflect this.

Taking equation (4.33) and a practical resistive load, R_L, then $h_{oc} \ll G_L$. Also, being an emitter-follower stage, $h_{rc} = 1$, so that the equation becomes:

$$R_{in} \simeq h_{ic} - h_{fc}R_L$$

or

$$R_{in} \simeq h_{ic} + h_{fe}R_L$$

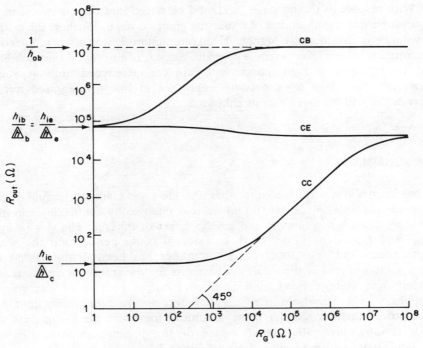

Figure 4.10. (*b*) Output resistance as a function of R_G.

and if $h_{fe}R_L \gg h_{ic}$

$$R_{in} \simeq h_{fe}R_L .$$

These equations should be compared with those derived by the hybrid-π method (equations (4.26*a*), (4.26*b*), and (4.26*c*)).

Taking equation (4.35), similar approximations yield:

$$G_{out} \simeq \frac{-h_{fc}}{R_G + h_{ic}}$$

or

$$R_{out} \simeq \frac{R_G + h_{ic}}{-h_{fc}} \simeq \frac{R_G + h_{ic}}{h_{fe}} \simeq \frac{R_G}{h_{fe}}$$

which should be compared with equations (4.25). These relationships are the basis of the use of an emitter-follower as an impedance changer, as will be seen later.

With respect to the CB stage, it should be noted that the input resistance is considerably less than that of either the CB or CC stage. Further, the output resistance is considerably higher. This combination is not often required in discrete audio amplifier stages, especially when it is also noted that although the voltage gain is high the current gain cannot exceed unity ($i_c < i_e$). However, the stage does become important in linear integrated micro-circuits, as will be explained in chapter 7.

4.4 SUMMARY

It has been shown in this chapter that a single-stage transistor amplifier can be made stable with respect to temperature changes by connecting external resistors in such a way as to reduce the effect of inherent changes in h_{FE}, V_{BE}, and I_{CBO}. The effect of these external components upon the small signal gains and resistances of the amplifier has been demonstrated by incorporating them in the equivalent circuits for the transistor and deriving current and voltage gains, and input and output resistances, for simple amplifiers. The equations derived are quite accurate providing that the transistor is operating under those conditions for which the parameters are known. Otherwise, the actual values of the parameters must first be obtained from curves such as those of figure 3.5.

For the purposes of the present introductory volume only resistive sources and loads have been considered, but it should be noted that reactive external values lead to the same equations, albeit in complex form.

This chapter has been concerned entirely with theoretical considerations, apart from a few notes regarding the particular merits of the three configurations. Later it will be seen how a more practical approach can simplify the expressions involved, but since not all practical circuits are of a stereotyped nature an insight into the theory behind their design will be found extremely useful.

4.4.1 Important Equations

For shunt feedback circuit of figure 4.2:

$$\frac{\partial I_C}{\partial I_{CEO}} = K = \frac{1}{1 + [h_{FE}R_C/(R_C + R_B)]}$$

$$\frac{\partial I_C}{\partial V_{BE}} = \frac{-Kh_{FE}}{R_C + R_B}$$

$$\frac{\partial I_C}{\partial h_{FE}} = KI_B \quad (\text{if } R_B \gg R_C).$$

For series feedback circuit of figure 4.3:

$$\frac{\partial I_C}{\partial I_{CEO}} = K = \frac{1}{1 + [h_{FE}R_E/(R_E + R_B)]}$$

$$\frac{\partial I_C}{\partial V_{BE}} = \frac{-Kh_{FE}}{R_E + R_B}$$

$$\frac{\partial I_C}{\partial h_{FE}} = KI_B$$

and

$$R_B = \frac{R_1 R_2}{R_1 + R_2}.$$

Small-signal gain and resistance formulae (*h*-parameters):

$$A_i = \frac{h_f}{1 + h_o R_L} \qquad R_{in} = h_i - \frac{h_r h_f}{G_L + h_o}$$

$$A_v = \frac{-h_f}{h_i(h_o + G_L) - h_r h_f} \qquad G_{out} = h_o - \frac{h_r h_f}{R_G + h_i}.$$

Small-signal gain and resistance formulae (hybrid-π parameters):

Common-emitter Stage

$$A_i \simeq g_m r_{b'e} \qquad\qquad R_{in} \simeq r_{bb'} + r_{b'e}$$

$$A_v \simeq \frac{-g_m R_L r_{b'e}}{r_{bb'} + r_{b'e}} \qquad R_{out} \simeq r_{ce}$$

Emitter-follower (*Common-collector*) *Stage*

$$A_i \simeq g_m r_{b'e} \qquad\qquad R_{in} \simeq r_{bb'} + r_{b'e} + h_{fe}R_L$$

$$A_v \simeq \frac{h_{fe}R_L}{r_{bb'} + r_{b'e} + h_{fe}R_L} \qquad R_{out} \simeq \frac{r_{bb'} + r_{b'e} + R_G}{h_{fe}}.$$

(Note: $-h_{fc} = (1 + h_{fe}) \simeq h_{fe}$ and this has been substituted for $-h_{fc}$ in the above expressions.)

QUESTIONS

1. Why must a transistor be biased before it can be used as a small-signal amplifier; and what is meant by the "quiescent operating point"?

2. In figure 4.3, the aim of the resistors R_1 and R_2 is to maintain V_B as constant as possible. Why cannot R_1 and R_2 be made very small to achieve this aim?

3. Explain how parameter variations tend to make the collector current of a transistor increase as a result of a rise in temperature. Which of these parameter variations is most important for silicon transistors?

4. In figure 4.7, the load has been shown as connected between the collector and the emitter nodes. Why is this a valid representation?

5. Keeping in mind that an emitter-follower produces no voltage gain, what are its applications?

6. Which part of figure 4.10(b) validates equation (4.25b), and why?

7. Which part of figure 4.10(a) validates equation (4.26c), and why?

8. Why is it not possible to directly measure all the hybrid-π parameters?

PROBLEMS

4.1. Derive the bias stability factor K for the shunt DC feedback circuit of figure 4.2 (given by equation (4.15)).

4.2. A germanium power transistor, of $h_{FE} = 29$, is connected into the circuit of figure 4.11. If I_{CBO} at 25°C is 10 μA, and doubles for every 9°C rise in temperature, by how much will V_{CE} change if the temperature rises to 52°C? (Neglect changes in V_{BE} and h_{FE}.)

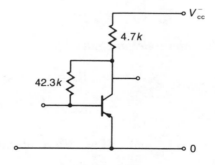

Figure 4.11.

4.3. If the half-voltage principle is to be used in a circuit like that of figure 4.11, recalculate the values of the two resistors if the transistor used has an h_{FE} of 100 and a V_{BE} of −0.6 V. Assume that the quiescent current is to be −1 mA and that the supply voltage is −12 V.

4.4. For an operating CE stage, the collector load is a 3.6 kΩ resistor, and under the biasing conditions involved, the h-parameters for the transistor are:

$$h_{fe} = 70 \quad h_{ie} = 1.8\,k\Omega \quad h_{oe} = 1.1 \times 10^{-3}.$$

Calculate the voltage gain, and check it using the hybrid-π voltage gain approximation, assuming that $I_C = 1\,mA$ and $kT/q \simeq 26\,mV$.

4.5. For the CE stage of problem 4.4, calculate the value of $r_{b'e}$ and compare it with h_{ie}.

4.6. A transistor is connected as an emitter-follower and works at a collector current of 1.6 mA with a 1.2 kΩ emitter load. Calculate its input resistance and voltage gain for small signals if $h_{fc} = -85$ and $h_{ie} = 1.4\,k\Omega$.

4.7. If a transistor exhibiting the "typical" h-parameters of table 4.1 were connected into a CE circuit, and a source having an internal resistance of 600 Ω were applied to the input, what would be the output resistance R_{out} at the collector? (Assume that any bias resistances are much greater than 600 Ω.)

4.8. Calculate the bias stability factor for a circuit like that of figure 4.3 given the following values:

$$R_1 = 86\,k\Omega \quad R_2 = 33\,k\Omega \quad R_E = 5.9\,k\Omega \quad h_{FE} = 100.$$

Is this a satisfactory result?

4.9. An emitter-follower has a load resistor $R_F = 1800\,\Omega$ and is properly biased. Calculate the mid-band gain accurately knowing that for the transistor used, the common-emitter h-parameters have been measured as follows:

$$h_{fe} = 225, \quad h_{ie} = 3600\Omega, \quad h_{oe} = 24\,\mu s.$$

5

Circuit Design and Performance

Having seen how transistors may be biased into an operating condition, then used as amplifiers of small signals, it is now possible to consider the practical performance of such stages, particularly with reference to their frequency responses.

Figure 5.1 repeats the standard series DC feedback bias circuit, and like figures 4.1(a), 4.2, 4.3, and 4.5(b), it includes a coupling capacitor C_C. This is necessary because the base is not usually at ground potential for biasing reasons, so the signal source must be isolated from the input point lest its internal resistance is low enough to change the bias voltage by shunting R_2. (Also, in some cases, such as a tape head or some pick-up cartridges, the resulting current through the source may well adversely affect it to the point of damage).

If gains greater than those available from a single stage are required, then it is possible, in principle, to *cascade* stages as shown in figure 5.2. This is obviously grossly wasteful in components, and in addition such networks are impossible to realise in integrated circuit (IC) form unless a multiplicity of external components is employed. Hence (apart from in a few high-frequency applications), such multistage amplifiers are now rarely found. It is possible, however, to omit the coupling capacitor and biasing resistors of a stage by arranging for its base to operate at the quiescent potential of the preceding collector; but even so, more modern design techniques lead not only to easy integration, but also to amplification down to DC, as will be seen later.

The single stage does remain useful, however, very often in signal processing related to sensors; and furthermore, its analysis conveniently illustrates the basic concepts of frequency response, for which reasons, it will now be treated in some depth.

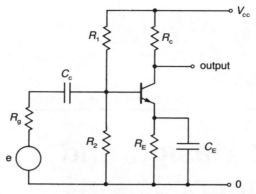

Figure 5.1. A simple common-emitter (CE) stage with series DC feedback biasing.

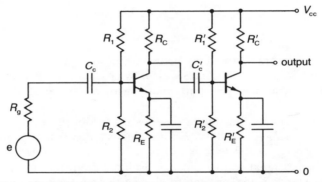

Figure 5.2. An R/C-coupled multistage amplifier (now largely obsolete).

5.1 FREQUENCY RESPONSE

So far it has been tacitly assumed that the amplifiers for which gain and impedance expressions have been derived were working at frequencies where neither the internal nor the coupling or emitter bypass capacitors had any effect upon the performance. A real amplifier is, however, markedly frequency sensitive and the (measured) voltage gain versus frequency curves of figure 5.7 exhibit this dependence.

The fall in voltage gain at low frequencies may be attributed to the effect of the capacitors external to the transistor, which means that the slope and position of this *low-frequency roll-off* may be defined by the designer. However, the fall at high frequencies is largely a function of the internal capacitances associated with the transistor. Thus, the slope and position of the *high-frequency roll-off* is only partly in the hands off the designer for any given set of transistor parameters.

The *pass-band* of the amplifier is that range of frequencies between the points where the voltage gain is 0.707 of its maximum value (or is 3 dB down—see Appendix I). These two points, the low- and high-frequency cut-off points, and the factors which determine them will now be discussed.

5.1.1 Low-frequency Response

The low-frequency performance of the CE stage of figure 5.1 can be predetermined by proper design of the input circuit components. Consider figure 5.3(a). Here the Thévenin equivalent of a signal source is shown feeding the input and biasing components of a transistor via a coupling capacitor C_c. As the frequency falls, X_{C_c} will become of increasing importance and the voltage drop across it will subtract from V_g and so lower V_{in}. This fall in input signal manifests itself as a reduction in gain and the result can be seen in the upper curve of figure 5.7.

To calculate the effect of C_c a simplified equivalent is used, and in figure 5.3(b) R_{BP} is made up of the parallel combination of R_1, R_2, and R_{in}. Using this circuit the fall in V_{in} can easily be found and a particularly important result appears when $R_g = 0$, that is, when a voltage source is used. Under these circumstances, and using sinusoidal quantities, $V_g = E$ and

$$\frac{V_{in}}{E} = \frac{R_{BP}}{\sqrt{(R_{BP}^2 + X_{C_c}^2)}} \tag{5.5}$$

and if $|X_{C_c}| = R_{BP}$, then

$$\frac{V_{in}}{E} = \frac{1}{\sqrt{2}} = 0.707 \,.$$

Expressed in decibels (see Appendix I) this becomes

$$20 \log_{10} 0.707 \simeq -3 \text{ dB} \,.$$

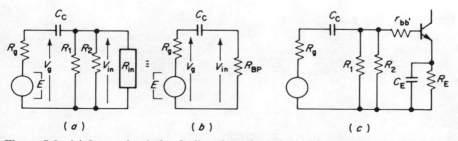

Figure 5.3. (a) Input circuit for finding the value of C_c; (b) equivalent circuit of (a) $1/R_{BP} = 1/R_1 + 1/R_2 + 1/R_{in}$; (c) circuit for finding C_E.

Thus the value of V_{in} is 3 dB below the value of E when $|X_{C_c}| = R_{BP}$ and $R_g = 0$. The frequency at which this occurs, f_L, can be found in this special case by equating $|X_{C_c}|$ to R_{BP}:

$$R_{BP} = |X_{C_c}| = \frac{1}{2\pi f_L C_c}$$

or

$$f_L = \frac{1}{2\pi R_{BP} C_c}. \tag{5.6}$$

When the input voltage is 3 dB below the source voltage then the gain will also be 3 dB down assuming that V_{in} is amplified linearly. This leads to the term "low-frequency cut-off", f_L, which is the frequency at which the gain is 3 dB down. It is also called the "half-power" point because it can be shown (see Appendix I) that the power transferred to the load is only one half of what it would be at higher frequencies where X_{C_c} becomes negligible. (Note that the treatment above refers only to sinusoidal voltages. If this were not so, the expression for X_{C_c} would have been invalid. The capitals V_{in}, E, etc. refer to root mean square (RMS) values.)

For practical circuits, of course, the source is unlikely to be a perfect voltage generator, and with a real source having an internal resistance R_g, the value of C_c for the 3 dB fall in gain will be different. Again using figure 5.3(b), let the Thévenin voltage be E and the generator resistance be R_g. The value of V_{in} at the middle of the pass-band where X_{C_c} is very small will be

$$V_{in(f_m)} = \frac{E}{R_g + R_{BP}} R_{BP}.$$

However, at f_L where X_{C_c} is significant

$$V_{in(f_L)} = \frac{E}{[(R_g + R_{BP})^2 + X_{C_c}^2]^{1/2}} R_{BP}.$$

From these two equations,

$$\frac{V_{in(f_L)}}{V_{in(f_m)}} = \frac{1}{\sqrt{2}} = \frac{R_g + R_{BP}}{[(R_g + R_{BP})^2 + X_{C_c}^2]^{1/2}}$$

or

$$\frac{1}{2} = \frac{1}{1 + [X_{C_c}^2/(R_g + R_{BP})^2]}$$

giving

$$|X_{C_c}| = (R_g + R_{BP}) \quad \text{at } f_L .\tag{5.7}$$

This equation shows that C_c can take a lower value for a source having a higher internal resistance yet still lead to the same low-frequency cut-off point.

An assumption made during the foregoing discussion is that C_E is so large that it has not been of importance in calculating the drop in gain at low frequencies. Were this not so, the combination of R_E and X_{C_E} would have to be taken into account according to equations (4.26) in their complex form. These show that an emitter load is reflected to the base and in this case Z_{in} would have been increased had X_{C_E} been significant.

A similar impedance transfer takes place when the value of C_E is being considered. In figure 5.3(c) the input circuit to the left of the base will have a resultant value Z_G which will be transferred to the emitter according to equations (4.25) also in their complex form. An analysis of the hybrid-π equivalent circuit shows that $r_{bb'}$ and $r_{b'e}$ are also transferred in like manner (see equation (4.25a)), so that the total impedance at the emitter is:

$$Z_{emitter} \simeq R_E \left\| \frac{Z_G + r_{bb'} + r_{b'e}}{h_{fe}} \right.$$

or, if Z_G is purely resistive,

$$R_{emitter} \simeq R_E \left\| \frac{R_G + r_{bb'} + r_{b'e}}{h_{fe}} \right. .$$

Three conditions are now of interest:

(i) If $|X_{C_E}|$ is very small compared with $R_{emitter}$, so that f_L is defined only by the input circuit, then C_E could turn out to be very large, particularly if $R_G \to 0$.

(ii) If C_E is not large enough to make this condition true, then *degeneration* or *negative feedback* will occur near f_L because some of the output signal will be dropped across the $R_{emitter} \| X_{C_E}$ combination *in opposition* to the input signal direction. This will result in the slope of the low-frequency roll-off increasing and with it the value of f_L.

(iii) If C_C is made very large, then the $R_{emitter} \| X_{C_E}$ combination alone will define f_L, which will be at the frequency where:

$$|X_{C_E}| = R_E \left\| \frac{R_G + r_{bb'} + r_{b'e}}{h_{fe}} \right.\tag{5.8}$$

If $|X_{C_c}|$ is genuinely zero down to DC (i.e., no coupling capacitor is present and biasing is applied at the inactive end of the source resistance) then the voltage gain will *not* fall to zero but tend to $-R_C/R_E$, an approximation which will be justified in the work leading to equation (5.27).

The situation may now be summarized.

If C_E is large enough, then C_c will define f_L; and if C_c is large enough, then C_E will define f_L. If the circuit components are such as to make the effects of C_E and C_c comparable, then a poorer low-frequency response may result and the rate of decrease of gain will be greater than when only one capacitor is dominant. (The actual rate of change is discussed later.) Usually, it is the magnitude of the source resistance which determines the situation representing the best design aim.

It is often profitable to make C_E the dominant capacitor because, owing to the impedance-changing properties of a stage using an emitter resistor, a much larger value of C_E is necessary to ensure negligible interference with f_L than for C_c. This is especially true for a very low value of f_L such as the 10 Hz or so which would be necessary if the amplifier were to accept a signal from a thermopile or Golay cell for example. These devices are often used to detect a mechanically chopped infrared light beam and their thermal delays are such as to prohibit a chopping frequency much above 15 Hz.

5.1.2 High-frequency Response

The high-frequency roll-off may be explained most conveniently by adding source and load resistances to the hybrid-π equivalent circuit of figure 3.6 to represent a single-stage CE amplifier. This is done in figure 5.4 where the external capacitances C_c and C_E have been omitted on the grounds that their reactances will be negligibly small at the frequencies of interest. Also, $r_{b'c}$ has been omitted because $|X_{C_{b'c}}| \ll r_{b'c}$ at these frequencies.

The active base voltage $v_{b'e}$ is developed across $C_{b'e} \| r_{b'e}$. Hence, as the frequency rises, the reactance of $C_{b'e}$ will become progressively smaller and $v_{b'e}$ will fall. With it will fall the output of the dependent current generator $g_m v_{b'e}$.

This effect will be more pronounced than is immediately apparent because an extra capacitance, proportional to $C_{b'c}$ and the voltage gain of the circuit can be shown to appear across $C_{b'e}$. This is the *Miller effect* and it is most conveniently demonstrated by writing an expression for the current flowing into $C_{b'c}$ from the active base region b'.

Using sinusoidal quantities,

$$I_{C_{b'c}} = (V_{b'e} - V_{ce})j\omega C_{b'c}$$

but

$$V_{ce} = -g_m V_{b'e} R_L \quad (\text{if } R_L \ll r_{ce})$$

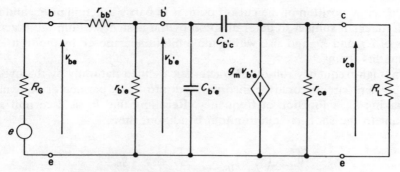

Figure 5.4. Loaded hybrid-π equivalent circuit.

so that

$$\frac{I_{C_{b'c}}}{V_{b'e}} = j\omega C_{b'c}(1 + g_m R_L).$$

This equation shows that an apparent capacitance of value $C_{b'c}(1 + g_m R_L)$ exists between b' and e and is therefore in parallel with $C_{b'e}$. The equivalent circuit for the input side of the stage may now be drawn as in figure 5.5, from which a high-frequency cut-off point f_H may be defined as follows.

Letting $(R_G + r_{bb'}) \| r_{b'e} = R$ and $C_{b'e} + C_{b'c}(1 + g_m R_L) = C_{in}$

$$f_H = \frac{1}{2\pi R C_{in}} \tag{5.9}$$

(the -3 dB or cut-off point for a parallel circuit is discussed in Appendix I).

This equation implies that $V_{b'e}$, and hence the dependent current generator $g_m V_{b'e}$, have values which are 3 dB below their mid-frequency values at f_H. Moreover, it implies that these quantities fall at a rate of 6 dB/octave; that is, they are inversely proportional to frequency at high frequencies.

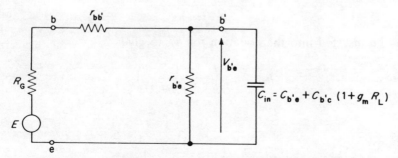

Figure 5.5. Equivalent input circuit incorporating the Miller effect.

The actual position of the cut-off point is in part a function of R_G and R_L, which makes it subject to determination by the designer within the practical limits of R_G and R_L and also within the limitations imposed by the transistor parameters themselves.

The high-frequency roll-off characteristic defined naturally by the hybrid-π equivalent circuit could be incorporated into the h-parameter equivalent by making h_{fe} a function of frequency. Recalling that h_{fe} is a current gain relevant to the short-circuited output condition, then,

$$h_{fe} = \left(\frac{I_{out}}{I_{in}}\right)_{(R_L=0)} = \frac{g_m V_{b'e}}{V_{b'e}/Z} = g_m Z$$

where Z is $r_{b'e}$ in parallel with $X_{C_{in}}$. When $R_L = 0$ then $C_{in} = C_{b'e} + C_{b'c}$ and because $C_{b'e} \gg C_{b'c}$ then $C_{in} \simeq C_{b'e}$. Hence,

$$h_{fe} \simeq \frac{g_m r_{b'e}(-jX_{C_{b'e}})}{r_{b'e} - jX_{C_{b'e}}} = g_m r_{b'e}\left(\frac{1}{1 + j(r_{b'e}/X_{C_{b'e}})}\right).$$

In this equation, if $X_{C_{b'e}} \gg r_{b'e}$, then $h_{fe} = g_m r_{b'e} = h_{feo}$ so that the equation can be rewritten

$$h_{fe} = h_{feo}\left(\frac{1}{1 + j(r_{b'e}/X_{C_{b'e}})}\right)$$

or

$$h_{feo}\left(\frac{1}{1 + j2\pi f C_{b'e} r_{b'e}}\right). \tag{5.10}$$

At one frequency, $|X_{C_{b'e}}| = r_{b'e}$ so that $|h_{fe}| = h_{feo}/\sqrt{2}$. This is the high-frequency cut-off (or -3 dB) point $f_{\alpha'}$ for the short-circuited output conditions and is obviously given by:

$$f_{\alpha'} = \frac{1}{2\pi r_{b'e} C_{b'e}}. \tag{5.11}$$

It may be inserted into the equation for h_{fe} to give:

$$h_{fe} = h_{feo}\left(\frac{1}{1 + j(f/f_{\alpha'})}\right) \tag{5.12}$$

or

$$|h_{fe}| = h_{feo}\left(\frac{1}{1 + (f/f_{\alpha'})^2}\right)^{1/2}. \tag{5.13}$$

Notice that when $f \gg f_{\alpha'}$, $|h_{fe}| \rightarrow h_{feo} f_{\alpha'}/f$. In this region, $|h_{fe}| \propto 1/f$; that is, it falls at a constant rate of 6 dB per octave. Also, $|h_{fe}|f = h_{feo} f_{\alpha'} = F_T$. The constant F_T is the *gain–bandwidth product* and from it the modulus of h_{fe} at any frequency higher than $f_{\alpha'}$ may be determined. In particular, notice that F_T is also f_1, the frequency at which the gain modulus has fallen to unity. These definitions are useful in high-frequency calculations and all depend upon the validity of the single time-constant model. There are in fact deviations from this 6 dB/octave slope and particularly from the associated "single lag" phaseshift. Treatment of these considerations is, however, outside the scope of this book.

Analogous to the CE cut-off $f_{\alpha'}$ is the CB cut-off, f_{α}, or $f_{h_{fb}}$. This frequency may be shown to much higher than $f_{\alpha'}$, which accounts for the superiority of the CB stage as a high-frequency amplifier. However, because the discrete CB stage has little value in audio-frequency work, this point will not be pursued.

5.1.3   A Simple Design Example

As an illustration of the use of the information so far presented, consider the design of a simple common-emitter amplifier stage using, for example, a small-signal silicon planar npn transistor having the limiting parameters given in table 3.2. Let the design value of I_C be 1 mA, which is the current at which the parameters of table 3.2 are valid. (For other currents, graphs such as those of figure 3.5 would be necessary.)

Since very low powers indeed are involved, it is not necessary to invoke the half-voltage principle. Instead, let $V_{CE} = 9$ V and let the drop across R_C also be 9 V. If a supply voltage of 24 V is used ($V_{CEM} = 40$ V) then the drop across R_E will be 6 V.

These values are shown in figure 5.6, and it is also assumed that the series biasing method is to be used. The 6 V drop across R_E also ensures that R_2 is not excessively low, which would otherwise make for a very low input resistance.

From these figures, R_E, R_C, R_1, and R_2 can be calculated:

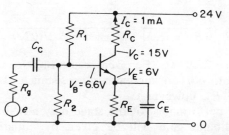

Figure 5.6. Practical amplifier stage using a small-signal silicon planar npn transistor.

$$R_E = \frac{6\,V}{1\,mA} = 6\,k\Omega$$

$$R_C = \frac{9\,V}{1\,mA} = 9\,k\Omega\,.$$

The maximum base current expected is that which will flow when the transistor is almost bottomed, that is, when $I_{Cmax} \simeq V_{CC}/(R_E + R_C)$ (neglecting V_{CE}). This gives,

$$I_{Cmax} \simeq \frac{24}{15} = 1.6\,mA\,.$$

At 1.6 mA, $h_{FE} \nless 35$, so that $I_{Bmax} \ngtr \simeq 1.6/35 \simeq 0.046\,mA$.

The current flowing through R_1 and R_2 should not be less than about five times I_{Bmax}, which in this case suggests 0.2 mA.

Hence, since V_{BE} during normal operation is about 0.6 V, the drop across $R_2 = V_B = 6.6$ V and $R_2 = 6.6\,V/0.2\,mA \simeq 33\,k\Omega$.

The current flowing through R_1 during normal operation will be 0.2 mA plus I_C/h_{FE} or $0.2 + 1/35 \simeq 0.23\,mA$.

Also the drop across R_1 is $24 - 6.6 = 17.4$ V. Hence, $R_1 = 17.4\,V/0.23\,mA = 75.6\,k\Omega$.

The nearest preferred value resistors to the calculated values for R_E, R_C, R_1, and R_2 are

$$R_E = 6.2\,k\Omega \quad R_C = 9.1\,k\Omega \quad R_1 = 75\,k\Omega \quad R_2 = 33\,k\Omega\,.$$

In an experimental amplifier, using these values of resistance and a randomly chosen transistor, the following values of current and voltage were observed:

$$I_C = 1.0\,mA \quad V_E = 6.3\,V \quad V_B = 6.96\,V \quad V_C = 14.8\,V\,.$$

These figures will be seen to compare very closely with the specified values for figure 5.6 and are well within the limits imposed by the 5% resistor tolerances.

The stability factor K may now be calculated:

$$K = \frac{1}{1 + [h_{FE}R_E/(R_B + R_E)]} = \frac{1}{1 + [35 \times 6.2/(22.9 + 6.2)]} \simeq 0.12$$

where $R_B = R_1 \| R_2 = 22.9\,k\Omega$. (For greater accuracy R_B should include $r_{bb'}$, since this is a real, not an incremental resistance. It is usually small compared with $R_1 \| R_2$, however.)

The small-signal performance of the stage may now be determined.

Firstly, the mid-band gain will be given by equation (4.23) in the hybrid-π system, or by equation (4.31) in the h-parameter system. Which system is used depends upon the nature of the known parameters; that is, those given by the manufacturer. For the purposes of the present example, consider the hybrid-π parameters. In order to determine the voltage gain, $r_{b'e}$ must first be found. This can be done by taking the maximum and minimum values of h_{fe} from table 3.2 and using them in equation (3.16b):

$$r_{b'e} \simeq \frac{26h_{fe}}{I_C}$$

which gives:

$$r_{b'e(min)} \simeq \frac{26 \times 50}{1} = 1300 \ \Omega$$

and

$$r_{b'e(max)} \simeq \frac{26 \times 200}{1} = 5200 \ \Omega \ .$$

Both of these values are much greater than $r_{bb'}$, which is quoted as about 50 Ω. Therefore, the simplified equation (4.23b) may be used to give the approximate voltage gain:

$$A_v \simeq -g_m R_L$$

where $g_m \simeq 39 I_C$ mA/V (by equation (3.15b)).
    Thus,

$$A_v \simeq -39 \times 1 \times 9.1 = -335 \quad \text{or} \quad 51 \ \text{dB} \ .$$

(Note that for a modern, general purpose, high-gain amplification/switching transistor (such as the Motorola 2N3903), $r_{bb'}$ will be very small, so that the more accurate equation, (4.23a), need not be used. Care should be exercised on this point, however, for some contemporary small-signal transistors exhibit values of $r_{bb'}$ up to about 1000 Ω.)
    Now consider the low-frequency cut-off characteristics of the stage. To determine f_L, R_G must first be found and this is given by R_g in parallel with $R_1 \parallel R_2$.
    Assuming that $R_g = 600 \ \Omega$ and that $r_{bb'}$ is a nominal 50 Ω for the purposes of this example,

$$R_G = 0.6 \parallel 75 \parallel 33 = 0.58 \ \text{k}\Omega \quad (\text{or} \ 580 \ \Omega) \ .$$

Hence,

$$R_{emitter} \simeq R_E \left\| \frac{r_{bb'} + r_{b'e} + R_G}{h_{fe}} \right.$$

gives

$$R_{\text{emitter(max)}} \simeq 6200 \left\| \frac{50 + 1300 + 580}{50} \simeq 38 \, \Omega \right.$$

$$R_{\text{emitter(min)}} \simeq 6200 \left\| \frac{50 + 5200 + 580}{200} \simeq 29 \, \Omega \right. .$$

If C_E were to define an f_L of 100 Hz, then

$$C_{E(\text{min})} \simeq \frac{10^6}{2\pi 100 \times 38} \simeq 42 \, \mu\text{F}$$

$$C_{E(\text{max})} \simeq \frac{10^6}{2\pi 100 \times 29} \simeq 55 \, \mu\text{F} .$$

A value of C_E large enough to ensure noninterference with f_L would be some ten times this larger figure, or 500 μF.

Now consider figures 5.3(a) and (b). Here, R_{in} can be readily calculated from either the relevant h-parameter or hybrid-π expression. Using the latter (equation (4.20)):

$$R_{\text{in}} \simeq r_{\text{bb}'} + r_{\text{b}'\text{e}}$$

giving

$$R_{\text{in(min)}} \simeq 50 + 1300 = 1350 \, \Omega$$

to

$$R_{\text{in(max)}} \simeq 50 + 5200 = 5250 \, \Omega$$

so that $R_{\text{BP}} = R_{\text{in}} \| R_1 \| R_2$ gives:

$$R_{\text{BP(min)}} \simeq 1275 \, \Omega$$

$$R_{\text{BP(max)}} \simeq 4271 \, \Omega .$$

Then,

$$R_{\text{g}} + R_{\text{BP(min)}} = 1874 \, \Omega$$

and

$$R_{\text{g}} + R_{\text{BP(max)}} = 4871 \, \Omega .$$

If C_C is made 0.5 μF then equation (5.7), in the form $f_L = 1/2\pi C_C (R_g + R_{\text{BP}})$ gives:

$$f_{L(max)} = \frac{10^6}{2\pi \times 0.5 \times 1874} \simeq 170 \text{ Hz}$$

and

$$f_{L(min)} = \frac{10^6}{2\pi \times 0.5 \times 4871} \simeq 65 \text{ Hz}.$$

Now consider the high-frequency performance of the stage. The high-frequency cut-off point, f_H, is dependent upon the interelectrode capacitances $C_{b'e}$ and $C_{b'c}$.

From table 3.2, the capacitances at $I_C = 1$ mA are:

$$C_{b'e} = 15 \text{ pF} \quad \text{and} \quad C_{b'c} = 2 \text{ pF}.$$

Using equation (5.9), where

$$\begin{aligned}
C_{in} &= C_{b'e} + C_{b'c}(1 + g_m R_L) \\
&= 15 + 2(1 + 39 \times 9.1) = 727 \text{ pF}
\end{aligned}$$

and

$$R = (R_G + r_{bb'})\|r_{b'e} = \begin{cases} (580 + 50)\|1300 = 424\ \Omega \text{ (min)} \\ (580 + 50)\|5200 = 562\ \Omega \text{ (max)} \end{cases}$$

gives $f_H = 1/2\pi R C_{in}$ as lying between

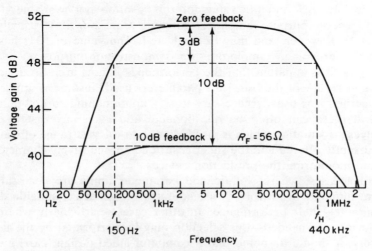

Figure 5.7. Frequency response of a common-emitter stage.

$$f_{H(max)} = \frac{10^{12}}{2\pi \times 424 \times 727} \simeq 516 \,\text{kHz}$$

and

$$f_{H(min)} = \frac{10^{12}}{2\pi \times 562 \times 727} \simeq 390 \,\text{kHz} \,.$$

Figure 5.7 is a measured gain/frequency plot of a practical amplifier stage using the resistor and capacitor values calculated above. Its performance may be compared with the calculated values:

	A_v	f_L	f_H	R_{in}
Calculated maximum	-355	170 Hz	516 kHz	5250 Ω
Measured	-355	150 Hz	440 kHz	2200 Ω
Calculated minimum	-355	65 Hz	390 kHz	1350 Ω

5.2   DISCUSSION ON SINGLE-STAGE CE DESIGN

The foregoing example has highlighted several points with respect to single-stage design in general.

Firstly, the performance of such a stage is very dependent upon the parameters of the particular device used and for a given type of transistor neither the frequency response nor the input resistance can be closely defined. The value of f_H in particular is very dependent upon $C_{b'c}$, and this is itself variable with I_C and V_{CE}. Also, the expression for voltage gain should strictly include $r_{bb'}$ and the value for this is often not available. (For a modern general purpose amplifier/switching transistor it is usually well below 50 Ω, however, and may be neglected. The value of 50 Ω has been used in the foregoing example only for demonstration purposes.)

Secondly, it is apparent that the performance figures are also dependent upon the *accuracy* of the transistor parameters used. Insofar as applications are concerned it is usually necessary to rely upon manufacturers' published data, and these can often be questionable and even inconsistent within themselves—a situation which is probably due to the vast range of transistor types currently offered and the sheer cost of comprehensively characterizing them and monitoring the production devices.

In some cases, where well-defined parameter limits are necessary and sufficient numbers are involved, a manufacturer will select suitable devices and guarantee their performance. In other cases—particularly where low-noise devices are needed—this selection may be performed by the user.

Thirdly, it should be noted that if computer-aided design (CAD) methods are justified by an application, it is imperative to guarantee the relevant limiting parameters of all the devices involved, otherwise the use of such exact methods becomes futile.

Finally, consider the effect of cascading two or more CE stages to produce greater gain.

In figure 5.2 the load on the collector of the first stage is R_C in parallel with the net input resistance to the second stage: that is (neglecting the reactance of C_C'),

$$R_L = R_C \| R_1' \| R_2' \| R_{in}' .$$

This can be quite a low resistance (such as a few hundred ohms), which implies that something approaching a short-circuit exists across the output of the first stage. Using this approximation the current gain of this stage can be said to be very close to h_{fe} and the input resistance to h_{ie}, since both are short-circuited-output parameters; that is, for the first stage,

$$R_{in} \simeq h_{ie} \quad \text{and} \quad A_i \simeq h_{fe} .$$

Since

$$A_v = \frac{v_{ce}}{v_{be}} \simeq \frac{-h_{fe} i_b R_L}{v_{be}}$$

and $v_{be}/i_b \simeq h_{ie}$, then,

$$A_v \simeq - \frac{h_{fe} R_L}{h_{ie}} .$$

(Note that this expression could be derived equally well by assuming the short-circuited-output condition $h_{ie} = r_{bb'} + r_{b'e}$ in equation (4.23a); and also inserting $h_{fe} = g_m r_{b'e}$.)

Using the parameters of table 3.2, and the values of R_{in} previously calculated,

$$R_{L(min)} = 9.1 \| 75 \| 33 \| 1.35 = 1.2 \text{ k}\Omega$$

and

$$R_{L(max)} = 9.1 \| 75 \| 33 \| 5.25 = 2.9 \text{ k}\Omega$$

giving

$$A_{v(min)} \simeq - \frac{50 \times 1.2}{1.8} = -33$$

and

$$A_{v(max)} \simeq - \frac{200 \times 2.9}{5.8} = -100 \; .$$

If the value of A'_v for the second stage is taken as being that calculated in the design example above, then the *overall* gain range of the two-stage amplifier can be taken as lying between the products of the gains:

$$A_{v(overall)} \simeq A_v A'_v \; .$$

That is,

$$A_{v(overall)(min)} \simeq (-33)(-355) = 11\,833$$
$$A_{v(overall)(max)} \simeq (-100)(-355) = 35\,500 \; .$$

This enormous gain range, and also the very wide range of possible input resistances, illustrates the need for design techniques which avoid reliance upon the transistor parameters themselves. Such techniques all involve the concept of *negative feedback*, and in fact this has already been met because the load resistor in an emitter-follower actually provides 100% negative feedback. An introduction to the subject of feedback now follows.

5.3 FEEDBACK

In order to understand the effects of feedback in particular cases it is useful to discuss first the general implications of the techniques, even though the analytical results may not always be easy to apply in their general form.

Feedback may take two basic forms, positive or negative. Positive feedback means that a portion of the output signal of an amplifier is fed back to the input in the same sense as the input signal. Thus the input signal is augmented, whereupon the output signal rises and an even larger feedback signal results. This cumulative process may continue until some component ceases to function; for example, the core of an inductance may saturate, or a transistor may "bottom." If the conditions are such that the device ceases to function momentarily, then builds up again, it is said to oscillate. Many free-running oscillators depend upon this principle and a formal treatment is given in chapter 10.

Certain forms of positive feedback can lead to the destruction of transistors, and one example has been met in the discussion on thermal runaway. Here the input signal is thermal, not electrical; this leads to an increase in I_C which feeds back more heat and the cycle repeats. For an amplifier the gain of the device measured *round* the complete loop, that is, from input to output then back via the feedback channel, is called the *loop gain*. Obviously if this is positive and greater than unity, then the build-up is cumulative and can only stop when some component ceases to function. For thermal runaway, the common ratio *r* is obviously analogous to loop gain. If, however, the loop gain is positive but less than unity the build-up will not be divergent.

Negative feedback has a rather different effect. Since it involves the application of a feedback signal which opposes the input signal, then the output signal is reduced, and so the *overall gain* of the amplifying system becomes less. As will be shown later, this gain also becomes much less dependent on the various parameters of devices within the amplifier. Other beneficial effects include the reduction of distortion and of noise arising within the amplifier and it also becomes possible to modify both input and output impedances.

One important aspect of negative feedback is that it may at certain frequencies become positive. This leads to the well known tendency of high-gain amplifiers with heavy feedback to burst into oscillation spontaneously. This effect is due to excessive phase shifts in such amplifiers, causing output signals at high frequencies to be shifted through phase angles approaching 360°. This, of course, brings them back into phase with the input signals and results in positive feedback.

Certain ways of describing the stability of amplifiers from this point of view will be discussed later. For the present only feedback which is genuinely 180° out of phase with the input signal will be considered.

There are four possible ways of applying negative feedback. The feedback signal itself may be a function of either the output voltage or the output current and it may be applied to the input in either series or parallel.

5.3.1 Voltage-derived Feedback

Consider the case of voltage feedback applied in series with the input. A perfect amplifier, represented in figure 5.8, may first be postulated and the *forward voltage gain* defined as $A = v_{out}/v_{in}$. The value of A may be positive, negative or complex, but according to the convention for circuit analysis previously used, the sense of the arrows representing input and output voltage must be as shown. This amplifier has an infinite input impedance and the output is the voltage source Av_{in}.

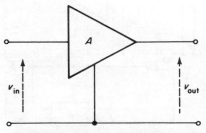

Figure 5.8. "Perfect" amplifier.

Now let a real source having an internal impedance of Z_G be connected to the input terminals and let a fraction B of the output voltage be applied in series with the input as shown in figure 5.9. This voltage fraction is given by:

$$v_f = Bv_{out}$$

where B is the *feedback fraction*.

Note that the box B may contain any network from a simple potentiometer to a transformer, but for the purposes of this discussion it will be assumed to result in no phase change between v_f and v_{out}.

Note also that the loop gain is given by the product AB.

The conventional signal polarities are shown by broken arrows and, using these, the gain equation for the general case may be derived.

Since $i_{in} = 0$, Z_G will have no effect. Therefore,

$$e = v_{in} - v_f = \frac{v_{out}}{A} - Bv_{out} \, .$$

The overall gain is $A_{ov(FB)}$ where $A_{ov(FB)} = v_{out}/e$, whence

$$A_{ov(FB)} = A/(1 - AB) \, . \tag{5.14}$$

Note that since $i_{in} = 0$, the terminal gain, that is the gain to the right of ab, is given by the same expression, or $A_{v(FB)} = A_{ov(FB)}$.

Several important results can be derived from this equation.

Firstly, if A is positive and $AB < 1$, then the overall gain will be greater than A. This is a stable case of positive feedback. However, if $AB > 1$, the circuit will become unstable and will burst into oscillation.

Secondly, if A has a negative value, then the signal polarities will be as shown by full arrows and v_f will oppose e, which is the condition for negative

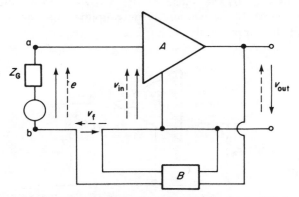

Figure 5.9. Amplifier with series voltage feedback.

feedback. Here it is apparent that the gains $A_{ov(FB)}$ and $A_{v(FB)}$ will be less than A by a factor $1 - AB$. This is therefore known as the *feedback factor* and is defined as the ratio of gain without feedback to gain with feedback.

As an example, consider an amplifier where $A = -100$ and $B = 0.1$. Then

$$A_{v(FB)} = \frac{-100}{1 + 10} = -9.1 .$$

If $|AB| \gg 1$, then $A_{v(FB)} \to -1/B$. This is of considerable importance since it means that the gain is no longer a function of the parameters of the amplifier components but only of the components in the feedback network. If this network is designed to be largely invariant with frequency, then the gain of the whole amplifier will be frequency independent, within the limits over which A is still large. Thus, $A_{v(FB)}$ can be defined by the designer and will cease to be affected by the parameters of the active elements.

Thirdly, any noise or distortion originating *within* the amplifier will be reduced by the value of the feedback factor. To show this assume that an amplifier without feedback has an input v_{in} and an output $v_{out} + v_N$. Here, v_N is that portion of the output generated within the amplifier and may be taken to represent either noise or distortion. Now let feedback be applied and let v_{in} be increased so that v_{out} is of the same magnitude as before. This means that the "forward" amplifier is working under identical conditions in both cases and will therefore generate the same unwanted signal v_N. However, v_N, owing to feedback, will change to v_N', where

$$v_N' = v_N + ABv_N' .$$

This gives the ratio of v_N to v_N' as $1 - AB$, which is the feedback factor. In other words, if noise or distortion is generated *within* an amplifier then the application of negative feedback will reduce its magnitude at the output point by the value of the feedback factor.

The improvement in gain stability can be demonstrated as follows. From equation (5.14),

$$A_{v(FB)} = \frac{A}{1 - AB}$$

so

$$\frac{dA_{v(FB)}}{dA} = \frac{1}{(1 - AB)^2} .$$

The fractional change is:

$$\frac{dA_{v(FB)}}{A_{v(FB)}} = \frac{dA}{A_{v(FB)}} \frac{1}{(1 - AB)^2}$$

or

$$\frac{\mathrm{d}A_{v(FB)}}{A_{v(FB)}} = \frac{\mathrm{d}A}{A}\frac{1}{1-AB}. \tag{5.15}$$

This means that the effect of changes in A is reduced by the value of the feedback factor.

To summarize, it may be said that a loss in gain has been traded for the advantages of an extended frequency response, improved linearity and lower distortion.

Consider now an amplifier having a finite input impedance Z_{in}, such as a transistor amplifier. This is represented in figure 5.10 as having an internal voltage gain of $A_v(=v_{out}/v_{in})$. If the source is represented by a voltage generator in series with the source impedance Z_G, the overall voltage gain is given by $A_{ov} = v_{out}/e$.

From figure 5.10

$$e = i_{in}(Z_G + Z_{in}) = \frac{v_{in}}{Z_{in}}(Z_G + Z_{in}).$$

Therefore

$$A_{ov} = \frac{v_{out}}{v_{in}[(Z_G + Z_{in})/Z_{in}]} = \frac{A_v}{(Z_G + Z_{in})/Z_{in}}. \tag{5.16}$$

This shows that the overall voltage gain is influenced by both the source and input impedances as would be expected since i_{in} is no longer zero. If the gain from the output terminals ab of the source were required, then by inspection this is obviously A_v. This is equivalent to putting $Z_G = 0$, when equation (5.16) reduces to A_v.

If negative feedback is applied to such an amplifier the circuit becomes as shown in figure 5.11 and A_v must always be negative.

If the overall voltage gain is again defined as $A_{ov(FB)} = v_{out}/e$ where

$$e = i_{in}(Z_G + Z_{in}) - Bv_{out}$$

$$= \frac{v_{in}}{Z_{in}}(Z_G + Z_{in}) - Bv_{out}$$

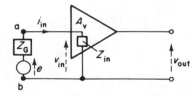

Figure 5.10. Amplifier with finite input impedance.

then

$$A_{ov(FB)} = \frac{v_{out}}{v_{in}[(Z_G + Z_{in})/Z_{in}] - Bv_{out}} = \frac{A_v}{[(Z_G + Z_{in})/Z_{in}] - BA_v}.$$
(5.17)

Note that if $Z_{in} \to \infty$ this expression reverts to equation (5.14), or if $B = 0$ (i.e., no feedback) it reverts to equation (5.16). If the gain from the output terminals of the source ab is required this is equivalent to putting $Z_G = 0$, when equation (5.17) again becomes the same as equation (5.14). This is once more obvious by inspection, since the source is then a perfect voltage generator and the value of Z_{in} can make no difference.

A feedback factor can be derived from equation (5.17) by using the definition:

$$\frac{\text{overall gain without feedback}}{\text{overall gain with feedback}} = \frac{(5.16)}{(5.17)} = 1 - \left(\frac{Z_{in}}{Z_G + Z_{in}}\right)BA_v.$$
(5.18)

The input impedance of interest in any amplifier is usually that "seen" by the source; here it is to the right of the terminals ab in figure 5.11. The qualitative effect of series feedback can be deduced by examination of the circuit. Since $A_v = v_{out}/v_{in} = v_{out}/Z_{in}i_{in}$, then for v_{out} to be the same as for an amplifier *without* feedback, i_{in} must be the same in both cases. For this to be so, *e* must *increase* if it is to make up for the series voltage v_f.

Quantitatively, let the input impedance at ab be $Z_{in(FB)}$, where

$$Z_{in(FB)} = \frac{v_{in} - v_f}{i_{in}} = \frac{Z_{in}}{v_{in}}(v_{in} - A_v Bv_{in})$$

or

$$Z_{in(FB)} = Z_{in}(1 - A_v B).$$
(5.19)

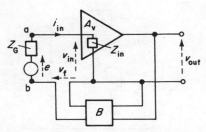

Figure 5.11. Real amplifier with series voltage feedback.

The factor $(1 - A_v B)$ is actually the feedback factor for the circuit to the right of ab; if the input impedance at the voltage generator itself is required, this is obviously given by $Z_G + Z_{in}(1 - A_v B)$.

This can be arranged as

$$(Z_G + Z_{in})\{1 - [Z_{in}/(Z_G + Z_{in})]A_v B\}$$

which is the input impedance of the circuit without feedback multiplied by the overall feedback factor.

Having determined the effects of series feedback in a general manner, these may be compared with the conditions imposed by the shunt feedback circuit of figure 5.12. Here, it will be seen that the source current, i_g, splits up into two branches, one flowing into the amplifier input impedance and the other round the feedback loop.

The terminal voltage gain is obviously unchanged:

$$A_{v(FB)} = \frac{v_{out}}{v_{in}} = A_v .$$

The overall voltage gain, however, may be determined as follows:

$$A_{ov(FB)} = \frac{v_{out}}{e} \quad \text{where} \quad e = v_{in} + i_g Z_G .$$

Now

$$i_g = i_{in} + i_f = \frac{v_{in}}{Z_{in}} + \frac{v_{in} - v_{out}}{Z_F}$$

$$= v_{in}\left(\frac{1}{Z_{in}} - \frac{A_v - 1}{Z_F}\right)$$

so

$$A_{ov(FB)} = \frac{A_v v_{in}}{v_{in} + v_{in} Z_G\{(1/Z_{in}) - [(A_v - 1)/Z_F]\}}$$

$$= \frac{A_v}{[(Z_{in} + Z_G)/Z_{in}] - (Z_G/Z_F)(A_v - 1)} . \quad (5.20)$$

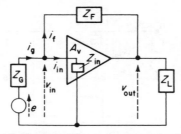

Figure 5.12. Amplifier with shunt feedback.

Note that the form of equation (5.20) is similar to that of equation (5.17), which means that by analogy $Z_G/Z_F = B$ providing that $|A_v| \gg 1$.

A most important form of equation (5.20) results when $Z_{in} \gg Z_G$ and $|A_v| \gg 1$. If these conditions are inserted, then

$$A_{ov(FB)} = -\frac{Z_F}{Z_G}. \tag{5.21}$$

This is the basic equation for the gain of the *operational amplifier*, so called since it is used to perform the mathematical operations required in analog computation. For example, if the impedances are pure resistances R_F and R_G, then any input voltage is multiplied by a constant, $-R_F/R_G$.

If Z_{in} is very high, practically all of the source current must flow round the feedback path, and if Z_F is a capacitor, then in order that this should happen, the output voltage must change according to the equation $v_{out} = (1/C_F) \int i_g \, dt$. This is the basis of the use of the operational amplifier as an integrator, often called a Miller integrator after the Miller effect described on page 94.

If A_v is very large, then the input point of the amplifier must be at a very low voltage since $v_{in} = v_{out}/A_v$. This has led to the concept of the *virtual earth*, for although the input point is not connected to the common rail, its voltage departs very little from this potential at any time. If, therefore, several input resistors are connected to the virtual earth point, their currents all flow to this point then around the feedback path *without interacting*. Thus, the feedback current is equal to the sum of the input currents, $v_{out} = R_F(i_1 + i_2 + i_3 \ldots)$ and the device can be used as an adding unit.

True operational amplifiers must will be considered further in chapter 7.

The actual value of the input impedance at the virtual earth point can be obtained as follows:

$$Z_{in(FB)} = \frac{v_{in}}{i_g} = \frac{v_{in}}{i_{in} + i_f}$$

$$= \frac{v_{in}}{(v_{in}/Z_{in}) + [(v_{in} - v_{out})/Z_F]} = \frac{1}{(1/Z_{in}) - [(A_v - 1)/Z_F]}. \tag{5.22}$$

If $|A_v| \gg 1$ and $1/Z_{in} \rightarrow 0$,

$$Z_{in(FB)} = -\frac{Z_F}{A_v}. \tag{5.23}$$

This is an extremely useful expression in practice and the negative sign is normally omitted—it arises in this case simply because the value of A_v is negative, making the impedance positive.

5.3.2   Current-derived Feedback

An analysis of current-derived feedback in the general case is much more complex than that for voltage feedback and its results are not so readily applicable to special cases. It is therefore more profitable to deal with specific examples as they arise. One such example is the emitter-follower of figure 4.5, the feedback analysis for this being given later.

Very often it is difficult to decide whether a given circuit represents voltage or current feedback. The most valid test is to ascertain whether the amount of feedback varies with the load impedance, though even this is not infallible. This is merely an expression of the inter-relationship between voltage and current in any real circuit.

5.3.3   Output Impedances

The provision of feedback from the output circuit of a real amplifier inevitably alters the output impedance, not by virtue of any loading effect but simply by changing the relationship of output current and voltage.

Consider figure 5.12, for example, and let the amplifier be replaced by one with an output impedance Z_{out}. Let the load be replaced by a voltage generator e_o and let the source generator be suppressed, as shown in figure 5.13. Then,

$$Z_{out(FB)} = \frac{e_o}{i_{out}} = \frac{e_o}{(e_o - A_v v_{in})/Z_{out}} .$$

Now

$$v_{in} = \frac{e_o Z}{Z_F + Z} \quad \text{where} \quad Z = Z_G \| Z_{in} .$$

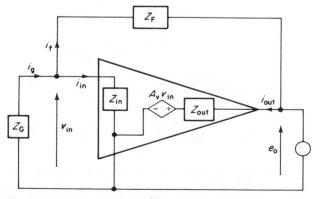

Figure 5.13. Circuit for calculation of output impedance.

If $Z_{in} \gg Z_G$, which is usual, $Z \simeq Z_G$ so that

$$Z_{out(FB)} = \frac{Z_{out}}{1 - [Z_G/(Z_F + Z_G)]A_v} . \qquad (5.24)$$

This means that for voltage-derived feedback, the output impedance is divided by a feedback factor defined by equation (5.24). (Notice that if a current source were used, so that $Z_G = \infty$, equation (5.24) would have had Z_{in} substituted for Z_G.)

If $Z_F = 0$ in equation (5.24), then

$$Z_{out(FB)} = Z_{out}/(1 - A_v) \simeq -Z_{out}/A_v$$

which is the output impedance of a *voltage-follower*, which is discussed in chapter 7.

It will have been observed that in the foregoing derivation the output impedance at the amplifier alone has been obtained. Strictly speaking the output impedance measured from the load itself should have been found, that is, account should have been taken of i_f as well as i_{out}. However, because the output impedance as calculated is very small for most real cases, the extra parallel path formed by $(Z_F + Z)$ affects R_{out} but little and so the extra complication is not justified.

For current-derived feedback, it is found that

$$Z_{out(FB)} = Z_{out}(1 - A_v B) . \qquad (5.25)$$

Here, however, it is usually better to treat each case on its own merits rather than to extract a generalized feedback factor.

5.3.4 Current and Power Gain

From figure 5.11 it can be seen that the current gain may change little when series feedback is applied. If a change occurs it is due to an alteration in the output impedance. On the other hand, shunt feedback would be expected to have a profound effect on the current gain, since i_g is augmented by i_f as shown in figure 5.13.

If the internal current gain is defined as $A_i = i_{out}/i_{in}$ then the current gain with feedback is

$$A_{i(FB)} = \frac{i_{out}}{i_g} = \frac{i_{out}}{i_{in} + i_f} .$$

If the feedback fraction is quoted in terms of current ratio at the output point, $B = -i_f/i_{out}$, then

TABLE 5.1. Summary of the effects of negative feedback.

	Voltage derived		Current derived	
	Series	Shunt	Series	Shunt
A_i		reduced		reduced
A_v	reduced		reduced	
R_{in}	increased	decreased	increased	decreased
R_{out}	decreased	decreased	increased	increased

$$A_{i(FB)} = \frac{i_{out}/i_{in}}{1 + (i_f/i_{in})} = \frac{A_i}{1 + (i_f/i_{out})(i_{out}/i_{in})} = \frac{A_i}{1 - A_i B} . \qquad (5.26)$$

This equation is of the same form as that for voltage gain and may be used in the same manner.

Expressions for power gain will not be derived since they may be obtained simply from the voltage and current equations.

Table 5.1 summarizes the effects of negative feedback as discussed above.

5.3.5 The Degenerate Common-emitter Stage

In figure 5.1 the shunting effect of C_E becomes progressively smaller as the frequency falls until at very low frequencies it may as well be absent. Under these circumstances the stage can be regarded as in part common-emitter and in part common-collector. Because such a circuit has some advantages, part of R_E is sometimes left unbypassed for normal operating frequencies as shown in figure 5.14(a). Here, the output signal current develops a voltage across R_F and this is applied in series with the input signal (because R_F is within the input loop). Further, this feedback signal is in opposition to the input signal so that the complete circuit becomes a case of *current-derived series feedback*. However, unlike the emitter-follower, less than 100% of the output voltage is fed back. In fact, if the total output signal voltage is developed across R_C and R_F in series, and if the cc gain from the base to R_F is nearly unity, then the fraction of the output signal fed back must be R_F/R_C. So, using the basic feedback equation,

$$A_{v(FB)} = \frac{A_v}{1 - A_v B} = \frac{A_v}{1 - A_v(R_F/R_C)} .$$

But the gain for $R_F = 0$ is approximately $-g_m R_C$, so

$$A_{v(FB)} \simeq \frac{-g_m R_C}{1 + g_m R_F} \qquad (5.27)$$

which reduces to $-R_C/R_F$ if $g_m R_F \gg 1$.

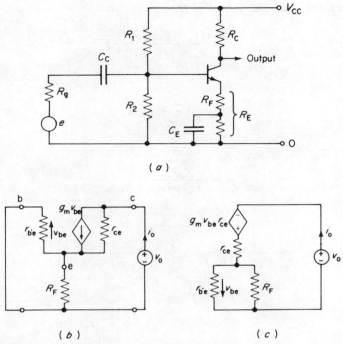

Figure 5.14. The degenerate CE stage: (*a*) practical circuit; (*b*) and (*c*) equivalents for analysis.

Because the stage acts in part as a common-collector, its input resistance is:

$$R_{in} \simeq r_{bb'} + r_{b'e} + h_{fe}R_F . \qquad (4.26b)$$

If $r_{bb'} \ll r_{b'e}$, which is usual and if h_{fe} is replaced by $g_m r_{b'e}$, this can also be written:

$$R_{in} \simeq r_{b'e}(1 + g_m R_F) \qquad (5.28)$$

which shows that the input resistance has been increased.

For current-derived feedback, the output resistance should also be increased and this may be demonstrated as follows. A hybrid-π equivalent circuit has been drawn in figure 5.14(*b*), where it has been assumed that $r_{bb'} \ll r_{b'e}$ and so may be neglected. The input has been short-circuited to AC and a voltage generator v_o has been applied to the output. Noting that $r_{b'e}$ now appears in parallel with R_F, the equivalent has been redrawn in figure 5.14(*c*) where a Norton-to-Thévenin transformation has also been made.

The current i_o at the output is now given by:

$$i_o = \frac{(v_o + g_m v_{be} r_{ce})}{r_{ce} + (r_{b'e} \| R_F)} \, .$$

But

$$v_{be} = -i_o(r_{b'e} \| R_F)$$

so

$$i_o = \frac{v_o - g_m r_{ce}(r_{b'e} \| R_F) i_o}{r_{ce} + (r_{b'e} \| R_F)}$$

so

$$\frac{v_o}{r_{ce} + (r_{b'e} \| R_F)} = i_o \left\{ 1 + \frac{g_m r_{ce}(r_{b'e} \| R_F)}{r_{ce} + (r_{b'e} \| R_F)} \right\}$$

or

$$R_{out} = \frac{v_o}{i_o} = r_{ce} + (r_{b'e} \| R_F) + g_m r_{ce}(r_{b'e} \| R_F) \, .$$

If $g_m r_{ce} \gg 1$, which is likely,

$$R_{out} \simeq r_{ce} \left\{ 1 + g_m \frac{r_{b'e} R_F}{r_{b'e} + R_F} \right\}$$

and if $R_F \ll r_{b'e}$, which is also likely, then

$$R_{out} \simeq r_{ce}(1 + g_m R_F) \, . \tag{5.29}$$

Notice that both the input and output resistances have been increased by a factor $(1 + g_m R_F)$, which must therefore be the feedback factor for this particular circuit, which is in accordance with the voltage gain expression (5.27).

5.3.6 A Design Example

For the simple common-emitter stage designed earlier, the voltage gain was shown to be −355 or 51 dB. If it is desired to apply 10 dB of series feedback, part of R_E may be left unbypassed, and this part, R_F, may be determined as follows.

The voltage gain will be reduced to 41 dB, which is −112, and using

$$A_{v(FB)} = \frac{A_v}{1 - A_v B}$$

$$-112 = \frac{-355}{1 + 355B}$$

where $B = R_F/R_C$ giving $R_F = 55.6\,\Omega$. (Note that this is much less than R_C, which is $9100\,\Omega$.)

When a $56\,\Omega$ section of R_E was left unbypassed, the lower curve of figure 5.7 resulted.

Figure 5.7 shows that the frequency response of the stage has been improved, as would be expected.

Should it be desired to construct a simple amplifier having a low output impedance, it would be appropriate to add an emitter-follower stage to the common-emitter stage as shown in figure 5.15(a).

Here, the bias voltage at the collector of Tr1 is the base bias voltage of Tr2, since no interstage coupling capacitor is involved. Tr2 presents an input resistance to Tr1 of approximately $h_{fe(2)}R'_F$, and this appears in parallel with R_C. This input resistance will normally be much greater than R_C, so that the overall voltage gain of the stage will be nearly $A_{v(1)}A_{v(2)}$. Here, $A_{v(1)}$ is the voltage gain of the simple degenerate CE stage, whilst $A_{v(2)}$ is that of an emitter-follower as given by equation (4.27).

This circuit consists of two stages, each with its own individual series feedback resistor, which implies a high input resistance at both bases. Figure 5.15(b) shows an example of *overall shunt* feedback leading to a *low* input resistance and an approximation to the *operational amplifier* condition.

Here, the load resistor of the emitter-follower has been split into two parts, R'_{F1} and R'_{F2}. That portion of the output voltage developed across R'_{F2} (being in phase opposition to the input voltage at the base of Tr1) has been used as a feedback signal and applied to the input via R_2. (Strictly, the value of R_2 should be reduced slightly because it no longer goes to ground but to a voltage $I_{E2}R'_{F2}$).

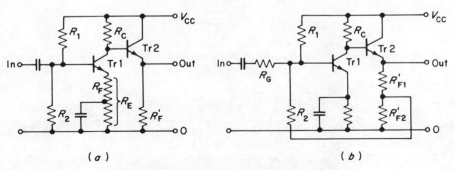

(a) (b)

Figure 5.15. Simple two-transistor amplifiers.

A simple way of predicting the performance of this amplifier is to think of the junction between R'_{F1} and R'_{F2} as the output point of a crude operational amplifier and R_2 as the shunt feedback resistor, R_F, of figure 5.12. Thus, if an input resistor R_G, is included (as shown in figure 5.15(b)), then equation (5.20) gives the voltage gain as:

$$A^\dagger_{ov(FB)} = \frac{A^\dagger_v}{[(R_{in} + R_G)/R_{in}] - (R_G/R_2)(A^\dagger_v - 1)} \ . \tag{5.30a}$$

Here, $A^\dagger_v$ is the gain measured from the base of Tr1 to the junction of R'_{F1} and R'_{F2}; that is,

$$A^\dagger_v = A_v \frac{R'_{F2}}{R'_{F1} + R'_{F2}} \tag{5.30b}$$

and the overall gain from the input to R_G to the emitter of Tr2 will be:

$$A_{ov(FB)} = A^\dagger_{ov(FB)} \frac{R'_{F1} + R'_{F2}}{R'_{F2}} \ . \tag{5.30c}$$

In the simple design example, the bias voltage at the collector of Tr1 was 15 V. This means that the base bias voltage of Tr2 is also 15 V, so that its emitter voltage is about $(15 - 0.6) = 14.4$ V. Hence, for $I'_C = 1$ mA,

$$(R'_{F1} + R'_{F2}) \simeq 14.4/I'_C = 14.4 \, \text{k}\Omega \ .$$

Suppose that this is made up from preferred values, so that

$$R'_{F1} = 12 \, \text{k}\Omega \quad \text{and} \quad R'_{F2} = 2.2 \, \text{k}\Omega$$

giving $(R'_{F1} + R'_{F2}) = 14.2 \, \text{k}\Omega$.
This implies that, from equation (5.30b),

$$A^\dagger_v = -355 \times \frac{2.2}{14.2} = -55 \ .$$

This can be put into equation (5.30a) to give:

$$A^\dagger_{ov(FB)} = \frac{-55}{[(2.2 + 4.7)/2.2] + (4.7/35)(56)} = -5.16 \ .$$

(Here R_2 has been increased by the small parallel combination of $R'_{F1} \parallel R'_{F2}$ for greater accuracy. It thus becomes 35 kΩ rather than 33 kΩ.)

Equation (5.30c) now gives the overall voltage gain for the complete amplifier:

$$A_{ov(FB)} = \frac{-5.16 \times 14.2}{2.2} = -33.3 \, .$$

(An actual measurement made upon an amplifier constructed in this way gave $A_{ov(FB)} = -33.5$.)

The assumption of the value of R'_{F2} in the above example was made in the interests of a clear sequence of calculations: in practice, the design steps would usually be carried out in reverse order to *determine* the value of R'_{F2} which would lead to a required voltage gain.

The examples given above illustrate that simple, though crude, approximations to amplifier performance can be deduced *provided* that the mode of operation of the circuits is thoroughly understood, which is *always* a desirable prerequisite to more sophisticated analyses.

Two transistors coupled together can often be treated as one unit and biased accordingly. They are referred to as *compound pairs* and exhibit characteristics which afford desirable design features. Some of these combinations will now be considered.

5.3.7 Compound Pairs

It has been shown in chapter 4 that the input resistance for a transistor is

$$R_{in} = h_i - \frac{h_r h_f}{G_L + h_o} \tag{4.33}$$

and this equation has been plotted for a common transistor in figure 4.10(a) where it is assumed that the signal frequency is such that only the resistive part of the h-parameters is significant and that the load is also a pure resistance. From this figure it is clear that the common-collector configuration will provide the highest input impedance and it is for this reason that the cc mode is chosen as the basic circuit for high-input-impedance amplifiers.

Over a large working range the R_{in} versus R_L curve for the cc or emitter-follower stage is almost at 45° to the logarithmic axes of the graph. This illustrates the resistance converter property of the emitter-follower, given previously as:

$$R_{in} \simeq r_{bb'} + r_{b'e} + h_{fe} R_L \tag{4.26b}$$

which becomes, when $h_{fe} R_L \gg (r_{bb'} + r_{b'e})$:

$$R_{in} \simeq h_{fe} R_L \, . \tag{4.26c}$$

When R_L becomes very large this approximation breaks down because (as will be seen from the h-parameter equivalent circuit of figure 5.16) R_L and

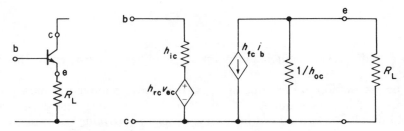

Figure 5.16. Emitter-follower with h-parameter equivalent circuit.

$1/h_{oc}$ are in parallel. Hence, the limiting value of the load is $1/h_{oc}$ (or r_{ce}) itself. In chapter 4, it was shown that when $R_L \to \infty$, equation (4.33) for R_{in} becomes:

$$R_{in} \to \frac{\triangle_c}{h_{oc}} = \frac{h_{ic}h_{oc} - h_{rc}h_{fc}}{h_{oc}}$$

or

$$R_{in} \to h_{ic} + h_{fe}\left(\frac{1}{h_{oc}}\right) \simeq h_{fe}\left(\frac{1}{h_{oc}}\right)$$

because

$$h_{re} = 1 \quad \text{and} \quad h_{fc} \simeq -h_{fe}$$

Comparing this approximation with (4.26c) shows that the limiting value of R_{in} is in fact determined by h_{oc} (or r_{ce}).

This elucidation is necessary when considering compound pairs, such as the *Darlington connection* of figure 5.17. Here the emitter current of Tr1 is the base current of Tr2 and the collectors are commoned. Hence the pair behaves as a single transistor and the overall incremental current gain can be found by determining the ways in which the various signal currents split up within the circuit.

Using the symbols $\alpha(=-h_{fb})$ and $\beta(=h_{fe})$ for convenience, it is apparent that if the emitter current of the compound pair is i_e then, because this is also the emitter current of Tr2, the collector current of Tr2 must be $\alpha_2 i_e$. Also, its base current (which is also the emitter current of Tr1) must be $i_{b_2} = i_{e_1} = (1 - \alpha_2)i_e$.

Hence, the base current of Tr1 (which is also the base current of the compound pair) must be:

$$i_b = (1 - \alpha_1)(1 - \alpha_2)i_e .$$

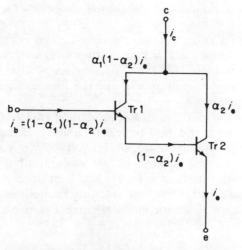

Figure 5.17. The Darlington connection (showing signal current magnitudes).

Also, the collector current of Tr1 must be:

$$i_{c_1} = \alpha_1(1 - \alpha_2)i_e$$

and this adds to i_{c_2} to give:

$$i_c = \alpha_1(1 - \alpha_2)i_e + \alpha_2 i_e .$$

These various signal currents are shown in figure 5.17 and can be used to write down the current gains for the compound pair:

$$\alpha = \frac{i_c}{i_e} = \frac{\alpha_1(1 - \alpha_2)i_e + \alpha_2 i_e}{i_e} = 1 - (1 - \alpha_1)(1 - \alpha_2) \simeq 1 \qquad (5.31)$$

$$\beta = \frac{i_c}{i_b} = \frac{\alpha_1(1 - \alpha_2)i_e + \alpha_2 i_e}{(1 - \alpha_1)(1 - \alpha_2)i_e} \simeq \frac{1}{(1 - \alpha_1)(1 - \alpha_2)} . \qquad (5.32)$$

If the two transistors were identical and working under identical bias conditions then if $\alpha_1 = \alpha_2 = 0.98$ for example, $\alpha = 0.996$ and $\beta = 2500$. However, it is apparent that because the direct or quiescent emitter current of Tr1 is also the (very small) quiescent base current of Tr2, then I_{E1} will be so small that h_{FE1} and hence the signal current gains of Tr1 will be much smaller than those of Tr2 (see figure 2.10), so that the apparent high gain of the compound pair is not fully realized. (It is possible to increase I_{E1} by including a resistor from its emitter to ground and this is often done when integrated compound pairs are fabricated, especially for power devices, as in figure 9.30.)

The Darlington pair is shown connected as an emitter-follower in figure 5.18 and if the composite β is known, then over a limited range $R_{in} \simeq \beta R_L$, which can be very high. However, it cannot exceed the limiting value placed upon it by the output resistance r_{ce} of the first transistor structure, as was initially explained.

Although the current gain of the compound emitter-follower is very high, its voltage gain must of course be less than unity. If the pair is connected in a common-emitter configuration, useful voltage gain can be obtained but it tends to be ill-defined because of the multiplicative effect of parameter variations. A better method is to use a pair of opposite-polarity transistors as shown in figure 5.19. Here, the small-signal analysis has been carried out as before and leads to an incremental current gain of:

$$\beta = \frac{i_c}{i_b} = \frac{\alpha_1}{(1 - \alpha_1)(1 - \alpha_2)} \ . \tag{5.33}$$

Consider now the voltage gain. Apart from the small signal voltage dropped across the base–emitter junction of the first transistor, the input voltage is given by

$$v_{in} = R_1 \left(\frac{1 - \alpha_2}{\alpha_1} + \alpha_2 \right) i_e \ .$$

Also

$$v_{out} = v_{in} + R_2 \alpha_2 i_e \ .$$

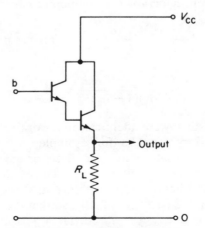

Figure 5.18. Compound emitter-follower.

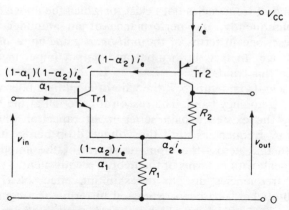

Figure 5.19. A complementary compound pair.

Hence

$$A_v = \frac{v_{out}}{v_{in}} = 1 + \frac{R_2}{R_1} \left(\frac{\alpha_2}{[(1 - \alpha_2)/\alpha_1] + \alpha_2} \right).$$

Since $(1 - \alpha_2)/\alpha_1$ is small, this becomes

$$A_v \simeq 1 + \frac{R_2}{R_1} . \tag{5.34}$$

This shows that the voltage gain of the amplifier is very well defined and that it may be controlled by making R_2 variable. (It is better to leave R_1 fixed, since R_1 is the basic current-limiting resistor.) Further, it shows that if $R_2 = 0$, the voltage gain is very near unity and in fact the approach to unity is much closer for this circuit than for the original Darlington connection.

Once again, the input resistance to the circuit is limited by the values of r_{ce} and of the biasing resistors. Positive feedback techniques (called "bootstrapping") may be employed to significantly raise the apparent values of these resistances, but it is easier to use field-effect transistors (see chapter 11) where input resistances of more than a few megohms are desired.

Large-signal *monolithic* (fabricated together on one silicon chip) compound pairs are now common and are used largely for power amplification and switching purposes. Their DC characteristics are of major importance in this context and will be considered in chapter 9 on power amplifiers.

5.3.8 Stability of Feedback Amplifiers

In the treatment of feedback amplifiers given so far, it has been assumed that the phaseshift within the amplifier has been exactly 180° and over the feedback path zero. In practice, this is not true for all frequencies, but normally a band of frequencies does exist for which the above conditions are obtained. Consequently, the performance of an amplifier at those frequencies is described in terms of the *mid-band* gain, input impedance and output impedance. In fact, both input and output impedances are largely resistive within this band.

As can be seen from figure 5.7, the gain of amplifiers falls at each side of the "working" frequency band. The reason for the fall at the low frequency end is usually the existence of a series input capacitor whose reactance increases at low frequencies until the voltage drop across it is significant compared with that across the input resistance of the amplifier. This has already been dealt with in terms of the hybrid-π equivalent, as has the fall in gain at high frequencies due to the shunting effect of C_{in}. The high-frequency roll-off has also been expressed in terms of a complex quantity formulation for h_f, and in fact at high frequencies all the h-parameters could be expressed in complex form. In practice, however, it is more convenient to use the admittance or y-parameters, which are all measured with short-circuited input or output (which are easier to realize practically than is an open-circuit input or output at high-frequency) and these are always presented in complex form.

In Appendix I it is shown that the voltage across a resistor in series with a capacitor falls at a rate of 6 dB per octave as the frequency is reduced. Such a circuit is similar to the input network of a transistor amplifier, where the capacitor is C_c and the resistor is given by the parallel combination of the inherent input resistance and the biasing components (see figure 5.3). If the low-frequency end of the gain versus frequency plot of figure 5.7 is examined, it will be seen that the gain does in fact fall at about 6 dB per octave, that is, as the frequency falls by a factor of two, so approximately does the gain. The region where this occurs is well below the -3 dB point since this slope must gradually decrease until the flat portion of the characteristic is reached. In Appendix I, it is shown that a simple geometrical construction based upon the 6 dB per octave slope will accurately give the -3 dB, or half-power, point.

At the high-frequency end of the characteristic, the gain also falls at 6 dB/octave, this time as the frequency *increases*. As has been explained, this is due largely to internal capacitances and the Miller effect.

As the gain of an amplifier changes, so does the phaseshift, and if only one capacitance and one resistance are involved, then the slope of the gain change approaches 6 dB per octave and the phaseshift never exceeds 90°. (This is also shown in Appendix I.) Consequently, if purely resistive feedback is applied around such an amplifier, it cannot become unstable, since the feedback can never be positive. This, of course, assumes that the feedback signal is derived from a point which is in antiphase with the input signal at the mid-band frequencies, and that the above discussion refers to phaseshifts which deviate from the ideal 180°. If this is the case, the feedback signal cannot come closer than 90° to the input signal at the extremes of the spectrum.

Such conditions obviously apply to single-stage amplifiers, which accounts for their inherent stability. The same is true for two-stage amplifiers, since two sets of simple *RC* networks in cascade can produce a phaseshift of up to 180° only. In practice, the phaseshift due to one *RC* network only approaches 90°, for in theory only a frequency of either zero or infinity would produce a true 90° shift. Similarly, for two *RC* networks, the phaseshift only approaches 180° and this accounts for the inability of two-stage amplifiers to oscillate, though they can easily exhibit transient instability.

To clarify this point, figure 5.20 has been included. Here the gain versus frequency and phaseshift versus frequency characteristics of single, double and triple *cut-off* networks have been drawn. A cut-off is defined as a

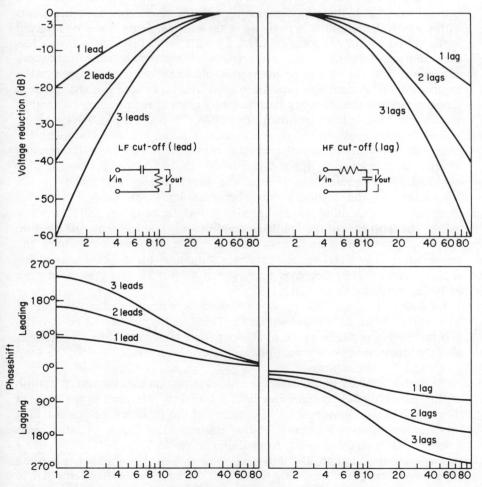

Figure 5.20. Gain versus frequency and phaseshift versus frequency curves for passive cut-offs.

capacitor combined with a resistor in such a manner as to cut off either low or high frequencies. A low-frequency cut-off (of which the input circuit of figure 5.3 is an example) is often termed a *single lead* since the output voltage leads the input. Similarly, a high-frequency *cut-off* is often termed a *single lag*.

Figure 5.20 shows that for more than two cut-offs, phaseshifts in excess of 180° are quite possible, and this explains the tendency of multi-stage feedback amplifiers to be unstable. It also shows that the slope of the gain versus frequency curve for two cut-offs approaches 12 dB per octave, and for three cut-offs approaches 18 dB per octave. Here lies one of the chief advantages of the dB method of handling gains. As a logarithmic scale is used gains are directly additive when components in cascade are involved.

The above discussion suggests one method of stabilizing an amplifier. If it is designed so that the gain remains high well outside the frequency limits within which the amplifier is to be used, then a single cut-off can be inserted in the circuit so that the loop gain falls by 6 dB per octave and becomes less than unity before the combined effects of the other cut-offs become important. This cut-off can be inserted in either the amplifier proper or the feedback loop. A common procedure is to shunt a feedback resistor with a capacitor so that the feedback becomes very great at high frequencies and so reduces the loop gain to less than unity before undesired phaseshifts occur.

The forward gain versus frequency diagram taken in conjunction with the phaseshift versus frequency diagram for an amplifier is useful in determining whether feedback can be applied, or whether the characteristics must first be modified. The requirement is that the forward gain multiplied by the attenuation of the feedback loop becomes less than unity before the phaseshift in both amplifier and feedback loop approaches 180°. In other words, the loop gain of the feedback amplifier must be less than unity before the loop phaseshift approaches 180°. If, at some frequency where the phaseshift is 180°, AB in the feedback equation for a perfect amplifier $A_{v(FB)} = A/(1 - AB)$ becomes unity (i.e., $AB = +1$), $A_{v(FB)} \to \infty$ and the amplifier becomes an oscillator.

Obviously the feedback equation should be written in complex form if frequencies other than those within the mid-band range are to be covered and this makes the relevant computations somewhat more difficult. There is also the question of how much phaseshift can be tolerated before the loop gain must fall below unity.

Both the complex nature of the feedback equation and the interpretation of the tolerable *phase margin* may be illustrated by reference to the Nyquist diagram. This is a graphical representation of the feedback equation taking into account its complex nature and is constructed as follows. Firstly, some length 0–1 is marked off on an Argand diagram as in figure 5.21. Then, using this as a standard length, the mid-band *loop* gain of the amplifier is marked off along the negative real axis OP. Thus, OP represents the factor AB in length and phase, for the loop gain must take a negative value if negative

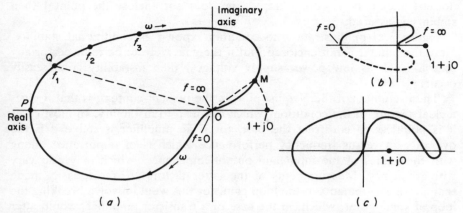

Figure 5.21. Nyquist plots for: (*a*) stable amplifier; (*b*) DC amplifier; and (*c*) conditionally stable amplifier.

feedback is being considered. When the frequency is of such a value that a phase change has taken place, the gain is represented by the vector OQ, and the phase change by the angle POQ. The frequency f, at which this occurs, may be inserted as shown, and the complete locus of the loop gain vector may be plotted. As it is known that AB must not exceed unity as the phaseshift approaches 180°, this means that the locus must never enclose the point 1, otherwise the amplifier will be unstable. This is known as the *Nyquist criterion*.

Note that the area above the abscissa contains the locus of the gain when the phase angle is lagging, which corresponds to a high-frequency cut-off. Conversely, the area below the abscissa contains the leading gain locus which corresponds to a low-frequency cut-off.

Since the vector 0–1 can be added to the loop gain AB, then the distance 1–Q represents the feedback factor $1-AB$ in vectorial form.

In the example depicted in figure 5.21(*a*), the fact that the gain is zero at zero frequency (DC) indicates that the amplifier in question is AC coupled. The gain also becomes zero just before a phaseshift of 180° lagging is reached, showing that only two lags have a significant effect.

At the point where the loop gain reaches unity, a line MO has been drawn. The angle 1OM gives the phase margin, this being the angle through which the phase would need to be shifted for instability to occur at a loop gain just above unity. It is desirable that, for good transient performance, the phase margin should not fall below 30°.

Figure 5.21(*b*) shows the form taken by the Nyquist diagram if a DC amplifier is being considered. Here, the gain has a real value at DC and reduces to zero at $f = \infty$, which means that the AB locus exists only in the first two quadrants of the diagram. To determine whether or not the amplifier will be stable, the conjugate of the locus must be drawn (shown dotted) and if the resulting closed line does not enclose the point 1 then stability is assured.

Figure 5.21(c) illustrates the condition known as conditional stability. Here the point 1 is not enclosed, but if the gain were to be reduced for any reason (such as low power supply voltages) then instability could easily occur.

The usefulness of the Nyquist plot, in the present context, is that it gives a clear picture of the conditions which can lead to instability. In most cases it is pointless to construct the plot, for if an amplifier is stable it is the overall gain versus frequency performance which is of importance, along with the values of the input and output impedances which may also vary with frequency. Measurements of the latter factors are more easily made than are measurements of the loop gain, for this would involve breaking the loop at some point, which, in the case of a transistor amplifier, would alter the loading on some point and so change the characteristics.

When an amplifier is unstable, the deductions from the Nyquist plot theory do allow valid attempts at stabilization to be made.

5.4 DC AMPLIFICATION

Strictly speaking, the term "DC" is a misnomer in this context, since there is obviously no informational content in DC and hence there is no point in amplifying it! In electronic terminology, however, it is used in reference to extremely slowly-varying signals which could not be amplified in a network containing capacitive elements which would produce a low-frequency roll-off.

In a later chapter it will be seen how the most common and useful DC amplifier stage—the long-tailed pair—is used in almost all operational amplifiers: the present section will provide an introduction to this configuration, which is arguably the most useful and versatile in the entire field of circuit design.

Consider first the problem of directly coupling two stages. In figure 5.2, if C_C' were omitted, then the base of the second transistor would be at the same bias level as the collector of the first. Hence, the values of R_C' and R_E' would have to be recalculated, and the collector of the second transistor would be at a higher bias voltage. Had further transistors been incorporated, their cumulative voltage drops would have made a high supply voltage necessary. This difficulty may be overcome in several ways. For example, Zener diodes may be used as coupling elements as in figure 5.22(a). Here, the diodes operate in their reverse breakdown mode and produce a voltage drop which is very well defined. Otherwise alternate pnp and npn transistors may be used as in figure 5.22(b), where their requirements of opposite polarity supplies effectively avoid any cumulative voltage drops.

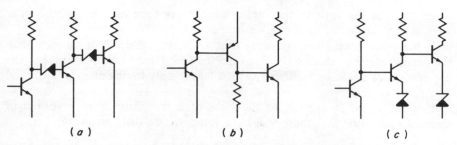

Figure 5.22. Direct coupling methods: (a) by use of Zener diodes; (b) by alternate polarity transistors; and (c) by raised emitter potentials.

Having seen how DC coupling may be achieved, consider the bypassed emitter resistors of figure 5.2. Obviously, at very low frequencies, the decoupling capacitors will cease to have any effect and the emitter resistors will give rise to considerable degeneration with a consequent severe reduction in gain. Therefore, either the circuit must be redesigned so that the decoupling capacitors are omitted altogether, or so that the emitter resistors themselves become unnecessary and the biasing of the entire amplifier becomes dependent upon the first stage. This situation can be improved by connecting a bias feedback loop around the entire amplifier, and this technique is often used for transformerless power amplifiers (see chapter 9).

If the quiescent emitter potential of a stage is to be raised without introducing degeneration, Zener diodes should be inserted as shown in figure 5.22(c). The constant voltage drop across the Zener diodes ensures that no degeneration takes places but no temperature stabilization can occur either, making necessary other biasing arrangements.

It has been pointed out by Hurley[1] that in a simple direct coupled common-emitter cascade the leakage currents act in series. If these leakage currents are identical (which would be rather unlikely), then, for transistors of one polarity, the current merely passes along the chain via the coupling line and eventually acts in one resistor, as would have been the case for a single transistor. This means that the bias stability is at least as good as for one stage alone.

If, however, transistors of mixed polarity are used, the leakage currents become additive, which leads to a poor bias stability.

From the above considerations, it is clear that a DC transistor amplifier can be built in principle, and it is now possible to define the performance of such a device.

If the amplifier is to accept low-level signals then the problem of drift can be very severe. For example, if the collector current of the first stage of a cascade changes because of a temperature variation, this change will be amplified and may assume considerable proportions at the output. In order to compare the drift performance of amplifiers having different gains, it is usual to refer the drift to the input, which may then be multiplied by the gain in any particular case to give output drift.

It has been seen (in chapter 4) that I_{CBO}, V_{BE} and h_{FE} each vary with temperature. The change in h_{FE} can be rendered unimportant by the application of negative feedback, which leaves only the changes in I_{CBO} and V_{BE} to be considered.

If a DC signal is applied to the base of a transistor from a source of internal resistance R_G, there will be a temperature drift given by

$$\delta I_{in} = \delta I_{CBO} + \frac{\delta V_{BE}}{R_G}$$

or

$$\delta V_{in} = \delta V_{BE} + R_G \delta I_{CBO} .$$

These equations show that if the source has a high resistance (a current source) the input current drift will tend to δI_{BCO}. For a low-resistance source (a voltage source) the input voltage drift will tend to δV_{BE}. This means that the form of drift compensation chosen may depend on the source resistance involved; for example, if a source of high resistance is involved, the effect of δI_{CBO} may be minimized by choosing a good quality silicon transistor having a very low leakage current.

If it is necessary to reduce further the effect of δI_{CBO}, a diode having similar leakage characteristics to the transistor may be connected in parallel with the base so that it is reverse biased. As the temperature varies, the leakage current through the diode is assumed to change at the same rate as I_{CBO} and will compensate for it.

A reduction of δV_{BE} is also easily attained. In figure 5.23, a diode is connected in series with the emitter circuit and is kept in the conducting

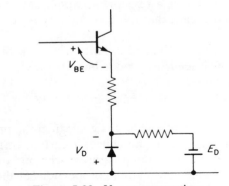

Figure 5.23. V_{BE} compensation.

state by the supply E_D. As is indicated in the diagram, the polarities of V_{BE} and V_D are in opposition and, if the temperature coefficients are similar, will lead to their effective cancellation. Actually, these coefficients are never exactly alike but by choosing a transistor and diode (or a similar transistor using only the emitter–base junction) with measured values of V_{BE} and V_D, a compensation can be achieved which will reduce drift by more than 90%.

A method of drift compensation which takes into account the overall change in output level due to all causes is afforded by the use of thermistors or silicon resistors. The former have a negative resistance versus temperature slope and the latter a positive one, which means that the ways in which they are used differ slightly. In order to achieve good compensation each amplifier must be subjected to a temperature test and a suitable network designed for it. However, if a high degree of compensation is not necessary, the same network may be used in any amplifier of similar design.

A thermistor may be incorporated in a base circuit as shown in figure 5.24(*a*). When a rise in temperature occurs I_C increases, but no compensation takes place since the emitter resistor is absent. However, the resistance of the thermistor R_2 decreases and so reduces the base current, thereby compensating for the rise in I_C.

A silicon resistor may be incorporated in the emitter circuit as shown in figure 5.24(*b*), and here the voltage drop across it opposes any increase in I_E by making the emitter potential go more positive with respect to the base. In this method the base potential remains constant but unfortunately some degeneration, which varies a little with temperature, is introduced.

In practice, both the thermistor and the silicon resistor must be coupled in a series–parallel network with fixed resistors because the resistance versus temperature law followed by the single components will not be suitable for accurate compensation over a wide temperature range. It is for this reason that the drift versus temperature curve for the amplifier under consideration must be plotted empirically, and a temperature-sensitive network designed

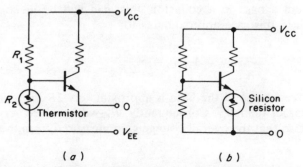

Figure 5.24. Principle of drift compensation by temperature-sensitive resistor of: (*a*) negative temperature coefficient; and (*b*) positive temperature coefficient.

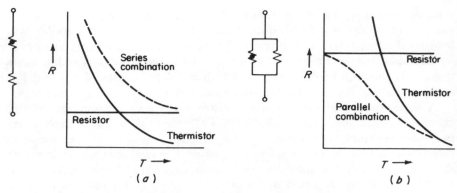

Figure 5.25. Characteristics of thermistor/resistor combinations.

to annul this over as wide a range as possible. Curves for a typical network are given in figure 5.25 and are self-explanatory.

All the attempts to facilitate DC amplification so far mentioned are aimed towards overcoming to some degree the essentially temperature-dependent nature of transistor parameters. By far the best approach to the problem is to use a pair of nearly identical transistors in such a way that their drift properties cancel out almost completely. The next section shows how this can be achieved.

5.4.1   The Difference Amplifier

Consider two identical transistors, each being connected as a common-emitter stage, and each with a large, unbypassed emitter resistor of value $2R_X$, as shown in figure 5.26(a). Let both collector resistors also be identical, and both input resistors, R_G. If $2R_X \gg R_G$, the bias stability will be very good; that is, the collector currents and hence the collector voltages will vary very little with temperature. On the other hand, the signal voltage gain of such a stage will be very poor, being that of a degenerate CE stage. Measured from a base to a collector, this gain will be given by equation (5.27), which in this case approximates to:

$$A_{v(FB)} \simeq -\frac{R_C}{2R_X}. \tag{5.35a}$$

The input resistance at the base is approximately $2R_X h_{fe}$, which is likely to be much larger than R_G. Consequently, R_G will produce very little signal voltage drop, so that the overall voltage gain measured from the input to R_G will also be:

$$A_{ov(FB)} \simeq -\frac{R_C}{2R_X}. \tag{5.35b}$$

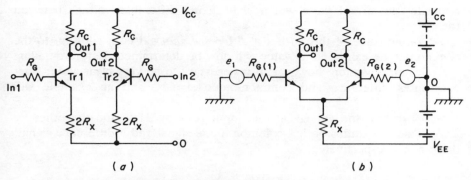

Figure 5.26. Development of the difference amplifier.

If the bias voltage V_B applied to each input is the same, then the quiescent collector currents and voltages will also be the same, the transistors being identical. Thus, if the temperature varies, the collector currents and voltages will also change by similar amounts because the temperature variations in h_{FE}, I_{CEO}, and V_{BE} must also correspond. Hence, if either collector voltage is measured with respect to the common line, it will be seen to vary very slightly with temperature, each stage being very stable; but if the voltage *between* the collectors is measured, it will be found to be zero under all conditions.

Further, if identical signals are applied at the inputs, the collector voltages will again change by the same amount, and again there will be zero voltage between these collectors. Such signals are called *common-mode signals*, and the voltage gain of equations (5.35) is called the *common-mode voltage gain*. Note that for these hypothetically identical transistors the common-mode gain exists only when the output is measured from one collector to the common line. If the output were measured *between* the collectors, then the common-mode gain would obviously be zero *unless the transistors were not truly identical*, when each would amplify the common-mode input signal by a slightly different amount from the other.

All the foregoing statements remain true if the two emitters are connected together as in figure 5.26(*b*) and a single *long-tail resistor* of value R_X is substituted so that the bias conditions remain the same as before. This circuit is now the classical *difference amplifier stage* or *long-tailed pair*.

One of the several attributes of this versatile stage is that whereas the common-mode gain and temperature drift are low, as has been shown, the difference gain is high, as will be demonstrated.

Suppose that the signals at each input differ (or a signal is applied at one input only, which is the same thing). The bias conditions will tend to hold the total current in R_X constant, so that the collector current will rise in one transistor and fall in the other. Hence, the collector voltages will change in opposite senses and there will now be a net difference voltage between them.

A consequence of this is that a *difference gain* exists in addition to the common-mode gain, and its value will now be determined. Before detailing the relevant deviation, however, it is necessary to be very clear about what constitutes difference and common-mode signals, and an example will illustrate this.

Suppose that the voltage at one input rises by 3 mV and the other by 2 mV. This constitutes both a common-mode signal and a difference signal, these being respectively:

$$e_{in(CM)} = \frac{e_1 + e_2}{2} = \frac{3+2}{2} = 2.5 \, mV$$

which will be amplified by the common-mode gain of equation (5.35); and

$$e_{in(diff)} = \frac{e_1 - e_2}{2} = \frac{3-2}{2} = 0.5 \, mV$$

which will be amplified by the difference gain, derived below. Notice that the difference input voltage quoted is that *at each input*. Measured *between* the two inputs it is obviously twice this value, or $(e_1 - e_2) = e_{in}$.

Figure 5.27 shows the low-frequency hybrid-π equivalent circuit relevant to the long-tailed pair of figure 5.26(*b*). The input generators have been represented by voltage generators e_1 and e_2 and their internal or Thévenin resistances R_{g1} and R_{g2} have been included in $R_{G(1)}$ and $R_{G(2)}$.

If the values of $r_{b'c(1)}$ and $r_{b'c(2)}$ are very large, and the value of R_X is made as large as possible, most of the input current will follow the path shown by the dotted loop. That is:

$$i_{b(1)} = -i_{b(2)} \approx \frac{e_1 - e_2}{R_{G(1)} + r_{bb'(1)} + r_{b'e(1)} + r_{b'e(2)} + r_{bb'(2)} + R_{G(2)}}.$$

If the transistors are identical, and $R_{G(1)} = R_{G(2)}$, then,

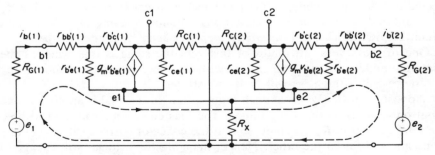

Figure 5.27. Low-frequency hybrid-π equivalent circuit for long-tailed pair.

$$i_{b(1)} = -i_{b(2)} \simeq \frac{e_1 - e_2}{2(R_G + r_{bb'} + r_{b'e})} \; .$$

The collector signal currents are given by $h_{fe}i_b$ and hence the collector signal voltages by $h_{fe}i_bR_C$, so that the difference voltage gain to *either* collector is:

$$A_{ov} = \pm \frac{v_c}{e_1 - e_2} \simeq \pm \frac{h_{fe}R_C}{2(R_G + r_{bb'} + r_{b'e})} \; . \qquad (5.36a)$$

If the difference input signal between the actual bases is used, the gain may be found simply by setting $R_G = 0$, giving,

$$A_v \simeq \pm \frac{h_{fe}R_C}{2(r_{bb'} + r_{b'e})} \; . \qquad (5.36b)$$

If the output voltage is measured *between* the collectors it will be *twice* the output voltage at either collector, since the two collector difference signals are equal in magnitude but opposite in sign. That is,

$$A_{ov(diff)} \simeq \frac{h_{fe}R_C}{R_G + r_{bb'} + r_{b'e}} \qquad (5.37a)$$

and

$$A_{v(diff)} \simeq \frac{h_{fe}R_C}{r_{bb'} + r_{b'e}} \; . \qquad (5.37b)$$

Knowing that $h_{fe} = g_m r_{b'e}$, this becomes,

$$A_{v(diff)} \simeq \frac{g_m r_{b'e} R_C}{r_{bb'} + r_{b'e}} \qquad (5.38)$$

which is the same as equation (4.23a) which gives the voltage gain of a simple common-emitter stage.

Thus, the difference voltage gain of a long-tailed pair is the same as the voltage gain of a single-transistor CE stage, but it offers a much improved drift characteristic and so can be used as a DC amplifier. Here, it should be noted that if a split power supply is used (e.g., ±15 V), then the input can be referred to the common rail of these supplies so that no input capacitor is needed. This is common practice and has been illustrated in figure 5.26(b).

The low common-mode gain of the long-tailed pair implies that common-mode input signals (temperature drifts included) are rejected in comparison with difference signals. This leads to a definition of a *common-mode rejection ratio* as the quotient of the difference and common-mode gains:

$$\text{CMR} = \frac{A_v}{A_{v(CM)}} \; .$$

Hence, both gains refer to the gains measured from the bases to either collector, so that equations (5.35) and (5.36b) are relevant:

$$\text{CMR} \simeq \frac{h_{\text{fe}}R_C}{2(r_{\text{bb}'} + r_{\text{b}'\text{e}})} \frac{2R_X}{R_C} = \frac{h_{\text{fe}}R_X}{r_{\text{bb}'} + r_{\text{b}'\text{e}}} . \tag{5.39}$$

The foregoing treatment is valid only for pairs of identical transistors with identical collector resistors. This is unattainable in practice but modern monolithic integrated circuit techniques do make for very close matching, so that an approach to the ideal is quite possible. This will be treated in detail later. If the transistors are *not* identical the derivations become complex and have been presented by Middlebrook[2] and Giacoletto[3].

Having seen how the long-tailed pair stage is biased, and how it performs from a small-signal point of view using the hybrid-π parameters, it is necessary to re-examine it in a more fundamental way.

5.4.2 The Difference Amplifier—DC Analysis

The common-mode rejection properties of a long-tailed pair are improved if R_X is large. However, this implies that the power supply voltage V_{EE} must also be high so that the quiescent current, I_X, can be provided. Clearly, it would be better to substitute a current source for R_X as shown in figure 5.28, and a close approach to this ideal can be realized in practice, as will be seen later.

Using this circuit, the basic transistor equations are

$$I_{C1} \simeq I_{ES1} \exp\left(\frac{q}{kT} V_{BE1}\right)$$

$$I_{C2} \simeq I_{ES2} \exp\left(\frac{q}{kT} V_{BE2}\right). \tag{5.40}$$

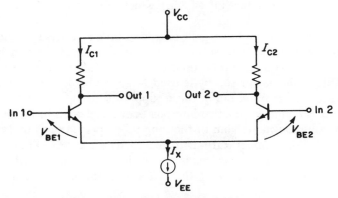

Figure 5.28. Difference amplifier stage.

If the transistors are identical, so that $I_{ES1} = I_{ES2}$, then,

$$I_X \simeq I_{C1} + I_{C2} \simeq I_{ES}\left[\exp\left(\frac{q}{kT}V_{BE1}\right) + \exp\left(\frac{q}{kT}V_{BE2}\right)\right]. \qquad (5.41)$$

Equations (5.40) can now be rewritten after extracting I_{ES} from equation (5.41):

$$I_{C1} \simeq \frac{I_X}{1 + \exp\left[q(V_{BE2} - V_{BE1})/kT\right]}$$

$$I_{C2} \simeq \frac{I_X}{1 + \exp\left[q(V_{BE1} - V_{BE2})/kT\right]}. \qquad (5.42)$$

These equations are plotted in figure 5.29 for 25°C, where $kT/q \simeq 26\text{ mV}$ and where each collector current has been normalized to I_X. This diagram shows that a linear region exists near the cross-over point of the two curves, which implies that the quiescent condition of the difference amplifier should in fact be defined by this cross-over point. It is also here where the maximum value of the transconductances occurs, the transconductances themselves being defined by I_X only, as would in fact be expected, having regard to equation (3.15).

Taking the input signal as $d(V_{BE1} - V_{BE2})$ and the output signal as the change in either collector current dI_C, the transconductance of the stage may be extracted from equations (5.42):

$$g_m = \frac{dI_C}{d(V_{BE1} - V_{BE2})} \simeq \frac{I_X \exp\left[(q/kT)(V_{BE1} - V_{BE2})\right]}{(kT/q)\{1 + \exp\left[(q/kT)(V_{BE1} - V_{BE2})\right]\}^2} \qquad (5.43)$$

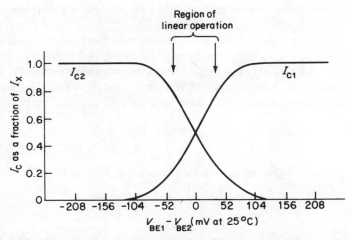

Figure 5.29. Plot of collector currents for a difference amplifier configuration.

and when $V_{BE1} = V_{BE2}$, this becomes

$$g_{m(max)} \simeq \frac{I_X}{4(kT/q)} . \tag{5.44}$$

If it is recalled that each quiescent collector current is one half of I_X, this can be written $g_{m(max)} = qI_Q/2kT$, which is one half of the transconductance of a simple CE-connected transistor as given by equation (3.15a). If, however, the output *voltage* is taken *between* the two collectors, it is clear that the voltage gain will be the same as for the simple CE stage, as has been shown.

The curve of g_m versus $(V_{BE1} - V_{BE2})$ is quite "peaky" for a simple difference amplifier, but can be flattened by the insertion of a pair of emitter degeneration resistors (see below).

Difference amplifier stages form the basis of almost all microcircuit operational amplifiers, both monolithic and hybrid, and this major application will be treated in chapter 7. However, certain applications may call for individual stages to be designed, and for this reason, closely matched transistor pairs are available, as are monolithic arrays of uncommitted transistors. The following simple example will highlight the basic design methods and will also introduce the important technique of common-mode feedback.

5.4.3 Simple Design Example

The choice of transistor pair to form the basis of a difference amplifier stage is dependent upon the constraints imposed by the design criteria. However, as an example, the CA3045 will be used, which is an array of five small-signal silicon npn transistors formed on a common monolithic substrate. The complete chip is encapsulated in a 14-pin dual-in-line package (DIL-pak), having connections as shown in figure 5.30. Two of the transistors are in very close proximity and have their emitters internally connected to form a difference amplifier configuration. At $I_C = 1$ mA, the difference between the values of V_{BE} for these transistors is guaranteed to be less than 5 mV, and the base bias current difference to be less than 2 μA. (These maxima are known as the offset base voltage and current, respectively.)

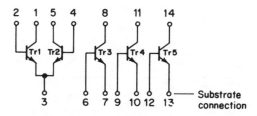

Figure 5.30. The CA3045 transistor array.

Suppose that a ± 15 V power supply is available, and that a long-tailed pair is to be designed having collector currents of 1 mA. Assuming that the input signals are referred to ground, the circuit becomes that of figure 5.26(b).

To calculate the value of R_X, it may be assumed that the bases are at ground potential (this implies a neglibibly small voltage drop across R_G), and that the current in R_X is $2I_C$, or 2 mA. (Both of these assumptions imply a high value of h_{FE}.)

Hence,

$$R_X = \frac{V_{EE} - V_{BE}}{I_X} = \frac{15 - 0.7}{2} \approx 7.2\,\text{k}\Omega$$

(because V_{BE} is given as typically 0.7 V at $I_C = 1$ mA).

If about half of the positive supply voltage is to be dropped across the transistor (V_{CE}) and the other half across its load, then,

$$R_C = \frac{\frac{1}{2}(V_{CC} + V_{BE})}{I_C} = \frac{15 + 0.7}{2} = 7.8\,\text{k}\Omega\,.$$

The design is now complete, and for purposes of illustration, input resistors will be chosen having values $R_G = 4.7\,\text{k}\Omega$.

The performance of the stage may now be calculated.

Firstly, the common-mode gain is, from equation (5.35):

$$A_{v(CM)} \approx \frac{-R_C}{2R_X} = \frac{-7.8}{2 \times 7.2} = -0.54\,.$$

This is the gain measured from either base to either collector for *identical* signals applied to the bases.

Secondly, the overall difference gain, defined as for equation (5.37a) is:

$$A_{ov(diff)} \approx \frac{h_{fe}R_C}{R_G + r_{bb'} + r_{b'e}}\,.$$

Here, h_{fe} is given as typically 110 in the relevant data sheet, and since $r_{bb'}$ is not given at all, the value for h_{ie} will be used in place of $(r_{bb'} + r_{b'e})$. Thus,

$$A_{ov(diff)} \approx \frac{110 \times 7.8}{4.7 + 3.5} = 105\,.$$

Had the difference gain been measured from *between* the bases to *between* the collectors then R_G would not have appeared, and equation (5.37b) would have been relevant:

$$A_{v(\text{diff})} \simeq \frac{h_{\text{fe}} R_{\text{C}}}{r_{\text{bb}'} + r_{\text{b}'\text{e}}} \simeq \frac{h_{\text{fe}} R_{\text{C}}}{h_{\text{ie}}}$$

$$= \frac{110 \times 7.8}{3.5} = 245 \, .$$

The input resistance to the stage may be approximated by considering the signal flow loop in figure 5.27. Taken as the resistance measured between the inputs, the overall input resistance is obviously

$$R_{\text{in}} \simeq 2(R_{\text{G}} + r_{\text{bb}'} + r_{\text{b}'\text{e}}) \, . \tag{5.45a}$$

Measured at one base (that is, omitting R_{G1} but retaining R_{G2}) it is:

$$R_{\text{in}} \simeq 2(r_{\text{bb}'} + r_{\text{b}'\text{e}}) + R_{\text{G2}} \tag{5.45b}$$

or, if the alternate base is grounded and the input resistance is measured from the other, it is:

$$R_{\text{in}} \simeq 2(r_{\text{bb}'} + r_{\text{b}'\text{e}}) \, . \tag{5.45c}$$

In this latter case, R_{in} is $2 \times 3.5 = 7 \, \text{k}\Omega$.

All three crude calculations are dependent upon the accuracy of h_{fe} and h_{ie}, which means that since "typical" values only are given, wide variations in gains and input resistances are likely.

A comparison between the calculated values and the measured values from an experimental stage was as follows:

	$A_{v(\text{CM})}$	$A_{\text{ov(diff)}}$	$A_{v(\text{diff})}$	R_{in}
Calculated	0.54	105	245	$7 \, \text{k}\Omega$
Measured	0.54	92	238	$6.2 \, \text{k}\Omega$.

The common mode rejection ratio (to one collector) is thus,

$$\text{CMR (calculated)} = \frac{245/2}{0.54} = 226.8 \quad (\text{or } 47.1 \, \text{dB})$$

$$\text{CMR (measured)} = \frac{238/2}{0.54} = 220.4 \quad (\text{or } 46.9 \, \text{dB}) \, .$$

If the drift in either collector voltage is of importance, then each side of the difference amplifier can be treated separately, as in figure 5.26(a). That is, the value of K can be calculated and the effects of temperature changes in V_{BE} and h_{FE} can be determined (for a small-signal silicon transistor, leakage current drift can usually be neglected, being small). In the present example K is very small:

$$K = \frac{1}{1 + [2R_X h_{FE}/(R_G + 2R_X)]} = \frac{1}{1 + [14.4 \times 100/(4.7 + 14.4)]} = 0.013$$

so that the drift will also be very small.

However, in the case of a difference amplifier, it is usually the drift *between* the collectors which is of importance, and this is related to differences between the transistors *and* between the values of nominally identical comonents.

In the present case, the difference between the values of V_{BE} (given as ± 5 mV maximum at $I_C = 1$ mA) *also* has a temperature coefficient of (typically) 1.1 $\mu V/°C$. This temperature-dependent *input offset voltage* constitutes a difference input signal, and so is amplified by the difference voltage gain.

Suppose, for example, that the temperature rises by 20°C. The change in input offset voltage is (typically),

$$\frac{\delta e_{os}}{\delta T} \Delta T = 1.1 \times 20 = 22 \ \mu V$$

so that the output drift voltage is,

$$\Delta V_{out} = 22 A_{v(diff)} = 22 \times 245 \simeq 5.4 \text{ mV} .$$

This small drift voltage is only one component of the total drift, albeit the most important one. It is small because of the excellent matching of transistors made possible by monolithic fabrication. Further considerations of drift will be postponed until complete integrated amplifiers are treated in chapter 7.

5.4.4 The "Long-tail" Current Source

In principle it is advantageous to make R_X as large as possible; equations (5.35) and (5.39) show that this reduces common-mode gain and so improves the common-mode rejection properties of the stage. Further, it makes the supposition inherent in figure 5.27 (that none of the difference signal input current flows in R_X) more valid, leading to greater accuracy in the calculation of difference gain. Unfortunately, any increase in R_X along with no change in I_X, must lead to a proportional increase in V_{EE}. However, when the biasing and signal concepts are divorced, a solution becomes obvious. If a transistor Tr5 is substituted for R_X as in figure 5.31, then its quiescent collector current becomes I_X and its collector presents a very high *incremental* resistance r_{ce} to the commoned emitters of the long-tailed pair.

The usual way of biasing the long-tail transistor is by means of the *current mirror* connection[4], also shown in figure 5.31. This technique is dependent upon the very close matching of transistors and so is ideal for monolithic

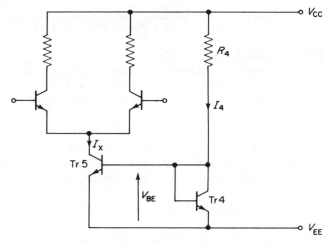

Figure 5.31. A long-tail current source ("current-mirror" connection shown).

integrated circuit fabrication, as will be explained in chapter 7. In figure 5.31, a matched transistor Tr4 is shown with its collector short-circuited to its base so that it becomes effectively a diode with only the base-emitter junction operative. Hence, its current is defined largely by the total voltage available and R_4:

$$I_4 = \frac{V_{CC} + V_{EE} - V_{BE}}{R_4}. \tag{5.46}$$

The voltage V_{BE} is applied between the base and emitter of Tr5 so that its collector current must become equal to I_4. In other words, Tr5 mirrors the defined current in Tr4, hence the term "current mirror."

This method of biasing provides extremely good temperature stability, because if $(V_{CC} + V_{EE}) \gg V_{BE}$, then the very small temperature variations in V_{BE} for Tr4 have a negligibly small effect on I_4. Because the transistors are matched, V_{BE} for Tr5 has identical temperature variations to that for Tr4, and so Tr5 continues to mirror I_4 exactly. So, I_X is essentially temperature-invariant.

There are numerous variations on this principle, and many involve the inclusion of an emitter resistor R_{E5} in the Tr5 circuit. If this resistor is more than a few ohms, it implies that the base of Tr5 must be driven by a voltage higher than V_{BE}, that is:

$$V_{B5} = V_{BE5} + I_X R_{E5}$$

giving

$$R_{E5} = \frac{V_{B5} - V_{BE5}}{I_X}. \tag{5.47}$$

An example of the use of such a modification occurs when a second stage is added. (Actually, if more than one stage of amplification is required, it is usually advisable to go to a complete integrated circuit, but this may well include common-mode feedback similar to that now to be described.)

5.4.5 A Second Stage and Common-mode Feedback

Figure 5.32 shows a second pair of transistors having their bases driven directly from the collectors of the first pair. This implies that only the voltage across R_{C1} (or R_{C2}) is available for division between V_{CE3} and R_{C3} (or V_{CE4} and R_{C4}). In the design example this was only 7.2 V, which implies that with equal division, V_{CE} for Tr3 and Tr4 is only some 3.5 V. Since the second stage will produce a much greater output voltage swing than will the first, it would be preferable to increase the voltage available to it. This can only be done by increasing the values of R_{C1} and R_{C2}, which must decrease the values of V_{CE1} and V_{CE2}.

Since the difference gains of the stages are equal to those of simple CE stages with similar values of R_C, clearly the gain of the complete two-stage amplifier will be as dependent upon the transistor parameter values as was the case for the two-stage CE amplifier. For some purposes this may be acceptable, particularly in high-frequency work, and some such amplifiers are available in monolithic form. Otherwise, negative feedback techniques again become necessary if it is desirable to closely define the gain and impedance properties.

Returning to figure 5.32, it will be noticed that the long-tail current, I_X, in the first stage is controlled by Tr5, which is driven from the second stage. This constitutes common-mode feedback, and its operation is as follows.

If a common-mode signal (or a temperature change) increases both collector currents (and hence I_X) in the first stage, then this *reduces* the

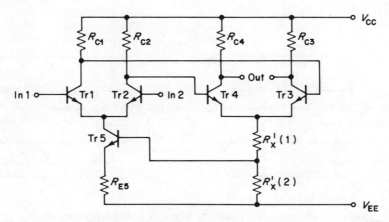

Figure 5.32. A two-stage difference amplifier with common-mode feedback.

voltage applied to both bases of the second stage. Hence, both collector currents (and hence I_x') in the second stage *fall*, so the voltage drop across $R_x'(2)$ also falls. This *reduces* the voltage applied to the base of Tr5, so that I_x is made to fall, so tending to counteract the initial rise. This constitutes negative feedback for common-mode signals *only*.

Notice that in the CA3045 monolithic transistor array used in the example, Tr5 is chosen as the long-tail transistor. This is because its emitter is part of the substrate on which the transistor structures are formed, so that it must be the most negative point in the circuit. Otherwise, a forward-biased junction might be formed with possibly disastrous results!

5.4.6 Level Shifting and Signal Feedback

One of the problems associated with the long-tailed pair is that the output voltage levels are different from those at the input. So, if the input signals are to be referred to a common line—very often ground—some means of level shifting the output signal must be devised. In chapter 7 methods of achieving this in integrated circuits are presented: for the present, one common technique in discrete transistor circuitry will be considered, because it also lends itself well to signal feedback.

Figure 5.33(*a*) shows a pnp transistor connected as a common-emitter stage and driven from one of a pair of difference amplifier transistors.

Notice that because only one of the difference pair is driving Tr3, the voltage-gain of this first stage is effectively halved. (Circuits which convert from the full difference output to single-ended working appear in chapter 7.) Hence, no voltage output is required from Tr2, so R_{C2} has been omitted.

The value of R_{C1} in the circuit is determined by the base-emitter voltage of Tr3. If I_{B3} in small in comparison with I_{C1},

$$R_{C1} \simeq \frac{V_{BE3}}{I_{C1}} .$$

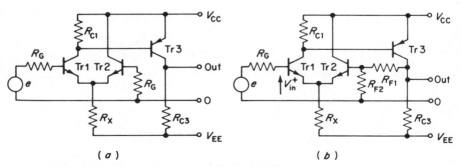

Figure 5.33. (*a*) a level shifting stage; (*b*) with signal feedback.

If I_{B3} is not very small compared with I_{C1}, then it can be taken into account by noting that the collector current of Tr1 must include I_{B3}:

$$I_{C1} = \frac{V_{BE3}}{R_{C1}} + \frac{I_{C3}}{h_{FE3}}$$

giving

$$R_{C1} = \frac{V_{BE3}}{I_{C1} - (I_{C3}/h_{FE3})}.$$

This expression involves several tolerances, not the least being the spread in values of V_{BE3}. These give rise to a range of values of I_{C3}, which is unfortunate, since it is the voltage drop in R_{C3} which shifts the level of output signal to ground potential in the circuit. Ideally, R_{C3} would be chosen so that

$$I_{C3} R_{C3} = V_{EE}$$

but since I_{C3} cannot be closely defined (and R_{C3} would be the nearest preferred value resistor anyway), this relation is rarely accurate. It could be made more so by giving Tr3 an emitter resistor, but this would reduce its voltage gain.

An improved situation is shown in figure 5.33(*b*) where the output point is tied to the base of Tr2 via R_{F1}. This technique completes a DC negative feedback loop which tends to hold the output point near ground potential, because the base of Tr1 is already held near ground by R_G and the base of Tr2 must follow it.

This circuit configuration also results in signal feedback, and by inspection it is seen that a fraction $R_{F2}/(R_{F1} + R_{F2})$ of V_{out} is applied to the base of Tr2. Letting the input signal to Tr1 be called v_{in}^+, because it leads to a *non-inverted* output signal, the *difference* signal *between* the bases is:

$$v_{in} = v_{in}^+ - \frac{v_{out} R_{F2}}{R_{F1} + R_{F2}}.$$

The voltage gain from the base of Tr1 to the output is therefore:

$$A_{v(FB)} = \frac{v_{out}}{v_{in}^+} = \frac{v_{out}}{v_{in} + [v_{out} R_{F2}/(R_{F1} + R_{F2})]}.$$

Here, $v_{out} = A_v v_{in}$, where A_v is the internal voltage gain of the difference amplifier stage in cascade with the CE stage, Tr3. Hence,

$$A_{v(FB)} = \frac{A_v v_{in}}{v_{in} + [A_v v_{in} R_{F2}/(R_{F1} + R_{F2})]} = \frac{A_v}{1 + [A_v R_{F2}/(R_{F1} + R_{F2})]}.$$

If $A_v R_{F2}/(R_{F1} + R_{F2}) \gg 1$, which is usual, then,

$$A_{v(FB)} \simeq \frac{R_{F1} + R_{F2}}{R_{F2}} \quad \text{or} \quad 1 + \frac{R_{F1}}{R_{F2}}. \qquad (5.48)$$

This shows that one of the prime targets of negative feedback has again been fulfilled: that the voltage gain of the amplifier has been closely defined.

A noteworthy property of this configuration is that because series feedback is involved (R_{F2} is in the input loop), the input impedance is *increased*. This can be most easily understood by observing that the output signal is in phase with the input signal (there being phase reversals in both Tr1 and Tr3) and that a portion of this output signal is fed back to the *alternate* input base, Tr2. Thus, *both* bases are driven by a signal which is similar in magnitude and phase, which means that it is a *common-mode signal*. For this reason, the input resistance is large, and may approach $2R_X h_{fe}$.

For reasons of good balancing, it is desirable to make the input resistance at the base of Tr1 similar to that at the base of Tr2, which can be taken as being R_{F2} if $(R_{F1} + R_{C3})$ is large compared with it.

The foregoing explanation is actually an introduction to the integrated circuit noninverting operational amplifier presented in chapter 7, and for this reason, it will not be treated further at this point. However, it is worthwhile observing that the circuit of figure 5.33(b) can be easily improved in two ways using techniques presented earlier. Firstly, a current mirror can be substituted for R_X, and secondly the output resistance can be made low by terminating the amplifier with an emitter-follower. The complete

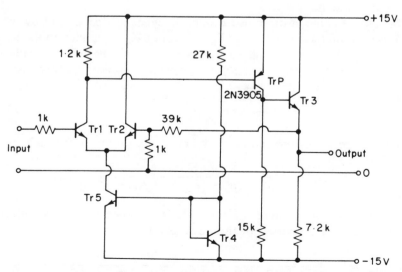

Figure 5.34. A complete amplifier based on the CA3045 integrated transistor array.

circuit is now as in figure 5.34 and is shown with typical component values. A CA3045 array has again been used, along with a single pnp transistor, and it is left as an exercise for the reader to calculate the performance figures.

5.4.7 Further Properties of the Difference Amplifier

When equations (5.42) were plotted, the graph of figure 5.29 resulted, which is essentially a transfer function for the long-tailed pair. Among other things, is showed that for reasonable linearity, the input voltage (V_{BE1} − V_{BE2}) had to be small (for example, not greater than $2kT/q$, or about 50 mV at room temperature).

The difference between two input voltages referred to the same datum, is the same as the difference between the two values of V_{BE}; that is,

$$(V_{in(1)} - V_{in(2)}) = (V_{BE(1)} - V_{BE(2)}) = V_{in} . \qquad (5.49)$$

The restriction of V_{in} to a few tens of millivolts can be removed by inserting emitter degeneration resistors as shown in figure 5.35, and this has the effect of linearizing the transfer curves as shown in figure 5.36 (where kT/q is now given the symbol V_T for convenience). This would be expected from considerations of series feedback, of which this is an example, and in fact the input voltage range is increased by approximately ($I_{C1}R_F + I_{C2}R_F$) or $I_X R_F$.

However, this increase in dynamic range is bought at the expense of difference gain, as would also be expected and again in accordance with feedback theory, the fractional fall in gain being similar to the fractional rise in dynamic range.

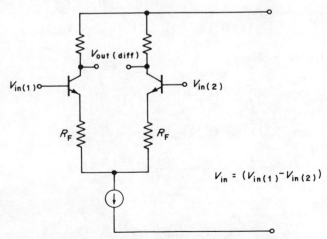

Figure 5.35. Difference amplifier with emitter degeneration.

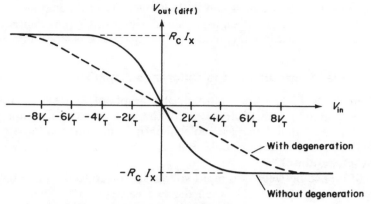

Figure 5.36. Effect of emitter degeneration.

The difference output voltage of a long-tailed pair of identical transistors having identical collector currents and identical load resistors is:

$$V_{\text{out(diff)}} = (I_{C1} - I_{C2})R_C = \Delta I_C R_C. \tag{5.50}$$

To plot ΔI_C and hence $V_{\text{out(diff)}}$, it is convenient to take the usual expressions for collector current (5.40) and rewrite them in the present context:

$$I_{C1} \simeq I_{ES} \exp{(V_{BE1}/V_T)}$$
$$I_{C2} \simeq I_{ES} \exp{(V_{BE2}/V_T)}$$

giving

$$I_{C1}/I_{C2} \simeq \exp{[(V_{BE1} - V_{BE2})/V_T]} = \exp{(V_{\text{in}}/V_T)}. \tag{5.51}$$

Now consider the following:

$$\frac{I_{C1} - I_{C2}}{I_X} = \frac{I_{C1} - I_{C2}}{I_{C1} + I_{C2}} = \frac{(I_{C1}/I_{C2}) - 1}{(I_{C1}/I_{C2}) + 1} \tag{5.52}$$

Putting equation (5.51) into (5.52) gives:

$$\frac{I_{C1} - I_{C2}}{I_X} = \frac{\exp{(V_{\text{in}}/V_T)} - 1}{\exp{(V_{\text{in}}/V_T)} + 1}. \tag{5.53}$$

Recalling that

$$\tanh x = \frac{\exp{(x)} - \exp{(-x)}}{\exp{(x)} + \exp{(-x)}} = \frac{\exp{(2x)} - 1}{\exp{(2x)} + 1}$$

equation (5.53) may be written:

$$\frac{I_{C1} - I_{C2}}{I_X} = \tanh\left(\frac{V_{in}}{2V_T}\right)$$

or

$$(I_{C1} - I_{C2}) = \Delta I_C = I_X \tanh\left(\frac{V_{in}}{2V_T}\right). \tag{5.54}$$

So, returning to equation (5.50),

$$V_{out(diff)} = R_C I_X \tanh\left(\frac{V_{in}}{2V_T}\right). \tag{5.55}$$

This expression can be plotted as in figure 5.36, and it shows (as does figure 5.29) that reasonably linear operation calls for small values of V_{in}. In fact, if it is recalled† that $\tanh x \to x$ when $x \ll 1$, it follows that,

$$V_{out(diff)} \to \frac{R_C I_X}{2V_T} V_{in} \quad \text{for} \quad V_{in} \ll 2V_T$$

whence

$$A_{v(diff)} \to \frac{R_C I_X}{2V_T}. \tag{5.56}$$

This may be compared with the expression (5.38) obtained using small-signals and balanced biasing:

$$A_{v(diff)} \simeq \frac{g_m r_{b'e} R_C}{r_{bb'} + r_{b'e}}.$$

Usually, $r_{bb'} \ll r_{b'e}$, so this becomes:

$$A_{v(diff)} \simeq g_m R_C. \tag{5.57}$$

Here, the value of g_m must be twice the value used for calculating the single-ended output gain; that is, using equation (5.44), $g_m = I_X/2V_T$, giving

$$A_{v(diff)} \simeq \frac{R_C I_X}{2V_T}$$

which is equation (5.56).

These insights into the performance of the long-tailed pair will now be used to explain the principle of the *transconductance multiplier*.

---

†This is because $\tanh x = x - (x^3/3) + (2x^3/15)\ldots$ so that when $x \ll 1$, then $\tanh x \to x$.

5.5   THE TRANSCONDUCTANCE MULTIPLIER

From equations (5.56) and (5.57),

$$V_{out(diff)} \simeq g_m R_C V_{in} = R_C I_X V_{in}/2V_T \tag{5.58}$$

which shows that if the transconductance is used as a variable by controlling I_X then, in principle, the multiplication of two variables can be achieved using the long-tailed pair. Moreover, if I_X is derived from a voltage V_m, then the product of two voltages results. Figure 5.37 shows how I_X can be derived from the voltage V_m using a current mirror, so that,

$$V_{out} \simeq \left(\frac{R_C}{2V_T R_m} \right) V_m V_{in} \quad (\text{if } V_m \gg V_{BE}) . \tag{5.59}$$

This equation implies a number of constraints and approximations, including the following.

(i) The transistors Tr1 and Tr2 and their loads R_{C1} and R_{C2} must be identical, which means the circuit is best realized using monolithic *IC* techniques.

(ii) The transistor structures and operating conditions must be such that $r_{bb'} \ll r_{b'e}$ (which is not difficult to achieve).

(iii) The value of V_m must be positive, and large enough to ensure that Tr5 remains in conduction.

(iv) V_{in} must be small compared with V_T, but can be of either polarity.

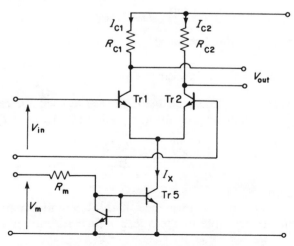

Figure 5.37. Basic two-quadrant multiplier circuit.

The problems involved with the inputs V_{in} and V_m are quite serious, for they imply that the two voltages may not only be of widely different magnitudes, but are referred to different voltage levels in the circuit. Furthermore, because V_m must be positive (unipolar) then only "two-quadrant" multiplication is possible.

If all these problems can be solved, then equations (5.54) and (5.55) become valid:

$$(I_{C1} - I_{C2}) = \Delta I_C = I_X \tanh \left(\frac{V_{in}}{2V_T} \right) \simeq \frac{V_m}{R_m} \tanh \left(\frac{V_{in}}{2V_T} \right)$$

which becomes, when $V_{in} \ll V_T$,

$$\Delta I_C \simeq \frac{V_m V_{in}}{2R_m V_T} .$$

To convert this circuit into a four-quadrant multiplier, and incidentally made the magnitudes of V_{in} and V_m similar, two more long-tailed pairs may be added, with cross-coupling as shown in figure 5.38. This circuit, due originally to Gilbert[5,6], is the basis for all modern monolithic *IC* transconductance multipliers.

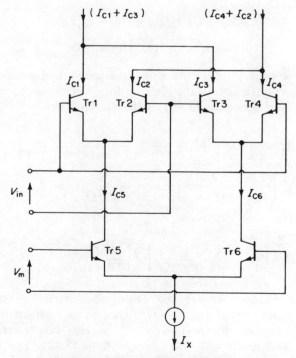

Figure 5.38. The basic Gilbert multiplier cell.

To understand the reasoning behind this cross-coupling, it is useful to observe what happens when only one signal is applied, which may be either V_{in} or V_m. Take $V_m = 0$ first. If V_{in} goes positive, I_{C1} increases, but I_{C3} decreases by the same amount. Therefore, the composite collector current $(I_{C1} + I_{C3})$ remains unchanged because of the cross-coupling. Also, I_{C4} increases, but I_{C2} decreases, so the composite current $(I_{C4} + I_{C2})$ is also unchanged. Hence, there is no output difference current ΔI, where $\Delta I = (I_{C1} + I_{C3}) - (I_{C4} + I_{C2})$.

Now suppose that $V_{in} = 0$. If V_m goes positive, I_{C5} and therefore I_{C1} and I_{C2} increase. Also, I_{C6} and therefore I_{C3} and I_{C4} decrease. But Tr1 and Tr3 have coupled collectors, so $(I_{C1} + I_{C3})$ remains unchanged. Similarly $(I_{C4} + I_{C2})$ remains unchanged.

These two arguments show that if either signal is zero, the other cannot produce an output, which is correct.

Again assume that V_m is positive. Because I_{C1} and I_{C2} have risen, then $g_{m(1)}$ and $g_{m(2)}$ must also have increased. Similarly, $g_{m(3)}$ and $g_{m(4)}$ will have decreased. So, if V_{in} now goes positive as well, I_{C1} must rise more than I_{C3} falls, so that there is a net *increase* in $(I_{C1} + I_{C3})$. Similarly, there will be a net *decrease* in $(I_{C4} + I_{C2})$. So now, a difference current exists, and it remains to show that this is proportional to $V_{in}V_m$.

Consider the difference current:

$$\Delta I = (I_{C1} + I_{C3}) - (I_{C4} + I_{C2})$$

$$= (I_{C1} - I_{C4}) - (I_{C2} - I_{C3})$$

Each of these collector currents arises in a long-tailed pair, so that each may be written as a fraction of the relevant long-tail current, as was done in the transition from equations (5.40) to (5.42). That is:

$$\Delta I = \left[\frac{I_{C5}}{1 + \exp\left(-V_{in}/V_T\right)} - \frac{I_{C6}}{1 + \exp\left(-V_{in}/V_T\right)} \right]$$

$$- \left[\frac{I_{C5}}{1 + \exp\left(V_{in}/V_T\right)} - \frac{I_{C6}}{1 + \exp\left(V_{in}/V_T\right)} \right]$$

or

$$\Delta I = \left[\frac{I_{C5} - I_{C6}}{1 + \exp\left(-V_{in}/V_T\right)} \right] - \left[\frac{I_{C5} - I_{C6}}{1 + \exp\left(V_{in}/V_T\right)} \right].$$

Because the numerators are the same, each term in this equation represents the collector current of a transistor in an equivalent long-tailed pair having a long-tail current of $(I_{C5} - I_{C6})$. This can be confirmed by comparing the two terms with the two equations (5.42). This being so, it must be possible to express the current difference ΔI by the hyperbolic tangent expression of equation (5.54):

$$\Delta I = (I_{C5} - I_{C6}) \tanh \left(\frac{V_{in}}{2V_T} \right). \qquad (5.60)$$

In this equation $(I_{C5} - I_{C6})$ is the difference between the collector currents of the Tr5–Tr6 long-tailed pair, so this can be written as:

$$(I_{C5} - I_{C6}) = I_X \tanh \left(\frac{V_m}{2V_T} \right). \qquad (5.61)$$

Combining equations (5.60) and (5.61) gives:

$$\Delta I = I_X \tanh \left(\frac{V_{in}}{2V_T} \right) \tanh \left(\frac{V_m}{2V_T} \right) \qquad (5.62)$$

and if V_{in} and V_m are both small compared with V_T, then,

$$\Delta I \simeq I_X V_{in} V_m / 2V_T = q I_X V_{in} / 2kT \qquad (5.63)$$

where V_{in} and V_m are now of similar (small) magnitude and are both difference input voltages, so that "four-quadrant" multiplication has now become possible. However, the coefficient in equation (5.63) is clearly temperature-dependent, which presents yet another problem.

Another desirable circuit modification would be to avoid the necessity for keeping V_{in} and V_m very small, and a method for doing this follows from a consideration of the actual operation of the multiplier[7]. In order that multiplication should occur at all implies that a nonlinear transfer function must exist somewhere in the circuit to which both input signals are applied. This actually occurs in the base–emitter junctions of the long-tailed pairs, which of course exhibit the usual well-defined exponential transconductance characteristics. So, if input networks are included which exhibit the inverse of this characteristic, then a linear input–output performance should result.

Referring again to figure 5.37, it is apparent that V_{in} should be derived from what amounts to a $\tanh^{-1}$ function circuit because V_{in} is applied between pairs of bases; whereas for the Tr5–Tr6 pair it is necessary only to establish that $(I_{C5}-I_{C6})$ is linearly related to V_m. Equation (5.60) will then be linearized.

Consider yet another modification of the long-tailed pair, shown in figure 5.39. Here, two fixed current sources of magnitude I are connected in the emitter leads and a resistor R is connected between the emitters. Neglecting base currents,

$$(I_{CA} - I_{CB}) = (I + \tfrac{1}{2}\Delta I) - (I - \tfrac{1}{2}\Delta I) = \Delta I$$

and it will now be shown that this difference current ΔI is proportional to

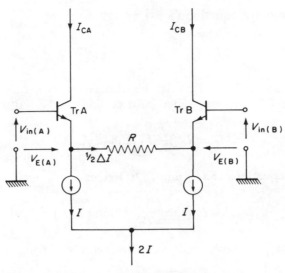

Figure 5.39. A voltage-to-current converter.

the input voltage and that the relevant relationship is largely temperature insensitive. Using the usual exponential transconductance expressions,

$$I_{CA} = (I + \tfrac{1}{2}\Delta I) = I_{ES} \exp\left(\frac{V_{BE(A)}}{V_T}\right) = I_{ES} \exp\left(\frac{V_{in(A)} - V_{E(A)}}{V_T}\right)$$

$$I_{CB} = (I - \tfrac{1}{2}\Delta I) = I_{ES} \exp\left(\frac{V_{BE(B)}}{V_T}\right) = I_{ES} \exp\left(\frac{V_{in(B)} - V_{E(B)}}{V_T}\right)$$

giving:

$$\frac{I - \tfrac{1}{2}\Delta I}{I + \tfrac{1}{2}\Delta I} = \frac{1 - (\Delta I/2I)}{1 + (\Delta I/2I)} = \exp\left(\frac{(V_{in(B)} - V_{in(A)}) + (V_{E(A)} - V_{E(B)})}{V_T}\right).$$

Now

$$(V_{in(B)} - V_{in(A)}) = -V_{in}$$

and

$$(V_{E(A)} - V_{E(B)}) = \tfrac{1}{2}\Delta IR$$

Using these, and re-writing,

$$V_T \ln\left(\frac{1 - (\Delta I/2I)}{1 + (\Delta I/2I)}\right) = \tfrac{1}{2}\Delta IR - V_{in}$$

If $\Delta I/2I \ll 1$, this becomes,

$$\tfrac{1}{2}\Delta IR \simeq V_{\text{in}} \quad \text{or} \quad \Delta I \simeq 2V_{\text{in}}/R \tag{5.64}$$

which shows that the difference of the two collector currents is indeed proportional to V_{in}, and that the relationship is largely insensitive to temperature variations. The remaining restriction is that the collector current difference ΔI should be small compared with the quiescent current I. Further, it is apparent that the excursions of V_{in} are now determined by the value of R and so can be large if R takes a high value.

This circuit can be applied to that of figure 5.38 by using the Tr5–Tr6 pair along with mirror transistors to form the current sources, as shown in figure 5.40.

To provide processed voltages for the multiplier base drive, another long-tailed pair like that of figure 5.39 is used, along with two more identical transistors connected as diodes in its collector leads, as is shown in figure 5.40. Again, mirror transistors are used as current sources and all four are driven from a single diode-connected transistor as shown, which implies that somewhere in the circuit a constant reference voltage must be generated. This technique, when suitably refined[5,6,7] can lead to good linearization, wherein the two input voltages can have wide ranges and be referred to the same datum. (They can also be single-ended and referred to ground.) Also,

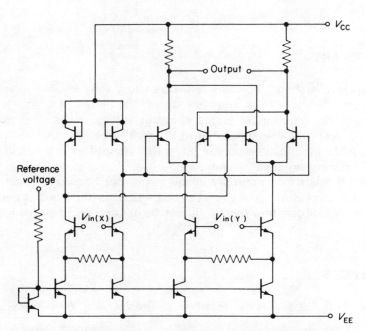

Figure 5.40. Basic circuit for integrated circuit multiplier.

the temperature dependence of the coefficient in equation (5.63) has now been largely overcome.

Implementation of the Gilbert multiplier is entirely dependent upon the very good matching of active structures and so lends itself well to monolithic processing. Figure 5.40 is the basic functional circuit only, and real ics (such as the Analog Devices A532 or the Motorola MC1595 ranges) are considerably more complex. For example, the output as it stands is the difference between two composite collector currents, and this must be converted to a single-ended output voltage, which can be accomplished using yet more current-mirror and difference amplifier circuitry as will be seen in chapter 7.

In practice, account has to be taken of the various offsets and scaling factors which relate to the circuit, and these are detailed by the relevant manufacturer. Insofar as the applications engineer is concerned a clear picture of the basic operation of the circuit is needed but more detailed knowledge of the ic design itself is unnecessary.

The multiplier as a microcircuit building block or subsystem is extremely important, for not only is it used in its mathematical sense where the product of two variables is needed, but it has applications in modulation (where one input is a square wave), and in phase measurement (where both inputs are square waves). Furthermore, the hyperbolic tangent properties of the basic long-tailed pair have been used in the conversion of triangular waves (which are easy to generate) into sine waves having frequency ranges of many orders of magnitude[8,9]. Some of these applications appear later in the book.

5.6 SUMMARY

This chapter has shown how discrete transistors can be used along with passive components to form amplifier stages, the design of which depends largely upon the appropriate juxtaposition of voltage drops. It then progressed to concepts of design peculiar to the high levels of active and passive structure matching, possible only in integrated circuit fabrication and which involved "current-oriented" techniques.

Chapter 7 follows on naturally at this point, but because the important topic of electrical noise arises several times later in the book a complete chapter on this subject now follows; it may be omitted on a first reading if so desired.

REFERENCES

1. Hurley, R. B., 1958, *Junction Transistor Electronics* (New York: Wiley) section 5.2.3.
2. Middlebrook, R. D., 1963, *Differential Amplifiers* (New York: Wiley).

3. Giocoletto, L. J., 1970, *Differential Amplifiers* (New York: Wiley).
4. Gray, P. R. and Meyer, R. G., 1977, *Analysis and Design of Integrated Circuits* (New York: Wiley) section 4.2.1.
5. Gilbert, B., 1968, A precise four-quadrant multiplier with sub-nanosecond response, *IEEE J. Solid St. Circuits*, **SC-3**, 365–373.
6. Gilbert, B., 1974, A high-performance monolithic multiplier using active feedback, *IEEE J. Solid St. Circuits*, **SC-9**, 364–373.
7. Huijsing, J. H., Lucas, P. and DeBruin, B., 1982, Monolithic analog multiplier-divider *IEEE, J. Solid St. Circuits*, **SC-17**, 9–15.
8. Gilbert, B., 1977, Circuits for the precise synthesis of the sine function, *Electron. Lett.* **13**, 506–508.
9. Evans, W. A. and Williams, J. S., 1979, The multi-tanh circuit as a triwave-to-sine convertor, *Electron. Circuits Syst.* **3**, 90–92.

QUESTIONS

1. What components are the causes of (*a*) low-frequency gain roll-off and (*b*) high-frequency gain roll-off in amplifiers?

2. What is meant by the "Miller Effect?"

3. The apparent resistance at the emitter of an operating transistor is usually very low. Why?

4. Define "forward voltage gain," "loop gain," "overall gain," "feedback factor" and "feedback fraction" for a feedback amplifier and include a relevant diagram.

5. The summing point or virtual ground at the inverting input of an operational amplifier working in the inverting configuration, presents a very low impedance and can accept only a very small signal. Explain these facts.

6. What is a "voltage-follower," and for what can it be used?

7. Explain why the voltage gain of a degenerate CE stage is nearly $-R_C/R_F$.

8. In Figure 5.17, for identical transistors the current gain of Tr1 is likely to be smaller than that for Tr2. Why is this, and could it be improved?

9. The voltage gain expression (5.34) for a complementary compound pair is given by $A_V \simeq 1 + R_2/R_1$ and is similar to that for the noninverting operational amplifier given in equation (7.32). Why?

10. What is the "Nyquist Criterion?"

11. What does "DC amplification" actually mean?

12. The difference amplifier is often called a "differential amplifier." Why is the former term preferable?

13. What is meant by the term "common mode signal?"

14. What is meant by the term "common mode rejection ratio?"

15. What is the main effect of emitter degeneration in a difference amplifier pair?

16. What is a "transconductance multiplier" and why is it so called?

PROBLEMS

5.1. A silicon planar transistor is connected into a common-emitter circuit with a very low resistance load. At a quiescent collector current of 1.6 mA the value of h_{fe} is 225. If the base spreading resistance is very small, what is the approximate value of the input resistance to the base at room temperature?

5.2. Using the hybrid-π equivalent circuit, determine the low-frequency cut-off point for the amplifier stage of figure 5.41. Show which capacitor defines this cut-off and prove that the other capacitor has a negligible effect at this frequency.

   Under the working conditions specified by the circuit, the following parameters apply to the transistor used: $V_{BE} = 0.6$ V, $r_{bb'} = 200\,\Omega$, $h_{fe} = 100$, and $kT/q \approx 26$ mV at 25°C. Assume that I_B is negligibly small.

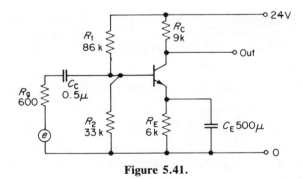

Figure 5.41.

5.3. Determine the mid-band gain and the high-frequency cut-off point for the amplifier stage of figure 5.41 if the following parameters apply: $V_{BE} = 0.6$ V, $r_{bb'} = 200\,\Omega$, $h_{fe} = 100$, $kT/q \approx 26$ mV at 25°C, $F_T = 300$ MHz, and $C_{b'c} = 6$ pF.

5.4   A small-signal transistor is connected as a single-stage CE amplifier. If $R_C = 12\,k\Omega$ and $R_E = 2.7\,k\Omega$, find the approximate voltage gains when:

(a) R_E is not bypassed;

(b) R_E is bypassed by a large capacitor, and a $350\,\Omega$ load is applied to the collector via another large capacitor;

(c) if a voltage generator drives the input via a $1\,\mu F$ coupling capacitor, how much of the R_E must remain unbypassed if the low-frequency cut-off point is to be 10 Hz?

Assume that $h_{fe} = 50$, $h_{ie} = 810\,\Omega$, and the transistor is properly biased, the resistors having values much greater than R_{in}.

5.5   In the circuit of figure 5.42 the transistor takes a quiescent collector current of $-1.0\,mA$ and the quiescent collector–emitter voltage is $-5.0\,V$. Calculate the values of R_C and $\frac{1}{2}R_E$ and the approximate mid-band gain. If C_E is very large, what is the low-frequency cut-off point? (Hint: take I_B into account).

For the transistor working at $-1\,mA$, $h_{FE} = 260$ and $h_{fe} = 300$. Also, $V_{BE} = -0.7\,V$.

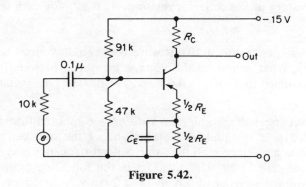

Figure 5.42.

5.6.   The two transistors in figure 5.43 are identical, and under the working conditions imposed by the circuit $V_{BE} = 0.7\,V$, $h_{FE} = 150$, $h_{fe} = 180$, and $r_{bb'} < 10\,\Omega$. Taking kT/q as 26 mV, calculate the following:

(i) the quiescent collector current in Tr1;

(ii) the quiescent collector current in Tr2;

(iii) the direct voltage at the output point;

(iv) the approximate input resistance at the base of Tr1 (to the right of the bias resistors);

(v) the approximate input resistance into the base of Tr2;

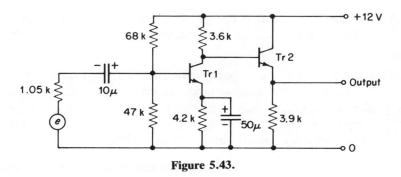

Figure 5.43.

(vi) the mid-band voltage gain A_v measured from the base of Tr1 to the output point;

(vii) the low-frequency cut-off point f_L.

5.7. Two matched npn transistors are connected as a difference amplifier with their bases taken to ground. The supply voltages are ± 15 V with respect to ground. If the long-tail resistor $R_X = 7.2$ kΩ, calculate:

(i) the current I_x in R_x;

(ii) the values of the collector resistors R_C if V_{CE} for each transistor is 6 V.

If a voltage generator of 10 mV were connected in series with one of the bases, what would be the voltage measured between the collectors? Assume that $V_{BE} = 0.6$ V; $r_{bb'} = 1000$ Ω; $h_{fe} = 100$; $kT/q \simeq 26$ mV.

5.8. Prove that the voltage gain for a difference input/difference output long-tailed pair is equal to that for one of the matched transistors operated as a CE stage with the same collector resistance as in the case of the long-tailed pair. Use this fact in the following example.

Two voltage sources of 1.6 mV and 2.0 mV, respectively, are connected from each base of a long-tailed pair to ground. Collector resistors of 7.2 kΩ are used along with a long-tail resistor of 10 kΩ. The stage is initially fully balanced with collector currents of 2 mA. Calculate: (i) the difference output voltage at each collector; and (ii) the common-mode output voltage at each collector. (Assume that $r_{bb'} \ll r_{b'e'}$).

5.9. The circuit of figure 5.44 is to be realized using the almost identical transistor structures Tr1 and Tr2 in a monolithic IC, and for each, $h_{fe} = 150$, $r_{bb'} = 100$ Ω, and V_{BE} can be taken as 0.7 V. If these parameters can be taken as applying also to the pnp structure Tr3, answer the following questions:

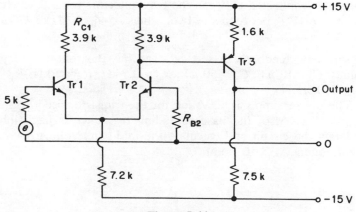

Figure 5.44.

(i) What would be a suitable value for R_{B2}, and why?

(ii) Is R_{C1} necessary? Explain your answer.

(iii) What are the quiescent currents in the collectors of Tr1 and Tr2?

(iv) What is the quiescent voltage at the output point?

(v) What is the approximate signal voltage gain of Tr3?

(vi) Does Tr3 contribute significantly to the load on the Tr2 collector?

(vii) What is the terminal voltage gain A_v for the complete amplifier?

5.10. For the circuit of figure 5.34, calculate the approximate value of the overall signal voltage gain and also the approximate quiescent current which the power supply must provide. (Take $V_{BE} \simeq 0.6\,V$ and neglect quiescent base currents.)

5.11. A difference amplifier stage is to be operated from $\pm 15\,V$ power supplies and biased using a 1 mA current mirror. The input to one base is derived from a vibration transducer having an internal resistance of $5\,k\Omega$, the other base being grounded via a matching resistor. If a noninverting output signal is taken from one collector and the overall voltage gain is 88.6, calculate the resistor values for the circuit and include them in an appropriate circuit diagram.

Assume that the two transistors are well matched and exhibit incremental current gains of 150 under room-temperature operating conditions, and that the base spreading resistances are very small.

What is the DC voltage level at the output point for room-temperature operating conditions?

5.12. A long-tailed pair consists of two identical transistors with a current source driving their commoned emitters. Each transistor takes a quiescent collector current of 1 mA via a $4.7\,k\Omega$ load resistor. If a voltmeter connected between the two collectors reads 3 V, what is the difference (V_{in}) between the two values of V_{BE}? (Take $kT/q = V_T = 26$ mV).

5.13. A series DC feedback biased CE stage has the following component values: $C_c = 10 \ \mu F$; $C_E = 50 \ \mu F$; $R_C = 2.7 \ k\Omega$; $R_E = 1 \ k\Omega$; $R_1 = 93 \ k\Omega$; $R_2 = 27 \ k\Omega$.

The power supply is 12 V and for the transistor used $h_{fe} = 200$ and $V_{BE} = 0.7$ V. Also, the base spreading resistance is negligibly small.

Draw the circuit and calculate the mid-band voltage gain and the low-frequency -3 dB point.

6

Noise

At the output of an amplifier, not only does the output signal appear, but so does a certain amount of electrical "noise", which is the term used to describe unwanted fluctuations which may exist in a frequency band from zero upwards.

Some of this noise enters the amplifier along with the input signal and is amplified by the same amount as that signal but the remainder is generated within the amplifier itself. Topics relevant to electrical "noise" generated in transistor audio amplifier stages have been clarified by several authors[1,2,3,4,5], so the point where it is now possible to present a simple coherent picture of the overall problem and its relation to the design process. To do this it is necessary only to accept the Nyquist and Schottky formulations for thermal and shot noise and to relate them to the relevant regions of the hybrid-π equivalent circuit. This model may then be reduced to a "two-noise generator" equivalent circuit, from which both the noise factor and the actual magnitude of the output noise may be calculated. Furthermore, the optimum magnitude of I_C for minimum signal-to-noise degradation through a bipolar transistor stage may be simply determined. Finally, it may be shown that a simple approach to the calculation of noise engendered by feedback resistors is provided by the noise-resistance concept.

6.1 SYMBOLS

Because the random noise sources in semiconductor devices have zero mean values, most calculations involve mean square or RMS values. This implies that noise power exists; that is, in the load resistor of an amplifier, power is dissipated which is proportional to the mean square value of the amplified noise voltage which appears at the input. These concepts are reflected in the necessity for the symbols listed below:

	Instantaneous	Mean Square	RMS
Noise current	i_n	$\overline{i_n^2}$	$\sqrt{\overline{i_n^2}}$
Noise voltage	v_n	$\overline{v_n^2}$	$\sqrt{\overline{v_n^2}}$
Total noise voltage at amplifier input	v_{ni}	$\overline{v_{ni}^2}$	$\sqrt{\overline{v_{ni}^2}}$
Total noise voltage at amplifier output	v_{no}	$\overline{v_{no}^2}$	$\sqrt{\overline{v_{no}^2}}$
Spot-noise voltage generator referred to amplifier input	—	$\overline{e_n^2}$	$\overline{e}_n$
Spot-noise current generator referred to amplifier input	—	$\overline{i_n^2}$	$\overline{i}_n$

6.2 NOISE SOURCES

When a current flows across a junction, the discrete nature of the charge carriers gives rise to a statistical fluctuation; that is, the number of carriers crossing a plane at a given instant is different from that at another instant. So, although the instantaneous value of the current may be higher or lower than the mean at different points in time, the average value over a long period will be the continuous current level, by definition. However, the fluctuations themselves will nevertheless be observable and so possess a measurable mean square value. Schottky gives the value of this mean square:

$$\overline{i_n^2} = 2qI\Delta f \tag{6.1}$$

where q is the electron charge, I is the current in question and Δf is the noise bandwidth.

Theoretically, this *shot noise* extends through an infinite bandwidth, and because it contains all frequencies is called "white noise." In practice, however, the bandwidth is limited by the power versus frequency characteristic of the circuit in which it arises. Note that the power/frequency characteristic is involved because the mean square value of the shot noise exists and so will produce noise power of value $\overline{i_n^2}R$ in a resistor R.

To simplify calculations involving white noise, a *noise bandwidth* Δf is postulated. This is the bandwidth of an ideal "square" power/frequency bandpass having the same height as the maximum value of the actual signal-power/frequency response of the system, and which passes the same noise power. This concept is illustrated in figure 6.1, and the way in which it arises naturally from the methematics of the situation is treated in Appendix III. For the present, it is sufficient to note that for a single-lead, single-lag amplifier, the noise bandwidth is:

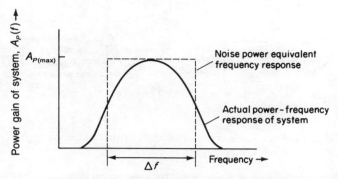

Figure 6.1. The noise bandwidth, Δf (Areas bounded by full and broken lines are equal.)

$$\Delta f \simeq \tfrac{1}{2}\pi(f_\mathrm{H} - f_\mathrm{L}) \tag{6.2}$$

where f_H and f_L are the high- and low-frequency cut-off points exhibited by the 6 dB/octave *voltage* gain characteristic.

The second important source of noise is thermal, or Johnson noise. This arises in any real (as opposed to dynamic) resistance and is due to thermal fluctuations of free electrons. Its mean square value is given by Nyquist as:

$$\overline{v_\mathrm{n}^2} = 4kTR\Delta f \tag{6.3}$$

where k is Boltzmann's constant, T is the absolute temperature and R is the resistance in question.

Here again, the noise is assumed to extend through an infinite bandwidth and to be limited only by the bandwidth of the associated circuit (including, of course, the self-capacitance of the resistance in question). Like shot noise, thermal noise is called "white noise."

Both shot and thermal noise may be assumed random in nature; that is, the probability of the noise amplitude lying within a given interval at a given instant follows the Gaussian probability function.

A third source of noise is the so called flicker or $1/f$ noise[6,7,8]. This is essentially a low-frequency phenomenon exhibited by almost all electronic devices including vacuum tubes and all forms of transistor. Its nature is still not properly understood, but from a design point of view this is less important that the fact that it can be represented by a rise in one or other of the two noise generators which appear in the two-noise-generator equivalent circuit as the frequency falls. (This point will be elaborated below.)

6.2.1 Noise Source Equivalents

Shot noise, having the dimensions of current, may be represented by a current generator in parallel with the (noiseless) current I, as shown in figure

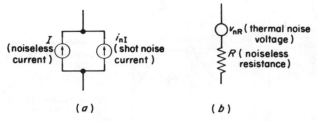

Figure 6.2. (*a*) Shot noise representation; (*b*) thermal noise representation.

6.2(*a*). Similarly, thermal noise may be represented by a voltage generator in series with the (noiseless) resistor producing it, as in figure 6.2(*b*).

If noise sources are additive, as in the case of thermal noises arising from two resistances in series, then the total instantaneous noise is:

$$v_{nR} = v_{nR_1} + v_{nR_2}.$$

The mean value of the sum will be zero, as were the individual mean values. However, the mean *square* value is:

$$\overline{v_{nR}^2} = \overline{\left(v_{nR_1} + v_{nR_2}\right)^2}$$
$$= \overline{v_{nR_1}^2} + \overline{v_{nR_2}^2} + \overline{2v_{nR_1}v_{nR_2}}.$$

If the two resistors are physically separate and distinct, there will be no correlation between the noise arising from each. Hence, because the mean values are separately zero, their product will also be zero, and

$$\overline{v_{nR}^2} = \overline{v_{nR_1}^2} + \overline{v_{nR_2}^2}.$$

The same will be true for two independent noise currents in parallel. In general, therefore, it may be said that if noise sources are *uncorrelated*, their mean square values may be added arithmetically.

6.3 BIPOLAR TRANSISTOR NOISE FORMULATION

Noise at the output of a transistor amplifier stage is the result of: (*a*) the amplification of noise entering the amplifier along with the signal and (*b*) the fact that any real amplifier stage must contain inherent noise sources. For the common-emitter bipolar transistor amplifier stage, both of these noise contributions may be included in the low-frequency hybrid-π equivalent circuit, as shown in figure 6.3(*a*). Here, the effective resistance at the input is labelled R_G, which includes the internal resistance of the signal source along with any bias and feedback resistances. The mean square

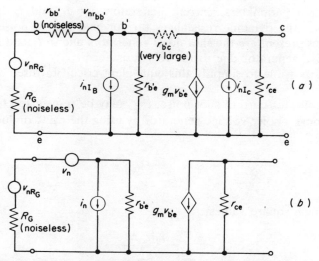

Figure 6.3. (a) Hybrid-π equivalent circuit with noise generators; (b) simplified equivalent circuit with only two noise generators. (*Note*: $r_{bb'}$ has been omitted because $r_{bb'} \ll R_G$; $r_{b'c}$ has been omitted because $r_{b'c} \gg R_G$.)

thermal noise generated in R_G is:

$$\overline{v_{nR_G}^2} = 4kTR_G\Delta f$$

and this has been represented in figure 6.3(a) by the noise voltage generator, v_{nR_G} in series with a noiseless resistor, $R_{G(\text{noiseless})}$.

Similarly, the noise due to the base-spreading resistance has been represented by a noiseless resistance, $r_{bb'(\text{noiseless})}$ in series with a noise voltage generator, $v_{nr_{bb'}}$, where,

$$\overline{v_{nr_{bb'}}^2} = 4kTr_{bb'}\Delta f \ .$$

Shot noise due to the base current crossing the emitter junction is represented by the noise current generator, i_{nI_B}, where,

$$\overline{i_{nI_B}^2} = 2qI_B\Delta f \ .$$

Finally, shot noise due to the collector current crossing the collector junction is:

$$\overline{i_{nI_C}^2} = 2qI_C\Delta f \ .$$

(Note that because the hybrid-π equivalent circuit refers only to incremental or small-signal parameters, current generators for I_B and I_C are not included. Also, the dynamic resistances do not give rise to noise; the only noise sources present are the shot noise generators due to I_B and I_C, and the thermal noise generator due to the real resistance $r_{bb'}$.)

It is now possible to simplify this equivalent circuit in three steps:

(i) The collector current shot noise, i_{nI_C}, may be transferred to the input side to become a noise voltage generator by using the transconductance, g_m; that is,

$$v_{nI_C} = \frac{i_{nI_C}}{g_m}$$

or, using mean square values,

$$\overline{v_{nI_C}^2} = \frac{2qI_C\Delta f}{g_m^2} .$$

Now $g_m = qI_C/kT$, so that,

$$\overline{v_{nI_C}^2} = \frac{2kT\Delta f}{g_m} . \tag{6.4}$$

(ii) This noise voltage generator may be added to the thermal noise due to $r_{bb'}$ to give a new composite noise generator. Assuming zero correlation and hence adding the mean square values gives:

$$\overline{v_n^2} = 4kT\left(r_{bb'} + \frac{1}{2g_m}\right)\Delta f . \tag{6.5}$$

This composite noise voltage generator, v_n, has been included in figure 6.3(b).

(iii) The base current shot noise generator may be simplified by putting $I_B = I_C/h_{FE}$:

$$\overline{i_{nI_B}^2} = \frac{2qI_C\Delta f}{h_{FE}}$$

and since $I_C = kTg_m/q$, then i_{nI_B} can be rewritten to become i_n, which is the only noise current generator remaining:

$$\overline{i_n^2} = \frac{2kTg_m\Delta f}{h_{FE}} . \tag{6.6}$$

The simplified equivalent circuit may now be completed as shown in figure 6.3(b).

Converting the Norton equivalent formed by i_n and R_G into its Thévenin dual $i_n R_G$ and summing it with the other noise voltage generators, gives:

$$v_{ni} = v_{nR_G} + v_n + i_n R_G \qquad (6.7)$$

or, using mean square values,

$$\overline{v_{ni}^2} = \overline{v_{nR_G}^2} + \overline{v_n^2} + \overline{i_n^2} R_G^2 . \qquad (6.8)$$

Inserting the values for v_n and i_n from equations (6.5) and (6.6) gives:

$$\overline{v_{ni}^2} = 4kTR_G\Delta f + 4kT\left(r_{bb'} + \frac{1}{2g_m}\right)\Delta f + \frac{2kTg_m\Delta f}{h_{FE}} R_G^2 . \qquad (6.9)$$

The noise voltage at the output as read by a true RMS meter will be:

$$\sqrt{\overline{v_{no}^2}} = A_v \sqrt{\overline{v_{ni}^2}} \qquad (6.10)$$

where A_v is the voltage gain of the amplifier defined in Appendix III. As will be shown later, it is often convenient to determine the unity bandwidth, or *spot* noise, and this may be found simply by dividing equation (6.9) by the noise bandwidth, Δf, to give:

$$\frac{\overline{v_{ni}^2}}{\Delta f} = \overline{v_{ni(spot)}^2} = 4kTR_G + \overline{e}_n^2 + \overline{i}_n^2 R_G^2 \qquad (6.11)$$

where

$$\overline{e}_n = \left(\frac{\overline{v_n^2}}{\Delta f}\right)^{1/2} = \left[4kT\left(r_{bb'} + \frac{1}{2g_m}\right)\right]^{1/2} \qquad (6.12a)$$

and

$$\overline{i}_n = \left(\frac{\overline{i_n^2}}{\Delta f}\right)^{1/2} = \left(\frac{2kTg_m}{h_{FE}}\right)^{1/2} . \qquad (6.12b)$$

Here, $\overline{e}_n$ and $\overline{i}_n$ are the *spot-noise generators*, and equations (6.12) explain their usual units of $\text{nV}/\sqrt{\text{Hz}}$ and $\text{pA}/\sqrt{\text{Hz}}$, respectively.

6.4 FLICKER AND HIGH-FREQUENCY NOISE

In the foregoing discussion, it has been tacitly assumed that the spot-noise generators are frequency-independent, as demonstrated for the bipolar transistor by equations (6.12). Otherwise, the wide-band noise equations containing Δf (equations (6.8) and (6.9)) would have been invalid and integral equations would be necessary. Actually, the frequency independence of $\bar{e}_n$ and $\bar{i}_n$ is a fairly well-justified assumption between 1 kHz and 100 kHz; that is, between the flicker noise and high-frequency noise regions.

In principle, both flicker and high-frequency noise could be taken into account by introducing further noise generators, but it is more convenient to make the existing spot-noise generators frequency dependent. This means that outside the range of frequency invariance, the spot-noise generators are no longer given by equations (6.12). However, if $\bar{e}_n$ and $\bar{i}_n$ are known (and both are subject to measurement), calculations of *spot* noise at any frequency may be made according to equation (6.11).

In the case of the bipolar transistor, flicker noise may be represented entirely by a rise in $\bar{i}_n$ as the frequency falls, while $\bar{e}_n$ remains sensibly constant. High-frequency noise is due largely to the onset of correlation between the noise generators, which is the result of noise being transmitted via the interelectrode capacitances at high frequencies. This form of noise is not of interest in the present audio-oriented treatment.

6.5 THE TWO-NOISE-GENERATOR EQUIVALENT

Figure 6.3(b) shows that the hybrid-π equivalent circuit now contains two noise generators at the input. If the transistor so characterized is used as the only stage in an amplifier, it can be represented as in figure 6.4(a), where the wide-band noise generators have been taken outside the system. This is the two-noise-generator equivalent, and from equation (6.10)

$$\sqrt{\overline{v_{no}^2}} = A_v(4kTR_G\Delta f + \overline{v_n^2} + \overline{i_n^2}R_G^2)^{1/2} . \tag{6.13}$$

For unity bandwidth noise calculations the two noise generators become $\bar{e}_n$ and $\bar{i}_n$ so that the spot noise at the output is, from equation (6.11):

$$\sqrt{\overline{v_{no(spot)}^2}} = A_{v(spot)}(4kTR_G + \bar{e}_n^2 + \bar{i}_n^2 R_G^2)^{1/2} . \tag{6.14}$$

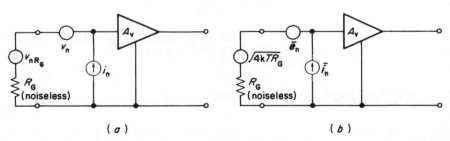

(a)   (b)

Figure 6.4. (a) Two-noise generator equivalent circuit; (b) two-spot-noise generator equivalent circuit. (*Note*: $\bar{e}_n = \sqrt{(\overline{v_n^2}/\Delta f)}$, $\sqrt{\bar{i}_n} = \sqrt{(\overline{i_n^2}/\Delta f)}$.)

If the noise at the output of a multi-stage amplifier is to be calculated, the two-noise-generator equivalent still applies provided that practically all of the noise is contributed by the first stage. This is the normal case for a properly designed amplifier, but some care should be taken to ensure that the second stage does not contribute significant noise in addition to amplifying that produced by the first stage[1].

6.6 NOISE FACTOR

It is often more useful to know by how much the signal-to-noise ratio is degraded as the signal passes through an amplifier, rather than to know the actual value of the noise at the output. To do this, it is convenient to compare the noise at the output of a "noisy" amplifier with that at the output of a hypothetical noiseless amplifier having the same voltage gain. Using figure 6.4(b), the mean square noise at the output is given by:

$$\overline{v^2_{no(spot)}} = A^2_{v(spot)}(4kTR_G + \overline{e^2_n} + \overline{i^2_n}R^2_G) \,. \tag{6.15}$$

For the noiseless amplifier, $\overline{e}_n$ and $\overline{i}_n$ would be zero, so that,

$$\overline{v^{2*}_{no(spot)}} = A^2_{v(spot)}4kTR_G \,. \tag{6.16}$$

Hence, the noise factor (NF) is:

$$NF = \frac{\overline{v^2_{no(spot)}}}{\overline{v^{2*}_{no(spot)}}} = 1 + \frac{\overline{e^2_n}}{4kTR_G} + \frac{\overline{i^2_n}R_G}{4kT} \tag{6.17}$$

$$= 1 + \frac{1}{4kT}\left(\frac{\overline{e^2_n}}{R_G} + \overline{i^2_n}R_G\right) \,. \tag{6.18}$$

This equation indicates that an optimum value of R_G will exist for which the noise factor is a minimum; that is, for which the signal-to-noise ratio is degraded as little as possible. This optimum value, $R_{G(opt)}$, can be extracted by differentiating equation (6.18) and equating to zero:

$$\frac{d(NF)}{dR_G} = \frac{1}{4kT}\left(-\frac{\overline{e^2_n}}{R^2_{G(opt)}} + \overline{i^2_n}\right) = 0$$

or

$$R_{G(opt)} = \frac{\overline{e}_n}{\overline{i}_n} \,. \tag{6.19}$$

The value of $R_{G(opt)}$ given in equation (6.19) may now be inserted back into equation (6.18) to give the minimum possible noise factor:

$$\mathrm{NF}_{\min} = 1 + \frac{\bar{e}_n \bar{i}_n}{2kT} .\qquad(6.20)$$

This equation simply confirms that a transistor having low values of spot-noise voltage and current will be capable of exhibiting a low noise factor *provided* the value of resistance at its input, R_G', is optimized. Because the choice of R_G is not usually the prerogative of the designer, however, it is appropriate to determine whether the transistor can be operated in such a way as to minimize the noise factor for any value of R_G.

6.6.1 Noise Factor Minimization

In the range where both $\bar{e}_n$ and $\bar{i}_n$ are fairly constant, equations (6.12) defining them may be inserted into the noise factor equation (6.17) to give:

$$\mathrm{NF} = 1 + \frac{r_{bb'}}{R_G} + \frac{1}{2g_m R_G} + \frac{g_m R_G}{2h_{FE}} .\qquad(6.21)$$

Putting $g_m = qI_C/kT$, this becomes:

$$\mathrm{NF} = 1 + \frac{r_{bb'}}{R_G} + \frac{kT}{2qI_C R_G} + \frac{qI_C R_G}{2kT h_{FE}} .\qquad(6.22)$$

To obtain the minimum possible noise factor for a given value of R_G, this equation may be differentiated with respect to I_C and equated to zero. Unfortunately, this process is complicated by the fact that h_{FE} is not constant but falls as I_C becomes smaller. However, for a good silicon planar low-level transistor, the following approximation may be made: for all values of I_C normally encountered in low-noise design, h_{FE} remains essentially constant, in which case:

$$\frac{d(\mathrm{NF})}{dI_C} = - \frac{kT}{2qI_{C(opt)}^2 R_G} + \frac{qR_G}{2kT h_{FE}} = 0$$

giving

$$I_{C(opt)} = \frac{kT}{qR_G} \sqrt{h_{FE}} .\qquad(6.23)$$

At room temperature, $kT/q \simeq 26$ mV, giving

$$I_{C(opt)} \simeq \frac{26}{R_G} \sqrt{h_{FE}} \, \mathrm{mA} \quad \text{if } R_G \text{ is in ohms} .\qquad(6.24a)$$

In Appendix III, it is shown that if the variation in h_{FE} with I_C for contemporary transistors is taken into account, a closer approximation is:

$$I_{C(\text{opt})} \simeq \frac{28}{R_G} \sqrt{h_{\text{FE}}} \text{ mA} \quad \text{if } R_G \text{ is in ohms} \tag{6.24b}$$

which is but little different from equation (6.24a).

In the frequency range below about 1 kHz, where flicker noise is important, it will be recalled that for a bipolar transistor, $\overline{i_n}$ is found to rise with falling frequency. As a crude approximation it may be assumed to follow a $(1/f)^{1/2}$ law, so that using equation (6.12b):

$$\overline{i_n^2} \simeq \frac{2kTg_m}{h_{\text{FE}}} \left(1 + \frac{f_F}{f}\right) \tag{6.25}$$

where f_F is a break-point in the curve of $\overline{i_n^2}$. If this is put into equations (6.21) through (6.24b), the optimum collector current for a minimal noise factor (equation (6.24b)) becomes:

$$I_{C(\text{opt})} \simeq \frac{28}{R_G} \left(\frac{h_{\text{FE}}}{1 + (f_F/f)}\right)^{1/2} \tag{6.26}$$

which clearly demonstrates that $I_{C(\text{opt})}$ becomes smaller as the spot frequency falls within the $1/f$ noise region.

6.6.2 Numerical Considerations

The factor $4kT$ and its root appear frequently in noise calculations and it is therefore useful to consider the numerical values involved:

$$\text{Boltzmann's constant, } k = 1.38 \times 10^{-23} \text{ J/K}$$

so that

$$4kT = 5.52 \times 10^{-23}(273 + {}^\circ\text{C})\text{J} .$$

At 25°C, this gives

$$4kT \simeq 1.65 \times 10^{-20} \text{ J}$$

whence

$$\sqrt{4kT} \simeq 1.28 \times 10^{-10} \text{ J}^{1/2} .$$

Calculation of thermal noise is now simple:

$$\overline{v_{nR}^2} = 4kTR\Delta f = 1.65 \times 10^{-20} R\Delta f \text{ V}^2 \quad \text{at 25°C} \tag{6.27}$$

and

$$\sqrt{\overline{v_{nR}^2}} = 1.28 \times 10^{-10} \sqrt{R \Delta f} \text{ V} \quad \text{at } 25°C. \tag{6.28a}$$

If R is in megohms, this becomes:

$$\sqrt{\overline{v_{nR}^2}} \simeq 0.13 \sqrt{R \Delta f} \ \mu\text{V} \quad \text{at } 25°C. \tag{6.28b}$$

Figures 6.5(a) and (b) are graphs of thermal noise voltage as functions of resistance, and are based on the foregoing numerical expressions. Figure 6.5(b) is of particular use when solving equations (6.13) and (6.14).

For a bipolar transistor both $\bar{e}_n$ and $\bar{i}_n$ are functions of I_C, which means that unless they are graphed in the manufacturer's data sheet they should be calculated for the relevant value of I_C using equations (6.12a) and (6.12b) or (6.25). These are:

$$\bar{e}_n = \left[4kT \left(r_{bb'} + \frac{1}{2g_m} \right) \right]^{1/2}$$

$$= \left[4kT \left(r_{bb'} + \frac{kT}{2qI_C} \right) \right]^{1/2} \tag{6.29}$$

and

$$\bar{i}_n = \left[\frac{2kTg_m}{h_{FE}} \left(1 + \frac{f_F}{f} \right) \right]^{1/2}$$

$$\simeq \left[\frac{2qI_C}{h_{FE}} \left(1 + \frac{f_F}{f} \right) \right]^{1/2} \tag{6.30}$$

where $q = 1.6 \times 10^{-19}$ C and $kT/q \simeq 26$ mV at 25°C.

In equation (6.30), the value of h_{FE} should be that pertaining to the relevant value of I_C, for it will be recalled that h_{FE} is a function of I_C. This often calls for the actual measurement of h_{FE} at the low levels of I_C which are necessary in low-noise design. (As a rule of thumb, it is good practice to establish that $h_{FE} \not< 100$ at $I_C = 1 \ \mu\text{A}$.)

Equations (6.24b) or (6.26) may be used to determine the optimum value of I_C, but their use is complicated by the dependence of h_{FE} on I_C. Taking equation (6.24b):

$$I_{C(opt)} \simeq \frac{28}{R_G} \sqrt{h_{FE}}$$

it will be seen that $I_{C(opt)}$ must first be known if the correct value of h_{FE} is to be inserted!

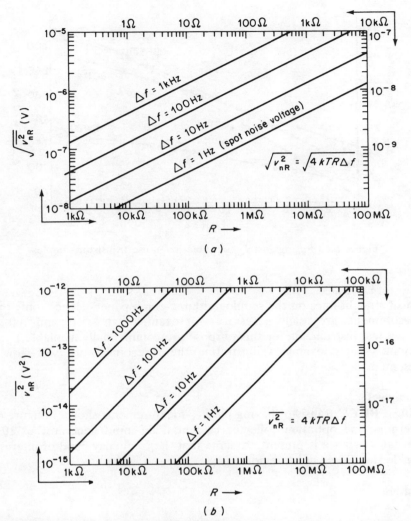

Figure 6.5. (*a*) Root mean square thermal noise voltage; (*b*) mean square thermal noise voltage.

This highlights the usefulness of measuring h_{FE} at various values of I_C, and this has been done for some representative transistors in figure 6.6. Clearly, the equation may be solved by taking an arbitrary value of h_{FE}, solving the equation, then using the value of $I_{C(opt)}$ so obtained to choose a better value of h_{FE}. A few successive approximations will quickly lead to the "best" value for $I_{C(opt)}$, remembering that the equation is approximate in the first place. (Note that below $I_C = 100\ \mu A$, the h_{FE}/I_C characteristics are

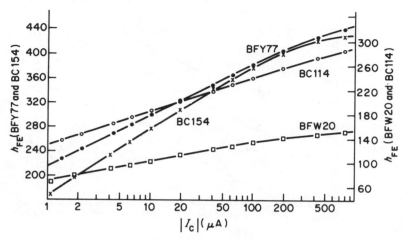

Figure 6.6. h_{FE} against I_C for some low-noise transistor samples.

almost straight lines on the semilogarithmic graph paper. Hence, only three measurements are really necessary—for example, at 1, 10, and 100 μA. Simple test instruments for this purpose are commercially available.)

Some simple examples illustrating the foregoing work will now be presented.

Example 1

Using a BFY77 transistor having the h_{FE} measurements shown in figure 6.6, determine the optimum collector current if an input generator of 10 kΩ internal resistance is present. Assume that the frequency band of interest is outside the flicker noise region.

Solution 1

$$I_{C(opt)} \simeq \frac{28}{R_G} \sqrt{h_{FE}} . \tag{6.24b}$$

Trial (a):
Assume that $I_C \simeq 70$ μA. From figure 6.6, at 70 μA $h_{FE} = 370$, giving, in this case

$$I_{C(opt)} \simeq \frac{28}{10^4} \sqrt{370} = 55 \ \mu A .$$

Trial (b):
Assume that $I_C = 55$ μA. From figure 6.6, at 55 μA $h_{FE} = 360$, giving, in this case

$$I_{C(opt)} \simeq \frac{28}{10^4} \sqrt{360} = 53 \ \mu A$$

This value is obviously satisfactory.

Example 2
Determine $\bar{e}_n$ and $\bar{i}_n$ for the transistor of Example 1.

Solution 2

$$\bar{e}_n = \left[4kT\left(r_{bb'} + \frac{kT}{2qI_C}\right) \right]^{1/2} \tag{6.29}$$

Assuming that $r_{bb'}$ is about 200 Ω, this is,

$$\bar{e}_n \simeq \left[1.65 \times 10^{-20}\left(200 + \frac{0.026}{2 \times 54 \times 10^{-6}}\right) \right]^{1/2}$$

$$\simeq 2.6 \ \text{nV}/\sqrt{\text{Hz}} \ .$$

$$\bar{i}_n \simeq \left(\frac{2qI_C}{h_{FE}}\right)^{1/2} \tag{6.30}$$

$$= \left(\frac{2 \times 1.6 \times 10^{-19} \times 54 \times 10^{-6}}{366}\right)^{1/2}$$

$$\simeq 0.22 \ \text{pA}/\sqrt{\text{Hz}} \ .$$

Example 3
Determine the noise factor for the BFY77 working as in the previous examples.

Solution 3

$$NF = 1 + \frac{1}{4kT}\left(\frac{\bar{e}_n^2}{R_G} + \bar{i}_n^2 R_G\right) \tag{6.18}$$

$$= 1 + \frac{1}{1.65 \times 10^{-20}}\left(\frac{6.8 \times 10^{-18}}{10^4} + 0.048 \times 10^{-24} \times 10^4\right)$$

$$= 1.07 \ .$$

That is, the *noise figure* (in decibels) is:

$$NF = 10 \log_{10} 1.07 = 0.3 \ \text{dB} \ .$$

Actually, the noise factor could have been calculated from the simpler equation (6.20) because choosing $I_{C(opt)}$ implies that $\bar{e}_n$ and $\bar{i}_n$ have become such that $R_G \simeq R_{G(opt)}$ for that value of I_C. That is,

$$R_{G(opt)} = \frac{\overline{e}_n}{\overline{i}_n} = \frac{2.6 \times 10^{-9}}{0.22 \times 10^{-12}} = 11.8 \, \text{k}\Omega \, ,$$

which is not far removed from the actual R_G of $10 \, \text{k}\Omega$. Hence, from equation (6.20):

$$\begin{aligned} NF_{min} &= 1 + \frac{\overline{e}_n \overline{i}_n}{2kT} \\ &= 1 + \frac{2.6 \times 10^{-9} \times 0.22 \times 10^{-12}}{0.5 \times 1.65 \times 10^{-20}} \\ &= 1.069 \, , \end{aligned}$$

the same value as before.

Figure 6.7(a) reproduces a set of measured noise contours for the BFY77

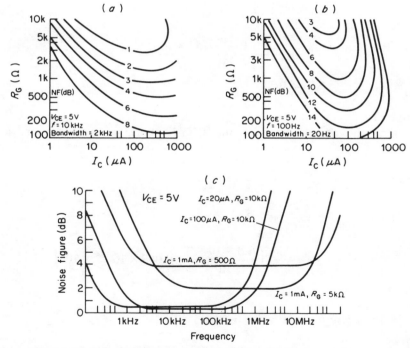

Figure 6.7. (a) Noise contours outside the $1/f$ noise region; (b) noise contours within the $1/f$ noise region; (c) noise figure against frequency for various operating conditions. (*Note*: All graphs are relevant to the BFY77 and are reproduced by permission of SGS THOMSON Microelectronics Ltd.)

and relevant to 10 kHz, which is well outside the flicker noise region. From this figure, it is clear that at $R_G = 10 \text{ k}\Omega$ and $I_C = 55 \mu\text{A}$, the noise figure is minimal as predicted and that 0.3 dB is a sound result.

Unfortunately, curves of $\bar{e}_n$ and $\bar{i}_n$ versus frequency are not presented (which is by no means unusual) so that f_F cannot be determined. Hence $I_{C(\text{opt})}$ for the flicker noise region cannot be determined either, but the measured noise contours of figure 6.7(b), which are relevant to 100 Hz, show that the noise factor will rise under the same operating conditions as above. In point of fact, for operation at 100 Hz, $I_{C(\text{opt})}$ should be about 25 μA.

Finally, figure 6.7(c) reproduces the measured noise figure as a function of frequency for several values of R_G and I_C. From this, it will be seen that at $I_C = 20 \mu\text{A}$ and $R_G = 10 \text{ k}\Omega$, the noise figure is about 3 dB up when the frequency has fallen to 100 Hz.

An alternative way of expressing noise generators, the noise resistance formulation, results in simpler expressions for noise factor and is also useful in demonstrating the variations in noise factor to be expected under nonoptimum working conditions. An account of this formulation follows.

6.7 THE NOISE RESISTANCE FORMULATION

Instead of representing the two spot-noise generators as a voltage and a current generator, respectively, it is equally valid to represent them as resistances that generate thermal noise.

Equation (6.3) states that a resistance R generates a mean square thermal noise voltage of

$$\overline{v_n^2} = 4kTR\Delta f$$

which implies a spot noise of

$$\frac{\overline{v_n^2}}{\Delta f} = 4kTR \, .$$

Thus a mean square noise generator could be represented as a resistor R, where,

$$R = \frac{\overline{v_n^2}}{\Delta f} \frac{1}{4kT} \, .$$

Now $\bar{e}_n^2$ is a mean square spot-noise voltage, so it may be represented by a *voltage noise resistance*, R_{nv}, where,

$$R_{nv} = \frac{\overline{e}_n^2}{4kT} . \tag{6.31}$$

Also, $\overline{i}_n$ being a current generator, it may be represented by a voltage generator divided by a resistance:

$$\overline{i}_n = \frac{\overline{e}_{ni}}{R_{ni}} = \frac{\sqrt{4kTR_{ni}}}{R_{ni}}$$

or

$$R_{ni} = \frac{4kT}{\overline{i}_n^2} . \tag{6.32}$$

The spot-noise generators for a transistor, in their square form, are now represented by the noise resistances R_{nv} and R_{ni}, and formulations (6.31) and (6.32) are such that it is easy to substitute these noise resistances into the equation for the noise factor previously derived. For example, equation (6.18) immediately becomes:

$$\mathrm{NF} = 1 + \frac{R_{nv}}{R_G} + \frac{R_G}{R_{ni}} . \tag{6.33}$$

This very simple expression clearly shows that a low noise factor will result if $R_{ni} \gg R_{nv}$, and this observation will be considered numerically later. The equation also leads to an expression for the optimum value of R_G which is given by:

$$\frac{d(\mathrm{NF})}{dR_G} = - \frac{R_{nv}}{R_{G(opt)}^2} + \frac{1}{R_{ni}} = 0$$

or

$$R_{G(opt)} = \sqrt{R_{nv}R_{ni}} . \tag{6.34}$$

Thus $R_{G(opt)}$ is the geometric mean of the two noise resistances, and it may be inserted back into equation (6.33) to give the minimum noise factor:

$$\mathrm{NF}_{min} = 1 + 2\left(\frac{R_{nv}}{R_{ni}}\right)^{1/2} . \tag{6.35}$$

Expressions (6.12), which give $\overline{e}_n$ and $\overline{i}_n$ for the bipolar transistor, may easily be converted into the noise resistance form by using formulae (6.31) and (6.32):

$$R_{nv} = \frac{\overline{e}_n^2}{4kT} = \left(r_{bb'} + \frac{1}{2g_m}\right) \tag{6.36a}$$

or

$$\left(r_{bb'} + \frac{kT}{2qI_C}\right) \tag{6.36b}$$

and

$$R_{ni} = \frac{4kT}{\overline{i}_n^2} = \frac{2h_{FE}}{g_m} \tag{6.37a}$$

or

$$\frac{2h_{FE}kT}{qI_C} . \tag{6.37b}$$

If R_{ni} is modified to include the crude approximation for the effect of flicker noise, it becomes:

$$R_{ni} = \frac{2h_{FE}kT}{qI_C[(1 + (f_F/f)]} . \tag{6.37c}$$

These expressions could be inserted into the noise factor equation (6.33) and would result in equation (6.22). This is nothing more than an alternative formulation of the same procedure and so will not be repeated, for it will again result in expressions for $I_{C(opt)}$ already derived (equations (6.23) through (6.26)).

6.7.1 Numerical Considerations

As has been pointed out in relation to the expression (6.33), a high R_{ni}/R_{nv} ratio will result in a low noise factor. If expression (6.34) for $R_{G(opt)}$ is combined with expression (6.33), plots of noise factor versus the ratio $R_G/R_{G(opt)}$ can be made for different R_{ni}/R_{nv} ratios. This has been done in figure 6.8, from which another important conclusion can be drawn. Not only

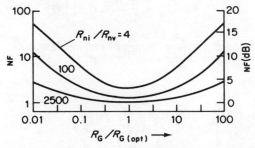

Figure 6.8. Effect of the R_{ni}/R_{nv} ratio on noise factor.

is the noise factor reduced as R_{ni}/R_{nv} increases, but it also remains low over a wider range of R_G values; that is, if R_{ni}/R_{nv} is high, then departures from the optimum value of input generator resistance are less important than if R_{ni}/R_{nv} were low.

The conclusions to be drawn are that: (*a*) for low-noise working over a wide range of R_G values, a transistor having a high R_{ni}/R_{nv} ratio should be chosen; and (*b*) if R_G is high enough, a FET should be used because inherently it has a very high R_{ni}/R_{nv} ratio, as will be shown in chapter 11.

If $\bar{e}_n$ and $\bar{i}_n$ are known either from manufacturers' data or by measurement using a noise test set, then R_{nv} and R_{ni} may be calculated using the previously established numerical values:

$$R_{nv} = \frac{\bar{e}_n^2}{4kT} = \frac{\bar{e}_n^2}{1.65 \times 10^{-20}}\ \Omega \quad \text{at } 25°C$$

and if $\bar{e}_n$ is in units of $nV/\sqrt{Hz}$, this becomes

$$R_{nv} \simeq \frac{\bar{e}_n^2}{16.5}\ k\Omega \quad \text{at } 25°C. \tag{6.38}$$

Similarly,

$$R_{ni} = \frac{4kT}{\bar{i}_n^2} = \frac{16.5}{\bar{i}_n^2}\ k\Omega \quad \text{at } 25°C \tag{6.39}$$

if $\bar{i}_n$ is in units of $pA/\sqrt{Hz}$.

Example 4
For the BFY77 transistor working under the conditions of Example 1, find the nearest relevant plot in figure 6.8.

Solution 4
Using the results of Solution 2,

$$R_{nv} = \frac{\bar{e}_n^2}{16.5} = \frac{6.7}{16.5} \simeq 0.41\ k\Omega$$

$$R_{ni} = \frac{16.5}{\bar{i}_n^2} = \frac{16.5}{0.048} \simeq 344\ k\Omega.$$

Hence,

$$\frac{R_{ni}}{R_{nv}} \simeq \frac{344}{0.41} = 840$$

which will be represented by a plot part-way between the lower two curves of figure 6.8; that is, the noise figure will remain sensibly constant over about a decade of variation in R_G.

6.8   FEEDBACK STAGES

The application of feedback will reduce the gain of an amplifier by the same amount as it reduces the internally generated noise and distortion. Consequently, the signal-to-noise ratio at the output *cannot* be improved by applying feedback. In fact, it may become worse, because (*a*) feedback resistors will contribute thermal noise to the system and (*b*) the bandwidth being increased by feedback, the wide-band noise will also be increased.

The noise contribution of feedback resistors can most easily be assessed by combining their values with the noise resistances, R_{nv} and R_{ni}. An equivalent circuit containing R_{nv} and R_{ni} is given in figure 6.9(*a*).

If series feedback is under consideration (as in a degenerate CE stage) the equivalent circuit will become that of figure 6.9(*b*). The feedback resistor appears in the input loop, so that it is in series with R_{nv}, and equation (6.33) for the noise factor becomes:

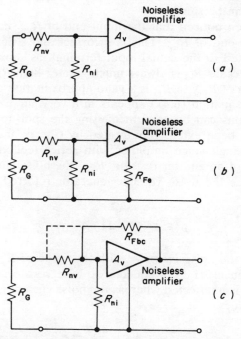

Figure 6.9. (*a*) The noise resistance equivalent circuit; (*b*) series feedback; (*c*) shunt feedback.

$$\mathrm{NF} = 1 + \frac{R_{\mathrm{nv}} + R_{\mathrm{Fe}}}{R_{\mathrm{G}}} + \frac{R_{\mathrm{G}}}{R_{\mathrm{ni}}}$$

which is usually written

$$\mathrm{NF} = 1 + \frac{R_{\mathrm{nv}}}{R_{\mathrm{G}}} + \frac{R_{\mathrm{G}}}{R_{\mathrm{ni}}} + \frac{R_{\mathrm{Fe}}}{R_{\mathrm{G}}} . \tag{6.40}$$

This expression shows that the noise factor is increased by the inclusion of a series feedback resistor, but if its value is much less than R_{G} this increase can be small. If the *actual* value of the noise is to be determined, however, the increased bandwidth *must* be included in expressions involving Δf.

If a shunt feedback resistor is present, as in figure 6.9(c), then R_{Fbc} and R_{ni} appear in parallel. Again, using equation (6.33), the noise factor becomes:

$$\mathrm{NF} = 1 + \frac{R_{\mathrm{nv}}}{R_{\mathrm{G}}} + \frac{R_{\mathrm{G}}(R_{\mathrm{Fbc}} + R_{\mathrm{ni}})}{R_{\mathrm{Fbc}} R_{\mathrm{ni}}}$$

$$= 1 + \frac{R_{\mathrm{nv}}}{R_{\mathrm{G}}} + \frac{R_{\mathrm{G}}}{R_{\mathrm{ni}}} + \frac{R_{\mathrm{G}}}{R_{\mathrm{Fbc}}} . \tag{6.41}$$

Here again the noise factor is increased but if R_{Fbc} is large compared with R_{G} the increase can be small.

It will have been noticed that the input end of R_{Fbc} has been shown to join the amplifier end of R_{nv}. This is incorrect and strictly speaking R_{Fbc} should be connected to the actual input terminal, as shown by the broken line. However, because R_{ni} is always much greater than R_{nv} (as in Example 4), the connection of R_{Fbc} to R_{ni} is a valid approximation and is justified on the grounds that equation (6.41) appears in a very simple form.

These conclusions can be confirmed using the spot-noise generator approach which has been extensively treated by Cherry and Hooper[4], who show that for a single-stage amplifier with series feedback (i.e., the unbypassed portion of an emitter resistor, R_{Fe}) the noise current generator is unaffected whereas the noise voltage generator referred to the input becomes:

$$\overline{v_{\mathrm{n}}^2} \simeq \overline{e_{\mathrm{n}}^2} \Delta f + 4kTR_{\mathrm{Fe}} \Delta f \tag{6.42}$$

which shows that the noise *increases* with the value of R_{Fe}.

For shunt feedback (i.e., a collector-to-base resistor, R_{Fbc}), the noise voltage generator is unaffected whereas the noise current generator referred to the input becomes:

$$\overline{i_{\mathrm{n}}^2} \simeq \overline{i_{\mathrm{n}}^2} \Delta f + \frac{4kT}{R_{\mathrm{Fbc}}} \Delta f \tag{6.43}$$

which shows that noise *decreases* as the value of R_{Fbc} rises.

Example 5

A BFW20 was connected into the circuit of figure 6.10, which is a shunt stabilized amplifier stage designed to operate the transistor at $I_C \simeq -40 \; \mu A$. Values of $\bar{e}_n$ and $\bar{i}_n$ were obtained at $R_G = 10 \; k\Omega$ using a commercial noise analyser and were found to compare well with the manufacturer's typical values taken from the graph of figure 6.11(a). At $f = 1 \; kHz$, these values were $\bar{e}_n = 2.83 \; nV/\sqrt{Hz}$ and $\bar{i}_n \simeq 0.79 \; pA/\sqrt{Hz}$. Equations (6.38) and (6.39) then give the noise resistance as

$$R_{nv} \simeq \frac{\bar{e}_n^2}{16.5} = \frac{2.83^2}{16.5} = 0.49 \; k\Omega \quad \text{and} \quad R_{ni} \simeq \frac{16.5}{\bar{i}_n^2} = \frac{16.5}{0.79^2} = 26.44 \; k\Omega$$

so that, using equation (6.41), the noise figure is

$$NF_{(1kHz)} \simeq 1 + \frac{0.49}{10} + \frac{10}{26.44} + \frac{10}{12 \times 10^3} = 1.43 \text{ or } 1.55 \; dB \;.$$

It will be seen from figure 6.11(b) that this value of noise factor is in

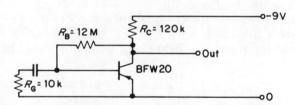

Figure 6.10. Circuit for sample noise figure calculation.

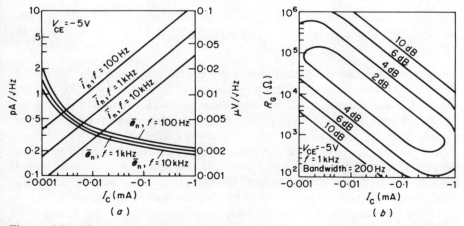

Figure 6.11. Noise characteristics for the BFW20 pnp planar diffused low-level transistor: (a) $\bar{e}_n$ and $\bar{i}_n$ as functions of I_C; (b) noise figure contours. (Reproduced by permission of SGS THOMSON Microelectronics Ltd.)

agreement with that predicted by the manufacturer for the same conditions of R_G and I_C.

The ratio of the noise resistances is $R_{ni}/R_{nv} = 26.44/0.49 = 54$, and if the graph of figure 6.8 is consulted it will be seen that this ratio relates to a curve somewhat above the $R_{ni}/R_{nv} = 100$ curve shown, so indicating that the noise figure may be expected to remain low only over about one decade of source resistance.

6.9 "PINK" NOISE

If the approximation is made that the mean square flicker noise follows a true $1/f$ law, this is called "pink" noise, and it enables calculations to be made of the mean square noise referred to the input within the flicker noise region. For modern low-noise planar bipolar transistors, flicker noise becomes important only below about 1 kHz. (For low-noise field-effect transistors flicker noise onset can be well below this, as will be described in chapter 11.)

As frequency falls, flicker noise in the field-effect transistor may be represented by a rise in the spot-noise *voltage* generator, whereas for the bipolar transistor it may be represented by a rise in the spot-noise *current* generator. Over a frequency band $(f_2 - f_1)$, the contribution of the flicker noise referred to the input of a field-effect transistor would be:

$$\overline{e}^2_{n(f_1)} f_1 \int_{f_1}^{f_2} \frac{1}{f} \, df = \overline{e}^2_{n(f_1)} f_1 \ln\left(\frac{f_2}{f_1}\right). \tag{6.44}$$

For a bipolar transistor, the mean square value of the pink noise contribution at the input would be:

$$\overline{i}^2_{n(f_1)} R_G^2 f_1 \int_{f_1}^{f_2} \frac{1}{f} \, df = \overline{i}^2_{n(f_1)} R_G^2 f_1 \ln\left(\frac{f_2}{f_1}\right). \tag{6.45}$$

These equations show that the pink noise contribution over any bandwidth defined by a frequency ratio (such as an octave or a decade) is the same as the pink noise contribution over any other bandwidth defined by the same frequency ratio. For example, the mean square pink noise voltage over the decade 1–10 Hz must be the same as that over 0.1–1 Hz or 10–100 Hz.

This observation makes it possible to calculate the pink noise contribution over any bandwidth if the value of the relevant spot-noise generator is known. Using mean square values, this may then be added to the noise contributions outside the $1/f$ region to give the wide-band noise.

Often, manufacturers give spot-noise generator values at spot frequencies inside and outside the $1/f$ region, particular in the case of field effect transistors. However, before embarking on a calculation, the designer must decide how low the frequency may become before flicker noise may be considered to be drift[9]. This problem is alleviated if the amplifier in question has a low-frequency roll-off, but if a DC amplifier is involved, the question devolves on how the load (or measuring instrument) will react to very low frequencies. A typical "transition frequency" between drift and noise could reasonably be taken as lying below about 0.1 Hz.

Examples relating to $1/f$ noise can be found in chapter 11.

6.10 "BURST" NOISE

This form of noise occurs in all semiconductor devices, and consists of sudden changes in the average level of the already existing noise. If an oscilloscope displays a typical noise trace consisting of a mêlée of random spikes within a fairly well-defined amplitude band, burst noise appears as a step function shift, up or down, of this band of spikes. If amplified and fed to a loudspeaker the result is a typical noise hiss plus crackling, hence the alternative term "popcorn noise!"

Burst noise is particularly common in monolithic integrated circuit amplifiers, and for a "poor" amplifier in this context, noise bursts of more than 50 μV (referred to the input) and lasting for periods of over 1 ms can be observed. Such bursts can occur at rates of several hundred per second, but burst rates are unpredictable. However, semiconductor devices and integrated circuits can be selected—and more recently, fabricated—to have minimal burst noise properties.

The origin of burst noise is still obscure, but is thought to be concerned with the modulation of current flow by the occupancy state of certain recombination centres[10].

6.11 CONCLUDING COMMENTS

The foregoing examples have shown that for good low-noise operation outside the $1/f$ region it is necessary to run a transistor at a collector current of a few tens of microamps. If frequencies below about 1 kHz are involved, where $1/f$ noise becomes important, then the optimum collector current may turn out to be only a few microamps, so that care must be taken to choose a transistor which will retain a high value of h_{FE} (>100) at such low levels. Also, because the noise figure is a weak function of V_{CE}, it is good practice to run the transistor at a low voltage.

Unfortunately, low noise and high input impedance are acquired at the expense of bandwidth, as is clearly demonstrated by equation (3.22): $C_{b'e} \simeq g_m/2\pi F_T$. That is,

$$F_T \propto g_m \propto I_C \quad (V_{BE} \text{ constant}) . \tag{6.46}$$

This proportionality is only approximate, but it does highlight the compromise which must be made between collector current and bandwidth when designing for low-noise performance.

In the $1/f$ noise region, field-effect transistors are particularly useful, and this point will be elucidated in chapter 11.

It will have been noticed that no examples of actual output noise calculations have been offered. This is because they are rarely useful: if the noise generated by the source resistance is determined (from the chart of figure 6.5(a)) and so is the minimum expected signal, this will give the worst-case signal-to-noise ratio. Thereafter, it is the noise factor which is of importance, because it is this which shows by how much the signal-to-noise ratio will be degraded by the amplifier. That is,

$$ \mathrm{s}/\mathrm{N}_{\mathrm{out}} = \frac{\mathrm{s}/\mathrm{N}_{\mathrm{in}}}{\mathrm{NF}} \qquad (6.47a) $$

which is actually a definition of the noise factor. Logarithmically,

$$ (\mathrm{s}/\mathrm{N}_{\mathrm{out}})\,\mathrm{dB} = (\mathrm{s}/\mathrm{N}_{\mathrm{in}})\,\mathrm{dB} - (\mathrm{NF})\,\mathrm{dB} . \qquad (6.47b) $$

When noise generator values are determined and used in noise calculations, the implication is that succeeding stages do not contribute significantly to the overall noise performance. This is true only if a proper design has been made from the noise point of view, and this is by no means common. For example, Darlington pairs are particularly poor[5] in this respect.

Long-tailed pairs with the output voltage taken as being between the collectors, exhibit about twice the mean square noise typical of a single CE stage working under the same conditions. This is reasonable, because each of the two transistors is contributing its own noise which is uncorrelated with that from the other. However, if the two transistors are not perfectly matched, some common mode noise (such as that arising in a long-tail mirror transistor) will be converted to difference noise and the mean square value of this will be added to the two existing mean square noise contributions. This connection normally forms the input stage of operational amplifiers in which succeeding stages may also contribute noise. Therefore, complete microcircuit operational amplifiers are characterized as single devices by the manufacturer, and this approach is introduced in the next chapter. The noise performance of the transconductance multiplier (and the Gilbert cell in particular) has also been investigated with some unexpected results, such that the two-quadrant multiplier is actually noisier than the four-quadrant[11].

Finally, the problem of effective noise matching must be considered. If the input devices can be chosen so that $R_{\mathrm{G(opt)}}$ is close to the existing R_G, this is the best solution. If not, then for AC signals, an input transformer having a ratio which will transform R_G to $R_{\mathrm{G(opt)}}$ may be used. For DC signals, the problem is more severe, but if it devolves on R_G being too low, then a number of input transistors may be operated in parallel[1]. The effect of this technique can be seen immediately by redrawing the noise resistance equivalent circuit of figure 6.9(a) to represent a number n of transistors, all connected in parallel (i.e., with their bases taken to a common input point, and the collectors taken to a common load resistor). This noise equivalent circuit is shown in figure 6.12.

The effective voltage noise resistance is now $R_{nv(eff)}$, where

$$\frac{1}{R_{nv(eff)}} = \frac{1}{R_{nv(1)}} + \frac{1}{R_{nv(2)}} + \frac{1}{R_{nv(3)}} + \cdots .$$

If all the transistors are identical, then

$$R_{nv(eff)} = \frac{R_{nv}}{n} . \tag{6.48}$$

Similarly,

$$\frac{1}{R_{ni(eff)}} = \frac{1}{R_{ni(1)}} + \frac{1}{R_{ni(2)}} + \frac{1}{R_{ni(3)}} + \cdots .$$

Again, if all the transistors are identical, then

$$R_{ni(eff)} = \frac{R_{ni}}{n} . \tag{6.49}$$

Equation (6.33) to obtain the noise factor now becomes:

$$NF = 1 + \frac{R_{nv(eff)}}{R_G} + \frac{R_G}{R_{ni(eff)}}$$

which for identical transistors is

$$NF = 1 + \frac{R_{nv}}{nR_G} + \frac{nR_G}{R_{ni}} . \tag{6.50}$$

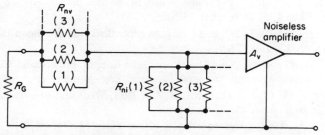

Figure 6.12. Noise resistance equivalent circuit for paralleled transistors.

The optimum value of R_G for noise matching is again given by equating the first differential to zero:

$$\frac{d(\text{NF})}{dR_G} = -\frac{R_{nv}}{nR_{G(\text{opt})}^2} + \frac{n}{R_{ni}} = 0$$

whence,

$$R_{G(\text{opt})} = \frac{(R_{nv}R_{ni})^{1/2}}{n}. \tag{6.51}$$

This expression implies that the optimum source resistance is reduced by a factor equal to the number of identical transistors connected in parallel.

If it were possible to use the same load resistor, R_L, for this arrangement as for a single transistor, then the voltage gain would obviously be n times that for the single transistor. However, the bandwidth would be proportionally reduced because of the increased collector–base capacitance. Also, the input resistance to the arrangement would be R_{in}/n where R_{in} is the input resistance of a single transistor.

The alternative case, where R_G is high, calls for the use of field-effect transistors, and the relevant techniques are introduced in chapter 11.

REFERENCES

1. Faulkner, E. A., 1968, The design of low-noise audio-frequency amplifiers, *Radio Electron. Eng.* **36**, 1, July.
2. Baxendall, P. J., 1968, Noise in transistor circuits, *Wireless World*, **74**, November and December.
4. Letzter, S. and Webster, N., 1970, Noise in amplifiers, *IEEE Spectrum*, **7**, 8, August.
4. Cherry, E. M. and Hooper, D. E., 1968, *Amplifier Devices and Low-Pass Amplifier Design* (New York: Wiley) chap. 8.
5. Motchenbacher, C. D. and Fitchen, F. C., 1973, *Low-Noise Electronic Design* (New York: Wiley).
6. Knott, K. F., 1968, $1/f$ voltage noise in silicon planar bipolar transistors, *Electron. Lett.* **4**, 5, December 13th.
7. Leuenbrger, F., 1968, $1/f$ noise in gate-controlled planar silicon diodes, *Electron. Lett.* **4**, 13, June 28th.
8. van Nie, A. G., 1968, Noise level and zero-drift of broad-band electrometers used for measuring small currents, *Electron. Eng.* September.
9. Bloodworth, G. G. and Hawkins, R. G., 1971, Drift and low-frequency noise, *Radio Electron. Eng.* **41**, 2, 61–64.
10. Ambrozy, A., 1982, Electronic Noise (New York: McGraw-Hill).
11. Bahnas, Y. Z., Bloodworth, G. G. and Brunnschweiler, A., 1977, The noise

properties of the linearized transconductance multiplier, *IEEE J. Solid St. Circuits*, SC-12, **5**, 580–584.

QUESTIONS

1. When calculating thermal noise or shot noise, the noise bandwidth must always be included. Why?

2. Why do the hybrid-π resistances $r_{b'e}$, $r_{b'c}$ and r_{ce} generate no thermal noise?

3. The actual resistance presented to the input of an amplifier stage R_G is less than the optimum value $R_{G(opt)}$ for minimum noise factor. Why is it fallacious to include a series resistor at the input to increase R_G up to the value of $R_{G(opt)}$?

4. What parameter in a transistor input stage may be adjusted to optimize the noise factor; and what disadvantage might accrue from this?

5. A transistor is found to exhibit an h_{FE} of 15 at $I_C = 10 \ \mu A$. Could this transistor be specified as a low-noise input device? Justify your answer.

6. Why is it possible to represent a noise generator by a simple resistor?

7. Idealized flicker noise is called "pink noise," and it rises at 3 dB/octave as the frequency falls. How may this be taken into account in a bipolar transistor noise formulation?

8. "Pink noise" has the same mean square value over any frequency interval with the same ratio of f_1/f_2. Why?

PROBLEMS

6.1. An amplifier exhibits 6 dB/octave roll-offs at both high and low frequencies, the -3 dB frequencies being 52 kHz and 1 kHz, respectively. Over this bandwidth, the RMS voltage and current noise generators referred to the input are 1.2 μV and 150 pA, respectively. If the internal resistance of a signal source applied to this amplifier is 40 kΩ, calculate the resulting noise factor.

   Is the amplifier well matched to the source? If not, suggest an alternative form of amplifier.

   Take $4 \, kT = 1.65 \times 10^{-20}$ J at 25°C.

6.2. A bipolar transistor exhibits a spot-noise voltage of 5 nV/$\sqrt{Hz}$ and a spot-noise current of 1 pA/$\sqrt{Hz}$, measured at 1 kHz. If it is used with a signal source of $R_G = 10$ kΩ, what would be the noise figure in dB?

What would be the value of the best possible noise figure, and at what value of R_G would this occur?

Would a junction FET having a similar spot-noise voltage and an R_G of 10 kΩ degrade the signal-to-noise ratio more or less than the bipolar transistor, and by how much? (See chapter 11.)

Assume that $4\,kT = 1.65 \times 10^{-20}$ J.

6.3. A transistor is connected into a CE amplifier stage. Its spot-noise voltage and current generator values are 20 nV/√Hz and 0.4 pA/√Hz, respectively, measured at 1 kHz. Using the noise resistance method, determine the spot-noise factor at 1 kHz at 25°C, assuming a net resistance at the input of 10 kΩ.

If a 20 kΩ shunt feedback resistor is connected from collector to base (with an appropriate blocking capacitor) by what percentage will the signal-to-noise ratio at the output decrease?

6.4. Why is the input generator resistance for optimum noise factor likely to be higher for a FET then for a bipolar transistor? A bipolar transistor is connected as a simple CE stage with a 3.2 MΩ shunt biasing resistor from collector to base. If the resistance of the input generator is 10 kΩ, what is the spot-noise factor for the stage at 10 kHz? Assume that $\bar{e}_n = 2$ nV/√Hz and $\bar{i}_n = 0.4$ pA/√Hz, both at 10 kHz. If the actual noise voltage at the output is measured using a true RMS meter which has a 100 Hz bandwidth centred at 10 kHz and 6 dB/octave roll-offs, what is the value of that noise voltage referred to the input?

6.5. The spot-noise generators of an amplifier input stage working at a given frequency are measured as 2.5 nV/√Hz and 0.25 pA/√Hz, respectively. Use the noise resistance method to calculate the noise figure in dB for that amplifier stage if the source resistance is 15 kΩ. (Neglect flicker and HF noise and assume that the temperature is 25°C.)

6.6. A bipolar transistor amplifier stage exhibits a spot-noise current generator value of 2 pA/√Hz at 100 Hz and a spot noise voltage generator value of 10 nV/√Hz. If a signal source with a 5 kΩ internal resistance is applied, calculate the contributions of the two spot-noise generators to the RMS noise referred to the input as measured over a single-lag, single-lead bandwidth of 100–500 Hz.

What is the total RMS contribution of the two noise generators, and how does this figure compare with the Johnson noise produced by R_G?

Is the stage well-noise-matched?

6.7. An inverting operational amplifier has a shunt feedback resistor of 110 kΩ and its input is derived from a signal generator of internal resistance 5.6 kΩ. If it has a spot-noise voltage generator value of 10 nV/√Hz and a spot-noise current generator value of 2 pA/√Hz, use the noise resistance method to calculate its noise factor.

Calculate also the minimum noise factor and determine whether the optimum input generator resistance so defined is close to the actual generator resistance of 5.6 kΩ.

6.8. An operational amplifier is connected in the inverting configuration with an input resistor $R_G = 470\,\Omega$ and a shunt feedback resistor $R_F = 9.4\,k\Omega$. The relevant data sheet gives an RMS noise voltage generator value of 3 μV and an RMS noise current generator value of 0.1 nA, both measured over a -3 dB bandwidth of 1–10 kHz.

If the input to R_G were short-circuited to ground, what noise voltage should be measured at the output by a true RMS meter with a -3 dB bandwidth of 1–10 kHz?

Assume the temperature to be 25°C at which the value of $4kT = 1.65 \times 10^{-20}$ J. Also assume that the relevant gain/frequency roll-offs are at 6 dB/octave.

6.9. An emitter-follower has a load resistor $R_F = 1800\,\Omega$ and biasing resistors $R_1 = 93\,k\Omega$ and $R_2 = 27\,k\Omega$, respectively. If a signal source having a signal-to-noise ratio of 80 dB is applied to the input, calculate the signal-to-noise ratio at the output.

Assume that the transistor used has a spot-noise voltage generator value of 5 nV/$\sqrt{Hz}$ and a spot-noise current generator value of 2 pA/$\sqrt{Hz}$. Disregard flicker and high-frequency noise.

7

Operational Amplifiers and Other Microcircuits

The early part of the decade 1960–1970 saw the gradual transition from discrete switching transistors to microcircuits in computers and other digital equipment. It was not until late in that decade, however, that linear microcircuits began to replace amplifiers and analog function modules comprised of discrete devices. This transition is virtually complete at the time of writing, and the practising circuit designer is currently incorporating linear microcircuits into his designs in place of the quantities of discrete components which would once have been required to perform the same functions.

Because the present volume is not concerned with digital electronics, this chapter will concentrate on the characteristics and applications of linear microcircuits with special reference to the operational amplifier, but will include a brief treatment of the design concepts behind the microcircuits themselves. This is because the majority of readers will be *users* of microcircuits rather than designers of them, but even so an insight into the very different approach dictated by integrated circuit, as opposed to discrete component, design will be found helpful in understanding the characteristics and limitations of current modules.

At the time of writing, two branches of microcircuit technology have been successful in producing linear amplifiers. One has resulted in the true *monolithic amplifier*, in which all the components—transistors and resistors—are formed by an epitaxial planar process on a single-crystal silicon chip, which is then mounted in a can (often TO-5 size), or in a flat-pack or dual in-line encapsulation. Many monolithic amplifiers also include stabilizing capacitors, though some still call for a simple external stabilizing circuit. Where several chips are used—for instance, when a monolithic amplifier chip is preceded by matched input field-effect transistors—the result is a *multi-chip microcircuit*.

Sometimes the active components are mounted on a substrate carrying thick- or thin-film passive components, and this is an example of the true *hybrid technique*. Whereas multi-chip microcircuits are encapsulated in the same way as single monolithic microcircuits, hybrids are usually sealed into small plastic boxes with an epoxy potting compound.

The input stage of a linear monolithic amplifier almost invariably consists of a difference amplifier or long-tailed pair. This configuration was treated in chapter 5 for bipolar transistors, and it will be readily appreciated that the requisite high degree of balance and symmetry is achieved naturally when both halves are fabricated concurrently by the same diffusion process. Further, because of the close proximity of the halves, temperature differences will be extremely small, making for excellent tracking over a wide range of ambient temperatures. A few monolithic amplifiers enhance this already good drift performance by incorporating fairly large transistor structures which heat the complete chip, and hold the temperature within a small range.

The hybrid amplifier usually incorporates a long-tailed pair input stage, either bipolar or FET, but is also available in single-ended form with a chopper-type input. This version can exhibit an extremely good voltage drift performance with respect to both temperature and time. Another version, the parametric amplifier, involves a cyclically varying capacitance (or varactor) "pumped" by an auxiliary generator. This module can exhibit a phenomenally low input bias current, which makes it eminently satisfactory as an electrometer amplifier.

7.1 SOME MONOLITHIC CIRCUIT DESIGN TECHNIQUES

The great advantage of monolithic methods is that close matching of the structures on the chip is possible, and it is on this premise that monolithic design techniques are based. Also, as Camenzind and Grebene[1] point out, although absolute values of resistance are not very easy to maintain in resistor-type structures, close matching of resistors—including ratios, one to the other—is not problematical.

Monolithic circuits are fabricated using extensions of the planar diffusion method mentioned with reference to figure 2.13; that is, successive diffusions of n- and p-type impurities are made through holes in layers of SiO_2 which have been grown sequentially over a single-crystal silicon slice, until the required structure is complete. It is not appropriate to describe such processes in detail in the present volume: excellent introductions to this complex subject will be found in other books[2,3]. For the present, it will suffice to note that figure 7.1 is an idealized representation (not to scale) of a cross-section through part of a silicon chip. A normal structure is shown on the left, along with a *lateral* pnp structure. The former has its base

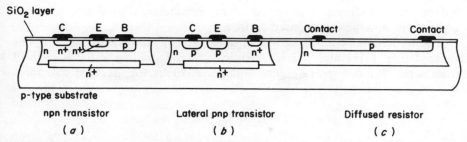

Figure 7.1. Idealized cross-sectional representation of part of a monolithic integrated circuit (*not* to scale).

sandwiched *between* the collector and emitter; whereas the latter has both emitter and collector diffused into the n-type base, so that the current flow is lateral to the chip. This is done in order to minimize and simplify the diffusion sequences, but it does result in the pnp structure having a poorer current gain and gain–bandwidth product.

In both cases, the major part of the current flow from emitter to collector is able to pass through a low-resistivity n-type region called the *buried layer*. This is actually the first diffusion in the fabrication process and it is usually achieved by doping with antimony or arsenic via windows in the initially-formed oxide layer. This oxide is then removed and an n-type epitaxial layer is grown over the entire surface of the silicon, including the buried layer. It is the thickness and resistivity of this layer which largely defines the eventual breakdown voltage of the transistor. The various structures are then isolated by boron diffusions through to the p-type substrate and these structures are completed by diffusing in the base and emitter regions. (Note the current paths in the base diffusions; these contribute the resistances $r_{bb'}$.) The final processes are the growth of an oxide passivating layer and the deposition of a metallization pattern for the contacts.

Resistors, as opposed to active structures, may be fabricated by several methods[2,3] including the comparatively simple diffusion of a p-type "stripe" into the epitaxial layer as shown in figure 7.1(c). It is not easy to control the absolute values of such resistors but alternative and better techniques are complex and require extra processing steps. However, good matching of diffused resistors is possible, which is of course desirable in integrated circuit design. Note that this form of resistor, like the active structures, involves a junction with the substrate, which implies not only that this junction must always be reverse-biased, but that the introduction of stray capacitance is inevitable.

In chapter 5 it was shown how monolithic analog integrated circuits relied on the good matching of both active and passive structures, and the current mirror was introduced. This circuit is very fundamental to analog IC design and so will now be investigated further.

The basic connection is shown in figure 7.2(a), and in chapter 5 it was claimed that if Tr1 and Tr2 are identical, then $I_{C2} \simeq I$, where I is derived from some well-defined reference voltage. Actually for identical structures $I_{C2} = I_{C1}$, so that

$$I = I_{C1} + I_{B1} + I_{B2}$$

$$= I_{C1} + \frac{I_{C1}}{h_{FE(1)}} \dotplus \frac{I_{C2}}{h_{FE(2)}}$$

$$= I_{C1}[1 + (2/h_{FE})]$$

or

$$I_{C2} = I_{C1} = I/[1 + (2/h_{FE})] . \qquad (7.1a)$$

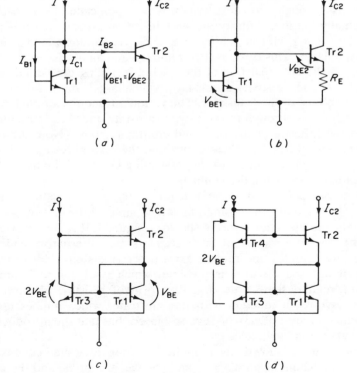

Figure 7.2. Various current mirror configurations.

If h_{FE} is large, obviously I_{C2} is very close to I, but in the case of transistor structures having low gain, such as lateral pnp types, the disparity might become significant.

In current mirrors the output resistance of the mirror transistor r_{ce} is often of importance, as for example when it is used as the long-tail transistor for a difference stage as in figure 5.31. Here, it is assumed that R_X (r_{ce} for the mirror transistor in this case) is very large, otherwise equation (5.38) for the difference gain would be invalid. Furthermore, a high value of R_X is seen to contribute directly to a low common-mode gain and hence a high CMR (equation (5.39)).

Ideally, the mirror transistor should be a true current source having infinite internal resistance, but the reality of the situation is again demonstrated by figure 3.8 and equation (3.20b). Figure 2.9 clearly shows that I_C varies somewhat with V_{CE} (for a constant I_B), especially at the higher current levels. The amount by which it does so can be determined from the Early voltage and the geometry of figure 3.8. Suppose that a collector current $I_{C(1)}$ flows when the collector–emitter voltage is $V_{CE(1)}$. Then, if $V_{CE(2)}$ is applied (whilst I_B remains constant), a new collector current $I_{C(2)}$ will arise:

$$\frac{I_{C(1)}}{V_{CE(1)} - V_A} = \frac{I_{C(2)}}{V_{CE(2)} - V_A}$$

or

$$I_{C(2)} = I_{C(1)} \frac{V_{CE(2)} - V_A}{V_{CE(1)} - V_A}. \qquad (7.1b)$$

As an example, take the Early voltage calculated below equation (3.20b), which is about $-106\,\mathrm{V}$ for $I_C = 1\,\mathrm{mA}$. Then suppose that V_{CE} rises from 10 V to 30 V. The change in I_C is thus:

$$I_{C2} = 1 \left(\frac{30 + 106}{10 + 106} \right) \simeq 1.17\,\mathrm{mA}.$$

That is, I_C has increased by 17%.

Equations (7.1) demonstrate two ways in which the simple current mirror departs from the ideal concept of a dependent current source.

To define a current I_{C2} which is different from I, then the two transistor structures can be diffused with different emitter areas, when $I_{E1}/I_{E2} = A_{E1}/A_{E2}$. Conversely a resistance may be included in the emitter connection of Tr2, as shown in figure 7.2(b). Here, the V_{BE} drop in Tr1 has to be shared between V_{BE2} and R_E. The value of R_E necessary to define a required current, I_{C2}, may be obtained as follows:

$$I_{C2} \simeq I_{ES} \exp\left(q V_{BE2} / kT \right).$$

Also,

$$I \simeq I_{ES} \exp\left(qV_{BE1}/kT\right)$$

giving

$$I_{ES} \simeq \frac{I}{\exp\left(qV_{BE1}/kT\right)} \, .$$

Putting this into the expression for I_{C2} gives:

$$I_{C2} \simeq I \exp\left[q(V_{BE2} - V_{BE1})/kT\right].$$

But $(V_{BE2} - V_{BE1}) \simeq -I_{C2}R_E$, so that

$$I_{C2} \simeq \frac{I}{\exp\left(qI_{C2}R_E/kT\right)} \, .$$

Hence,

$$qI_{C2}R_E/kT \simeq \ln\left(I/I_{C2}\right)$$

whence

$$R_E \simeq (kT/qI_{C2}) \ln\left(I/I_{C2}\right) \tag{7.2}$$

where the value of I_{C2} is a design parameter.

In this circuit, the resistance presented at the collector of Tr2 will be greater than r_{ce} because R_E, being unbypassed, acts as an emitter degeneration resistor. Therefore, equation (5.29) becomes valid, and $R_{out} \simeq r_{ce}(1 + g_m R_E)$ in this case.

Note that if another resistor of the same value were included in the emitter circuit of the diode-connected transistor I_{C2} would take the same value as I. However, if two unequal resistors R_{E1} and R_{E2} were included in the two emitter circuits, then I_{C2} would become $I(R_{E1}/R_{E2})$.

A modification to the basic circuit which improves the output resistance and also makes I_{C2} much more closely equal to I, is the *Wilson current mirror*[4] shown in figure 7.2(c). If the transistors Tr1 and Tr3 are identical, then the basic current equality is:

$$I_{C3} = I_{C1}$$

or

$$I_{C2} + I_{B2} - I_{B3} - I_{B1} = I - I_{B2}$$

giving

$$I_{C2} = I + (I_{B1} + I_{B3}) - 2I_{B2} \qquad (7.3a)$$

which shows that the current matching is determined by any small differences in the base currents around the circuit. An alternative way of expressing this relationship for monolithic transistor structures where the current gains are assumed to be identical, is to rewrite equation (7.3a) as follows:

$$I_{C2} = I + \frac{I_{C1}}{h_{FE}} + \frac{I_{C3}}{h_{FE}} - \frac{2I_{C2}}{h_{FE}}.$$

Knowing that $I_{C1} = I_{C3}$,

$$I_{C2}\left(1 + \frac{2}{h_{FE}}\right) = I + \frac{2I_{C3}}{h_{FE}} = I + \frac{2[I - (I_{C2}/h_{FE})]}{h_{FE}}$$

so

$$I_{C2}\left(1 + \frac{2}{h_{FE}} + \frac{2}{h_{FE}^2}\right) = I\left(1 + \frac{2}{h_{FE}}\right)$$

giving

$$I_{C2} = I\left(1 + \frac{2}{2h_{FE} + h_{FE}^2}\right)^{-1} \qquad (7.3b)$$

which clearly shows how little I_{C2} differs from I.

The inclusion of Tr2 in the Wilson current mirror effectively buffers Tr1 against collector voltage variations in addition to converting the circuit into a feedback configuration, with the resulting close match between I_{C2} and I. Consequently, it is useful not only within monolithic networks, but is also available as a small subsystem in its own right (e.g., the Texas Instruments TL011–TL021 range, in which different current ratios are offered depending upon the type number chosen).

In figure 7.2(c) it will be noticed that V_{CE3} is held at $2V_{BE}$, whereas V_{CE1} is only V_{BE}. Another current mirror (described by another Wilson[5]) is shown in figure 7.2(d), and includes a fourth transistor which equalises V_{CE1} and V_{CE3} to a single V_{BE}, so establishing that they are working under truly identical conditions. The performance of this circuit, in both current ratio accuracy and output resistance, is even better than that of the basic Wilson mirror.

When a diode is required, as in the current mirror configuration, it is convenient to take a normal transistor structure which has been diffused along with all the other transistors in the integrated circuit, and simply connect it as a diode. In principle, the diode connection can take five different forms:

(1) base–emitter junction (collector connected to base);
(2) base–emitter junction (collector open-circuited);
(3) base–collector junction (emitter connected to base);
(4) base–collector junction (emitter open-circuited);
(5) collector and emitter connected together.

Each of these connections has its own advantages and disadvantages[2], notable amongst which are that the connections involving the base–emitter junction (1, 2 and 5) have reverse breakdown voltages of between 5 and 8 V, whereas the base–collector junction (3 and 4) is designed for breakdown usually over 30 V. However, if figure 7.1(*a*) is examined, it will be seen that the base–collector–substrate sequence forms a so-called parasitic pnp structure. This can lead to the diversion of current into the substrate. For this and other reasons, the base–emitter connections are more common, their low breakdown voltages being adequate for more low-level work. Also, such structures can be used as Zener diodes having a V_Z of about 6 V.

The current mirror can also be used as an active load for a difference amplifier as shown in figure 7.3. This technique not only increases the gain of the stage markedly, but also acts to convert the difference output to a single-ended output. These attributes come about as follows.

Firstly, the mirror transistor, Tr4, being biased by the diode-connected transistor Tr3, presents a large *dynamic* load resistance to Tr2, its value being nearly equal to r_{ce} for Tr4. This leads to a high voltage gain.

Secondly, the output signal current ΔI is given by

$$\Delta I = \Delta I_{C4} - \Delta I_{C2}$$

where $\Delta I_{C4} \simeq \Delta I_{C1}$ (because of current mirror action) and $\Delta I_{C2} \simeq -\Delta I_{C1}$

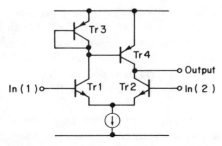

Figure 7.3. The current mirror as an active load.

(because Tr1, Tr2 is a difference pair). Hence, $\Delta I \simeq 2\Delta I_{C1}$ which shows that full advantage is taken of *both* collector current changes.

Unfortunately, this circuit is no longer symmetrical because the voltage across Tr3 is now smaller than across Tr4, leading to unequal collector voltages on Tr1 and Tr2. The resulting different amounts of base-width modulation can lead to an excessive base voltage difference (called the "offset" voltage). A technique for overcoming this, and other, problems is shown in figure 7.4(*a*). Here, two emitter-followers, Tr1 and Tr2, drive a *common-base* difference amplifier comprised of Tr3, Tr4 and a current source Because the two pnp transistors Tr3 and Tr4 are operated in the CB mode, their frequency response is at its best, so compensating for the comparatively poor performance of the lateral pnp structure.

The arrangement provides current gain in Tr1 and Tr2 and voltage gain in Tr3 and Tr4, which is large because of the current mirror active load Tr6.

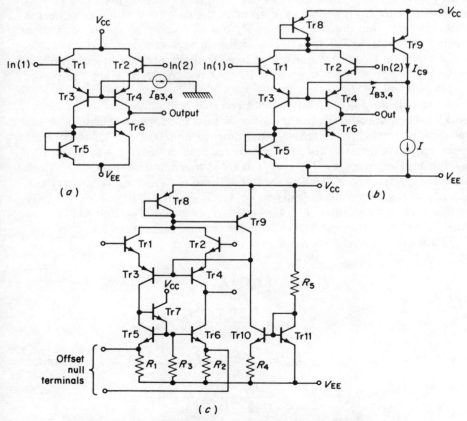

Figure 7.4. Development of a monolithic difference input stage.

Further, the difference input resistance, R_{in}, is now increased by the two $r_{\text{bb}'}$ and the two $r_{\text{b}'\text{e}}$ resistances of Tr3 and Tr4.

Figure 7.4(b) shows a method of biasing this composite stage. The collector currents of Tr1 and Tr2 pass through the diode-connected transistor Tr8, and the resulting voltage drop drives the mirror transistor Tr9. The resulting collector current I_{C9} plus the commoned-base current $I_{\text{B3,4}}$ comprises the constant current I. That is,

$$I_{\text{B3,4}} = I - I_{\text{C9}} . \tag{7.4}$$

If the currents through Tr1 and Tr2 rise because of a rise in temperature, the voltage drop across Tr8 also rises, leading to an increase in I_{C9}. Hence, from equation (7.4) above, it is seen that $I_{\text{B3,4}}$ must fall, since I is constant. This results in a reduction in the currents through Tr1 and Tr2, which is the desired result. The whole system thus constitutes a common-mode feedback network.

In practice, the current generator, I, is again a mirror transistor Tr10 as shown in figure 7.4(c), and its current is defined by the diode-connected transistor Tr11 and the two resistors R_4 and R_5 (according to the derivation leading to equation (7.3)). Tr10 is a high-gain npn structure, and will act as a good practical constant current source, though the current generated is dependent upon that in R_5, which in turn is dependent upon the power supply voltages.

Also in figure 7.4(c), it will be noticed that a modification to the active load has been introduced. In the earlier circuits the collector currents of Tr3 and Tr4 were slightly different, according to equation (7.1), which must introduce some offset. This difference is reduced in figure 7.4(c) by supplying the base currents of Tr5 and Tr6 from the new transistor Tr7. Since these base currents are small the resulting base current into Tr7 is very low indeed, so that the collector current of Tr3 is greater than that of Tr4 only by this very small amount, I_{B7}. However, R_3 is included so that the current through Tr7 is not so small that its h_{FE} becomes excessively low.

This form of current mirror is easily analysed as follows:

$$I_{\text{C3}} = I_{\text{C5}} + I_{\text{B7}} = I_{\text{C5}} + \frac{I_{\text{E7}}}{h_{\text{FC}}}$$

$$= I_{\text{C5}} + \frac{I_{\text{B5}} + I_{\text{B6}} + I_{\text{R3}}}{(h_{\text{FE}} + 1)}$$

Now $I_{\text{B5}} = I_{\text{C5}}/h_{\text{FE}}$ and $I_{\text{B6}} = I_{\text{C6}}/h_{\text{FE}}$ and also $I_{\text{C5}} = I_{\text{C6}}$ if $R_1 = R_2$, therefore

$$I_{\text{C3}} = I_{\text{C6}} + \frac{2I_{\text{C6}}}{h_{\text{FE}}(h_{\text{FE}} + 1)} + \frac{I_{\text{R3}}}{(h_{\text{FE}} + 1)}$$

giving

$$I_{C6} = \frac{I_{C3} - [I_{R3}/(h_{FE} + 1)]}{1 + [2/h_{FE}(h_{FE} + 1)]}$$

or

$$I_{C6} = I_{C3}\left(1 + \frac{2}{h_{FE} + h_{FE^2}}\right)^{-1} - I_{R3}\left(\frac{h_{FE}}{h_{FE^2} + h_{FE} + 2}\right)$$

The first part of this expression is little different from that for the Wilson current mirror (equation (7.3b)); and the second part reduces to I_{R3}/h_{FE} for high values of h_{FE}.

This form of current mirror also allows the inherent mismatch offset to be removed by slight adjustment of the relative values of the two quiescent currents in the input chain, and this is done by connecting a potentiometer across R_1 and R_2 with its slider coupled to the negative supply.

Tr7 acts as a (nearly unity-gain) emitter-follower which transfers the signal at the collector of Tr5 to the base of Tr6 so maintaining the difference-to-single-ended facility offered by the simpler circuit of figure 7.4(a).

The final circuit of figure 7.4(c) is that used as the first stage of the well-known 741 monolithic operational amplifier, and in modified form in many others. It can attain a voltage gain in excess of 1000, which makes possible the design of operational amplifiers having only two stages, but still exhibiting gains above 10^5. It does, however, require input base biasing currents to Tr1 and Tr2 of 0.1 μA or more. Further, the difference input resistance is rarely more than a few megohms. Several techniques are available to improve this aspect of the circuit's performance. For example, the input transistors Tr1 and Tr2, could be replaced by a Darlington pair (see chapter 5), which could take the difference input resistance up to more than 10 MΩ. However, the input offset voltage would be higher than for the single input transistor and other problems due to difficulties in matching would arise, such as marked differences in collector current changes with temperature.

An alternative solution is to use *super-gain* (or "punch-through") transistor[6] structures at the input. Such transistors have very thin base diffusions so that the transport factor, b (equation (2.1)), approaches unity very closely. This results in an extremely high value for h_{FE}—typically 2000–5000 at a collector current of a few microamps—but it also means that the CE breakdown voltage is only 2–3 V. Breakdown under these conditions occurs when the collector depletion layer "punches" through the base layer to the emitter (which accounts for the alternative name of the structure). Further, the value of r_{ce} is lower than for a normal structure because the base-width modulation is comparatively large.

However, because of the very high gain, the input bias currents are low—typically less than 1 nA (10^{-9} A) at room temperature—with a corresponding differential input resistance approaching 100 MΩ.

To prevent breakdown the voltage across a super-gain transistor must be kept low, and figure 7.5(a) shows how such a structure (denoted by the "open-base" symbol) can be operated from the V_{BE} of a lateral pnp structure. The composite structure then resembles a very high-gain npn transistor, but with the high breakdown voltage characteristic of the lateral pnp structure.

Figure 7.5(b) illustrates the basic input circuit of the LM 108 monolithic operational amplifier[7]. Here, it will be seen that the voltage from the emitter of each super-gain transistor to the base of its complement is held at $2 V_{BE}$ by Tr3 and Tr4. Hence, although the common-mode input voltage to Tr1 and Tr2 may be large, this voltage change is also applied to the bases of Tr5 and Tr6 so that the voltage *across* each combination Tr1/Tr5 and Tr2/Tr6 remains at $2 V_{BE}$. Further, because the collector–base voltage across the super-gain transistors is so small, leakage is reduced to negligible proportions. The two super-gain transistors are protected against large *difference* input voltages by the diode structures across the inputs.

The foregoing material is a brief introduction to some approaches to the design of first stages for monolithic operational amplifiers. It is by no means comprehensive, nor does it cover the highly sophisticated techniques used in hybrid or discrete amplifier design[8], which can still result in amplifier modules that are superior in many respects to their monolithic counterparts.

Remaining with monolithic techniques, however, it is appropriate to consider the design of the second and output stages. Usually, modern monolithic operational amplifiers effectively contain only two stages, albeit complex ones, the reason being that the phaseshift through the amplifier is limited (see chapter 5) and unconditional stability can be simply achieved, as will be seen later. In this context, there will be an *output* configuration,

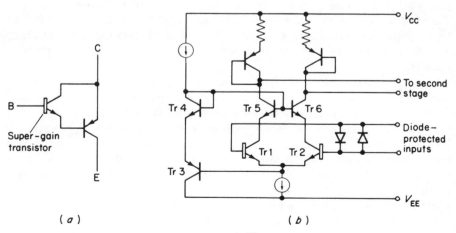

Figure 7.5. Use of the super-gain transistor[7]. (Warning: in some literature, the "open-base" symbol is used for "normal" npn transistors.)

whose duty it is to provide adequate power to the load along with a good voltage swing; and a *driver* configuration which provides an appropriate signal for this output stage. By judicious design, all of the voltage gain can be contained within the driver stage, and this can also provide the high-frequency limitation (or dominant time constant) for the amplifier. The output stage will then be simply an impedance transfer network.

In many monolithic operational amplifiers, the driver stage is simply a Darlington pair having an active load as shown in figure 7.6. In principle, this stage could simply drive an emitter-follower output stage, which would result in a low output impedance. The emitter-follower has disadvantages, however, including the fact that it takes significant power under quiescent conditions. This can easily be seen if such a stage is visualized as having its load resistor connected to V_{EE} and its collector connected to V_{CC}. Under no-signal conditions the emitter must be at zero volts, when the load resistor will dissipate V_{EE}^2/R_L W. If $V_{CC} = V_{EE}$, then the transistor must also dissipate this power.

A better solution is to use a *complementary pair* emitter-follower, as shown in figure 7.7(*a*). This stage has long been used in discrete transistor power amplifiers, and is described in detail in chapter 9. For the present, it need only be noted that a positive-going signal will turn Tr14 further ON, and Tr20 further OFF. A negative-going signal will do the converse.

At zero signal, both transistors will be OFF, and will remain so until the signal at the commoned bases approaches $\pm V_{BE}$. In order that conduction should commence for signals lower than this it is usual to hold the two output transistors in a slightly conducting state by separating their bases by a voltage approaching $2V_{BE}$, and one method of doing this is shown in figure 7.7(*b*). Here Tr18 ensures that the voltage across R_8 is V_{BE}, and if the current gain of this transistor is high then the current through R_7 must be similar, $I_{B(18)}$ being negligible. That is,

$$I_7 \simeq I_8 = \frac{V_{BE(18)}}{R_8}.$$

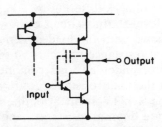

Figure 7.6. A Darlington second stage with active load and compensation capacitor (shown by the broken lines).

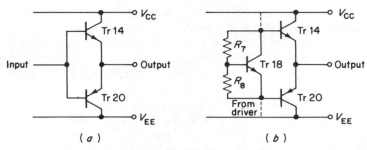

Figure 7.7. Complementary pair output stage: (*a*) principle; (*b*) with level-shifting network for crossover distortion reduction.

Hence,

$$V_{R_7} = I_7 R_7 \simeq I_8 R_7 = V_{BE(18)} R_7 / R_8 \ .$$

The total voltage across Tr18 is simply the sum of the voltages across R_7 and R_8:

$$V_{CE(18)} = V_{R_8} + V_{R_7} = V_{BE(18)} \left(1 + \frac{R_7}{R_8} \right). \tag{7.5}$$

Note that this expression involves the *ratio* of two resistances, which is good design practice insofar as monolithic integrated circuits are concerned, and it has already been mentioned that consistent resistor ratios are much easier to achieve than are consistent absolute values.

Equation (7.5) shows that if $R_7 = R_8$, then the voltage across Tr18 is $2 V_{BE(18)}$. The current through Tr18 can be chosen to give an appropriate value of V_{BE} for purposes of crossover distortion minimization, a problem which is fully explained in chapter 9. This circuit is sometimes known as a V_{BE} *multiplier*.

Actually, the network is an example of a *level-shifting circuit* in that it transmits the signal through a direct voltage drop with little attenuation; that is, in this particular case, identical drive signals are applied to the bases of Tr14 and Tr20 at datum voltages differing by $2 V_{BE(18)}$.

A basic method for a level-shifting circuit is shown in figure 7.8(*a*), where an emitter-follower is shown with the load in series with a current source. Recalling that the internal resistance of a current source is infinitely high so that it passes its current I under all load conditions, it can be seen that whereas the two ends of the load resistance R are at potentials differing by IR, nevertheless the same signal appears at both ends. In practice a current source has a high but finite internal resistance, and it is necessary only that this be large compared with R. Hence, a current mirror transistor will suffice, and in figure 7.8(*b*) its dynamic collector–emitter resistance r_{ce} will

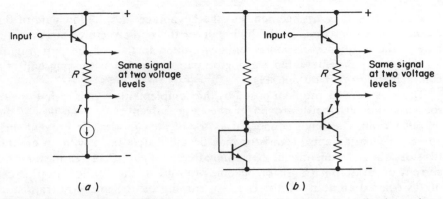

Figure 7.8. (*a*) Principle of the level shifter; and (*b*) practical realization using a current mirror.

be much greater than R. The arrangement thus constitutes a level shifter, the circuit function of which is to translate signals from the output of one stage to the input of another at a different voltage level.

Returning to the output stage depicted in figure 7.7, it will be noticed that the collector of the pnp transistor is at the most negative level in the circuit. This makes possible the use of an alternative method of fabrication, the *substrate pnp transistor*. An idealized cross-section of this device is shown in figure 7.9, from which it will be seen that unlike the lateral pnp structure, the substrate transistor employs depthwise planar diffusions like the normal npn structure, and indeed it is this which makes it possible to fabricate both polarity transistors during thes same sequence of processes.

Figure 7.9 shows that an n-island is diffused in to form the base region, along with a heavily-doped (n^+) region to minimize $r_{bb'}$. A p-diffusion forms the emitter, and the p-substrate itself becomes the collector. This is why the substrate must be the most negative part of the integrated circuit; otherwise a forward-biased junction could result.

Like the lateral pnp structure the substrate pnp has a comparatively low gain, but it does have a high current capability resulting from the large substrate collector area, which makes it appropriate as a large-signal stage.

An alternative way of realizing the pnp transistor is to combine a lateral pnp structure with a npn structure to form a Darlington pair as shown in

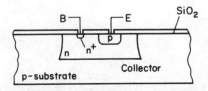

Figure 7.9. The substrate pnp structure.

figure 7.10(*a*). This arrangement has the advantage that the low gain of the lateral pnp is overcome by the high gain of the npn structure; but unfortunately, the frequency response remains limited to that of the pnp. Figure 7.10(*b*) shows the integrated Darlington pair connected as the pnp half of a complementary pair output stage.

The most serious problem posed by the complementary pair is that one of the structures can be destroyed by the large current which can flow if the output terminal becomes connected to one of the supplies. To prevent such an occurrence, an extra transistor may be fabricated along with an emitter resistor, the relevant circuit becoming that of figure 7.11. If the emitter current of an output transistor becomes too great, the voltage drop across the associated emitter resistor rises, so turning ON the ancillary transistor, which diverts the base current of the output transistor, so limiting the collector current which can pass.

That is, if

$$I_E R_E = V_{BE}$$

then

$$I_{E(max)} = \frac{V_{BE(max)}}{R_E} .$$

This form of protection is very common in modern linear integrated circuits and often the lower, pnp, ancillary transistor is omitted where the output pnp is a high-current low-gain substrate transistor whose current gain falls rapidly when I_C becomes large.

Having introduced some of the basic techniques underlying monolithic (as opposed to discrete) circuit design, it is appropriate to illustrate some of them with reference to a complete monolithic operational amplifier.

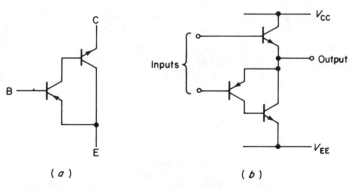

Figure 7.10. (*a*) A Darlington pnp combination; (*b*) used in a complementary output stage.

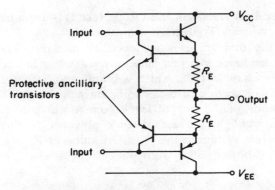

Figure 7.11. Overcurrent protection in a complementary pair output stage.

7.1.1 THE 741 Monolithic Operational Amplifier

Figure 7.12 illustrates the circuit of the 741. The input stage, consisting of npn emitter-followers Tr1, Tr2 driving the common-base pnp difference transistors Tr3, Tr4, corresponds exactly to the input stage of figure 7.4(c), where the various components have been annotated as in figure 7.12.

The Tr5, Tr6, Tr7 active load configuration supplies the Darlington driver stage Tr16, Tr17 which is actively loaded by Tr13. The complementary output pair Tr14, Tr20 is driven via the level-shifting network involving Tr18, which operates as described previously with reference to figure 7.7(b).

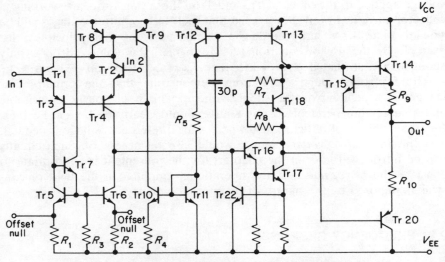

Figure 7.12. The 741 monolithic operational amplifier circuit.

Tr15 is an ancillary transistor used to protect Tr14 from overcurrent. The substrate pnp transistor Tr20 is left unprotected.

Biasing of the first stage is achieved as previously described, as is common-mode feedback via Tr9. The only transistor not accounted for in the previous discussion is Tr22, which is simply another ancillary transistor, this time for the protection of the Darlington pair driver Tr16, Tr17.

As an indication of the current levels common in monolithic integrated circuits, and to highlight the use of some previously-defined expressions, consider the biasing system in the 741. The value of R_5 is about 39 kΩ, so that, using ±15 V power supply rails, its current is:

$$I_{R5} = \frac{V_{CC} - V_{EE} - V_{BE11} - V_{BE12}}{R_5}$$

$$= \frac{30 - (2 \times 0.7)}{39} \simeq 0.73 \text{ mA} .$$

Using equation (7.2) to calculate I_{C10} implies a trial-and-error method because I_{C10} appears in both sides effectively. That is, if $R_4 = 5$ kΩ:

$$I_{C10} = \frac{kT}{qR_4} \ln\left(\frac{I_{R5}}{I_{C10}}\right) = \frac{0.026}{5} \ln\left(\frac{0.73}{I_{C10}}\right)$$

which $I_{C10} = 0.019$ mA will be found to satisfy.

Now $I_{C8} = I_{C9} \simeq I_{C10}$ if $(I_{B3} + I_{B4})$ is small, so this must also be 19 μA. Hence, assuming symmetry, the collector currents of Tr5 and Tr6 must be 9.5 μA each. This gives the voltage drops in R_1 and R_2 (which are about 1 kΩ each) as 9.5 mV and this must be added to V_{BE5} or V_{BE6} to arrive at the voltage applied to the bases of Tr5 and Tr6. These transistors are working at collector currents of only 9.5 μA, so that their base–emitter voltages will be smaller than 0.7 V and should be calculated from first principles, for example by the use of equation (2.11). In fact, this voltage drop is only about 0.55 V, so that if $R_3 \simeq 50$ kΩ, then $I_{R3} \simeq I_{C7} \simeq 0.55/50 \simeq 11$ μA.

The foregoing rough calculations demonstrate that the primary bias current is dependent on the supply voltage. This is obviously undesirable and a supply-independent current source would clearly be better. Further, for circuits like that of figure 5.40 where a fixed reference voltage is needed, a supply-insensitive *voltage* source would also be necessary. Such circuits will be introduced later in the chapter; for the present, it is appropriate to consider the *performance* of the operational amplifier in the light of constraints imposed by the nature of its internal circuitry.

7.2 FREQUENCY RESPONSE

An amplifier such as the 741 has two input points, and it is the extremely small difference between the signal voltages at these inputs which is amplified to produce most of the output signal. The relevant transfer function will be frequency dependent, as was the case for the simple amplifier stages described in chapter 5. That is, the open-loop (i.e., without feedback) difference-input voltage gain is:

$$A_v(f) = \frac{v_{\text{out}}}{(v_{\text{in}}^+ - v_{\text{in}}^-)}(f). \qquad (7.6)$$

If the amplifier is to be stable under all conditions of resistive feedback, then the internal phaseshift (over and above the normal phase reversal) should not be allowed to approach 180°. If the gain can be tailored to fall at approximately 6 dB/octave, then, according to chapter 5, the phaseshift will approach only 90°. This roll-off characteristic is common to most operational amplifiers and is achieved by the inclusion (externally or internally) of the compensating circuits mentioned previously.

Figure 7.13 shows idealized gain and phase characteristics of a single-lag operational amplifier. Here, the noninverting input point is assumed to be grounded, so that the gain from the inverting input is

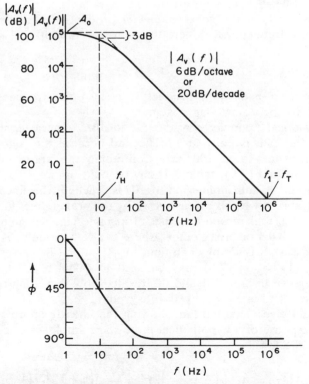

Figure 7.13. Idealized gain and phaseshift example for a "single-lag" operational amplifier.

$$A_v(f) = \frac{A_0}{1 + j(f/f_H)} \tag{7.7}$$

A_0 being the DC gain and f_H the high-frequency cut-off point where the phaseshift is (by definition) 45°.

In general, the phaseshift is given by

$$\phi = \tan^{-1}\left(\frac{f}{f_H}\right) \quad \text{(lagging)} \tag{7.8}$$

and the modulus of the gain is

$$|A_v(f)| = A_0\left(\frac{1}{1 + (f/f_H)^2}\right)^{1/2}. \tag{7.9}$$

For frequencies much in excess of f_H, (7.9) may be rewritten

$$|A_v(f)| = A_0\frac{f_H}{f} \quad \text{if} \quad \left(\frac{f}{f_H}\right)^2 \gg 1$$

so that $|A_v(f)|f = A_0 f_H = f_T$, a constant.

The constant f_T is the *gain–bandwidth* product of a single-lag system. Note that at a frequency f_1, where the gain falls to unity

$$f_1 = f_T$$

that is, the gain–bandwidth product f_T is also the frequency at which $|A_v(f)| = 0\,\text{dB}$ for a single-lag system.

As a numerical example, consider the idealized characteristics of figure 7.13. Here the gain is unity at 1 MHz, and because the roll-off is 20 dB/decade it rises to 10 at 100 kHz. Continuing the argument and using equations (7.8) and (7.9), table 7.1 may be built up.

When the DC or maximum gain of 10^5 is reached, the linearly approximated slope abruptly levels off (as shown by the broken lines in figure 7.13). If, however, the real response is that of a single lag, the point where this occurs is the $-3\,\text{dB}$ or break point, where the gain modulus is 0.707×10^5. Also, the phaseshift is 45° by definition, for equation (7.8) gives $\phi = \tan^{-1} 1$ when $f = f_H$. Equation (7.8) may be used to calculate ϕ at each of the frequencies given in table 7.1 and the results show that a lag of 90° is very rapidly approached after the 10 Hz break point.

Figure 7.14 gives a dotted line showing the linearly approximated gain/frequency response of a hypothetical three-stage amplifier. Such a response would have the form

$$A_v(f) = A_0\left[\left(1 + j\frac{f}{f_{HA}}\right)\left(1 + j\frac{f}{f_{HB}}\right)\left(1 + j\frac{f}{f_{HC}}\right)\right]^{-1}.$$

TABLE 7.1

Frequency (Hz)	Idealized Open-loop Internal Gain Modulus $\lvert A_v(f) \rvert$			
	Nominal Value	dB	Phase-lag	
10^6	1	0	90°	
10^5	10	20	90°	
10^4	10^2	40	90°	
10^3	10^3	60	89°26′	
10^2	10^4	80	84°17′	
10	10^5	100	45°	
	(actually 0.707×10^5)			
1	10^5	100	0	

If resistive feedback were applied, this amplifier would become unstable. A compensating network could be designed, however, which would give the amplifier a single break point instead of three; that is, by severely restricting the bandwidth, a 6 dB/octave roll-off could be produced as shown by the full line. (More sophisticated methods of phase compensation exist[9], such as pole cancellation, but are outside the scope of this volume.)

Having considered some elementary aspects of the frequency response question, it is appropriate to relate the conclusions to the monolithic operational amplifier, and in particular to the classic configuration of the 741. If the output stage has unity gain and a good high-frequency cut-off

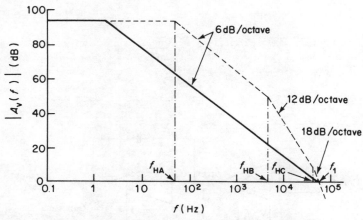

Figure 7.14. "Three-lag" operational amplifier gain characteristic (– – –) compensated for "single-lag" performance (——).

point, the input stage and the second stage only can be assumed to define the frequency-dependent voltage gain[10]. Figure 7.15 illustrates this approximation, and shows a compensating capacitor C connected from the input to the output of the second stage. This situation will be seen to correspond with that relating to the Miller effect produced by stray capacitance across a single transistor, and it can be conveniently analysed using sinusoidal quantities.

If the reactance of the capacitor is smaller in magnitude than the input resistance to the second stage (which is a reasonable assumption at the frequencies of interest), then $|I_{in}| \ll |I_C|$. Hence, if the gain of the second stage is large so that $V_{in(2)}$ is small,

$$V_{out} = \frac{I_C}{j\omega C} \simeq \frac{I_{G(2)}}{j\omega C}. \tag{7.10}$$

Also,

$$I_{G(2)} \simeq g_{m(1)}V_{in} \tag{7.11}$$

where $g_{m(1)}$ is the transconductance of the first stage.

Putting equation (7.11) into equation (7.10) gives,

$$V_{out} \simeq \frac{g_{m(1)}V_{in}}{j\omega C}$$

or

$$\frac{V_{out}}{V_{in}} \simeq A_v(f) \simeq \frac{g_{m(1)}}{j\omega C} \tag{7.12a}$$

or

$$|A_v(f)| \simeq \frac{g_{m(1)}}{\omega C}. \tag{7.12b}$$

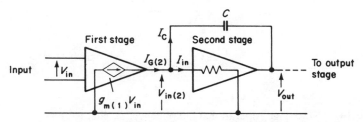

Figure 7.15. The two gain stages of an operational amplifier.

This equation expresses the fact that the inclusion of a capacitor across the second stage results in a single-lag slope of 6 dB/octave provided that neither the first stage nor the output stage contribute significantly to the frequency response.

The unity-gain (or the gain–bandwidth product) for this simple case may be extracted from equation (7.12b) by letting $|A_v(f)| = 1$:

$$f_1 = f_T \simeq \frac{g_{m(1)}}{2\pi C} .\tag{7.13}$$

In a real amplifier such as the 741, f_T is typically about 1 MHz for a C of about 30 pF, and it is this value which is shown in figure 7.12.

If a 741 silicon chip is examined, it will be noticed that the capacitor occupies about 15% of the total area. If it is desired to fabricate more than one amplifier upon a single chip (for example, a "quad" of amplifiers), then the area occupied by the capacitors contributes greatly to the overall size of the chip needed, and hence to the cost of fabrication.

Looking at equation (7.13) it will be seen that to reduce the value of C whilst still maintaining the same gain–bandwidth product implies also a reduction in g_m. In principle, this could be achieved by reducing the value of the collector currents in the input pair, but this would also restrict the bandwidth of the pair and introduce further frequency response problems in the current mirror. Solomon[10] has introduced a method of transconductance reduction which is applicable to lateral pnp transistors, and which has been used in the pnp common-base transistors in the input circuit of figure 7.4 without reduction in bandwidth.

The technique involves the diffusion of more than one collector into the lateral pnp structure so that the output current divides in the ratio of the areas presented by the collectors. For example, if two collectors exist having relative effective areas of A_1 and A_2, then (neglecting I_B) I_E divides into two currents I_{C1} and I_{C2} where

$$I_E = I_{C1} + I_{C2} \quad \text{and} \quad \frac{I_{C2}}{I_{C1}} = \frac{A_2}{A_1} = n .$$

If only I_{C1} is passed through the load, the apparent transconductance is:

$$g_{m(app)} \simeq \frac{q}{kT} I_{C1} .$$

The normal transconductance would be:

$$g_{m(1)} \simeq \frac{q}{kT} (I_{C_1} + I_{C_2})$$

so that the reduction effected by this technique is,

$$\frac{g_{m(app)}}{g_{m(1)}} = \frac{I_{C1}}{I_{C2} + I_{C2}} = \frac{1}{1 + n} . \tag{7.14}$$

As a numerical example, if the area of collector (2) is five times greater than the area of collector (1) and only I_{C1} is passed through the load, then from equation (7.14),

$$\frac{g_{m(app)}}{g_{m(1)}} = \frac{1}{1 + 5} = \frac{1}{6} .$$

For an unchanged bandwidth, equation (7.13) will give the new value of compensating capacitor, C':

$$\frac{g_{m(1)}}{2\pi C} = \frac{g_{m(app)}}{2\pi C'}$$

or

$$C' = C \frac{g_{m(app)}}{g_{m(1)}} = \frac{C}{6} .$$

In the case of the 30 pF capacitance used in the 741, this implies that a reduction to 5 pF would result if a collector area ratio of 1:5 were chosen.

Figure 7.16(a) illustrates the basic input circuit using collector-split pnp transistors as the CB difference pair, and it will be seen that the second pair of collectors "waste" current into the base-biasing network.

If lateral pnp transistors are used as the input devices themselves, then they can also be collector-split types as shown in figure 7.16(b). It is

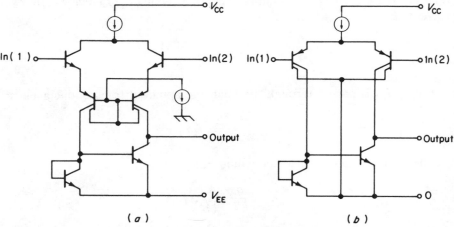

(a) (b)

Figure 7.16. Input circuits using lateral, multicollector pnp structures as: (a) a CB difference pair[10]; and (b) input transistors for a single-supply amplifier[11].

sometimes useful to employ a pnp input stage where a single, positive power supply is involved so that the input signal can swing down to nearly ground level. This is the case where amplifiers for automotive purposes are involved and only a 12-V battery is available[11].

The collector-split pnp lateral transistor is also useful as a source of more than one constant current for biasing purposes. Clearly, if several collectors are included, then each will produce a constant current proportional to its area; all resulting from a single base–emitter voltage.

One of the problems which continue to beset monolithic linear integrated circuits is that of *thermal feedback*, where the heat dissipated by the comparatively high-current output pair causes a small temperature gradient to appear across the silicon chip, the highest temperature being in the "output" region. This temperature gradient can result in one of the input transistors being at a slightly higher temperature than the other, so that an apparent input voltage is produced which is proportional to the output power.

This can modify the overall gain of the complete amplifier quite significantly[10], so that any expression derived for the DC gain on purely electrical grounds will be in error. However, insofar as the user is concerned, the gain limits of the amplifier will be specified by the manufacturer.

7.2.1   Slewing Rate and Settling Time

When a large, fast rising signal is applied to the input of an operational amplifier it is observed that the output voltage does not follow it exactly, but rises more slowly. This means that a large step input results in a ramp output; and that large, high-frequency sinusoidal signals become distorted. The parameter which defines this phenomenon is the *slew rate*, which is usually quoted in volts per microsecond at the output for a step input voltage.

A prime reason for this effect may be found in the first stage of an operational amplifier of the 741 type. In figure 7.12, the slew rate SR is determined largely by the rate at which the compensating capacitor, C, can be charged by the current in the collector lead of the Tr4/Tr6 combination. If the current mirror is assumed to transfer the (equal) collector current changes of Tr3 and Tr4 to the capacitor[12], then

$$SR \simeq \frac{dV_C}{dt} = \frac{2I_{C4}}{C} .$$

Insofar as sinusoidal signals are concerned, the maximum frequency for large output signals of good quality is usually taken as being limited to[12]:

$$f_{max} \simeq \frac{SR}{2\pi V_{Pk}}$$

where V_{Pk} is the peak output voltage.

For the normal 741 structure, the SR is about $1 \text{ V}/\mu\text{s}$, whereas for a modern fast monolithic amplifier[13] it can exceed $50 \text{ V}/\mu\text{S}$, and may reach some $600 \text{ V}/\mu\text{S}$ in specialized cases.

Related to slewing rate is the *settling time*. After a maximal slew resulting from a step input, the output voltage tends to overshoot and eventually recover to settle at the appropriate voltage level. The time taken for the amplifier to settle to within a small percentage of the final value is called the settling time, and it does, of course, include the slewing time. However, because it is a function of both the amplifier *and* the feedback circuit parameters, it is difficult to define[14].

Having presented the salient points of monolithic amplifier construction and frequency-response behavior, it is appropriate to consider it from the user viewpoint and establish an equivalent circuit, as was done for the discrete transistor in chapter 3.

7.3 THE OPERATIONAL AMPLIFIER EQUIVALENT CIRCUIT

In principle, it is possible to draw an equivalent circuit for an operational amplifier which will contain elements describing all the imperfections encountered. This circuit is, however, very complex and fortunately is rarely required. Figure 7.17 shows a simplified equivalent circuit, to which the following parameters apply.

(1) The differential input resistance, measured *between* the inputs, is termed R_{in}; whereas the common-mode input resistance, measured by applying the same signal to *both* inputs, is represented at each input by $R_{\text{in(CM)}}^{-}$ and $R_{\text{in(CM)}}^{+}$. (Both these terms will be familiar from the treatment of the difference amplifier in chapter 5.)

(2) When the inputs are short-circuited, the output should be zero, but in practice there will be a deviation from this ideal. This is due to asymmetry within the amplifier (for example, the two values of V_{BE} for the input transistors will not be exactly matched) and the zero impedance voltage applied between the inputs, which is necessary to zero the output voltage, is called the *offset bias voltage* e_{os}. It consists of several components, namely:

(a) An initial offset E_{os}. This is the value of e_{os} for zero output voltage at room temperature and at a given time.

(b) An *average* temperature-dependent component $\Delta e_{\text{os}}/\Delta T \ \mu\text{V}/°\text{C}$. In actual fact, de_{os}/dT is a function of T, which is why an average value is given in most data sheets.

(c) Average variations with supply voltages, $\Delta e_{\text{os}}/\Delta V_{\text{CC}}$ and $\Delta e_{\text{os}}/\Delta V_{\text{EE}} \ \mu\text{V}/\text{V}$.

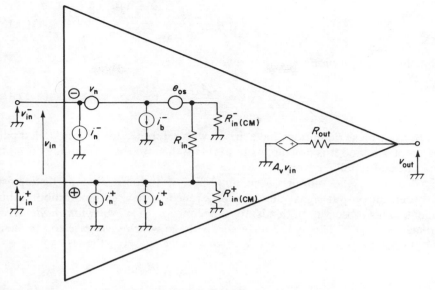

Figure 7.17. Operational amplifier equivalent circuit. $\ominus$ = inverting input; $\oplus$ = noninverting input.

(*d*) A random drift with time, having a guaranteed maximum value $\Delta e_{os}/\Delta t$.

(3) For the output to be zero when the inputs are open-circuited, bias currents i_b^+ and i_b^- must be supplied to the input bases from current sources. These bias currents consist of initial values I_B^+ and I_B^- plus temperature, supply voltage and time-varying components. (This applies even for the FET input stage where minute gate leakage currents must be supplied.) The difference between the bias currents is called the *differential offset current* i_d, where $i_d = (i_b^+ - i_b^-)$. This also has an initial value, I_d, plus temperature, supply voltage and time-varying components.

(4) The noise generators referred to the input are quoted in the same way as described in chapter 6 except that two separate noise current generators i_n^+ and i_n^- appear. i_n^+ and i_n^- are included to take account of cases where the difference-input operational amplifier is used with both inputs active; that is, when neither input is grounded.

(5) Any real amplifier must have an output resistance R_{out}, as shown in the equivalent circuit.

(6) If signals identical in magnitude and phase are applied to the inputs of a real operational amplifier, an output voltage will appear due to the *common-mode gain*. This is, if $v_{in}^- = v_{in}^+ = v_{in(CM)}$, then

$$A_{v(CM)} = \frac{v_{out(CM)}}{v_{in(CM)}} \qquad (7.15)$$

and the *common-mode rejection ratio* is defined as

$$CMR = \frac{A_v}{A_{v(CM)}} . \qquad (7.16)$$

Both of these definitions have arisen with reference to the long-tailed pair in chapter 5, and it is now apposite to relate them to the terminology used in operational amplifier work.

Clearly, at the output of an amplifier, it is impossible to distinguish between an output signal due to a difference or to a common-mode input signal. This observation leads to the concept of a *common-mode error voltage*, $v_{in(ECM)}$. This is the *difference* voltage which gives rise to the same output signal as a *common-mode input voltage*, $v_{in(CM)}$, that is,

$$v_{in(ECM)} = \frac{v_{out(CM)}}{A_v} = \frac{v_{in(CM)}A_{v(CM)}}{A_v} .$$

Hence, the common-mode rejection ratio is also given by

$$CMR = \frac{A_v}{A_{v(CM)}} = \frac{v_{in(CM)}}{v_{in(ECM)}} = \frac{\text{common-mode input voltage}}{\text{common-mode error voltage}} . \qquad (7.17)$$

The existence of a common-mode output voltage is really an expression of the fact that the signals at the two inputs are not amplified equally, but that some mismatch is present. (It may, however, also be the result of the external input circuitry to a difference amplifier being unbalanced in such a way as to present a portion of a common-mode signal to the amplifier inputs as a difference signal[15].)

If the gain from the inverting input is A_v^- and that from the noninverting input is A_v^+, then if a common-mode signal $v_{in(CM)}$ is applied to both inputs, the output will be

$$v_{out(CM)} = v_{in(CM)}(A_v^- + A_v^+)$$

that is,

$$A_{v(CM)} = A_v^- + A_v^+ . \qquad (7.18)$$

As an example, suppose that $A_v^- = -100\,000$ and $A_v^+ = 99\,990$. If 0.1 V were applied at each input, then

$$v_{out(CM)} = 0.1(-100\,000 + 99\,990) = -1.0\,\text{V} .$$

The common-mode rejection ratio is given by the ratio of the average difference gains to the common-mode gain:

$$\text{CMR} = \frac{(A_v^- - A_v^+)/2}{A_{v(\text{CM})}} \tag{7.19}$$

which for the present example is

$$\text{CMR} = \frac{-99\,995}{-10} = 9999.5$$

or approximately 80 dB.

Note also that, in the present example, the common-mode error voltage is

$$v_{\text{in(ECM)}} = \frac{-1.0}{-99\,995} \text{ V} \simeq 10 \ \mu\text{V}$$

and can be represented by a 10 μV voltage generator in series with the inverting input.

Normally, operational amplifiers are not used in the open-loop condition but employ various forms of feedback. Consequently, the effects discussed above must be re-examined for various conditions of feedback. This will now be done for the most common configurations.

7.3.1   The Inverting Configuration

Figure 7.18(a) shows an ideal amplifier connected as an operational amplifier, with the noninverting input taken to the common line and input and feedback resistors connected as shown.

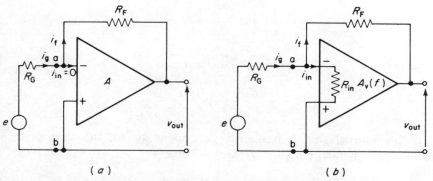

Figure 7.18. (a) "Perfect" inverting amplifier; (b) inverting amplifier with finite R_{in} and frequency-dependent gain.

Because the input resistance is infinite, the current delivered by the voltage generator will flow around the feedback loop, so that $i_g = -i_f$. If A is very large (ideally infinite), then

$$v_{in} = \frac{v_{out}}{A} \to 0 .$$

Hence

$$i_g = \frac{e}{R_G} = -i_f = \frac{-v_{out}}{R_F}$$

or

$$\frac{v_{out}}{e} = -\frac{R_F}{R_G} = A_{ov(FB)} . \tag{7.20}$$

This equation for the voltage gain from the input generator has already been obtained (equation (5.21)) in terms of impedances. Notice that not only are the input impedance and internal voltage gain A assumed to be infinitely high, but the bandwidth is also assumed to range from DC to infinity! Because any real amplifier cannot aspire to such a prodigious performance it is necessary to take into account the modifications imposed by practical limitations.

If the differential input resistance of the amplifier is R_{in}, then the relevant modification to the operational amplifier equation is given by equation (5.20) in its resistive form:

$$A_{ov(FB)} = A_v \left(\frac{R_{in} + R_G}{R_{in}} - \frac{R_G}{R_F} (A_v - 1) \right)^{-1} \tag{7.21}$$

where the internal gain A_v is no longer infinite, but is nevertheless invariant with frequency. It may be written as follows:

$$A_{ov(FB)} = -\frac{R_F}{R_G} \left[1 - \frac{1}{A_v} \left(1 + \frac{R_F}{R_G} + \frac{R_F}{R_{in}} \right) \right]^{-1} . \tag{7.22a}$$

Note that if $R_{in} \gg R_F$, which is usual, then

$$A_{ov(FB)} = -\frac{R_F}{R_G} \left[1 - \frac{1}{A_v} \left(1 + \frac{R_F}{R_G} \right) \right]^{-1} . \tag{7.22b}$$

If the internal gain is frequency dependent and of the form given in equation (7.7), then equation (7.22a) becomes

$$A_{ov(FB)}(f) = -\frac{R_F}{R_G} \left[1 - \left(\frac{1 + j(f/f_H)}{A_0} \right) \left(1 + \frac{R_F}{R_G} + \frac{R_F}{R_{in}} \right) \right]^{-1} \tag{7.23}$$

which is more conveniently written by letting

$$\frac{1}{A_0}\left(1 + \frac{R_F}{R_G} + \frac{R_F}{R_{in}}\right) = \frac{1}{A_0^*} \tag{7.24}$$

so that

$$A_{ov(FB)}(f) = -\frac{R_F}{R_G}\left[1 - \frac{1}{A_0^*}\left(1 + j\frac{f}{f_H}\right)\right]^{-1}$$

$$= -\frac{R_F}{R_G}\left[\left(1 - \frac{1}{A_0^*}\right) - j\left(\frac{f}{A_0^* f_H}\right)\right]^{-1}. \tag{7.25}$$

This expression is now of the form

$$-\frac{R_F}{R_G}\left(\frac{1}{a - jb}\right)$$

so that the phaseshift may be written simply as $\tan^{-1}(b/a)$, which is

$$\phi = \tan^{-1}\left(\frac{f}{f_H(A_0^* - 1)}\right) \quad \text{(lagging)} \tag{7.26}$$

and the modulus of the gain becomes

$$|A_{ov(FB)}(f)| = \mathcal{A} = \frac{R_F}{R_G}\left[\left(1 - \frac{1}{A_0^*}\right)^2 + \left(\frac{f}{A_0^* f_H}\right)^2\right]^{-1/2}. \tag{7.27}$$

This equation now relates to the system drawn in figure 7.18(b), and will be valid when the load on the amplifier is sufficiently light that there is little voltage drop in the output impedance.

7.3.2 Example

For a lightly loaded operational amplifier with a high-resistance input stage and the gain and phase response of figure 7.13, determine the forward gain modulus and phaseshift at 10 kHz if $R_G = 1\,\text{k}\Omega$ and $R_F = 100\,\text{k}\Omega$.

Because R_{in} is large, R_F/R_{in} may be neglected in equation (7.24), so that

$$\frac{1}{A_0^*} = \frac{1}{-10^5}(1 + 100) = -\frac{101}{10^5}$$

which may be inserted into equation (7.27):

$$\mathcal{A} = 100\left[\left(1 + \frac{101}{10^5}\right)^2 + \left(\frac{101 \times 10^4}{10^5 \times 10}\right)^2\right]^{-1/2}$$

or

$$\mathcal{A} \simeq \frac{100}{\sqrt{(1 + 1.0201)}} = 70.34 \ .$$

This is a 30% departure from the "ideal" gain of 100, but it should be noted that had the value of f been only one decade lower (or the value of A_0^* been only one decade higher), the departure from the ideal would have been much less than 1%. However, use of equation (7.26) to calculate the phaseshift in the present example gives

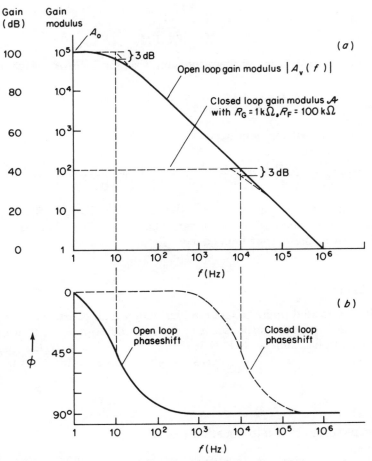

Figure 7.19. Effect of input and shunt feedback resistors on the gain and phaseshift of an operational amplifier.

$$\phi = \tan^{-1}\left(\frac{10^4}{10[-(10^5/101)-1]}\right) \simeq 45° \quad \text{(lagging)}$$

which is to be expected, because a gain of 70.34 is not far removed from the $-3\,\text{dB}$ point relevant to the forward gain modulus of the feedback amplifier. The gain $\mathcal{A}$ and the phaseshift ϕ for this example are shown by the broken lines in figure 7.19. Note that these curves clearly show how gain has been traded for an increased bandwidth.

The foregoing example does in fact demonstrate a rule of thumb, that if negligible phaseshift through such an amplifier is desired, the working frequencies should not approach nearer to the break frequency then approximately two decades.

7.3.3 Offset, Drift, Noise, and CMR for the Inverting Configuration

The input offset voltage of an amplifier is defined as that voltage which, applied in series with the input, would produce a voltage e_{os} across R_{in}. This voltage, e_{os}^*, may be determined by removing the hypothetical generator e_{os} in figure 7.20(a) and substituting a voltage generator e_{os}^* in series with the input, as in figure 7.20(b). This generator must now produce a voltage e_{os} across R_{in}, which is, of course, in parallel with R_F. That is,

$$e_{os} = e_{os}^* \frac{R_{in} \| R_F}{R_G + R_{in} \| R_F}.$$

If $R_{in} \gg R_F$, which is usual, this becomes,

$$e_{os} \simeq e_{os}^* \frac{R_F}{R_G + R_F}$$

or

$$e_{os}^* \simeq e_{os} \frac{R_G + R_F}{R_F} \tag{7.28}$$

which is the value shown in figure 7.20(b).

Consider now the two bias currents i_b^+ and i_b^-. Because the noninverting input is taken directly to the common line, i_b^+ will be supplied from this line and will have no effect on the output. It is therefore omitted in figure 7.20(c).

Conversely, i_b^- is in parallel with R_G, and if a Norton–Thévenin transformation is applied, it can be replaced by a voltage generator in series with R_G:

$$e_{i_b^-} = i_b^- R_G \tag{7.29}$$

which is the value shown in figure 7.20(d).

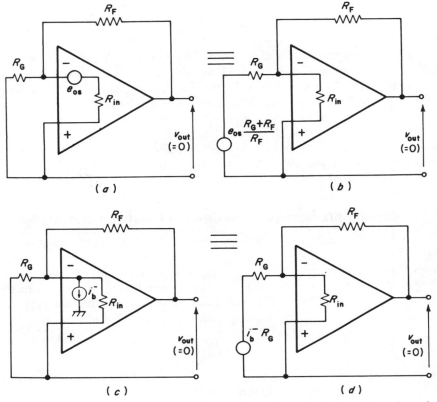

Figure 7.20. Offset and bias components referred to the input of an inverting operational amplifier.

Having referred both e_{os} and i_b^- to the input circuit in the form of voltage generators, they may now be multiplied by the closed-loop DC voltage gain to give the total drift component at the output. Using equation (7.22a) along with the DC open-loop gain A_0, this is:

$$\Delta V_{out} = -\frac{R_F}{R_G}\left[1 - \frac{1}{A_0}\left(1 + \frac{R_F}{R_G} + \frac{R_F}{R_{in}}\right)\right]^{-1}\left(\Delta e_{os}\frac{R_G + R_F}{R_F} + \Delta i_b^- R_G\right).$$
$$(7.30a)$$

Normally, for a high-gain amplifier, the correction factor in the first bracket will tend to unity and may be ignored, so that

$$\Delta V_{out} \simeq -\frac{R_F}{R_G}\left(\Delta e_{os}\frac{R_G + R_F}{R_F} + \Delta i_b^- R_G\right). \qquad (7.30b)$$

A modification which leads to a better current-drift performance is shown in figure 7.21. Here, the noninverting input is taken to the common line via

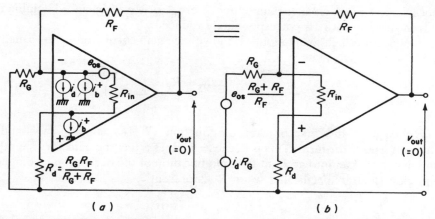

Figure 7.21. (*a*) Incorporation of R_d to reduce current drift; (*b*) equivalent circuit showing both drift components referred to input (inverting configuration).

a resistor R_d, which has a value equal to the parallel combination of R_G and $R_{F'}$:

$$R_d = \frac{R_G R_F}{R_G + R_F}.$$

This means that both i_b^+ and i_b^- produce voltage drops in equal resistances, so that if $i_b^+ = i_b^-$, an extremely small common-mode voltage appears at each input. This smallness, coupled with the high CMR of such an amplifier, means that almost no output due to the common-mode voltage appears. For this reason the inverting amplifier may be assumed to be inherently free of common-mode errors even when R_d is non-zero.

However, if $i_b^+ \neq i_b^-$, which is usual, then the difference between these bias currents will result in a difference input voltage which can be expressed as shown in figure 7.21. In diagram (*a*), the required bias current at the noninverting input is termed i_b^+ as usual, whilst that at the inverting input is given by $i_b^- = i_b^+ \pm i_d$, which is shown as two separate generators. If $R_d = R_G \| R_F$, then both i_b^+ and the i_b^+ component of i_b^- can be expressed as common-mode voltage generators as previously explained, whilst the difference input voltage generator replacing $\pm i_d$ is simply $\pm i_d R_G$ as usual. This transformation is shown in figure 7.21(*b*).

Under these circumstances, equation (7.30*a*) (slightly modified by R_d) becomes

$$\Delta V_{\text{out}} = -\frac{R_F}{R_G} \left\{ 1 - \frac{1}{A_0} \left[1 + \frac{R_F}{R_G} + \frac{R_F}{R_{\text{in}}} + \frac{R_d}{R_{\text{in}}} \left(1 + \frac{R_F}{R_G} \right) \right] \right\}^{-1}$$

$$\times \left(\Delta e_{\text{os}} \frac{R_G + R_F}{R_F} + \Delta i_d R_G \right). \tag{7.31a}$$

Because i_d is much smaller than either i_b^+ or i_b^- for a well-balanced amplifier, current-induced drift is correspondingly reduced.

Again, for a high-gain amplifier, the voltage gain correction factor usually tends to unity, so that

$$\Delta V_{out} \simeq -\frac{R_F}{R_G}\left(\Delta e_{os}\frac{R_G + R_F}{R_F} + \Delta i_d R_G\right). \qquad (7.31b)$$

Notice that the first bracket in equation (7.31a) contains a slightly changed gain-error term. This is because R_d does affect the gain very slightly and may be taken into account by a modification of the derivation leading to equation (5.20). Specifically, when R_d is present

$$e = i_g R_G + i_{in}(R_{in} + R_d)$$

and

$$i_g = i_{in} + i_f = \frac{v_{in}}{R_{in}} + \frac{(v_{in} + i_{in}R_d) - v_{out}}{R_F}.$$

However, the extra term in this first bracket is a product involving R_d/R_{in}, so that for amplifiers where $R_{in} \gg R_d$ (which are the majority) this term may be neglected.

The noise generators appear in similar positions to the bias and offset generators in the equivalent circuit, which is not surprising when it is observed that drift is merely very low-frequency noise.

7.3.4 The Noninverting Configuration

An "ideal" operational amplifier is shown connected in a noninverting configuration in figure 7.22(a). The DC voltage gain may be simply found as follows:

$$\begin{aligned}
A_{ov(FB)} = A_{v(FB)} &= \frac{v_{out}}{v_{in} + [v_{out}R_{F2}/(R_{F1} + R_{F2})]} \\
&= \frac{Av_{in}}{v_{in} + [Av_{in}R_{F2}/(R_{F1} + R_{F2})]} \\
&= \left(\frac{1}{A} + \frac{R_{F2}}{R_{F1} + R_{F2}}\right)^{-1}.
\end{aligned}$$

If A is very large,

$$A_{v(FB)} \simeq \frac{R_{F1} + R_{F2}}{R_{F2}} \quad \text{or} \quad 1 + \frac{R_{F1}}{R_{F2}}. \qquad (7.32)$$

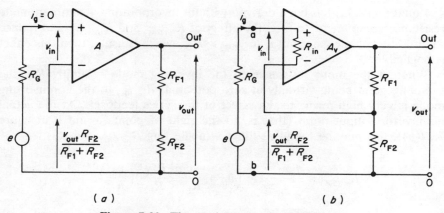

Figure 7.22. The noninverting configuration.

Notice that the voltage gain in this configuration cannot be less than unity, and that the amount by which it exceeds unity is dependent only on the ratio of the two resistors R_{F1} and R_{F2} provided that A remains large.

A gain expression for a real amplifier must take into account not only finite open-loop gain A_v and finite input resistance R_{in} but also the fact that i_g will flow through R_G, R_{in}, and the parallel combination $R_{F1} \parallel R_{F2}$, producing a voltage drop in each.

If $R_{F1} \parallel R_{F2} = R_P$, then this single resistor may be assumed to terminate at a voltage generator given by $v_{out} R_{F2}/(R_{F1} + R_{F2})$. Hence, by using figure 7.22(b) along with this simplification, $A_{ov(FB)}$ may be determined:

$$A_{ov(FB)} = \frac{v_{out}}{e} \tag{7.33}$$

$$A_{ov(FB)} = \frac{v_{out}}{i_g R_G + i_g R_{in} + i_g R_P + [v_{out} R_{F2}/(R_{F1} + R_{F2})]} \tag{7.34}$$

and because $i_g = i_{in}$, this becomes

$$A_{ov(FB)} = A_v v_{in} \left(v_{in} \frac{R_G}{R_{in}} + v_{in} + v_{in} \frac{R_P}{R_{in}} + \frac{A_v v_{in} R_{F2}}{R_{F1} + R_{F2}} \right)^{-1}$$

which reduces to

$$A_{ov(FB)} = \frac{R_{F1} + R_{F2}}{R_{F2}} \left[1 + \frac{1}{A_v} \left(1 + \frac{R_G}{R_{in}} + \frac{R_P}{R_{in}} \right) \left(\frac{R_{F1} + R_{F2}}{R_{F2}} \right) \right]^{-1}. \tag{7.35}$$

Note that this reduces to equation (7.32) when A_v is large.

Equation (7.35) has been derived without incorporating a common-mode resistance component. In fact, the effect of $R^-_{in(CM)}$ and $R^+_{in(CM)}$ is to reduce slightly the voltage gain, but if $R_{in(CM)} \gg R_G$, which is usual, then this effect is negligible.

Whereas the input resistance to the inverting mode is simply R_G, the summing point being virtually at zero potential, $R_{in(FB)}$ to the noninverting mode is very high owing to the effect of the series feedback. At the actual noninverting input point (that is, to the right of points a and b in figure 7.22(b)), this may be found as follows (neglecting $R_{in(CM)}$ and R_P):

$$R_{in(FB)} = \frac{v_{in} + v_f}{i_{in}} = \left(v_{in} + \frac{A_v v_{in} R_{F2}}{R_{F1} + R_{F2}} \right)\left(\frac{v_{in}}{R_{in}} \right)^{-1}$$

or

$$R_{in(FB)} = R_{in}\left(1 + \frac{A_v R_{F2}}{R_{F1} + R_{F2}} \right). \tag{7.36}$$

Because A_v is normally large and has a positive numerical value, $R_{in(FB)}$ can be very high. However, it must be noted that $R^+_{in(CM)}$ is in parallel with $R_{in(FB)}$, so that the value of $R_{in(FB)}$ cannot exceed $R^+_{in(CM)}$ and will tend to this value if A_v becomes very large. Because $R_{in(FB)}$ is very high it may be assumed that i_g is extremely small, so that unless R_G is phenomenally large, $A_{v(FB)} = A_{ov(FB)}$.

7.3.5   Offset, Drift, Noise, and CMR for the Noninverting Configuration

Figure 7.23(a) clearly shows that the offset voltage generator e_{os} acts in series with R_G, and is therefore also the offset generator referred to the input.

From figure 7.23(b) it will be seen that the bias current at the noninverting input flows through R_G and can therefore be represented by a voltage generator $i^+_b R_G$ in series with R_G. The bias current at the inverting input, on the other hand, is derived largely from the parallel combination $R_{F1} \| R_{F2}$. If a resistor R_d is now inserted in series with the inverting input as shown in figure 7.24 and

$$R_d = R_G - \frac{R_{F1}R_{F2}}{R_{F1} + R_{F2}}$$

then the two bias currents flow through identical resistances, so producing a common-mode signal which is amplified only by $A_{v(CM)}$, which is small. The difference current drift, however, may be referred to the input (see figure 7.24), becoming $i_d R_G$. The drift at the output may now be obtained using

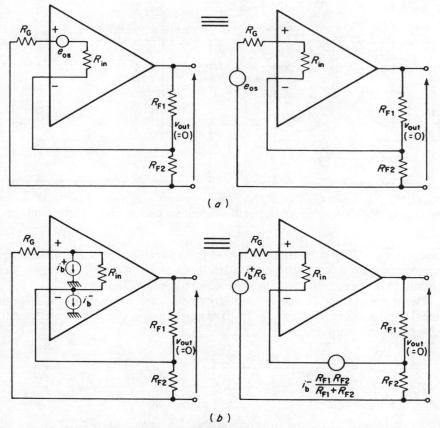

Figure 7.23. Offset and bias components referred to the input of a non-inverting operational amplifier.

(7.35), slightly modified by writing R_G in place of R_P, and by writing $A_v = A_o$, the DC gain:

$$\Delta V_{out} = \frac{R_{F1} + R_{F2}}{R_{F2}} \left[1 + \frac{1}{A_o} \left(1 + \frac{2R_G}{R_{in}} \right) \left(\frac{R_{F1} + R_{F2}}{R_{F2}} \right) \right]^{-1} (\Delta e_{os} + \Delta i_d R_G) .$$

$$(7.37a)$$

For a high-gain amplifier, the gain error term usually tends to unity, so that

$$\Delta V_{out} \simeq \frac{R_{F1} + R_{F2}}{R_{F2}} (\Delta e_{os} + \Delta i_d R_G) . \qquad (7.37b)$$

One of the disadvantages of the noninverting configuration is that the input signal is also the common-mode signal, and will lead to a common-

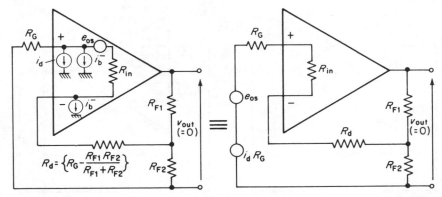

Figure 7.24. Incorporation of R_d to reduce current drift; and equivalent circuit showing both drift components referred to input (non-inverting configuration).

mode output. This is easily understood by observing that if A_v is large the signals appearing at each input with respect to the common line are essentially identical; that is, they are common-mode signals.

The common-mode error voltage is, by definition, in phase with the input signal v_{in}^+ (and hence with v_{in}^- and v_{in}). Therefore, $v_{in(ECM)}$ may be added to v_{in}^+ to give the output signal:

$$v_{out} \simeq \frac{R_{F1} + R_{F2}}{R_{F2}} \left(v_{in}^+ + v_{in(ECM)} \right)$$

$$= \frac{R_{F1} + R_{F2}}{R_{F2}} \left(1 + \frac{1}{CMR} \right) v_{in}^+ .$$

This last equation shows clearly that, for the noninverting configuration, the true gain is given by

$$A_{v(FB)} = \frac{v_{out}}{v_{in}} \simeq \frac{R_{F1} + R_{F2}}{R_{F2}} \left(1 + \frac{1}{CMR} \right) . \qquad (7.37c)$$

(The accurate expression must of course include the relevant gain error factor, as in equations (7.35) and (7.37a). It has been omitted here only for reasons of clarity.)

For most amplifiers the CMR is very large, which in practice means that the tolerances in R_{F1} and R_{F2} usually have a much more marked effect on the gain than does the common-mode signal.

Note that in the foregoing discussion, it has been tacitly assumed that the CMR is constant; that is, $v_{in(ECM)}$ is directly proportional to $v_{in(CM)}$. This assumption is usually true for a limited range of $v_{in(CM)}$ and any serious deviation is sometimes given in the relevant data sheet.

The noise generators, which appear in similar positions to the offset generators, may be treated in the manner described in chapter 6, keeping in mind that i_n^+ and i_n^- are not correlated.

7.3.6 Comparison of the Inverting and Noninverting Configurations

The approximate expressions for voltage gain, input resistance and output voltage drift derived in the foregoing sections may be tabulated (table 7.2) to provide a convenient basis for a comparison of the two configurations.

For both configurations, the voltage gains are well defined by the external circuitry, but in the case of the inverting configuration it is reasonable to connect the noninverting input to the common line ($R_d = 0$) so that a single-input amplifier may be used. This would mean that Δi_b^- would appear in the drift equation as opposed to Δi_d (that is, equation (7.30b) becomes valid), but if a good chopper-type amplifier is used the drift performance might well be better than that possible with a differential-input amplifier.

The drift equations tabulated below show that if the inverting configuration is used with a large value of R_G, then the voltage-drift component can be high as compared with the case for the noninverting configuration. In both cases, however, the current-drift component is the same. If voltage amplification with low drift is of interest, therefore, it is usually best to specify the noninverting configuration if a high impedance source generator is involved. This conclusion could also be reached simply by observing that $R_{in(FB)}$ for the noninverting configuration is very high, whereas a virtual earth ($Z_{in(FB)} \to 0$) point exists for the inverting configuration. It must be remembered, however, that there will be a common-mode component of output voltage present in the case of the noninverting configuration. An example will serve to illustrate the foregoing points.

7.3.7 Design Example

An operational amplifier is to be used as a low-frequency amplifier having a gain of 100, and is to be driven from a source generator having an internal

TABLE 7.2.

	Inverting	Noninverting
$A_{ov(FB)}$	$-\dfrac{R_F}{R_G}$	$1 + \dfrac{R_{F1}}{R_{F2}}$
$R_{in(FB)}$	$\to 0$	$\to R_{in(CM)}$
	to right of points a and b in figure 7.18(b)	
ΔV_{out}	$-\dfrac{R_F}{R_G}\left(\Delta e_{os}\,\dfrac{R_G + R_F}{R_F} + \Delta i_d R_G\right)$	$\dfrac{R_{F1} + R_{F2}}{R_{F2}}\,(\Delta e_{os} + \Delta i_d R_G)$
	for $R_d = \dfrac{R_G R_F}{R_G + R_F}$	for $R_d = R_G - \dfrac{R_{F1} R_{F2}}{R_{F1} + R_{F2}}$

resistance of $1000\,\Omega$. What will be the worst-case output drift over a temperature range of 10–$50°C$, the relevant parameters being as follows:

$$A_0 = 2.5 \times 10^5 \text{ (min)} \qquad\qquad R_{in} = 1\,M\Omega$$

$$R_{in(CM)} = 1000\,M\Omega \qquad\qquad \text{Rated output} = 5\,mA \text{ at } \pm 10\,V$$

$$\frac{\Delta e_{os}}{\Delta T} = \pm 5\,\mu V/°C \qquad\qquad \frac{\Delta i_d}{\Delta T} = \pm 0.05\,nA/°C .$$

If this amplifier were to be used in the inverting configuration, the value of R_G would have to be much greater than $1000\,\Omega$. If, for example, $R_G = 1\,M\Omega$, then $R_F = 100\,M\Omega$, which is most unsatisfactory. It is therefore clearly better in this case to use the non-inverting configuration which will present a very high input resistance.

If the required load current is less than $4\,mA$ at the maximum output voltage of $\pm 10\,V$, then up to $1\,mA$ can be spared for supplying R_{F1} and R_{F2}; that is,

$$R_{F1} + R_{F2} = \frac{10}{1} = 10\,k\Omega$$

and because from equation (7.32),

$$A_{v(FB)} = 1 + \frac{R_{F1}}{R_{F2}}$$

$$\frac{R_{F1}}{R_{F2}} = 100 - 1 = 99 .$$

Then,

$$R_{F1} = 9900\,\Omega \quad \text{and} \quad R_{F2} = 100\,\Omega .$$

A full design would take into account the tolerances of the preferred value resistors to be used, and a gain-controlling pre-set potentiometer would be used in series with R_{F1}, that is, $R_{F1} = 9.2\,k\Omega + 1\,k\Omega$ potentiometer.

To minimize current-induced drift, a balancing resistor R_d should be included, as in figure 7.24:

$$R_d = R_G - \frac{R_{F1}R_{F2}}{R_{F1} + R_{F2}} = 1000 - \frac{9900 \times 100}{10\,000} = 910\,\Omega$$

(to the nearest preferred value). (*Note*: Here, R_G is simply R_g for the purposes of the calculation, where R_g is the internal resistance of the source generator.)

The total changes in e_{os} and i_d are

$$\Delta e_{os} = \frac{\Delta e_{os}}{\Delta T}\, \Delta T = \pm 5 \times 40 = \pm 200\ \mu V$$

$$\Delta i_d = \frac{\Delta i_d}{\Delta T}\, \Delta T = \pm 0.05 \times 40 = \pm 2\ nA\ .$$

(Note that variations due to changes in supply voltage and those to be expected over a period of time should be *added* to these values to give the full worst-case environmental changes in e_{os} and i_d.)

Equation (7.37b) now gives the output drift:

$$\Delta V_{out} \simeq \frac{R_{F1} + R_{F2}}{R_{F2}}\, (\Delta e_{os} + \Delta i_d R_G)$$

$$= 100(200 \times 10^{-6} + 2 \times 10^{-9} \times 10^{3})$$

$$= 20.2\ mV\ .$$

Notice, in this calculation, that because R_g ($=R_G$) is low, the value of $\Delta i_d R_G$ is only 2 μV compared with the 200 μV contributed by Δe_{os}. Clearly, the current-induced drift would have been greater had a source generator with a higher input resistance been involved. The configuration could, in fact, have accepted a much higher value of R_g because, from equation (7.36),

$$R_{in(FB)(min)} = R_{in}\left(1 + \frac{A_v R_{F2}}{R_{F1} + R_{F2}}\right)$$

$$= 10^{6}\left(1 + \frac{2.5 \times 10^{5} \times 100}{10\,000}\right)$$

$$= 2501\ M\Omega\ .$$

This value is of course unrealistic, because $R_{in(CM)}$ for the amplifier is 1000 MΩ, which is therefore the actual value of $R_{in(FB)}$.

Because of the nature of the noninverting configuration, the signal voltages at the two inputs, v_{in}^{+} and v_{in}^{-}, are almost equal and constitute also the common-mode input voltage.

The gain error due to the common-mode signal is given by equation (7.37c), and if the data sheet value for the CMR is 20 000:

$$CM\ gain\ error = \frac{1}{CMR} \times 100 = 0.005\%\ .$$

In other words, $v_{in(ECM)}$ appears in series with v_{in}^{+} and has a value v_{in}^{+}/CMR. For example, in the present case, if $v_{in}^{+} = 0.1$ V, then $v_{in(ECM)} = 0.1/20\,000 = 5\ \mu V$. This results in a common-mode output signal of only $5 \times 100 = 500\ \mu V$ compared with a required output signal of $0.1 \times 100 = 10$ V.

7.3.8 Basic Applications of the Inverting Configuration

Because of the existence of a virtual earth point at the input to an operational amplifier connected in the inverting configuration, it is valid to treat this input point as a *current sink*. This leads to the realization that the transfer function of major interest for this configuration is the *transresistance*, v_{out}/i_g. If i_{in} is small (which is usual) then $i_g \simeq -i_f$, so that

$$v_{out} = -i_f R_F = i_g R_F \tag{7.38}$$

and the transresistance is simply $v_{out}/i_g = R_F$.

Equation (7.38) demonstrates that the system is now a current-to-voltage converter, and figure 7.25 shows some typical applications of this concept. Diagram (*a*) illustrates how a small current i_g from a high impedance source may be measured (or utilized) by converting it to a proportional voltage $i_g R_F$. Diagram (*b*) shows the operational amplifier used in its most common role, as a summing element. Here, currents i_{g1}, i_{g2}, i_{g3}, etc. are fed to the virtual earth point, and the current sum may be assumed to flow through R_F, producing an output voltage given by

$$v_{out} = (i_{g1} + i_{g2} + i_{g3} + \cdots)R_F$$

$$= \left(\frac{-e_1}{R_{G1}} + \frac{-e_2}{R_{G2}} + \frac{-e_3}{R_{G3}} + \cdots \right) R_F .$$

If $R_{G1} = R_{G2} = R_{G3} = \cdots$

$$v_{out} = -\frac{R_F}{R_G}(e_1 + e_2 + e_3 + \cdots) . \tag{7.39}$$

The summing property is put to use extensively and it will be appreciated that the drift problem is one to which much attention must be given if results of high accuracy are demanded.

If a capacitor instead of a resistor is used as the feedback element, as shown in figure 7.25(*c*), an *integrator* results. This circuit will be recognized as being identical to the Miller effect equivalent circuit for an active device, as discussed in previous chapters. If A_v is high, so that $i_g = -i_f$, then

$$v_{out} = v_{C_F} = \frac{1}{C_F} \int i_f \, dt = -\frac{1}{C_F} \int i_g \, dt$$

and because

$$i_g = \frac{e}{R_G}$$

$$v_{out} = -\frac{1}{C_F R_G} \int e \, dt . \tag{7.40}$$

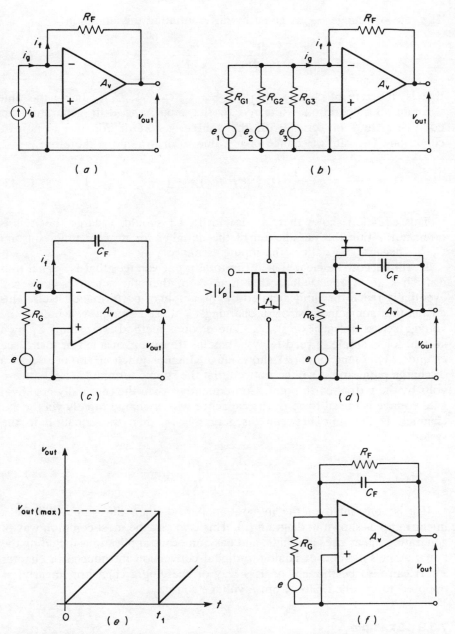

Figure 7.25. Basic applications of the inverting configuration. (*a*) Transresistance amplifier; $v_{out} = \simeq i_g R_F$. (*b*) Summing amplifier; $v_{out} \simeq -R_F[(e_1/R_{G1}) + (e_2/R_{G2}) + (e_3/R_{G3}) \ldots]$. (*c*) Integrating amplifier; $v_{out} \simeq -(1/C_F R_G)\int e\, dt$. (*d*) Integrator with FET discharge path; $v_{out} \simeq -(1/C_F R_G)\int_0^{t_1} e\, dt$. (*e*) Integrator output for constant input; (*f*) AC integrator.

The rate of change v_{out} is given by differentiating this equation:

$$\frac{dv_{out}}{dt} = -\frac{e}{C_F R_G} \tag{7.41}$$

that is, the rate of change of v_{out} is $-e/C_F R_G$ V/s, $C_F R_G$ being the time constant of integration. Referred to the input, the drift generators act additively, and in series with the source generator for the worst-case condition. The rate of change of v_{out} due to drift alone is therefore

$$\frac{v_{out}}{T} = -\frac{1}{C_F R_G}(e_{os} + i_b^- R_G) = -\left(\frac{e_{os}}{C_F R_G} + \frac{i_b^-}{C_F}\right). \tag{7.42}$$

This equation shows that, for low drift, C_F should be large, but this is often not within the jurisdiction of the designer for a given time constant $C_F R_G$, because R_G defines the input resistance.

It will be now be clear that if the total input voltage, that is, signal plus drift, has an *average* value other than zero, then the output voltage will eventually reach the limit dictated by the amplifier performance itself. This means that some method of discharging C_F must be employed before the voltage across it reaches $v_{out(max)}$. The obvious method is to shunt C_F by a switch; for example, a reed relay, a bipolar transistor, or a FET or IGFET (see chapter 11). Figure 7.25(d) shows an n-channel junction FET in use as a discharge path, and it will be noticed that the source terminal is held at zero volts by the virtual earth point. Consequently, when the (normally negative) gate voltage is raised to zero, the capacitor will discharge rapidly via the FET channel. If the time between discharges is t_1, then the equation to the system is

$$v_{out} = -\frac{1}{C_F R_G}\int_0^{t_1} e\,dt. \tag{7.43}$$

If e is constant over the integration period, then the output will be a linear ramp as shown in diagram (e). This is in fact the most common way of generating linear rises and falls, and has numerous applications including the "dual slope" method of analog-to-digital conversion introduced in chapter 15. It can also be thought of as a way of charging a capacitor linearly, as opposed to exponentially, from a voltage source.

7.3.9 Example

A 2-V DC voltage source with an internal resistance of $10\,k\Omega$ is applied to an integrator. Calculate the value of feedback capacitor which would be needed to produce a positive-going ramp which reaches 10 V in 0.5 s. What polarity input signal would be needed to achieve this?

Using equation (7.43)

$$C_F = -\frac{1}{V_{out}R_g} \int_0^{0.5} (-2)\, dt$$

$$= \frac{2 \times 0.5}{10 \times 10^4} = 10\ \mu F.$$

Note that the input polarity must be negative if the output ramp is to be positive-going, which is evident by inspection because the relevant operational amplifier is connected in the inverting mode.

Figure 7.25(f) shows an AC integrator; that is, a system where C_F is shunted by a permanent resistance R_F. For frequencies much lower than $1/C_F R_G$, the gain will be simply $-R_F/R_G$, but at frequencies much higher than $1/C_F R_G$ the system becomes an integrator. In this latter case the input signal e will be integrated, and will appear at the output superimposed on the voltage produced by the drift. That is, the total output voltage is

$$v_{out} = -\frac{R_F}{R_G}\left(e_{os}\frac{R_G + R_F}{R_F} + R_G i_b^-\right) - \frac{1}{C_F R_G}\int e\, dt. \tag{7.44}$$

By choosing an operational amplifier with a low drift characteristic, the first term may be rendered very small. If, however, an AC output only is called for, a capacitive coupling to the load will render the drift plateau unimportant *providing* that the maximum value of v_{out} given by equation (7.44) does not exceed the capability of the amplifier.

One interesting property of the AC integrator is that if the input is sinusoidal, it can be used to shift the phase by 90°. Neglecting the drift component, and using equation (7.44), let $e = E \sin \omega t$ so that,

$$v_{out} = -\frac{1}{C_F R_G}\int E \sin \omega t\, d(\omega t)$$

$$= \frac{E}{C_F R_G}\cos \omega t$$

$$= \frac{E}{C_F R_G}\sin\left(\omega t + \frac{\pi}{2}\right)$$

which clearly constitutes a 90° phaseshift, and this is particularly useful for the generation of quadrature signals.

The foregoing treatment of basic applications has assumed an infinite R_{in}, and a very large and frequency independent value of A_v. Neither of these assumptions is true, but state-of-the-art operational amplifiers are such as to render the relevant equations very close approximations. For marginal applications, however, full analyses should be performed using the tenets outlined in the early part of the chapter, otherwise errors will occur.

Because of the very large number of applications appropriate to operational amplifiers, however, such analyses cannot be included in the present volume and the reader is referred to the list of publications given at the end of the chapter.

7.3.10 Basic Applications of the Noninverting Configuration

Fundamentally, the noninverting connection is a voltage amplifier having a gain of $1 + R_{F1}/R_{F2}$ and (usually) an input resistance of $R_{in(CM)}$. The standard circuit is that of figure 7.26(a), while figure 7.26(b) shows an important modification, the voltage follower. Here, $v_{out} = e$ (provided that $R_G \ll R_{in(CM)}$) and the circuit is essentially a high-accuracy buffer. If the input is a precision DC voltage reference (as described in section 7.6) the circuit becomes a useful low-impedance voltage source.

The output impedance of the voltage follower has already been shown in chapter 5 to be $-Z_{out}/A_v$, where A_v is the internal gain of the amplifier used. This is clearly a very small quantity, so that the circuit does closely approach the ideal voltage source concept.

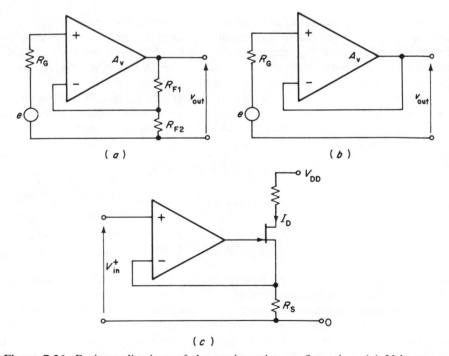

Figure 7.26. Basic applications of the noninverting configuration. (a) Voltage amplifier; $v_{out} \simeq e[1 + (R_{F1}/R_{F2})]$. ($b$) Voltage follower; $v_{out} = e$. (c) Current source; $I_D = V_{in}^+/R_S$.

A close approach may also be made to the ideal current source by including an auxiliary active element such as the FET (see chapter 11) shown in figure 7.26(c). Here, because the voltages at each of the inputs are almost identical, and are of value V_{in}^+, the output current is simply

$$I_D = \frac{V_{in}^+}{R_S}.$$

7.4 THE DIFFERENCE AMPLIFIER

Because both inverting and noninverting inputs are available, pure difference operation of the balanced-input operational amplifier is possible. The simplest configuration is the open-loop circuit of figure 7.27(a) where, neglecting drift and noise components, the output voltage is

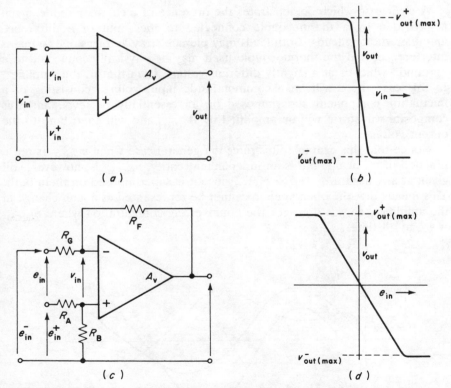

Figure 7.27. Difference amplifiers and characteristics. (a) Simple difference amplifier (or voltage comparator if A_v is large); $v_{out} \simeq A_v v_{in}$. ($b$) Voltage comparator transfer characteristic. (c) Difference amplifier with feedback; $v_{out} \simeq -(R_F/R_G)(e_{in}^- - e_{in}^+)$. ($d$) Transfer characteristic of feedback difference amplifier.

$$v_{\text{out}} = A_v(v_{\text{in}}^- - v_{\text{in}}^+) + A_{v(\text{CM})} \frac{v_{\text{in}}^- + v_{\text{in}}^+}{2}. \qquad (7.45)$$

At this point, it is convenient to examine more closely the meaning of the common-mode input voltage. Figure 7.28 shows two input voltage waves v_{in}^- and v_{in}^+ and the resultant difference and common-mode voltages. When the two input voltages are equal, $v_{\text{CM}} = v_{\text{in}}^- = v_{\text{in}}^+$, but when they are not equal, v_{CM} takes the median value between the two. This results in the expression for the common-mode voltage being $(v_{\text{in}}^- + v_{\text{in}}^+)/2$.

This being so, the two input voltages measured with respect to the common line must be

$$v_{\text{in}}^- \text{ or } v_{\text{in}}^+ = \frac{v_{\text{in}}^- + v_{\text{in}}^+}{2} \pm \frac{v_{\text{in}}^- - v_{\text{in}}^+}{2}$$

$$= v_{\text{in(CM)}} \pm \frac{v_{\text{in}}}{2}. \qquad (7.46)$$

A clear case which demonstrates the presence of a common-mode signal is provided by a thermocouple connected to the input of a difference amplifier via long leads. Both leads may pick up stray voltages due to mains interference, and the thermocouple itself may be physically connected to a "ground" which is at a slightly different voltage from that of the amplifier. In this case there will be a common-mode input voltage consisting of a fluctuating component superimposed on an essentially DC level, and this composite waveform will be amplified by $A_{v(\text{CM})}$ and will contribute to the output voltage.

Notice that this example illustrates the general case when $v_{\text{in(CM)}}$ is not a function of v_{in}^+ or v_{in}^-, but is an independent entity. $v_{\text{in(CM)}}$ is, however, still given at any instant by $(v_{\text{in}}^+ + v_{\text{in}}^-)/2$ since it is superimposed on them both. This means that the phenomenon cannot be represented as a gain change in the amplifier, as was done for the noninverting configuration where $v_{\text{in(CM)}} = v_{\text{in}}^+$ at all times.

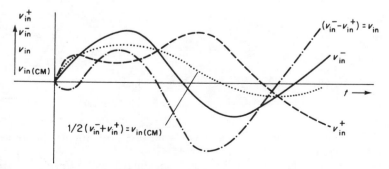

Figure 7.28. Input, difference and common-mode voltages.

Returning to the open-loop difference amplifier of figure 7.27(a), it will be apparent that because A_v is very large, then v_{in} must be minute. Hence, for practical purposes, the system has ceased to be an amplifier, and has become a *voltage comparator*; that is, when v_{in}^- differs from v_{in}^+ even very slightly, v_{out} will swing to the maximum value of which the amplifier is capable, $v_{out(max)}$. The polarity of $v_{out(max)}$ will, of course, depend on whether v_{in}^- is greater or less than v_{in}^+, as is evident from the transfer function of figure 7.27(b).

Figure 7.27(c) shows a difference-input circuit which will accept comparatively large input voltages. The difference-input voltage gain may be determined simply by letting e_{in}^- and e_{in}^+ be zero separately, then summing the two outputs so determined:

for $e_{in}^+ = 0$,

$$v_{out} = e_{in}^-\left(-\frac{R_F}{R_G}\right)$$

and for $e_{in}^- = 0$,

$$v_{out} = \frac{e_{in}^+ R_B}{R_A + R_B}\,\frac{R_G + R_F}{R_G}$$

so that, in general,

$$v_{out} = e_{in}^-\left(-\frac{R_F}{R_G}\right) + e_{in}^+\,\frac{R_B}{R_A + R_B}\,\frac{R_G + R_F}{R_G}\,.$$

If resistors are chosen so that $R_A = kR_G$ and $R_B = kR_F$, where k is a constant,

$$v_{out} = e_{in}^-\left(-\frac{R_F}{R_G}\right) + e_{in}^+\left(\frac{R_F}{R_G}\right)$$

$$= -\frac{R_F}{R_G}\,(e_{in}^- - e_{in}^+)\,. \tag{7.47}$$

This equation shows that useful input signal differences may be linearly amplified by choosing a suitable R_F/R_G ratio.

The chief disadvantages of this circuit both devolve upon the difficulty of establishing that $(R_A + R_B) = k(R_G + R_F)$ *exactly*. Firstly, a variable-gain facility is virtually impossible to achieve without introducing a mismatch resulting from variable-resistor tracking problems; and secondly, if a mismatch *is* present, then any common-mode signal will be amplified in part as a difference signal. This latter point becomes obvious if a common-mode input signal e_{cm} is added to both e_{in}^- and e_{in}^+, and the foregoing derivation is repeated. For exact resistor matching, the common-mode component of the output becomes zero: if a mismatch is present, it does not. Various techniques have been developed to overcome such disadvantages, the results being found in the range of *instrumentation amplifiers* now available.

7.4.1 The Instrumentation Amplifier

The development of instrumentation amplifiers resulted from the need for amplifiers able to select millivolt difference signals from volts of common-mode interference—that is, present very high common-mode rejection ratios—along with pre-settable gains and high input impedances at both inputs. Such amplifiers are often necessary for use with transducers such as biomedical electrodes, difference-connected thermocouples or twin photo-detectors in double-beam-in-space optical systems.

Instrumentation amplifiers usually incorporate more than one operational amplifier and figure 7.29 illustrates a typical three-amplifier system, which itself indicates the need for multi-amplifier integrated circuits.

The system shown includes two input amplifiers, each of which accepts an input signal at its noninverting input. The outputs are applied to a difference-coupled amplifier which has four very closely matched, fixed resistors of value R_0. (These are often individually trimmed thick-film resistors.)

A common-mode input signal e_{cm} has also been included, and the following analysis is similar to that used for the single difference amplifier. The output voltages of the two input amplifiers may be found by letting only one signal be nonzero at a time.

If $e_{in(2)}$ and e_{cm} are zero, then $e_{in(1)}$ is amplified by the noninverting gain of amplifier A_1:

$$e_3 = \left(1 + \frac{R_2}{R_1}\right)e_{in(1)} . \tag{7.48a}$$

If $e_{in(1)}$ and e_{cm} are zero, then because the difference input signal of amplifier A_2 is negligibly small, essentially the whole of $e_{in(2)}$ is applied to

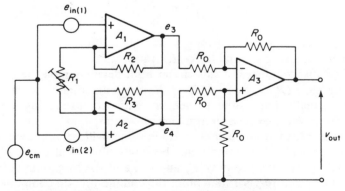

Figure 7.29. An instrumentation amplifier system.

the inverting input of A_1:

$$e_3 = -\frac{R_2}{R_1} e_{in(2)} . \tag{7.48b}$$

If $e_{in(1)}$ and $e_{in(2)}$ are zero, then

$$e_3 = \left(1 + \frac{R_2}{R_1}\right) e_{cm} - \frac{R_2}{R_1} e_{cm} = e_{cm} . \tag{7.48c}$$

Equations (7.48) can now be added by the Theorem of Superposition to give:

$$e_3 = \left(1 + \frac{R_2}{R_1}\right) e_{in(1)} - \frac{R_2}{R_1} e_{in(2)} + e_{cm} . \tag{7.49a}$$

Also, by symmetry,

$$e_4 = \left(1 + \frac{R_3}{R_1}\right) e_{in(2)} - \frac{R_3}{R_1} e_{in(1)} + e_{cm} . \tag{7.49b}$$

The output amplifier performs the subtraction $(e_4 - e_3)$ and if $R_2 = R_3$ then,

$$v_{out} = \left(1 + \frac{2R_2}{R_1}\right) (e_{in(2)} - e_{in(1)}) . \tag{7.50}$$

Note that in this equation, there is no common-mode component in the output. This would also be the case if R_2 were not exactly equal to R_3, and only a gain error would result. Equations (7.49) show that whereas the difference gains of the input amplifiers are $[1 + (R_2/R_1)]$ the common-mode gains are unity. Hence, the common-mode rejection ratios of these amplifiers are simply the ratio of the two gains, or $[1 + (R_2/R_1)]$. The inherent CMR of the output amplifier is therefore multiplied by this factor, which accounts for the excellent overall CMR of the system.

Note also that R_1 appears in the final gain expression (equation (7.50)), so that if it is made variable, a single-resistance pre-set voltage gain facility is afforded.

Finally, the third requirement of an instrumentation amplifier—that high impedances should be presented at both inputs—is achieved by both input signals being taken to noninverting input points.

The problem of common-mode voltage can be very severe; for example, if small voltage variations at the terminal of a high-voltage generator need to be measured, the effective common-mode voltage can be some thousands of volts. Conversely, in certain electro-medical applications, a patient must be isolated from possible leakage currents down to $50 \mu A$ or so, while still being monitored using bioelectrodes. Both applications call for complete unilateral isolation of the amplifier and its power supply from the input.

Various methods of accomplishing this are possible, the most popular being the opto-isolator (discussed in chapter 12) and the transformer. The latter technique involves modulation of a carrier frequency by the input signal so that it can be transmitted across the highly insulated windings of a transformer.

The technique of modulating a carrier by a low-frequency signal has long been used in the so-called *chopper amplifier*, where it was used to avoid the drift problems encountered in direct-coupled amplifiers. However, while modern instrumentation amplifiers are challenging the chopper amplifier in all areas of performance it is nevertheless appropriate to consider this form of amplifier since it remains widely used, and this will be done after introducing some modern techniques of offset, bias and drift reduction.

7.5 SOME TECHNIQUES OF LOW OFFSET, BIAS AND DRIFT DESIGN

Consider the difference input stage, as shown in figure 7.30, and assume that the input offset voltage is simply the difference between the two base–emitter voltages for zero output voltage:

$$e_{os} = V_{BE1} - V_{BE2} . \tag{7.51}$$

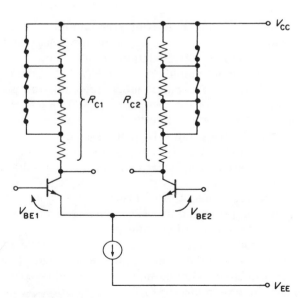

Figure 7.30. Offset voltage trimming method showing fusible links.

This can be expressed in terms of R_{C1} and R_{C2}, which implies that e_{os} may be minimized if R_{C1} and/or R_{C2} can be appropriately adjusted. To show this, first principles may be invoked (as usual):

$$I_C \simeq I_{ES} \exp \left(\frac{qV_{BE}}{kT} \right)$$

$$\frac{qV_{BE}}{kT} \simeq \ln \left(\frac{I_C}{I_{ES}} \right).$$

Hence

$$e_{os} \simeq \frac{kT}{q} \left[\ln \left(\frac{I_{C1}}{I_{ES1}} \right) - \ln \left(\frac{I_{C2}}{I_{ES2}} \right) \right]$$

$$= \frac{kT}{q} \ln \left(\frac{I_{C1}}{I_{C2}} \frac{I_{ES2}}{I_{ES1}} \right). \tag{7.52}$$

For the difference output voltage to be zero, $V_{CE1} = V_{CE2}$, so that $(I_{C1}/I_{C2}) = (R_{C2}/R_{C1})$, giving:

$$e_{os} \simeq \frac{kT}{q} \ln \left(\frac{R_{C2}}{R_{C1}} \frac{I_{ES2}}{I_{ES1}} \right).$$

Thus, for $e_{os} = 0$,

$$\left(\frac{R_{C2}}{R_{C1}} \frac{I_{ES2}}{I_{ES1}} \right) = 1 \tag{7.53}$$

which can be attained by trimming one or both of the collector load resistors even if the two transistors are not quite identical (I_{ES1} and I_{ES2} were retained in this derivation to highlight the latter point).

In practice, each collector load can be made up from chains of several resistors of varying (usually binary-weighted) magnitudes. Some of them are shunted by fusible aluminum links or equally fusible Zener diode structures[16], represented by fuse symbols in figure 7.30. Then, during testing, e_{os} is measured, then minimized by electrically fusing an appropriate number of the trimming links.

As an example of the practicality of this technique, the Precision Mono-lithics Inc. type OP–07A operational amplifier, is quoted as having a maximum offset voltage of 25 μV (typical value 10 μV) and a temperature drift of 0.6 (typically 0.2) μV/°C. Such low offset and drift voltages are useful in numerous applications involving low "DC" signals such as those from thermocouples, and also where integration is involved, as indicated by equation (7.42).

In chapter 11 it will be shown that the field-effect transistor can be used in the design of extremely low input bias current amplifiers and the super-gain bipolar transistor has already been mentioned in this context. Another technique is that of *bias-current cancellation*, which once again uses the properties of the current mirror.

The idea behind bias-current cancellation is to supply the base bias current to the two input transistors from somewhere other than the signal source. Therefore, it becomes necessary to determine the magnitude of these currents, generate them, and apply them to the two input bases in parallel with the signals.

Figure 7.31 shows how this may be achieved. Taking the left side of the input pair, the collector current of Tr1 is also the emitter current of Tr3. Thus, if the two structures are well matched, the base current of Tr3 will be the same as that of Tr1, and it is this current which establishes the V_{BE} of the diode-connected pnp structure Tr5. Therefore, Tr6 mirrors this current and applies it to the base of Tr1 (at a high impedance level).

The same happens on the right-hand side, and if all the npns and all the pnps are well matched and have high current gains, the signal source will be called upon to provide very little bias current. It will never be zero, however, because the current gains are not infinite, and in fact those of the pnp structures may be quite low because of the low collector current levels at which they operate. Also, the input *offset* current is likely to be high because it now depends upon mismatches not only between Tr1 and Tr2 but between their associated current-cancellation circuits. In fact it is this which limits the bias current reductions to about a twentieth of their uncompensated values at best.

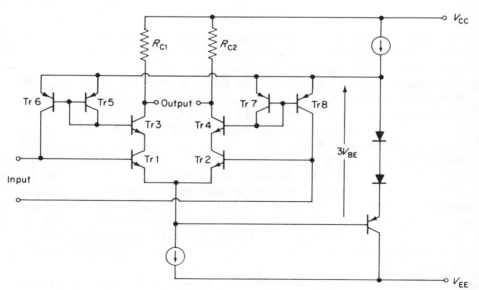

Figure 7.31. A base current cancellation technique.

The current source on the extreme right of the circuit energises the diode string shown. This establishes that a voltage of $3V_{BE}$ is available to drive the input transistors and their cancellation circuits. Therefore, looking again at the left side of the input circuit, one V_{BE} exists across Tr5, one across Tr3 (B to E) and hence only one V_{BE} is left to supply Tr1.

The OP–07A exemplifies this technique also, and exhibits a maximum input bias current of ± 2 nA (typically ± 0.7 nA); and an input offset current of 2 nA maximum (typically 0.3 nA).

Continued progress in solid-state design and fabrication is allowing continuous-current operational amplifiers to challenge the performance of chopper amplifiers, but new approaches to these latter techniques continue to afford extra benefits at very low frequencies, as will now be described.

7.5.1 Chopper Amplifier Techniques

By definition, AC-coupled amplifiers do not drift, because they cannot operate at DC. Therefore, if a "DC" (i.e., very slowly changing) signal can be converted to AC, then amplified and finally reconstituted, the result is a drift-free "DC" amplifier. The inherent proviso in this statement is that the processes of *modulation* and *demodulation* do not introduce drift (or other spurious signals) of their own.

Figure 7.32(*a*) shows a system where the modulator and demodulator are represented by an oscillating change-over switch, and figure 7.32(*b*) shows the resulting waveforms. The input signal e_{in} is chopped by the modulator switch, and appears as AC at the amplifier output, with a mean value of zero. The function of the demodulator switch is to clamp the top (or bottom) of this waveform to zero; that is, to effect *DC restoration*.

Early chopper amplifiers used electromechanical switching, which limited both their frequency response and life. Modern choppers are solid state, using bipolar or field-effect transistors in ways which will be described in later chapters, but the frequency response is still limited. In fact, the highest signal frequency which can be amplified must be less than half the chopping frequency, otherwise *aliasing* will result. (This is a form of frequency mixing between the signal and the chopper which gives rise to apparent signals at the difference frequency between the two.)

The frequency restriction imposed by a chopper amplifier is not an insuperable problem, for the higher frequencies can be amplified by an associated amplifier. This is the principle of the Goldberg amplifier[17], shown in block diagrammatic form in figure 7.33.

The upper amplifier is a standard direct coupled difference amplifier having two inputs, so that the gain may be $+A_2$ or $-A_2$. The lower amplifier is an AC coupled unit connected between a modulator and demodulator. Its gain is $-A_1$ and its drift is assumed to be very low. The whole arrangement carries input and feedback resistors R_G and R_F, making it an operational amplifier of gain $-R_F/R_G$.

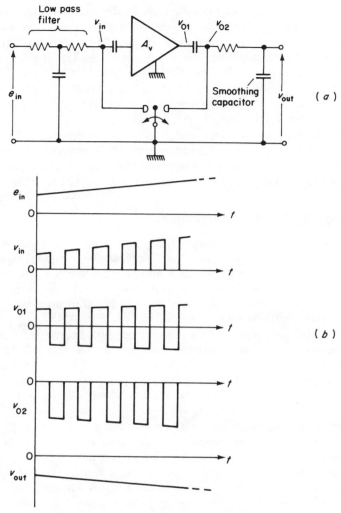

Figure 7.32. (*a*) Chopper amplifier and (*b*) associated (idealized) waveforms.

When a signal v_{in} is applied, a current will flow into input 1 of A_2 if the frequency is high enough for the reactance of C_1 to be small. The output will be $-A_2 v_{in}$. However, if the signal is of a very low frequency, then it will be blocked by C_1 but it will be able to enter A_1 via the low-pass filter and the chopper. Here it will be amplified to $-A_1 v_{in}$ and applied to the alternate input of A_2. When this input point is in use the gain involved is $+A_2$, so the net output is $-A_1 A_2 v_{in}$.

This means that when a DC or very low-frequency signal is applied, it passes through the two amplifiers in cascade, while if the frequency is

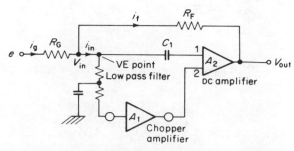

Figure 7.33. Basic form of chopper stabilized (Goldberg) amplifier.

higher, only A_2 is operative. This situation is not deleterious to the linearity of the system, however, for the overall gain is always given by $-R_F/R_G$ providing that the forward gain at all working frequencies is great enough to make the operational amplifier relationship true. In practice, the filters present in the circuits are so arranged that a uniform rate of decrease of gain is obtained, and this is tailored to be always less than 10 dB/octave so that the system can never become unstable.

Numerically, A_2 might be $\pm 10^5$ while A_1 is -10^3, so that for DC the internal gain is -10^8, falling to unity at about 10 MHz.

Drift reduction is effected in the following manner. Inherently, the chopper amplifier A_1 is almost drift free, so that any drift at the output may be assumed to arise in A_2. If the drift at the output is e_{od}, then this represents a signal at one of the inputs, e_{od}/A_2. However, this input drift will be much less than any DC or very low-frequency signal since A_2 has been preceded by A_1. Thus, the drift signal at the virtual earth point to the left of C_1 will be e_{od}/A_1A_2. In other words, the drift has been reduced by a factor equal to the gain of the amplifier A_1.

Actually, this argument is not affected even if C_1 is absent (as it was in the original Goldberg amplifiers), for the DC signal which now appears at input 1 of A_2 is only $1/A_1$ of that appearing at input 2, which makes it of negligible importance.

An analysis giving the gain and drift performance of such an amplifier may be made by considering the voltage and currents indicated in figure 7.33. Let the drift referred to the input of A_2 be e_d. The shunt feedback system may be approached as in chapter 5 where an analysis of figure 5.12 was made, which resulted in equation (5.20). Equating currents at the virtual earth point gives:

$$i_{in} = i_g - i_f$$

$$= \frac{e - v_{in}}{R_G} - \frac{v_{in} - v_{out}}{R_F}$$

or

$$i_{in}R_G R_F = R_F(e - v_{in}) - R_G(v_{in} - v_{out}) \, .$$

Rearranging,

$$i_{in}R_G R_F = eR_F + v_{out}R_G - v_{in}(R_G + R_F) \, . \tag{7.54}$$

Also,

$$v_{out} = -A_2(v_{in} + e_d) - A_1 A_2 v_{in}$$
$$= -A_2 e_d - A_2 v_{in}(1 + A_1)$$

giving

$$v_{in} = \frac{-A_2 e_d - v_{out}}{A_2(1 + A_1)} \, . \tag{7.55}$$

Inserting (7.55) in (7.54) gives

$$i_{in}R_G R_F = eR_F + v_{out}R_G + \frac{A_2 e_d + v_{out}}{A_2(1 + A_1)}(R_G + R_F) \, .$$

Rearranging to find v_{out} gives:

$$v_{out} = \left(-e\,\frac{R_F}{R_G} + i_{in}R_F - \frac{e_d}{(1 + A_1)}\,\frac{R_G + R_F}{R_G} \right)$$
$$\times \left(\frac{1}{1 + (R_G + R_F)/R_G A_2(1 + A_1)} \right) \, .$$

$$\tag{7.56}$$

Consider the implications of equation (7.56), ignoring the second bracket for the present. The first bracket has three terms which contribute to the output voltage as follows:

(a) The first term implies the gain of a perfect operational amplifier, its contribution to v_{out} being a linear function of e.

(b) The second term introduces an error proportional to the input leakage current i_{in} and the feedback resistor. Normally, i_{in} is very small since the relevant driving voltage, v_{in}, is also very small owing to its appearance at the virtual earth point. Note that i_{in} may be assumed to act through R_F, though in fact it does not pass through this resistor, but only into the internal input impedance of the amplifiers.

(c) The final term shows that the qualitative argument given previously is essentially true. The drift referred to the input of A_2 is reduced by a factor equivalent to the gain of A_1 (where $A_1 \gg 1$). If $e_d/(1 + A_1)$ is assumed to act across R_G, then e_{od} must be given by the product of this quantity and the resistor ratio $(R_G + R_F)/R_G$.

All these terms must now be multiplied by the second bracket, and the meaning of this is as follows. From the algebra leading to equation (7.55), it is apparent that $-A_2(1 + A_1)$ is the gain of the whole amplifier, v_{out}/v_{in}, when the drift term is neglected. Also, the factor $R_G/(R_G + R_F)$ is the attenuation ratio at the virtual earth point for any signal passing from the output to the input via R_F and R_G. Therefore, the product of these two factors is the gain around the loop through A_1, A_2, and R_F; that is, the loop gain

$$A_L = -A_2(1 + A_1) \frac{R_G}{R_G + R_F} .$$ (7.57)

The second bracket now becomes

$$\left(\frac{1}{1 - (1/A_L)} \right) .$$

In order to show that for "normal" gains, this bracket tends to unity, take the case where $A_1 = 10^3$, $A_2 = 10^5$, $R_G = 1000 \, \Omega$ and $R_F = 1 \, \text{M}\Omega$.

From equation (7.57), $A_L = 10^5$, so the second bracket takes a value very nearly unity and its effect on the first bracket is negligible.

From a practical viewpoint, the input signal has been split up into low and high frequencies, the former being handled by the chopper amplifier A_1 and the latter by the "normal" DC amplifier A_2. This means that the low bandwidth limitation of the chopper amplifier has been overcome and the drift associated with a "normal" DC amplifier of gain A_1A_2 has been reduced by the gain of A_1.

An alternative form of chopper stabilization is shown in figure 7.34. Here, the two amplifiers are so arranged that the drift at the output of A_2 is sensed by A_1, amplified, and applied to the input of A_2 so as to cancel out the original output drift. This is a straightforward application of negative feedback principles, the basic problem being to separate the drift from the signal in order that the original aim may be accomplished. The way in which this is done may be shown as follows.

If the gain A_2 is large, and negative

$$v_{out} = -\left(\frac{R_F}{R_G} e + e_{od} \right)$$ (7.58)

where e_{od} is the drift at the output. Therefore

$$v_{in} = -\frac{R_F}{R_G} \frac{e}{A_2} - \frac{e_{od}}{A_2} .$$ (7.59)

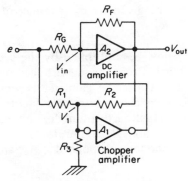

Figure 7.34. Alternative form of chopper stabilization.

This means that the voltage at the virtual earth point must be changed by e_{od}/A_2 in order that the drift might be cancelled out.

If $R_1 \gg R_3$ and $R_2 \gg R_3$, then

$$v_1 = e \frac{R_3}{R_1} + v_{out} \frac{R_3}{R_2}$$

$$= e \frac{R_3}{R_1} - \frac{R_3}{R_2} \left(\frac{R_F}{R_G} e + e_{od} \right)$$

from (7.58), or

$$v_1 = e \frac{R_3}{R_1} \left(1 - \frac{R_1 R_F}{R_2 R_G} \right) - \frac{R_3}{R_2} e_{od}.$$

If

$$\frac{R_2}{R_1} = \frac{R_F}{R_G}$$

this becomes

$$v_1 = - \frac{R_3}{R_2} e_{od}. \tag{7.60}$$

If $A_1 = -R_2/(R_3 A_2)$, then its output voltage becomes e_{od}/A_2, which is the voltage required to cancel out the drift according to equation (7.59).

A modern form of drift sensing is afforded by the Intersil Commutating Auto Zeroing amplifier (the CAZamp)[18]. This device uses solid-state switches called CMOS transmission gates (described in chapter 11), which are able to transmit analog signals of either polarity with very little distortion.

The CAZamp contains two similar operational amplifiers, both having normal drift properties, and figures 7.35 show the two alternating states of these operational amplifiers.

In figure 7.35(*a*), capacitor C_2 is charged only by the offset and noise voltage of amplifier 2, which is connected in the unity-gain configuration. The various solid-state switches then change over, so that figure 7.35(*b*) becomes valid. Now, capacitor C_2 is connected in a direction such as to *oppose* the offset and noise voltage of amplifier 2; and the signal is in series with it. At this time, C_1 is being charged to the drift and noise level of

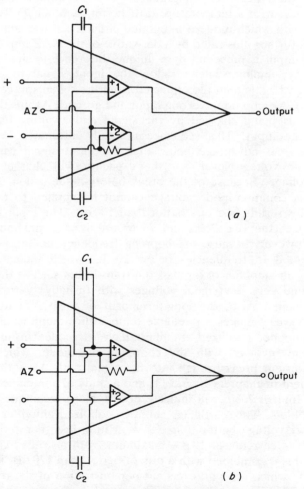

Figure 7.35. The two switched states of the Intersil CAZamp (reproduced by permission of Harris Semiconductor Inc.).

amplifier 1; and after the next change-over, C_1 will null out this voltage and also appear in series with the signal.

Hence, the CAZamp switches and their associated capacitors act to cancel out drift and noise sequentially rather than continuously; that is, whilst one amplifier is processing the signal, the drift and noise voltage of the other is being stored ready for oppositional application at the next switch-over.

Such amplifiers have an auto-zero connection which is available for referencing by the user and the only extra circuitry required are the two capacitors C_1 and C_2, which are typically 1 μF polycarbonate units. The chopping frequency is normally 160 Hz, but can be externally modified if so desired. For the types ICL 7600 and 7601, the maximum offset voltage is quoted as 5 μV, with a temperature drift better than 0.1 μV/°C.

The ease with which modern integrated circuits can use analog switching techniques is further illustrated by a derivative of the CAZamp, in which the noninverting input is preceded by a unique difference-to-single-ended converter. Briefly, analog switches allow the input signal to charge up a capacitor C_3, which is then disconnected from the signal source and applied between the CAZamp noninverting input and ground. At this time a second capacitor C_4 is being charged by the signal and awaiting its turn to be applied to the input. This sequence normally alternates 160 times per second, the great advantage being that the input signal can be at any common-mode voltage level, for the capacitor will charge only to the difference voltage. Because of the break-before-made action of the analog switches, this common-mode voltage cannot be applied to the CAZamp input, and this results in a CMR better than 100 dB. The input bias currents for this device (the ICL 7605 and 7606) are 1.5 nA maximum (typically 0.15 nA). However, because of chopping frequency limitations, the input signal is limited to frequencies below 20 Hz, which indicates its prime usefulness as an amplifier of inputs from transducers such as thermocouples which combine very low signal voltages with (usually) very low rates of amplitude change. Also, the long-term drift of the CAZamp (quoted at about 0.5 μV/year) makes for reliable temperature monitoring.

Modern chopper stabilized amplifiers have succeeded in minimizing the basic problem involved with the chopping technique, which is that of switching transient breakthrough into the signal path. How this arises will become evident in chapters 10 and 11, and a study of contemporary devices, such as the Intersil 7650, will indicate how modern techniques have coped with the problem. This particular chopper-stabilized amplifier exhibits not only an offset voltage better than 5 μV, drifting less than 0.01 μV/°C but also offers bias currents of 10 pA maximum, with an offset current below 0.5 pA. All this is combined with a CMR of better than 120 dB, and a voltage gain over 10^6, which offer an excellent demonstration of the results obtainable using modern monolithic integrated circuit techniques combining both analog and digital networks.

7.6 SUPPLY-INSENSITIVE CURRENT AND VOLTAGE SOURCES

The current provided by the collector of the mirror transistor in the simple pair of figure 7.2(a) is directly proportional to the diode current; and if this is provided by a voltage source (such as V_{CC}) and a resistor, then the mirror current becomes proportional to that voltage. In other words a voltage-to-current converter has been produced, and such devices have many uses in electronics. However, if an independently constant current is required, then either a constant-current circuit *per se* can be designed, or else a constant voltage source can be used.

The form of current source which uses a degenerated mirror transistor, as in figure 7.2(b), does itself provide some degree of independence. Consider equation (7.2) again, and put $kT/q = V_T$ for convenience:

$$I_{C2} = \frac{V_T}{R_E} \ln\left(\frac{I}{I_{C2}}\right).$$

The rate of change of I_{C2} with I is:

$$\frac{dI_{C2}}{dI} = \frac{V_T I_{C2}}{R_E I} \frac{d}{dI}\left(\frac{I}{I_{C2}}\right)$$

$$= \frac{V_T I_{C2}}{R_E I}\left(\frac{I_{C2} - I(dI_{C2}/dI)}{I_{C2}^2}\right)$$

$$= \frac{V_T}{R_E}\left(\frac{1}{I} - \frac{dI_{C2}}{I_{C2}\,dI}\right).$$

So

$$\frac{dI_{C2}}{dI}\left(1 + \frac{V_T}{R_E I_{C2}}\right) = \frac{V_T}{R_E I}$$

giving

$$\frac{dI_{C2}}{dI} = \left[\frac{R_E I}{V_T}\left(1 + \frac{V_T}{R_E I_{C2}}\right)\right]^{-1}$$

$$= \frac{I_{C2}}{I}\left(\frac{R_E I_{C2}}{V_T} + 1\right)^{-1}.$$

The fractional rate of change of I_C with I is:

$$\frac{dI_{C2}/I_{C2}}{dI/I} = \frac{dI_{C2}}{dI}\frac{I}{I_{C2}} = \left(1 + \frac{R_E I_{C2}}{V_T}\right)^{-1} \tag{7.61}$$

which is always less than unity. This fractional rate of change improves as the voltage drop across the emitter resistor increases, which is to be expected because this drop represents the amount of degenerative feedback applied to the mirror transistor. Hence, the circuit provides greater stability for greater ratios of I to I_{C2}. (Notice also that the form of equation (7.61) is similar to that of equation (4.6) for the bias stability of a simple CE stage with an emitter resistor.) Finally, if a voltage source along with a resistor were used to supply I, then any fluctuations in that voltage source would result in lower fractional fluctuations in I_{C2}, because $I \propto V_{ref}$.

7.6.1 A Design Example

A mirror transistor is to provide a current of 10 μA, and a reference current is available using a voltage source (which can vary between 5.5 and 6.5 V) along with a 5300 Ω series resistor. Calculate the value of the emitter resistance needed and determine the percentage fluctuation possible in the mirror current. Assume that V_{BE} for the diode-connected transistor is 0.7 V.

Using the current mirror of figure 7.2(b), the mean reference current is given by the mean value of the reference voltage as:

$$I_{mean} = \frac{6 - 0.7}{5300} = 1 \text{ mA} .$$

Hence, using equation (7.2),

$$R_E = \frac{V_T}{I_{C2}} \ln \left(\frac{I}{I_{C2}} \right)$$

$$= \frac{0.026}{10^{-5}} \ln \left(\frac{10^{-3}}{10^{-5}} \right) \simeq 12 \text{ k}\Omega .$$

From equation (7.61) the fractional change in I_{C2} relative to that in I is:

$$\frac{dI_{C2}/I_{C2}}{dI/I} = \left(1 + \frac{R_E I_{C2}}{V_T} \right)^{-1}$$

$$= \left(1 + \frac{12 \times 10^3 \times 10^{-5}}{0.026} \right)^{-1} \simeq 0.18 .$$

The percentage change in V_{ref} (and hence in I) is:

$$\left(\frac{6.5 - 5.5}{6} \right) 100 \simeq 16.7\%$$

giving the percentage change in I_{C2} as $16.7 \times 0.18 \simeq 3\%$.

Because there are several methods of producing constant voltage sources (some of which are detailed later), the need for a linear and temperature-stable voltage-to-current converter exists. The current-mirror variations described above are simple examples of such $V-I$ converters, as is the difference pair circuit of figure 5.39. More accurate and temperature-stable circuits exist which provide both unilateral and bilateral (i.e., one and two-quadrant) voltage-to-current conversion in both discrete and monolithic form[19,20,21].

Voltage sources are derived from three convenient references, the Zener diode, the V_{BE} drop and V_T. The Zener diode is treated further in chapter 14 in terms of its operation and use in power supplies, but in the context of integrated circuits two special forms of it are of particular interest.

Firstly, the base–emitter junction of a standard monolithic transistor structure can be reverse-biased, and will break down near to 6 V. By limiting the current through this junction to a level which will cause neither permanent damage nor significant heating, this reverse breakdown voltage can be used as a reference.

An idealized cross-section through a normal npn structure is shown in figure 7.36(a), and when reverse-biased, this breaks down at about 6.7 V, and exhibits a temperature coefficient of about 2 mV/°C. Figure 7.36(b) shows how the p-region can be diffused down to the buried n-layer, and the junction between this isolation region and the emitter tends to break down at a somewhat lower voltage, typically 5.6 V, and also exhibits a lower temperature coefficient.

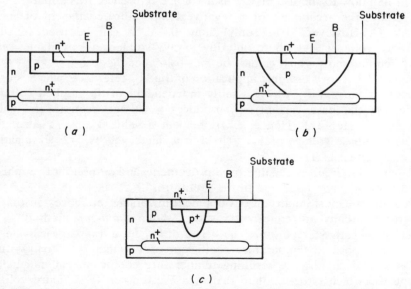

Figure 7.36. Monolithic Zener diode structures (idealized cross-sections). (a) Standard emitter-base structure; (b) emitter-isolation Zener and (c) buried Zener structure.

However, both types of Zener structure produce considerable noise and long-term drift, largely because they are essentially surface-phenomena devices. Also, their internal dynamic resistances can be several tens of ohms.

The second alternative is provided by the buried Zener structure shown in figure 7.36(c). Here, a deep, but small-area p^+-diffusion is made as shown and breakdown (at about 6.3 V) occurs between this and the n^+ emitter, that is below the surface. The current to the p^+-region passes laterally through the p-type base surrounding it. This results in lower noise, temperature coefficient and dynamic resistance.

The National Semiconductor Corp. type LM199 monolithic IC[22] uses a buried Zener structure along with a heating and temperature stabilizing circuit which maintains the die at 90°C. The buried Zener itself is associated with ancilliary circuitry (on the same chip) which controls its operating current, affords some temperature compensation, and buffers the output. The result is a device which will operate between 0.5 and 20 mA, and provide a 6.95 V reference with an internal resistance of 0.5 Ω and a temperature coefficient given as 0.3 ppm/°C. Also, its total noise voltage is about 7 μV over a bandwidth of 10 Hz to 10 kHz.

The LM199 is actually a complete IC replacement for a discrete Zener diode, but buried Zener structures of the type it uses are built into other integrated circuits which need reference voltages, such as digital-to-analog converters, by several manufacturers.

Turning now to the use of V_{BE} as a voltage reference, it is apparent that this implies the acceptance of a negative temperature coefficient of about -2 mV/°C. However, being only about 0.7 V, it can be used for low reference voltages and where only low supply voltages are available, both being requirements which cannot be met by Zener diodes.

Figure 7.37 shows a simple application of the V_{BE} reference, which is used to define a current I_{C2} which is fairly independent of the applied voltage. Neglecting the base current, I_{C2} produces a voltage drop in R_E which provides V_{BE} for Tr1. Hence, I_{C2} is held at a value V_{BE1}/R_E. Also, the collector-to-base voltage of Tr1 is held at a single V_{BE} by Tr2, so negating the Early effect in Tr1.

Again, this circuit is capable of improvement, and several more sophisticated versions have appeared in the literature.

Possibly the most important development in voltage references, insofar as integrated circuits are concerned, is the Widlar *band-gap* method[23], in which the negative temperature coefficient of V_{BE} for a transistor is balanced against the positive temperature coefficient of V_T (that is kT/q). With a well-designed circuit, the resulting temperature coefficient can fall below 20 ppm/°C which is better than that of a Weston cell (a standard voltage reference).

To obtain V_T, two transistors are run at two different current densities and the difference between their values of V_{BE} is extracted. In practice, this

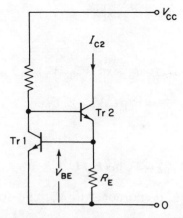

Figure 7.37. A current source using V_{BE} as a reference.

means using either two identical transistors with different values of I_C or two transistors of different emitter areas, and similar values of I_C. For reasons of simplicity, consider the former and refer to figure 7.38.

If Tr1 and Tr2 are identical, but carry different collector currents, then the difference between their values of V_{BE} is obviously the voltage drop across R_E, which may be found as follows:

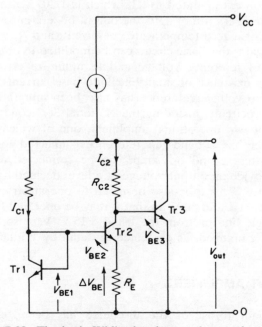

Figure 7.38. The basic Widlar band-gap voltage reference.

$$I_{C1} \simeq I_{ES} \exp \left(V_{BE1}/V_T \right)$$

and

$$I_{C2} \simeq I_{ES} \exp \left(V_{BE2}/V_T \right)$$

$$\frac{I_{C1}}{I_{C2}} \simeq \exp \left(\Delta V_{BE}/V_T \right)$$

or

$$\Delta V_{BE} \simeq V_T \ln \left(I_{C1}/I_{C2} \right)$$

which shows that ΔV_{BE} is proportional to V_T, and so exhibits a positive temperature coefficient.

Now ΔV_{BE} exists across R_E, so the voltage across R_{C2} must approach $\Delta V_{BE} R_{C2}/R_E$. This, plus V_{BE3}, is the output voltage:

$$V_{out} \simeq V_{BE3} + \left(\Delta V_{BE} R_{C2}/R_E \right). \tag{7.62}$$

The negative temperature coefficient of V_{BE3} is greater than the positive temperature coefficient of ΔV_{BE}, so if they are to cancel out, R_{C2}/R_E must be carefully chosen. In fact it can be shown that[24] for a net temperature coefficient of zero the output voltage should be equal to the band-gap voltage of silicon extrapolated to 0 K which is 1.205 V.

At a room temperature of 25°C the output or reference voltage turns out to be 1.262 V for a zero temperature coefficient, and R_{C2}/R_E is adjusted at about 10:1. Again, this basic circuit can be modified to obtain, for example, higher values of reference voltage and to minimize errors such as those incurred by the existence of small but finite base currents[25].

The band-gap voltage reference has also been integrated into complete monolithic microcircuits including the National Semiconductor Inc. LM10. This consists of two operational amplifiers, one of which is uncommitted, whilst the other has a band-gap reference connected to its noninverting input. Hence this second op. amp. can be connected as a noninverting amplifier to provide an output voltage at a level decided by the user, and at a low impedance. This IC is also designed to operate from power supplies ranging between 1.1 and 40 V, so that it may be operated from a single cell, from a 5 V logic line or from the usual ±15 V. Very sophisticated design techniques are employed, as explained in detail by Widlar[26].

7.7 CURRENT AMPLIFIERS

So far the main thrust of the discourse has been related to voltage amplification, for it is this in which the designer is normally interested as an end result. However, it will have become apparent that voltage can be treated as a secondary variable; for example, it is voltage which is produced across a collector load resistance by the current in that load, which can be regarded as being derived from a current source. Diodes and bipolar transistors can be thought of as current-operated devices and design procedures based on this outlook have much to offer. In particular, if it is recalled that for a capacitor $i = C \, dv/dt$, then if there is little change in the voltage applied there will be little change in the charge inserted and extracted from that capacitor. Such a situation occurs at the virtual ground point of an inverting operational amplifier; but it does not occur at the point from which an internal compensating capacitor is driven. (This capacitor determined the slew rate in the op. amp. of figure 7.12.)

If capacitive current can be minimized then, in general, the high-frequency response improves. A case in point is afforded by the common-base stage, and this and its equivalent circuit are shown in figures 7.39. The full analysis will be left to the reader as an exercise, but the main point to note is that $C_{b'c}$ now appears in parallel with the output (if $r_{bb'}$ is small) as

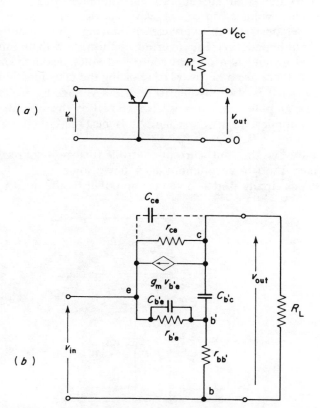

Figure 7.39. (*a*) Common base configuration and (*b*) hybrid-π equivalent circuit.

opposed to appearing from output to input as in the CE stage. Hence there is no Miller effect resulting from this capacitance (though a little "negative" Miller capacitance is contributed by the very small value of C_{ce} along with the noninverting voltage gain of the stage). This is why the lateral pnp transistors in figures 7.4 and 7.12 do not limit the frequency response of the network. These lateral pnps are interposed between the npn emitter-followers and their loads, and one of these loads is a diode-connected transistor. Therefore, very little voltage swing occurs in the system, except at the commoned collectors of Tr4 and Tr6, thus making for a good frequency response.

It is also possible to isolate the load of a CE stage using a CB stage as shown in figure 7.40. This is called the *cascode* configuration, in which voltage excursions at the collector of Tr1 are held at almost zero by the V_{BE} drop in Tr2 so negating the Miller effect in Tr1. The load appears in the collector of the CB stage Tr2, which itself has a negligible Miller capacitance. Hence the high-frequency capability of the cascode stage is very good indeed[27].

Returning to the complete operational amplifier, this is essentially a voltage-controlled voltage-source, or vcvs, device and the noninverting configuration demonstrates this very clearly, having a high input impedance and a low output impedance. The inverting connection with an input directly into the virtual ground is a current-controlled voltage-source or ccvs amplifier. (In fact, there are other ways of realizing the ccvs function as will be described below.) It is also possible to realize a voltage-controlled current-source or vccs amplifier and this is also known as the operational trans-conductance amplifier, or OTA, because it is best described by a Norton equivalent circuit.

Finally, there is the full current-controlled current-source, or cccs, amplifier which though uncommon does have some useful features. In particular, a cccs circuit used as a voltage amplifier by the inclusion of input

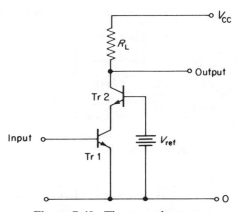

Figure 7.40. The cascode stage.

and feedback resistors has a value of f_H which is sensibly independent of the gain[28], whereas a normal op. amp. exhibits a gain–bandwidth product type of response in which gain must be traded for bandwidth. Figure 7.41 shows a symbol for a cccs amplifier; the filled-in arrow at the output denotes a high-impedance current source output and the circle-and-arrow version of a current-source symbol between the inputs denotes a current-difference characteristic:

$$A_i = \frac{i_{out}}{(i_{in}^+ - i_{in}^-)} \ .$$

The output voltage produced by such an amplifier is:

$$V_{out} = i_{out} R_F \quad (\text{if } R_L = \infty)$$
$$= A_i(i_{in}^+ - i_{in}^-)R_F$$
$$= A_i\left[\frac{e_{in}^+}{R_g^+} - \left(\frac{e_{in}^-}{R_g^-} + \frac{V_{out}}{R_F}\right)\right]R_F$$
$$= A_i\left(e_{in}^+\frac{R_F}{R_g^+} - e_{in}^-\frac{R_F}{R_g^-} - V_{out}\right)$$

giving

$$V_{out}(A_i + 1) = A_i\left(e_{in}^+\frac{R_F}{R_g^+} - e_{in}^-\frac{R_F}{R_g^-}\right)$$

or

$$V_{out} = \frac{A_i}{A_i + 1}\left(e_{in}^+\frac{R_F}{R_g^+} - e_{in}^-\frac{R_F}{R_g^-}\right). \qquad (7.63a)$$

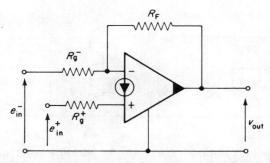

Figure 7.41. A current-controlled current-source (cccs) amplifier.

If $R_g^+ = R_g^-$ this becomes,

$$V_{out} = \frac{A_i}{(A_i + 1)} \left(\frac{R_F}{R_g} (e_{in}^+ - e_{in}^-) \right). \tag{7.63b}$$

The important point here is that the bracket contains no frequency-dependent terms, so that the voltage gain as a function of frequency is dependent upon the A_i term alone. The current amplifier may consist of a modified Wilson current mirror or various similar configurations[28]. It is worthy of note that the basic Gilbert cell is a current amplifier, and was originally treated as such[29].

An example of a ccvs or Norton amplifier[30] is the National Semiconductor Inc. type LM159/359, the symbol and input circuit for this being given in figures 7.42. Figure 7.42(a) shows the current symbol at the input, and also an arrow on the non-inverting terminal to show that this is the main input

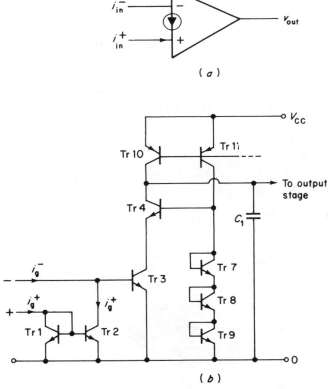

Figure 7.42. Symbol (*a*) and input circuit (*b*) for National Semiconductor Inc. type LM159/359 Norton amplifier (reproduced by permission).

current point. Figure 7.42(b) explains this; most of the noninverting input current flows into the diode-connected transistor Tr1 and is closely mirrored in the collector of Tr2. The inverting input current is thereby forced to supply this mirror current, and the small difference between the two currents provides the base drive for Tr3.

The CE transistor Tr3 forms part of a cascode pair with the CB transistor Tr4, the base of which is held at three V_{BE} drops by the diode-connected chain Tr7, 8, and 9. This chain is supplied by current from the mirror transistor Tr11, and another mirror transistor Tr10 (driven from the same source) forms the active load for the cascode pair[31]. The current drive to this amplifier, along with the cascode gain stage, results in a gain–bandwidth product of some 400 MHz.

Other advantages include a small chip area, because of the simplicity of the circuit and the necessity for only a 12 pF roll-off capacitance C_1, which means that multiple units can be fabricated on one chip. Also, the amplifier can be conveniently operated from a single supply (such as a standard 5 V logic rail) when the output swing will range from a few millivolts to about 2 V below V_{CC}.

The design approach to circuits employing the CCVS amplifier is best illustrated by reference to the basic biasing system of figure 7.43[31]. Here the input bias current to the noninverting terminal I_{in}^+ is mirrored by that into the inverting terminal, which can only be provided via R_F. This means that V_{out} must rise to a level sufficient to produce this current, which defines the quiescent output voltage.

Keeping in mind that a V_{BE} drop (of about 0.6 V) will appear at each input, the output voltage is then given by:

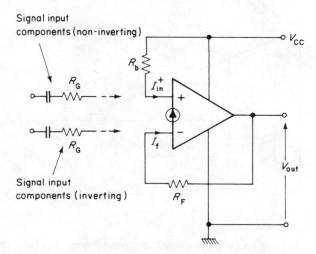

Figure 7.43. Biasing and signal input scheme for the LM159/359 Norton amplifier.

$$V_{\text{out}} = I_f R_F + V_{\text{BE}}^- \qquad (7.64)$$

and the input bias current must be:

$$I_{\text{in}}^+ = \frac{V_{\text{CC}} - V_{\text{BE}}^+}{R_b} \approx I_f. \qquad (7.65)$$

Signals may now be injected into either input as shown on the left of figure 7.43.

The converse of the Norton amplifier is the voltage-controlled current-source or operational transconductance amplifier (OTA) for which figure 7.44(a) is a symbol. The input circuit of this amplifier is a normal long-tailed pair; but the output is taken from the commoned collectors of a complementary pair of npn/pnp structures. This means the output is at a high impedance, so simulating a current source. Hence, the output side of the device is best represented by a balanced Norton equivalent circuit as shown in figure 7.44(b). The transconductance g_m of this equivalent is determined

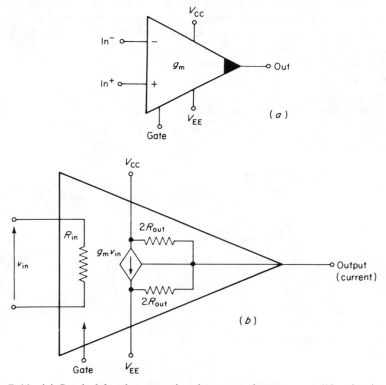

Figure 7.44. (a) Symbol for the operational transconductance amplifier (OTA) symbol and (b) simplified equivalent circuit.

by the long-tail current I_x of the input difference amplifier, and this may be determined by the user because the base of the long-tail transistor is accessible via the "Gate" electrode. The OTA is therefore a variable-gain device, but the input pair can also be switched ON or OFF, so vastly increasing the range of its applications[32].

The OTA is particularly useful where linear capacitance charging is desired, such as in time-bases and sample-and-hold circuits (see chapter 13). It is also applicable where it is necessary to connect outputs in parallel as in the case of multiplexers (see chapter 11). (Remember that current sources can be connected in parallel, but not in series, and vice-versa for voltage sources.)

7.8 NOISE IN OPERATIONAL AMPLIFIERS

As was described in chapter 6, the various sources of noise can be combined into hypothetical noise generators having lumped values referred to the input. Figure 7.17 shows where such generators appear in an equivalent circuit for the operational amplifier. Figure 7.45(a) shows those generators alone at the input of an ideal amplifier.

If sufficient information is available on the magnitude of these generators—and particularly if spot noise values are available in graphical form—the noise figure for the amplifier can be determined, as was shown in chapter 6. However, most operational amplifier specification sheets specify noise generators in terms of peak-to-peak or RMS values over various bandwidths.

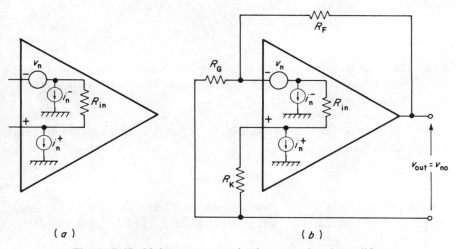

Figure 7.45. Noise generators in the operational amplifier.

Should it be necessary to determine noise levels at the output this can be done simply by multiplying each noise generator by the appropriate gain for the circuit involved. For example, in figure 7.45(b) the contributions to the output noise of the internal noise generators and the thermal noise generated in the various resistors is as in the following list (using instantaneous quantities).

(a) Because v_n is in a similar position to the drift voltage generator e_{os}, it can be treated in the same way; that is, it may be referred to the input to become $v_n(R_G + R_F)/R_F$. The output noise due to this source is now:

$$v_{no(a)} = \frac{v_n(R_G + R_F)}{R_F} \frac{R_F}{R_G} = v_n\left(1 + \frac{R_F}{R_G}\right).$$

(Note: The factor $[(1 + (R_F/R_G)]$ in an inverting amplifier is sometimes called the "noise gain.")

(b) The noise-current generator at the inverting input appears across R_G. Therefore, by a Norton–Thévenin transformation, it becomes a voltage $i_n^- R_G$, which must be multiplied by the inverting gain:

$$v_{no(b)} = i_n^- R_G\left(\frac{R_F}{R_G}\right) = i_n^- R_F .$$

(c) The noise-current generator at the noninverting input appears across $R_K{}^*$, so that a Norton-Thévenin transformation gives a voltage $i_n^+ R_K$, and this must be multiplied by the noninverting gain:

$$v_{no(c)} = i_n^+ R_K\left(1 + \frac{R_F}{R_G}\right).$$

(d) The thermal noise in R_G is amplified by the inverting gain to give:

$$v_{no(d)} = v_{nR_G}\left(\frac{R_F}{R_G}\right).$$

(e) The thermal noise generated in R_F exists between the output and the virtual earth (or summing) point, so that its contribution is simply,

$$v_{no(e)} = v_{nR_F} .$$

(f) Finally, the noise generated in R_K is amplified by the noninverting gain to give:

$$v_{no(f)} = v_{nR_K}\left(1 + \frac{R_F}{R_G}\right).$$

Each of the above expressions has involved instantaneous noise values; according to the material presented in chapter 6, the resultant RMS output

*Often, R_K will be R_d as in figure 7.21(b).

noise may be expressed by taking the root of the sum of the squares of the individual components, provided that they are uncorrelated. Thus,

$$\sqrt{(\overline{v_{no}^2})} = \sqrt{[\overline{v_n^2}(1 + R_F/R_G)^2 + \overline{i_n^{-2}}R_F^2 + \overline{i_n^{+2}}R_K^2(1 + R_F/R_G)^2}$$
$$+ \overline{v_{nR_G}^2}(R_F/R_G)^2 + \overline{v_{nR_F}^2} + \overline{v_{nR_K}^2}(1 + R_F/R_G)^2]. \qquad (7.66)$$

Because the terms under the root sign are squares, it is possible to simplify the expression at the expense of a little accuracy. Suppose, for example, that the RMS value of a term is more than three times the RMS value of another. This means that the square of the former is nine times larger than the square of the latter. Therefore, neglecting the latter would introduce an error of only some 5%, if only these two terms were involved. Hence, no great error will be introduced if any term which has a numerical value below $\frac{1}{3}$ of any other term is neglected.

The thermal noise terms may be calculated using the methods of chapter 6 but the values of v_n, i_n^+, and i_n^- must be either extracted from manufacturers' data sheets or measured. The former calculation depends upon the noise bandwidth being known (equation (6.2)) and the latter depends upon both the noise bandwidth and the spot-noise generator values being known. Frequently, manufacturers do not quote spot-noise values but present the RMS (or peak) noise values themselves over given bandwidths.

For instance, the Analog Devices 118 operational amplifier noise generators referred to the input are given as:

$$v_n(0.01 \text{ Hz to } 1 \text{ Hz}) = 1 \text{ } \mu\text{V peak-to-peak (max)}$$
$$v_n(10 \text{ Hz to } 10 \text{ kHz}) = 2 \text{ } \mu\text{V RMS (max)}$$
$$i_n^- = i_n^+(0.01 \text{ Hz to } 1 \text{ Hz}) = 20 \text{ pA RMS (max)}.$$

Notice firstly that in the low-frequency band where flicker noise would be expected to predominate, the peak-to-peak (P-P) noise voltage is given. This is because in certain critical applications this maximum instantaneous P-P noise voltage may be added to the maximum difference input signal expected plus the maximum drift voltage so that the maximum output excursion can be determined. Such a calculation can determine the maximum gain conditions in which the amplifier should be used.

Keeping in mind that noise is random, it is, strictly, only possible to specify a *probability* that the noise will never exceed the P-P maximum given. Often, a 99.9% probability is used, which according to the Gaussian distribution implies a P-P value of $6.6 \times$ RMS value. Using this factor, noise calculations over the lower bandwidths using RMS values may also be carried out.

As an example, the Analog Devices 118 operational amplifier may be used along with the noise data given above.

7.8.1 Noise Calculation Example

In the noninverting configuration, the noise generators would appear as in figure 7.46.

To determine the output noise in the bandwidth 10 Hz to 10 kHz, the list of input noise components may be made up as in the above discussion, using the noise value for the 118 of $v_n = 2\ \mu V$ over the frequency band 10 Hz to 10 kHz. This frequency band is largely outside the $1/f$ region, where i_n^+ and i_n^- are too small to matter (which is why they are not given). However, to illustrate how i_n^+ and i_n^- would be treated if they were significant, the value given for the lower band (0.01–1.00 Hz), which is 20 pA RMS, will be used in the appropriate calculations. Further, a sequence of calculations for this noninverting amplifier will be given in the same order as for the inverting-amplifier sequence given on page 272–3. (Note that had the lower frequency band been used, then the P-P value of v_n would have been divided by 6.6 to convert it to RMS.)

(a)
$$v_{no(a)} = v_n\left(1 + \frac{R_{F1}}{R_{F2}}\right) = 2(100) = 200\ \mu V\ .$$

(b) In this case i_n^- appears across the combination of R_d and $R_{F1} \parallel R_{F2}$; that is, across a resistance value R_G. Hence, a Norton–Thévenin transformation gives a noise voltage generator, $i_n^- R_G$ and this appears in series with R_{in} and R_d etc. Hence it must be multiplied by the noninverting gain to give

$$v_{no(b)} = i_n^- R_G\left(1 + \frac{R_{F1}}{R_{F2}}\right) = (20 \times 10^{-12})(10^3)(100) = 2\mu V$$

(c) The noise-current generator at the noninverting input appears across R_G so that,

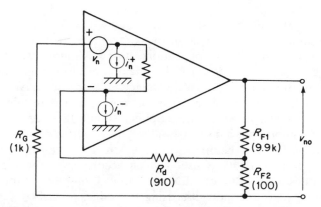

Figure 7.46. Noise generators in the noninverting configuration ($A_{v(FB)} = 100$; $R_K = R_G = 1\ k\Omega$).

$$v_{no(c)} = i_n^+ R_G \left(1 + \frac{R_{F1}}{R_{F2}} \right) = 2 \ \mu V \ .$$

(d) The thermal noise generated in R_G may be determined from figure 6.5(a) as about $4 \ nV/\sqrt{Hz}$. The RMS value over the bandwidth 10 Hz to 10 kHz is given by multiplying this spot noise by the square root of that bandwidth. The contribution to the output noise will therefore be:

$$v_{no(d)} = (4 \times 10^{-9})(\sqrt{10^4})(100) = 40 \ \mu V \ .$$

(e) There is no inverting feedback resistor R_F in this example.

(f) The thermal noise generated in the $R_d + R_{F1} \| R_{F2} = R_G$ combination is exactly as in (d) above, and is also multiplied by the noninverting gain

$$v_{no(f)} = 40 \ \mu V \ .$$

The root of the sum of the squares of these components will give the output noise according to equation (7.66). However, only items (a), (d), and (f) are of importance so that,

$$v_{no} = \sqrt{200^2 + 40^2 + 40^2} = 208 \ \mu V \ \ \text{RMS} \ .$$

The sequence of noise contributions can also be used to determine the noise factor. Recalling that this is the ratio of the output noise power of the amplifier to the output noise power of an amplifier of the same gain but without sources of internal noise, then, using mean square values,

$$NF = \frac{\Sigma_a^f v_{no}^2}{\Sigma_d^f v_{no}^2} \ .$$

Here again, (b) and (c) can be ignored, so that,

$$NF = \frac{200^2 + 40^2 + 40^2}{40^2 + 40^2} = 13.5$$

or, in decibels

$$NF \ (dB) = 10 \log_{10} 13.5 = 11.3 \ dB \ .$$

(Note that these values refer to the 10 Hz to 10 kHz frequency band, not to the "spot-noise" condition.)

Finally, it is occasionally found that spot-noise values *are* given in data sheets. In such cases, the flicker noise generators can be dealt with as in chapter 11 on field effect transistors. Also, it has been shown[33] that for ccvs or Norton amplifiers, most of the noise arises in the input current mirror and can be represented by a single noise current generator connected between the inverting input and ground[33].

REFERENCES

1. Camenzind, H. R. and Grebene, A. B., 1969, An outline of design techniques for linear integrated circuits, *IEEE J. Solid St. Circuits*, **SC-4**, 110–121.

2. Grebene, A. B., 1972, *Analog Integrated Circuit Design* (New York: Van Nostrand Reinhold).

3. Till, W. C. and Luxon, J. T., 1982, *Integrated Circuits: Materials, Devices, and Fabrication* (New Jersey: Prentice-Hall).

4. Wilson, G. R., 1968, A monolithic junction f.e.t–n.p.n. operational amplifier, *IEEE J. Solid St. Circuits*, **SC-3**, 341–348.

5. Wilson, B., 1981, Current mirrors, amplifiers and dumpers, *Wireless World*, 47–50 (December).

6. Widlar, R. J., 1969, Super-gain transistors for I/C's, *IEEE J. Solid St. Circuits*, **SC-4**, 249–251.

7. National Semiconductor Inc, 1969, I/C Op. Amp. beats FETs on input current, *Application Note*, AN 29–1.

8. Graeme, J. G., Tobey, G. E. and Huelsman, L. P., 1971, *Operational Amplifiers—Design and Applications* (New York: McGraw-Hill).

9. Bowers, D., 1979, Monolithic op. amps. in high-speed applications, *Electronics* (October 16th).

10. Solomon, J. E., 1974, The monolithic operational amplifier—a tutorial study, *IEEE J. Solid St. Circuits*, **SC-9**, 314–332.

11. Russell, R. W. and Fredericksen, T. M., 1972, Automotive and industrial building blocks, *IEEE J. Solid St. Circuits*, **SC-7**, 446–454.

12. Hearn, W. E., 1971, Fast-slewing monolithic amplifier, *IEEE J. Solid St. Circuits*, **SC-6**, 20–24.

13. Davis, P. C., Mayer, S. F. and Saari, V. R., 1974, High slew-rate monolithic amplifier using compatible complementary PNP's, *IEEE J. Solid St. Circuits*, **SC-9**, 340–347.

14. Komath, B. Y., Meyer, R. G. and Gray, D. R., 1974, Relationship between frequency response and settling time of operational amplifiers, *IEEE J. Solid St. Circuits*, **SC-9**, 347–352.

15. Demrow, R. I., 1968, Narrowing the margin of error, *Electronics*, **41**, No 8, April 15th.

16. Erdi, G., 1975, A precision trim technique for monolithic analog circuits, *IEEE, J. Solid St. Circuits*, **SC-10**, 412–416.

17. Goldberg, E. A., 1950, Stabilization of wideband amplifiers for zero and gain, *RCA Review*, June 298.

18. Intersil Staff 1980, *Data Acquisition Handbook* (California-Intersil Inc.) 293–296.

19. Hart, B. L. and Barker, R. W. J., 1975, A precision bilateral voltage-current converter, *IEEE J. Solid St. Circuits*, **SC-10**, 501–503.
20. Hart, B. L., 1976, Circuit technique for wide-range, linear, voltage-to-current converters, *Electron. Lett.*, **12**, 298–299.
21. Pookaiyaudom, S. and Surakampontorn, W., 1978, An integrable precision voltage-to-current converter with bilateral capability, *IEEE J. Solid St. Circuits*, **SC-13**, 411–413.
22. Dobkin, R. C., 1976, On-chip heater helps to stabilize monolithic reference Zener, *Electronics*, 106–112, Sept. 16th.
23. Widlar, R. J., 1971, New developments in IC voltage regulators, *IEEE J. Solid St. Circuits*, **SC-6**, 2–7.
24. Kujik, K. E., 1973, A precision reference voltage source, *IEEE J. Solid St. Circuits*, **SC-8**, 222–226.
25. Brokaw, A. P., 1974, A simple three-terminal IC bandgap reference, *IEEE J. Solid St. Circuits*, **SC-9**, 388–393.
26. Widlar, R. J., 1978, Low voltage techniques, *IEEE J. Solid St. Circuits*, **SC-13**, 838–846.
27. Horowitz, P. and Hill, W., 1980, *The Art of Electronics* (Cambridge: Cambridge University Press) section 13.05.
28. Allen, P. E. and Terry, M. B., 1980, The use of current amplifiers for high performance voltage applications, *IEEE J. Solid St. Circuits*, **SC-15**, 155–162.
29. Gilbert, B., 1968, A new wide-band amplifier technique, *IEEE J. Solid St. Circuits*, **SC-3**, 353–365.
30. Frederiksen, T. M., Davis, W. F. and Zobel, D. W., 1971, A new current-differencing single-supply operational amplifier, *IEEE J. Solid St. Circuits*, **SC-6**, 340–347.
31. Regan, T., Sevastopoulos, N., Kohan, C. and Lange, J., 1979, Dual current-differencing amplifier permits high-frequency designs, *EDN* 99–105, Sept. 20th.
32. Wittlinger, H. A., 1974, Application of the CA3080 and CA3080A high-performance operational transconductance amplifiers, *RCA Application Note*, ICAN 6668.
33. Haslett, J. W., 1974, Noise performance of the new Norton op. amps., *IEEE Trans. Electron. Dev.* ED-21, 571–577.

QUESTIONS

1. What are the main disadvantages of a lateral transistor in a monolithic integrated circuit, compared with a vertical structure?

2. Describe two methods of realizing a current mirror in which the mirror current is smaller than the reference current.

3. What is the simplest method of realizing a Zener diode structure in an integrated circuit containing bipolar transistors?

4. Explain how a current mirror can act as the load for a difference amplifier, and effect a double-to-single-ended conversion.

5. What is a "super-gain" transistor?

6. For what is a "V_{BE}-multiplier" used?

7. In figure 7.12, which component defines the 6 dB/octave roll-off of the amplifier; and which defines the quiescent power consumption?

8. In figure 7.12, explain the functions of (*a*) the Tr5/6/7 combination, and (*b*) Tr15 or Tr22.

9. In figure 7.15, why can the first stage of the operational amplifier be regarded as a transconductance stage?

10. Assuming that $R_d = 0$ in both cases, explain why the inverting operational amplifier configuration of figure 7.18 never has a common-mode input signal, whereas the noninverting configuration of figure 7.22 always experiences a common-mode input signal.

11. What is the basic concept behind the technique of reducing drift by incorporating a resistor R_d into an operational amplifier circuit?

12. If the signal voltage to an integrator (figure 7.25(*c*)) is zero, why does the output rise or fall until limited by the power supply voltage?

13. How may a sinusoidal signal be phaseshifted by 90°?

14. List the attributes of an instrumentation amplifier.

15. What are the advantages of the cascode amplifier stage?

16. Using the equivalent circuit of figure 7.39, explain why the lateral CB difference pair used in the 741 operational amplifier does not severely restrict its frequency response.

PROBLEMS

7.1. An integrated circuit contains three identical uncommitted transistor structures of $h_{FE} = 50$. A current mirror is to be realized using these structures which will produce a mirror current I_C which is within 0.1% of a reference current I. Can this be achieved using only two of the structures or should a Wilson current mirror configuration be utilized? Justify the answer numerically.

7.2. The defined current I for a current-mirror pair of matched npn transistor structures is 2 mA. If a current of 0.4 mA is to flow in the collector of the second transistor, calculate the value of the relevant emitter resistor. Assume that $kT/q = 26$ mV.

7.3. If the mirror current I_{C2} provided by a Wilson current mirror is to be within 99.5% of the defining current I in an integrated circuit, what must be the minimum values of h_{FE} for the (identical) transistors used?

7.4. A V_{BE}-multiplier is to be used to provide a voltage drop of $3V_{BE}$. If 1 mA is available to flow through the two necessary resistors, calculate their values if V_{BE} is taken as 0.6 V, and draw the relevant circuit. Assume that I_B may be neglected.

7.5. An operational amplifier has an open-loop gain of 10^5 and exhibits a high-frequency -3 dB point at 10 Hz. The input resistance is very high and the roll-off is at 6 dB/octave.

If it is connected in the inverting mode with an input resistor of $100\,\Omega$ and a feedback resistor of $4900\,\Omega$, calculate the new high-frequency cut-off point.

7.6. The slewing rate (SR) of an operational amplifier is to be at least $10\,V/\mu s$. If the chip area available limits the compensating capacitor to 15 pF, how much current must be made available from the transistor collector which charges that capacitor?

7.7. A high-gain integrated operational amplifier has a bipolar transistor long-tailed pair as its first stage, the noninverting input being grounded. The input and feedback resistors are $1\,k\Omega$ and $47\,k\Omega$, respectively. If the offset voltage and input bias current drifts are 20 $\mu V/°C$ and 0.1 nA/°C, respectively, by how much will the output voltage change (to a close approximation) if the ambient temperature rises from 18°C to 35°C?

7.8. An inverting amplifier is to have a voltage gain of -100 and an input resistance of not less than $1\,k\Omega$. The output drift voltage must not exceed 8 mV over a temperature range of 10°C to 35°C. An operational amplifier having the following drift characteristics is available:

$$\frac{\Delta e_{os}}{\Delta T} = 1\ \mu V/°C\,, \quad \frac{\Delta i_b}{\Delta T} = 10\ nA/°C\,, \quad \frac{\Delta i_d}{\Delta T} = 2\ nA/°C\,.$$

Assuming that the open-loop gain of this amplifier is very large, can it be used in the required design?

Prove your answer by designing the necessary circuit and calculating the (approximate) maximum possible drift at the output. Also, determine whether a resistor R_d is necessary at the noninverting input, and if so, give its value.

7.9. An operational amplifier has an offset voltage drift of 25 $\mu V/°C$ and an offset current drift of 0.5 nA/°C referred to the input. It is connected in the noninverting configuration where $R_{F1} = 10\,k\Omega$ and

$R_{F2} = 1 \, k\Omega$. The external resistance at the input is $10 \, k\Omega$ and R_d is included. If the output voltage is zero at 20°C, what can it change to at 50°C?

7.10. A pressure transducer has an internal resistance of $20 \, k\Omega$ and its open-circuit voltage is expected to lie between 0.3 and 1.0 V. An operational amplifier is used to amplify this signal by a factor of ten without significantly loading the transducer. Design an appropriate circuit knowning that the temperature drift of the amplifier must never be larger than 1% of the signal magnitude, and that the circuit must operate from 10°C to 60°C.

The maximal drift parameters for the amplifier are as follows:

$$\frac{\Delta e_{os}}{\Delta T} = 20 \, \mu V/°C, \quad \frac{\Delta i_b}{\Delta T} = 8 \, nA/°C, \quad \frac{\Delta i_d}{\Delta T} = 2 \, nA/°C.$$

7.11. An inverting operational amplifier has an input resistor $R_G = 1.1 \, k\Omega$ and a feedback resistor $R_F = 110 \, k\Omega$. A transducer is connected to the input which may be modeled as a linear pressure-dependent voltage generator. At 25°C the output voltage swing of the amplifier is 2.4 V when a pressure change of $10 \, N/m^2$ is applied to the transducer.

If the temperature of the amplifier rises to 45°C after which $8 \, N/m^2$ is applied to the transducer, by what percentage will the output swing be in error if its drift characteristics are maximal as given below?

If the maximum acceptable error is 2.5%, can this be achieved using the same amplifier. If this can be done, justify the method used.

The maximum drift parameters are:

$$\frac{\Delta e_{os}}{\Delta T} = 10 \, \mu V/°C, \quad \frac{\Delta i_b}{\Delta T} = 0.05 \, nA/°C, \quad \frac{\Delta i_d}{\Delta T} = 0.01 \, nA/°C.$$

7.12. The common mode rejection ratio (CMR) for an operational amplifier connected in the noninverting mode is 60 dB. What will be the percentage error in the overall gain which results from this common mode factor? Comment upon the magnitude of this result in the light of other approximations in the expression used.

7.13. A temperature sensor can be modeled as a direct current source which generates $60 \, \mu A$ at 0°C, increasing linearly at $10 \, \mu A/°C$ thereafter. Design an operational amplifier system which will provide an output of zero at 0°C and 10 V at 50°C. Use a $\pm 15 \, V$ power supply, one high-gain operational amplifier with a very high input resistance, and two external resistors. Assume that the sensor needs polarising by connection across one of the 15 V supplies.

7.14. An operational amplifier is connected as a simple integrator to which is applied a 1 kHz sine wave of amplitude 1 V peak. What is the phase shift of the output sine wave and what is its peak value if the input resistor $R_G = 1 \, k\Omega$ and the feedback capacitor $C_F = 0.1 \, \mu F$?

7.15. A simple integrator uses an operational amplifier having an offset voltage of 50 mV and bias currents of 2 μA at the working temperature. If the input resistor $R_G = 1 \, M\Omega$ and the feedback capacitor $C_F = 2.2 \, \mu F$, how long would it take for the output to reach its maximum of ± 14 V if the input were short-circuited at switch-on? Take the value of R_{in} to the amplifier as being very large.

7.16. For the integrator of problem 7.14, what would be the value of the feedback resistor R_F necessary to keep the output voltage down to a magnitude of not more than 0.1 V?

7.17. An instrumentation amplifier has a difference voltage gain of $+100$ and a common mode rejection ratio of 80 dB. If "DC" signals of $v_{in}^+ = 4.010$ V and $v_{in}^- = 3.990$ V are applied at the inputs, determine:

(a) the difference input voltage v_{in};
(b) the common mode input voltage $v_{in(CM)}$;
(c) the exact output voltage.

What percentage is the contribution of the common mode part of the output to the difference part of the output for this amplifier?

7.18. If the defined current I in problem 7.2 changed by 10%, by what percentage would the mirror current I_C change? (Assume that $kT/q = 26$ mV at 25°C).

7.19. An inverting operational amplifier has input and feedback resistors of 4.7 kΩ and 47 kΩ, respectively, and the noninverting input is grounded. At 25°C the output is trimmed to 1 V for a 100 mV DC input. What is the maximum percentage by which the output could be in error if the ambient temperature rose to 55°C?

Demonstrate whether it would be justified to insert a resistor R_d in the noninverting input lead to minimise this error.

For the amplifier used:

$$\frac{\Delta e_{os}}{\Delta T} = 10 \, \mu V/°C \, , \quad \frac{\Delta i_b}{\Delta T} = 20 \, nA/°C \, , \quad \frac{\Delta i_d}{\Delta T} = 4 \, nA/°C \, .$$

8

Frequency-dependent Circuits

Strictly speaking all circuits are frequency dependent, as has been seen already, but there are many cases in electronics where the frequency response of a network forms the main design criterion. In the case of *filters* for example, it may be required to pass signal components at frequencies up to a certain value (low-pass filters), or only above a certain value (high-pass filters), or within a narrow band around a center frequency (bandpass filters). The design and analysis of filters is a very well-documented topic and various powerful mathematical methods have been developed in relation to it. In particular, frequency-plane methods have led to transfer functions in which the relevant constants are fully tabulated[1]. Filters utilizing passive components allied with transistors or complete integrated circuits are called *active filters*, and whilst their theory has been assembled into a coherent whole[2] new techniques continue to develop, such as state variable and digital methods[3]. Also, the use of rapidly-switched capacitors to simulate resistors has made possible the complete integration of *switched-capacitor filters* (see chapter 11).

It is only a short step from filter design to oscillator design, but because the purpose of this brief chapter is not to introduce either topic *per se* but only to show how frequency-dependent passive components may be combined with active elements, a few examples of each will be described. These will highlight the concept of tailoring overall frequency responses, and indicate how input and output resistances (and bias requirements) may be accommodated by various circuit techniques.

Finally, brief introductions to waveform generators, voltage-controlled oscillators (vcos) and phase locked loops (PLLs) conclude the chapter.

8.1   FREQUENCY-SELECTIVE AMPLIFIERS

It is often necessary to select a sinusoidal signal of a particular frequency from some composite waveform. For example, if a light beam is chopped at a mechanical frequency f_0 using a simple segmented chopper wheel and the waveform from a subsequent photocell (see chapter 13) is observed, it will be found to be trapezoidal. Mathematically, such a waveform can be expressed as the sum of a Fourier series containing only sinusoidal terms, and in practice it is possible to select the fundamental (lowest frequency) or any harmonic by using a narrow bandpass filter.

The selection of a fundamental is also sometimes useful in reducing noise in a system. Chapter 6 indicated that bandwidth appears in the expressions for both shot and thermal noise, which implies that any circuit selecting a narrow frequency band will also reduce the noise in the system. It will, of course, also reduce the signal power; for all parts of a complex waveform, other than those due to noise alone, represent signal power, and if an available signal is very small the decision to reduce it further by frequency selection should be carefully considered.

The most obvious way of selecting a narrow band of frequencies is to insert a suitable filter network between two amplifier stages as in figure 8.1(*a*). In principle this can be done by including a set of lags and a set of leads to form ladder networks, but it would result in considerable attenuation, apart from being highly complex and wasteful of components. A more satisfactory method is to use a frequency-selective network (FSN) around a feedback loop as in figure 8.1(*b*). Such a system suggests several modes of operation.

(1) If A_v is negative, then the FSN must have a rejection characteristic as shown in figure 8.2(*b*). This means that zero feedback will be applied at the center frequency f_0, but increasing amounts of negative feedback will appear as the frequency departs from f_0 at either side. The result will be a gain versus frequency curve as in figure 8.2(*c*).

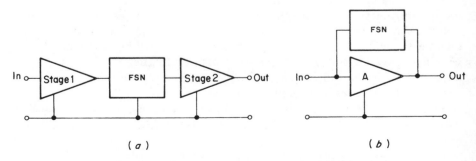

Figure 8.1. Application of frequency-selective networks.

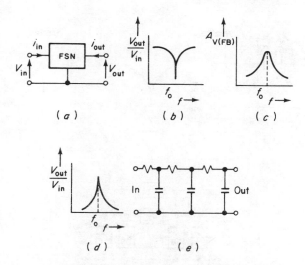

Figure 8.2. FSN characteristics.

(2) If A_v is positive, then either a phase-reversing component (such as a transformer) must exist in the FSN, or else the loop gain A_vB must be less than unity. In the latter case, positive feedback should be a maximum at the center frequency and fall off at either side, as in the acceptance characteristic of figure 8.2(d). It is apparent that with such a system the stability will depend on A_v if A_vB approaches unity.

(3) It is possible to utilize the phase shifting properties of certain networks rather than their selective attenuation characteristics. (Actually, all practical selective amplifiers use a combination of both.) Consider for example a set of three lags forming a ladder network as in figure 8.2(e). The frequency response curves of figure 5.20 for three lags show that the phaseshift curve passes through 180° at one frequency. If such a ladder were connected into the feedback path of an amplifier having a negative gain, then at one frequency the feedback would become positive and, providing that $A_vB < 1$, the forward gain would be a maximum. In practice, the gain versus frequency curve would be severely modified by the attenuation characteristic of the ladder.

These three methods may be illustrated by reference to the following examples.

Firstly, when an amplifier having a negative gain is available, a network such as the *twin-T* may be used. This is illustrated in figure 8.3(a) and its gain and phaseshift characteristics are given in figure 8.3(b). This is an example of a passive *notch filter*.

The frequency for zero transmission from input to output and the relations between component values can be derived by performing a star-

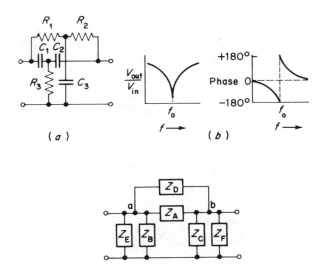

Figure 8.3. The twin-T network.

delta transformation on each set of three elements to give the circuit of figure 8.3(c). From this circuit it can be seen that if $Z_A = -Z_D$, then the admittance between points a and b is zero:

$$1/Z_A + 1/Z_D = 0$$

and so no current will flow and no voltage drop across Z_C and Z_F will appear. The values of Z_A and Z_D are given by the following equations:

$$Z_A = -\frac{j}{\omega C_1} - \frac{j}{\omega C_2} + \frac{(-j/\omega C_1)(-j/\omega C_2)}{R_3}$$

$$= -j\left(\frac{1}{\omega C_1} + \frac{1}{\omega C_2}\right) - \frac{1}{\omega^2 C_1 C_2 R_3} \tag{8.1}$$

also

$$Z_D = R_1 + R_2 + \frac{R_1 R_2}{(-j/\omega C_3)}$$

$$= R_1 + R_2 + j\omega R_1 R_2 C_3. \tag{8.2}$$

Equating the real part of (8.2) to the sign reversed real part of (8.1) gives

$$\omega^2 C_1 C_2 R_3 (R_1 + R_2) = 1 .$$ (8.3)

For the imaginary parts,

$$\omega R_1 R_2 C_3 = \frac{C_1 + C_2}{\omega C_1 C_2} \quad \text{or} \quad \omega^2 R_1 R_2 C_3 = \frac{C_1 + C_2}{C_1 C_2}$$

and substituting for ω^2 from (8.3) gives

$$\frac{C_3}{C_1 + C_2} = R_3 \left(\frac{1}{R_1} + \frac{1}{R_2} \right) ..$$ (8.4)

This analysis assumes that a voltage source is used and that the network feeds into an infinitely high load impedance. Neither of these assumptions can be realized in practice, but may be interpreted to mean that the load impedance must be much higher than the source impedance. A good voltage amplifier approximates to this condition, because the output (which drives the twin-T) has a very low impedance and the input point (which loads the twin-T) exhibits a very high impedance. Consider the design of the network itself. By inspection, the relation of equation (8.4) will obviously hold if $C_3 = C_1 + C_2$ and $R_1 = R_2 = 2R_3$. This simplification gives a rapid method of calculating the center frequency, since equation (8.3) becomes

$$\omega^2 = \frac{1}{4 C_1^2 R_3^2} \quad \text{or} \quad \omega = \frac{1}{2 C_1 R_3}$$

and since $\omega = 2 \pi f_0$

$$f_0 = \frac{1}{4 \pi C_1 R_3} .$$ (8.5)

If a twin-T for about 400 Hz is to be designed, for example, the following components can be used:

$$R_1 = R_2 = 3.9 \, \text{k}\Omega , \quad R_3 = 1.8 \, \text{k}\Omega , \quad C_1 = C_2 = 0.1 \, \mu\text{F} , \quad C_3 = 0.2 \, \mu\text{F} .$$

These values give a center frequency of 442 Hz by equation (8.5) and an actual experiment gave the rejection curve of figure 8.4, which shows that the f_0 is approximately 420 Hz. In order to achieve a good rejection curve it is necessary to match the components to within 2% and this was in fact done, with the result shown in figure 8.4.

This twin-T circuit could be incorporated into the feedback loop of an operational amplifier as in figure 8.1(b), but in order to illustrate various design factors previously introduced, a circuit like that of figure 5.34 will be utilized, along with various modifications.

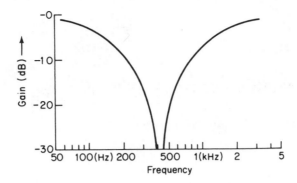

Figure 8.4. Twin-T response curve.

Figure 8.5 shows such a circuit, and here a CA 3045 transistor array has been used, along with a discrete pnp transistor. Collector currents of 1 mA have been chosen for the difference pair Tr1–Tr2, so that Tr5 must act as a 2 mA constant-current source. (Tr5 has its emitter connected to the substrate in the CA 3045 array.) This transistor is driven by the diode-connected Tr4, so that Tr4–Tr5 form a current mirror. The 6.8 kΩ resistor defines a current of about 2 mA in Tr4, which is mirrored in Tr5.

The output from Tr1 is applied to a pnp common-emitter stage and Tr3 acts as an emitter-follower buffer which drives the twin-T from the necessary

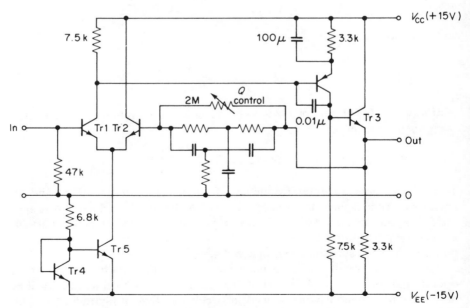

Figure 8.5. A twin-T selective amplifier.

low-impedance source. If the Tr1–Tr2 difference amplifier is to work in the linear mode, then the inputs must be near-identical. This implies that the two input impedances are high, being largely common-mode, so that the output of the twin-T is applied to the requisite high-impedance point.

Figure 8.6 gives the frequency response of the amplifier and its performance can be described by assigning a Q-factor, which in this context is defined by:

$$Q = \frac{f_o}{f_H - f_L} \tag{8.6}$$

where f_H and f_L are the upper and lower -3 dB, or half-power points. The Q can be controlled by a variable bridging resistor connected across the twin-T as shown in figure 8.5, and its effect is also shown in figure 8.6.

With the full 2 MΩ in circuit, the Q is very high—of the order of several hundred.

A network often used in a *positive* feedback loop is the Wien bridge or one of its modifications. The original Wien network is a four-arm AC bridge used for the measurement of capacitance or frequency. Two of the arms contain resistance only, and the remaining two have the form shown in figure 8.7(a). In this form the transfer function V_{out}/V_{in} can be obtained as follows:

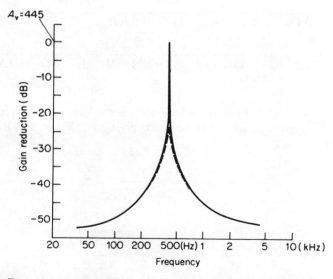

Figure 8.6. Frequency response of the circuit of figure 8.5: full curve, Q-control = 2 MΩ; broken curve, Q-control = 100 kΩ.

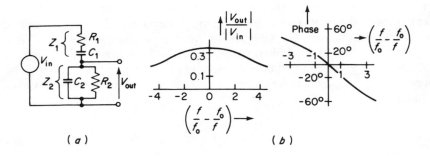

(a) (b)

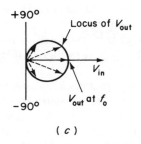

(c)

Figure 8.7. Wien bridge characteristics.

$$\frac{V_{out}}{V_{in}} = \frac{Z_2}{Z_1 + Z_2} = \frac{(-jX_{C_2})R_2/(R_2 - jX_{C_2})}{R_1 - jX_{C_1} - (jX_{C_2}R_2)/(R_2 - jX_{C_2})}$$

$$= \frac{1}{[(X_{C_1}/X_{C_2}) + (R_1/R_2) + 1] + j[(R_1/X_{C_2}) - (X_{C_1}/R_2)]} \cdot$$
(8.7)

Equation (8.7) is of the form $1/(a + jb)$, and when V_{out} is in phase with V_{in} then the imaginary part must be zero, giving $R_1/X_{C_2} = X_{C_1}/R_2$. Therefore

$$R_1 R_2 = \frac{1}{\omega^2 C_1 C_2}$$

or

$$\omega^2 = \frac{1}{R_1 R_2 C_1 C_2} \quad \text{whence} \quad f_o^2 = \frac{1}{4\pi^2 R_1 R_2 C_1 C_2} \cdot$$
(8.8)

Let

$$R_1 = R_2 = R \quad \text{and} \quad C_1 = C_2 = C .$$

Equation (8.8) now becomes

$$f_o = \frac{1}{2\pi RC} . \tag{8.9}$$

Having determined the center frequency at the point where zero phaseshift occurs, consider the attenuation at this frequency when $R_1 = R_2 = R$ and $C_1 = C_2 = C$. Equation (8.7) reduces to $V_{out}/V_{in} = \frac{1}{3}$, which shows that if the network is used as a positive feedback element, then the value of A_v must not exceed three, otherwise instability will result. Since at the center frequency $R = 1/\omega_o C$, the general form of equation (8.7) will be

$$
\begin{aligned}
\frac{V_{out}}{V_{in}} &= \frac{1}{3 + j[(\omega C/\omega_o C) - (\omega_o C/\omega C)]} \\
&= \frac{1}{3 + j(f/f_o - f_o/f)} = \frac{1}{3 + jb} .
\end{aligned} \tag{8.10}
$$

When $f \gg f_o$, $b \gg 3$ which means that the output voltage lags the input by nearly 90°, for $V_{out}/V_{in} = -j(1/b)$. Conversely, when $f_0 \gg f$, b becomes a large negative number and V_{out} leads by nearly 90°.

The amplitude and phase characteristics are plotted in figure 8.7(b), from which it can be seen that the variation in the magnitude of V_{out} is not great at frequencies near f_o, but that the phaseshift is quite marked. One method of plotting the gain and phase responses which illustrates the performance of the circuit very graphically is the circle diagram of figure 8.7(c). Here V_{out} has been plotted on an Argand diagram, and the locus of V_{out} ($=V_{in}/(a + jb)$) is, of course, a circle.

When the Wien bridge is connected as a positive feedback element, both the gain and phaseshift characteristics contribute to the overall performance, and in the form so far discussed, a voltage source drive should be used, and the bridge should work into a very high impedance. An approach to ideal conditions can be achieved by the use of the pnp/npn composite of figure 5.19. Here, the input impedance is high, the output impedance low, and the gain can be adjusted to be slightly less than three. A suitable circuit is given in figure 8.8, from which it can be seen that the bias resistors apply a bias voltage to the first transistor base via the Wien bridge itself, so that they do not shunt the actual input point. In fact, it is necessary to decouple R_D as shown if the input signal to the amplifier is injected at point a, otherwise R_2 will alter the position of f_o by contributing to the parameters of the Wien network. Point a will accept a high-impedance source, which means that the amplifier should be preceded by a suitable high output impedance stage in order that the performance should not depend on the source impedance. However, a low-impedance source can be connected at point b after

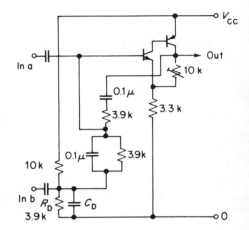

Figure 8.8. Use of Wien bridge in a simple selective amplifier (400 Hz).

removing C_D, but here the position of f_o will change and some compensation for this must be made. The performance of the circuit using a low-impedance source and a 33 kΩ series input resistor connected to a is shown in figure 8.9 and the Q-factor for this arrangement is 7.6.

An advantage of the Wien bridge is that the position of f_o may be controlled by using pre-set potentiometers in the arms, and a ganged component is useful here, since only two elements are needed compared with three for the twin-T network.

At this point it will be convenient to discuss the alternative to tailoring an amplifier circuit to suit a frequency selective network; that is, modifying the FSN to suit the amplifier.

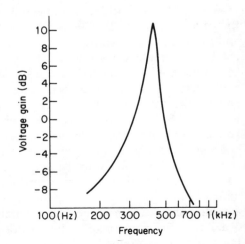

Figure 8.9. Response curve for circuit of figure 8.8 ($Q = 7.6$).

A current source in parallel with an internal impedance is equivalent to a voltage source in series with the same impedance, the former being a Norton equivalent and the latter being a Thévenin equivalent. This is a simple demonstration of the existence of dual networks, the former being considered from the point of view of current flow and the latter from the voltage aspect. In the present context, it is desired to design a network whose *current* transfer function is of the same form as the *voltage* transfer function of the Wien bridge so far considered. In other words, the transfer function of equations (8.7) and (8.10), namely

$$\frac{V_{out}}{V_{in}} = \frac{Z_2}{Z_1 + Z_2} = \frac{1}{a + jb}$$

must become

$$\frac{I_{out}}{I_{in}} = \frac{Z_2}{Z_1 + Z_2} = \frac{1}{a + jb}.$$

This may be accomplished by using the circuit of figure 8.10, which is therefore the *current dual* of figure 8.7(a). Consequently, the relations giving the center frequency (equation (8.8)) and the attenuation (equation (8.10)) still hold, and only the form of the circuit has changed.

It is now the current gain of the associated amplifier which is of interest and one suitable circuit, which has been shown to give a Q of several tens[4], is given in figure 8.11. In this circuit, the Q can be controlled by adjusting the input resistance to the amplifier.

It is also possible to construct a network requiring a voltage source and a low resistance load so that the transfer function becomes an admittance I_{out}/V_{in}, and this is done as follows.

For the circuit of figure 8.12,

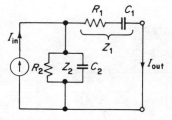

Figure 8.10. Wien bridge current dual.

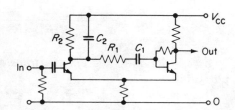

Figure 8.11. Application of current dual (biasing components not shown).

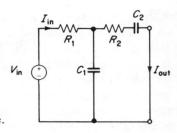

Figure 8.12. Admittance equivalent of Wien bridge.

$$\frac{I_{\text{out}}}{V_{\text{in}}} = \frac{I_{\text{out}}}{I_{\text{in}}} \frac{I_{\text{in}}}{V_{\text{in}}}$$

and

$$\frac{I_{\text{out}}}{I_{\text{in}}} = \frac{-jX_{C_1}}{R_2 - j(X_{C_1} + X_{C_2})} \ . \tag{8.11}$$

Also

$$\frac{I_{\text{in}}}{V_{\text{in}}} = \left[R_1 + \left(\frac{(-jX_{C_1})(R_2 - jX_{C_2})}{R_2 - j(X_{C_1} + X_{C_2})} \right) \right]^{-1} . \tag{8.12}$$

Therefore

$$\frac{I_{\text{out}}}{V_{\text{in}}} = (8.11) \times (8.12) = \frac{(-jX_{C_1})}{R_1R_2 - jR_1(X_{C_1} + X_{C_2}) + (-jX_{C_1})(R_2 - jX_{C_2})}$$

$$= \frac{1}{R_1 + R_2 + R_1(X_{C_2}/X_{C_1}) + j(R_1R_2/X_{C_1} - X_{C_2})} .$$

This is again of the form $1/(a + jb)$, and at zero phaseshift the imaginary part becomes zero and

$$\frac{I_{\text{out}}}{V_{\text{in}}} = \frac{1}{R_1 + R_2 + R_1(C_1/C_2)} \ . \tag{8.13}$$

If $R_1 = R_2$ and $C_1 = C_2$, the *transfer conductance is* $1/3R$. Also at zero phaseshift

$$\frac{R_1R_2}{X_{C_1}} = X_{C_2} \quad \text{or} \quad R_1R_2 = X_{C_1}X_{C_2} = \frac{1}{\omega_0^2 C_1 C_2} \ .$$

Therefore

$$\omega_0^2 = \frac{1}{R_1 R_2 C_1 C_2}.$$

Again, if

$$R_1 = R_2 \quad \text{and} \quad C_1 = C_2$$

then

$$f_o = \frac{\omega_o}{2\pi} = \frac{1}{2\pi R_1 C_1}. \tag{8.14}$$

A simple two-stage amplifier can be used to demonstrate a method of utilizing this form of the Wien bridge. In figure 8.13, the network is used to couple unbypassed parts of the emitter resistors, producing selective positive feedback. The second emitter resistor is made variable in order to control the degree of feedback, and with a setting which enables the amplifier to operate just below the point of oscillation will result in a high Q-factor. At this setting, however, the stability of the amplifier is very sensitive to changes in gain. A circuit of this form gave the results plotted in figure 8.14, curve (*b*) representing the condition for a maximum Q-factor of 16, while curve (*a*) represents a more stable condition with a Q-factor of about 5.

Note that as the frequency increases far beyond f_o, the Wien network becomes a low-pass filter, and the high frequencies are shunted to ground via C_1. This results in the approach of the curves of figure 8.14 to an asymptotic gain of about 22 dB, for negligible positive feedback will occur at these frequencies and the amplifier will revert to its basic gain figure.

The continued fall at the low frequencies is again due to the increasing reactance of the input capacitor.

The third technique used in frequency-selective amplifiers, that is, the use of a phaseshift characteristic as the major factor in a FSN, is well illustrated by the circuit of figure 8.15. This is of a form suggested by Butler[5], and

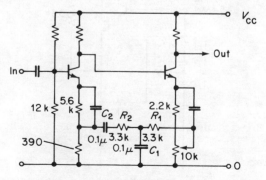

Figure 8.13. Application of admittance equivalent.

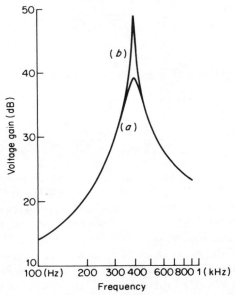

Figure 8.14. Response curve for circuit of figure 8.13. Q for curve $(a) = 4.7$ and for curve $(b) = 16$.

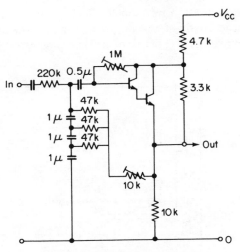

Figure 8.15. Very low-frequency selective amplifier ($f_o = 10 \, \text{Hz}$, $Q = 3.1$).

combines the characteristics of a ladder network with those of a high-input-impedance compound pair. This pair uses a degree of bootstrapping from the output to both collectors and to the shunt bias resistor. The ladder network is also fed from the low-impedance output point and is lightly loaded by the high-impedance input point. As this point is also fed from the source, it is apparent that a high impedance, or current source, is necessary. The degree of feedback is controlled by the $10 \, k\Omega$ variable resistor, and it is apparent that here the high current gain of the compound pair is being used, the voltage gain being, of course, approximately unity.

The three-stage ladder network has gain and phaseshift characteristics as shown in figure 5.20 but these will be modified by the associated circuitry, making calculations of the value of f_o difficult. (The values shown in figure 8.15 give an f_o of about 10 Hz.) This is often the case where low-gain discrete-component amplifiers are involved, for their input and output impedances and gains remain functions of the components used, since large amounts of feedback cannot be employed. Other than the employment of digital filtering methods, the alternative is to use operational amplifier techniques, where the amplifier parameters can be made to approach the requirements of the passive frequency-selective networks much more closely. Techniques of this sort will be introduced in the next section, where the three frequency-selective passive networks described above will be used in sinusoidal oscillator circuits.

8.2 SINUSOIDAL OSCILLATORS

Consider again the basic feedback equation:

$$A_{v(FB)} = \frac{A_v}{1 - A_v B} .$$

For *positive* feedback to occur, one of two conditions must be fulfilled:

(*a*) The forward gain, A_v, should have a positive numerical value and the feedback path B should have zero phaseshift. This condition is fulfilled by a noninverting amplifier and a resistive feedback network. It is also fulfilled if the feedback network is frequency-sensitive, and at some frequency exhibits zero phaseshift, such as is the case for a Wien bridge circuit at f_o.

(*b*) The forward gain, A_v, should have a negative numerical value and the feedback path B should exhibit 180° phaseshift. This condition is fulfilled by an inverting amplifier and a signal transformer in the feedback loop so connected that the feedback signal is positive. It is also fullfilled if the feedback network is frequency-sensitive, and at some frequency exhibits 180° phaseshift, such as is the case for a twin-T or a three-section ladder network.

If either of these conditions is fulfilled, then the amplifier will only be stable if at *all* frequencies, $A_vB < 1$.

Suppose that $A_vB > 1$. When such a circuit is switched on, any small disturbance at the input (such as noise) will be amplified and the large output will be fed back additively to the input. This means that in a real amplifier the maximum possible output will very quickly be reached, the usual result being that the active devices saturate and cease to amplify. If, after a specific dwell-time, the circuit quickly reverts to normal and the cycle continues to repeat, then this is a *relaxation oscillator*, the output of which is a square wave. Such oscillators, which are essentially very high gain positive feedback circuits, are described in chapter 10.

Oscillators having a controlled degree of feedback can produce sinusoidal output (as will be described below) and the only other principle on which oscillators work involves a *negative resistance slope*. These are described in chapter 12.

Now suppose that an amplifier with a frequency-selective feedback path conforms with criterion (a) or (b) above and that at the center frequency, f_o, the loop gain, A_vB, is very close to unity. Under these conditions oscillations will be maintained at a constant level, and the waveform generated will be sinusoidal. This is because only a sine wave is unchanged in shape by a circuit containing passive reactances, and is therefore the only waveform which can be transmitted round and round the feedback loop; that is, part of the power from the DC power supply has been converted to a sinusoidal signal.

This situation adds a third criterion if the positive feedback loop is to lead to stable sinusoidal oscillations:

(c) That some method must be employed for maintaining the loop gain at unity, resulting in a stable output waveform magnitude.

Design criterion (a) suggested a noninverting amplifier with a Wien bridge network in the feedback loop, and this method, coupled with circuitry to establish criterion (c), has formed the basis of most oscillator designs in the 1.0 Hz to 10 MHz frequency range, including the famous original Hewlett-Packard design[6] and many excellent modern realizations[7,8]. The popularity of the Wien circuit is in part because only two elements need to be varied simultaneously to tune the oscillator over a wide frequency range. Depending on the frequencies involved, either the two capacitors may be switched for coarse control, and fine control achieved by ganged potentiometers (for LF) or vice versa (for HF).

The basic Wien oscillator circuit is drawn in two ways in figure 8.16. In figure 8.16(a) a full bridge network is shown in which the operational amplifier is seen amplifying the difference output from the bridge between points a and b, and so energizing the entire bridge at points c and d. (Power supply rails have been omitted in the interests of clarity.) This diagram

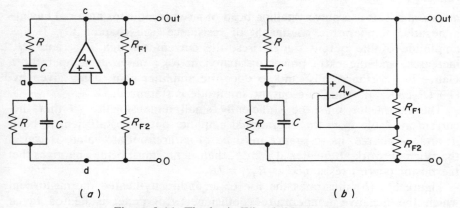

Figure 8.16. The basic Wien bridge oscillator.

stresses the point that there must be *some* output between points a and b, otherwise the bridge will not be energized at all. If the amplifier gain were not well controlled, the amplitude of the oscillations could build up until some nonlinearity appeared, which itself could be used to limit the gain in the system. Methods of gain control involving both nonlinearities and secondary feedback loops all depend upon the gain of the amplifier being determined by external elements, and in the redrawn circuit of figure 8.16(b) it becomes clear that resistors R_{F1} and R_{F2} form the gain-determining elements of a noninverting operational amplifier, as in figure 7.22. Control of the value of R_{F1}, or R_{F2}, will now control the gain of this amplifier.

There are several ways of achieving such control, three of which are shown in figure 8.17. Diagram (a) shows the replacement of R_{F1} using a

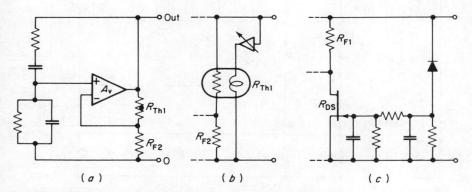

Figure 8.17. Amplitude stabilization of a Wien oscillator by: (a) a directly-heated thermistor; (b) an indirectly-heated thermistor and (c) a FET VCR.

thermistor. This is a polycrystalline bead of a semiconductor material having a negative temperature coefficient of resistance (see chapter 13). If the amplitude of the output signal rises, the current through R_{Th1} and R_{F2} increases, and the extra power dissipated in R_{Th1} raises its temperature. Hence its resistance falls, and so does the amplifier gain, being given by $[1 + (R_{Th1}/R_{F2})]$. Thus, the output amplitude is stabilized.

The thermistor must be sufficiently small dimensionally so that the current available from the operational amplifier output is sufficient to heat it; and, of cource, its resistance must be of an appropriate value. Because the Wien network attenuates at f_o by $\frac{1}{3}$, then $A_{v(FB)}$ must be 3. This gives the thermistor (warm) resistance as $R_{Th1} = 2R_{F2}$.

Figure 8.17(b) illustrates the use of an indirectly-heated thermistor, in which the negative temperature coefficient (NTC) pellet is heated by a separate filament. Similar to this is the calistor, in which a filament lamp is encapsulated with a photosensitive resistor (see chapter 13). For either device, an extra amplifier—which may be simply an emitter-follower—may be necessary to energize the filament.

Finally, a FET may be used as a voltage-control resistor (VCR) as described in chapter 11. Here, the output magnitude is converted to a direct voltage using a rectifier and low-pass filter and applied to the gate of the FET, the resistance of which is a function of this voltage.

The FET is also useful as a very high input impedance voltage amplifier, and a complete Wien oscillator using it is shown in figure 8.18. The passive components shown have been chosen for about 50 kHz, and the design technique for the active part will become evident after reading chapter 11.

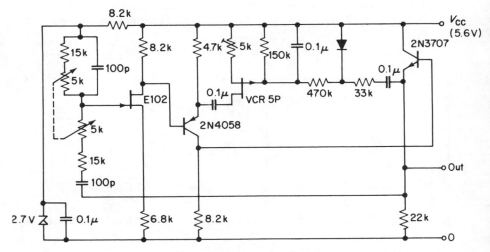

Figure 8.18. A Wien oscillator using FET VCR amplitude control. (Values shown lead to oscillations at about 50 kHz.)

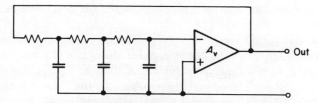

Figure 8.19. The basic phaseshift oscillator.

In all these circuits a second feedback loop has been added, which itself may be subject to instability. In the case of thermistor control, however, the thermal delay inherent in the bead provides a dominant lag to stabilize the system; while in the case of the FET ·VCR, the low-pass circuit for obtaining the DC gate level provides this dominant lag.

The alternative design criterion (*b*) suggests a twin-T network around an inverting amplifier. It also suggests that a simple three-section ladder network would form a suitable feedback path because it can provide a phaseshift of 180°, as shown in figure 5.20. This is in fact the well-known phaseshift oscillator shown in figure 8.19.

Because the attenuation of a three-section ladder having equal capacitance values and equal resistance values is 29, the three sections are often graded so as to produce a phaseshift of 60° per section, but with lower overall attenuation. In either case, the (inverting) gain of the amplifier must be controlled so that the loop gain is very close to unity.

For both the three-section ladder and the twin-T oscillator methods, variation in the center frequency, f_o, is difficult, because it implies varying either all three resistances or all three capacitances equally, which accounts for the popularity of the Wien circuit, as has been mentioned.

A more modern sinusoidal oscillator technique is the so-called *state-space* method, which is derived from analog computing concepts. Figure 8.20 gives the block diagram for an oscillator based on this approach, and it contains a

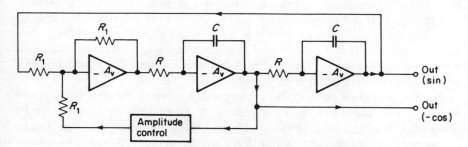

Figure 8.20. The basic quadrature oscillator.

unity-gain inverting amplifier and two integrators (as described in relation to figure 7.25(c)).

If a sinusoidal signal is applied to the input of the first amplifier it is phaseshifted through 180° but not otherwise changed. The middle amplifier then integrates it with a time constant $1/CR$; that is, a 90° phaseshift occurs at some frequency f_o. The signal then passes to the final amplifier, which again integrates it, and since R and C have the same values as in the central amplifier the signal is shifted through a further 90° at f_o.

The total phaseshift is now 360°, but the amplitude of the signal is unchanged, because the gain of each integrator is unity at f_o. Hence, the output signal can be applied to the input of the first amplifier, so that the complete system becomes a positive feedback loop with unity gain, which is a condition for stable oscillation.

In practice, some method of amplitude stabilization is necessary, as for the Wien oscillator, and various techniques are available[1,10]. In figure 8.20, such a network block is shown around the first two amplifiers, and the main output is taken from the third. If this constitutes the sine wave, v_0, then a second output is available from the central amplifier which constitutes a wave at $-90°$ to this, that is, a cosine wave. This is why this form of network is also called a *quadrature oscillator*.

8.2.1 Waveform Generators

In linear electronics it is frequently necessary to use waveforms other than sinusoidal, mainly in the checking of networks which are intended to accommodate such waveforms and for specialized applications such as the provision of scanning waveforms in cathode-ray oscillographs. In addition to sine-wave generators such as those described above, square-wave generators are introduced in chapter 10, and pulse and ramp generators appear in chapter 12. However, commercial generators exist which produce a series of waveforms, all derived from a common oscillator, and which are therefore automatically synchronized. Further, single integrated circuits which also perform this function are readily available.

A basic block diagram relating to the generation of the most common waveforms—the square, triangle and sine—is shown in figure 8.21. Here, it is supposed that the upper current source charges the capacitor, C, linearly for a pre-determined time, after which the lower current source discharges it, also linearly. If the charge and discharge times are equal, a triangular wave such as the one shown results. A hysteresis circuit (see figure 10.26) is able to convert this triangle to a square wave.

Finally, a shaping circuit can be designed to convert the triangle to a sine wave[11]. This can be done by a piecewise approximation method, or by using the nonlinear characteristic of field-effect transistors[12] or bipolar difference amplifiers. Sine waves built up by these methods are usually somewhat distorted compared with those obtained from linear oscillators,

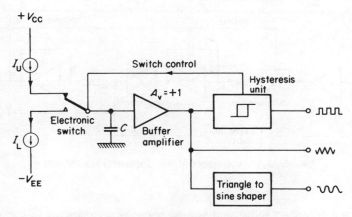

Figure 8.21. A basic waveform generator block diagram.

but recent techniques have improved this situation considerably. In particular the tanh characteristics of the difference amplifier (derived in chapter 5) have been used to produce extremely accurate triangle-to-sine shaping[13].

In many commercial waveform generators, the frequency can be set up by a series of push-buttoms or thumbwheels which program the frequency to several significant figures. This is possible because the currents supplied by the current generators in the linear charging system may be accurately defined and also because the hysteresis switch and triangle-to-sine converter can be very fast acting compared with the periods of the waveforms generated. This constitutes an excellent example of a combination of digital and analog techniques.

Returning to figure 8.21, if the discharging time is very short and the discharging current very high then a sawtooth wave results in which a slow linear rise is followed by a rapid fall. Conversely, a rapid rise will be followed by a slow linear fall if the charging time is short and the charging current is very high.

8.2.2 The Voltage-Controlled Oscillator

If the frequency generated by an oscillator is some function of an applied voltage, it is called a *voltage-controlled oscillator* (vco). The output may be of any regular waveshape, and if its frequency is directly proportional to the applied voltage it becomes a *voltage-to-frequency converter* (vfc).

In data acquisition and transmission, it is often desirable to convert an analog voltage from a transducer (see chapter 13) into a series of pulses, the repetition rate of which is proportional to the initial voltage. These pulses can be transmitted in digital form so that noise is less important, and demodulated at an appropriate receiver. This is one typical application of a vfc.

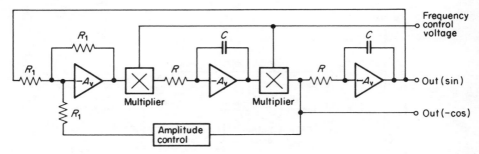

Figure 8.22. A voltage-controlled quadrature oscillator.

Voltage-controlled pulse and square-wave generators are not difficult to design, and arise naturally from the relevant circuitry, as will become apparent in chapters 10 and 12. To realize a sine wave vco is more difficult, though figure 8.21 does indicate one method. If the current sources are made voltage dependent (as in the voltage-dependent current source of figure 7.26(c)), then the frequency of the triangular wave is also voltage-dependent and hence so is the shaped sine wave.

Another approach is to substitute the fixed resistors in a frequency-selective circuit, such as the Wien bridge, by voltage-controlled resistors (vcrs, see chapter 11). This can also be done in the case of the two integrator resistors in the quadrature oscillator of figure 8.20. The problem in all such applications is that of tracking, however, and a better solution is to incorporate a pair of multipliers in the quadrature oscillator circuit. The principle of operation is that the incoming signal to an integrator resistor R in figure 8.20 is changed in amplitude if multiplied by a constant. This is the same thing as changing the conductance of R, upon which the frequency of the loop is dependent. The diagram now appears as in figure 8.22, and the multipliers in question (such as the transconductance type described at the end of chapter 5) are available in integrated circuit form, along with the operational amplifiers themselves.

8.3 THE PHASE LOCKED LOOP

The phase locked loop is an electronic system which can be thought of as a feedback loop in which the error signal at the input is a minimized phase difference rather than a voltage difference. Its operation can be explained with reference to figure 8.23, where an input waveform (which is shown as a sinusoid, though other waveforms are also used) enters a *phase detector*. This is a circuit which produces an output signal which is a function of the phase difference between two input signals.

The phase-detector output is fed to a low-pass filter, the output of which

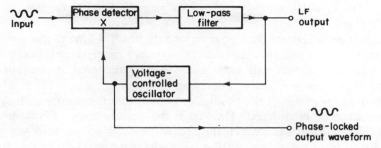

Figure 8.23. The phase locked loop.

is used to determine the frequency generated by a voltage-controlled oscillator. The output from this VCO then enters the phase detector, which completes a negative feedback loop, the result being that the phase difference is reduced to near zero; that is, the VCO frequency is held the same as the input signal frequency.

In this system, the LF output from the low-pass filter may be extracted, this being a voltage level proportional to the difference in frequency between the input signal and the natural or center frequency of the VCO. Also, the signal from the VCO may be extracted, which can be thought of as a "cleaned-up" version of a noisy and/or phase-fluctuating input signal.

If the output of the VCO is of interest, then the linearity of the phase detector is of little importance. Conversely, if a LF output from the low-pass filter is required to represent the frequency difference then clearly the phase detector should be a linear device. A case in point is where demodulation of an FM signal is desired; the low-pass filter will remove the carrier and produce a LF signal proportional to the frequency modulation.

The foregoing description of the PLL implies that the phase detector must also operate as a frequency-difference detector and pass an appropriate signal to the VCO via the low-pass filter, otherwise the VCO could not lock in to the same frequency as the input signal. This function can be achieved using a multiplier, when the loop operates as follows.

Suppose that there is no input and the VCO is oscillating at its *free-running frequency*, f_o. When a signal of frequency f_i appears at the input, the two inputs will be multiplied to give a pair of sum and difference frequencies:

$$(\hat{V}_i \sin \omega_i t)(\hat{V}_o \sin \omega_o t) = \tfrac{1}{2}\hat{V}_1\hat{V}_o[\cos(\omega_i - \omega_o)t - \cos(\omega_i + \omega_o)t].$$

Hence two signals, a low-frequency difference component and a high-frequency sum component, result. The difference frequency component is transmitted by the low-pass filter to the VCO, the frequency of which is thereby shifted to that of the incoming signal. When these two signals are at the same frequency it becomes valid to consider the phase difference between them, and the multiplier then performs the following computation:

$$(\hat{V}_i \sin \omega_i t)\hat{V}_o \sin (\omega_i t - \phi) = \tfrac{1}{2}\hat{V}_i\hat{V}_o[\cos \phi - \cos (2\omega_i t - \phi)] .$$

The first term represents a DC component, which is always transmitted by the low-pass filter and used to pull the VCO very nearly into phase with the input signal. The VCO will now *track* the input signal in frequency by maintaining a minimal phase difference.

This dual operation of the multiplier implies that there are two frequency differences to be considered. The first is the *capture range*, which describes the frequency band over which the PLL will lock in to an incoming signal. The second frequency range is that over which the PLL will remain locked. This is called the *lock range* or *tracking range*, and it is larger than the capture range.

The capture range is limited in part by the fact that the low-pass filter must be designed to pass the difference frequency $(f_i - f_o)$, but not the sum frequency $(f_i + f_o)$. Hence, for larger differences, $(f_i - f_o)$ may reach the cut-off frequency of the filter and so not be passed. When the PLL is locked, however, the output from the phase detector always contains a DC component, which will therefore always be transmitted by the filter. This accounts for the tracking range being greater than the capture range.

An aspect of PLL utilization which is of considerable importance when considering circuit stability is that the loop behaves overall as an *integrator*. This is because it is a *phase difference* which produces the control signal, whereas it is a *frequency* which is adjusted (and phase is the integral of frequency). Hence a 90° lagging phaseshift is produced, just as is the case for integrating operational amplifiers. If little extra phaseshift were introduced then there would be no stability problem, but in most (but not all) phase locked loops, the low-pass filter is mandatory, which itself produces a lag. Thus, over at least part of the frequency spectrum a lag approaching 180° and the concomitant 12 dB/octave roll-off will appear. However, it is normally possible to design the low-pass filter so that the roll-off reverts to 6 dB/octave near unity gain so that spurious oscillation is avoided.

The PLL has been extensively treated in both books[14] and papers[15] and has been used in a very wide range of applications[16,17]. The integrated circuit form—which normally involves square-wave operation—has become another basic analog component, along with operational and instrumentation amplifiers, multipliers, dividers, VCOs and sundry more exotic functional blocks[11].

REFERENCES

1. Wait, J. V., Huelsman, L. P. and Korn, G. A., 1975, *Introduction to Operational Amplifier Theory and Applications* (New York: McGraw-Hill).
2. Bowron, P. and Stephenson, F. W., 1979, *Active Filters for Communications and Instrumentation* (New York: McGraw-Hill).

3. Huelsman, L. P. (ed.) 1970, *Active Filters: Lumped, Distrubuted, Integrated Digital and Parametric* (New York: McGraw-Hill).

4. Hutchins, R., 1961, Selective RC amplifiers using transistors, *Electron. Eng.* (February).

5. Butler, F., 1962, Transistor RC oscillators and selective amplifiers, *Wireless World* (December).

6. Hewlett, W. R., 1942, *US Patent*, No. 2, 268, 872, January 6th.

7. Ewins, A. J., 1971, Wien bridge audio oscillator, *Wireless World*, 104–107 (March).

8. Williams, P., 1971, Wien oscillators, *Wireless World*, 541–546 (November).

9. Bajen, G., 1976, Design transistor oscillators, *Electron. Des.*, **24** (8) 98–102.

10. Grame, J. G., Tobey, G. E. and Huelsman, L. P., 1971, *Operational Amplifiers—Design and Applications* (New York: McGraw-Hill) pp. 389–391.

11. Sheingold, D. H. (ed.), 1974, *Non-Linear Circuits Handbook* (Analog Devices Inc.) chapter 2.

12. Middlebrook, R. D. and Richer, I., 1965, Non-reactive filter converts triangular waves to sines, *Electronics*, 96–101 (March 8th).

13. Evans, W. A. and Williams, J. S., 1979, The multi-tanh circuit as a triwave-to-sine converter, *Electron. Circuits Syst.*, **3**, 90–92.

14. Gardner, F. M., 1979, *Phaselock Techniques*, 2nd. edn. (New York: Wiley).

15. Gupta, S. C., 1975, Phase-locked loops, *Proc. IEEE*, **63**, 291–306.

16. *Linear Phase-locked Loops Applications Book*, 1972 (Signetics, Inc.).

17. Kronpa, V. F., 1973, *Frequency Synthesis* (London: Griffin).

QUESTIONS

1. The voltage gain of a frequency-selective amplifier which uses the two reactive arms of a Wien bridge as its feedback element, must not be more than three. Explain this.

2. What are the center-frequency phase shifts of (*a*) the Wein bridge network and (*b*) the twin-T network? How do these influence the choice of amplifier?

3. Redraw the selective amplifier circuit of figure 8.5 using an operational amplifier instead of discrete components and explain its operation.

4. Why do oscillators like the Wein bridge or phaseshift types always produce a sine wave?

5. Explain the purpose of the multipliers in the VCO of figure 8.22.

6. Why is the tracking range of a PLL greater than its capture range?

PROBLEMS

8.1. Design a 10 kHz Wein oscillator using an operational amplifier. Specify (realistic) component values for the Wein network and for the gain-defining amplifier resistors. Do you anticipate any stability problems? If so, suggest a solution.

8.2. The voltage-controlled oscillator in a phase locked loop has just frequency-locked to an incoming signal, and the phase angle between the two waveforms is 30°. If the phase detector can be thought of as a unity-gain multiplier, what is the DC component of its output if the RMS magnitude of the input and VCO waves are 1 V and 3 V, respectively?

9

Power Amplifiers

The material presented so far has been concerned with small voltages and currents, with the implication that these will be measured at the output of a stable amplifier or oscillator. In many cases it is necessary to use such output signals to actuate mechanisms such as relays, AC or DC servo-motors and loudspeakers. Here, the signal power required is much greater than is available from any of the amplifiers so far considered and may, in fact, be anything from a few hundred milliwatts to several hundred watts. In this section a review of the design of small power amplifiers will be presented.

The necessity for a high output power dictates that the signal excursions should be large; in other words calculations based on small-signal parameters are likely to be inaccurate owing to severe changes in these parameters. Further, although the excursions should be large, they should not take the transistor outside the boundaries of safe operation, otherwise damage may result, or at best, considerable distortion may be encountered. This distortion criterion is of less importance in single-frequency power amplifiers intended for motor operation than in, for example, audio amplifiers, but it will nevertheless be considered here since it provides a graphic illustration of the effects of large signal operation where the parameters vary with signal amplitude.

9.1 OPERATING BOUNDARIES

Figure 9.1(a) shows a set of output characteristics for the common-emitter mode using a small power transistor. The nature of the various lines bounding the operating area is as follows.

Firstly, the maximum power hyperbola is given by the product $I_C V_{CE} = P_{tot}$. This expression recognizes that, since most of the voltage drop in a transistor occurs at the collector junction, it is here that most of the internal

309

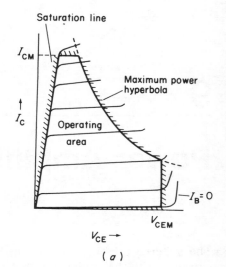

Figure 9.1. (*a*) Operating limits for a
large signal stage.

(*a*)

power will be dissipated. The small dissipation at the emitter junction $I_B V_{BE}$
is therefore ignored.

If the transistor is operated beyond this hyperbola, other than transiently,
its temperature will rise beyond the destruction level. Obviously, this
temperature rise will depend on the construction of the transistor and its
physical surroundings. Usually, the collector of a power transistor is bonded
to the metal case, which is in turn firmly fixed to a suitable heat sink. The
maximum power hyperbola may therefore take several different positions,
depending on the efficacy of this heat sink. The relevant design considerations will be discussed later.

The second limitation is observed at low collector voltages, and is the line
where all the output curves coalesce. The inverse slope of this saturation
line is designated the *collector saturation resistance*, R_{CS}. The actual bulk
resistance of the collector material contributes most of R_{CS}, which implies
that it may be controlled to some extent. That R_{CS} should be as low as
possible in a power transistor is apparent, because at high currents the
collector voltage at saturation can be several volts, which implies a severe
limitation of the operating area for a high R_{CS}. For modern silicon power
transistors R_{CS} can be as low as a few ohms, while their germanium
counterparts may exhibit an R_{CS} of only a small fraction of an ohm.

The third limitation occurs at high collector current, where the values of
h_{fe} and h_{FE} fall so low as to introduce *clipping*, which means that any
changes in input signal amplitudes at these levels are amplified only negligibly, so that the tops of output signal waves are severely clipped. However,
for some transistors, the maximum permissible value of I_C occurs before the
clipping limitation becomes too serious for certain applications.

Fourthly, if the base current reverses, clipping at the bottom of the signal wave will occur, for the transistor ceases to conduct, or is *cut off* in this region.

Finally, the breakdown voltages must be considered and figure 9.1(*b*) clarifies the meaning of the various breakdown voltage abbreviations. In each case the term $V_{(BR)CE}$ (sometimes written BV_{CE}) refers to the voltage which breaks down the transistor under conditions specified by a final subscript. Hence, $V_{(BR)CEO}$ refers to the breakdown voltage with the base open-circuited; $V_{(BR)CER}$ to that with a (specified) base–emitter resistance in circuit; $V_{(BR)CES}$ to that with the base and emitter short-circuited and $V_{(BR)CEV}$ to that with a (specified) base–emitter voltage applied. As the diagram shows, the various breakdown voltages differ somewhat, the lowest being $V_{(BR)CEO}$.

After breakdown the collector current rises rapidly, and there exists a region of negative resistance slope as V_{CE} falls while I_C continues to rise. Eventually, a virtually constant *sustaining voltage* is reached, at which I_C will rise catastrophically unless limited by external circuitry. The value of the sustaining voltage is again a function of the base connection and is minimal for the open-circuit condition. Notice that the curve corresponding to this condition is that to which at the I_C/V_{CE} curves coalesce.

There will, of course, be a family of breakdown and sustaining voltages unique to each individual transistor, which means that a manufacturer must specify a maximum operating voltage (herein called V_{CEM}) which is some-

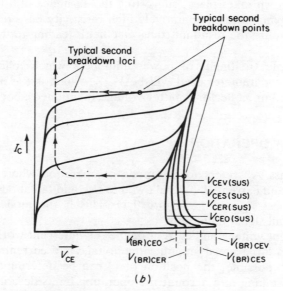

(*b*)

Figure 9.1. (*b*) Voltage breakdown characteristics.

what lower than the lowest value of $V_{CEO(SUS)}$ expected for each type number.

The processes involved in breakdown include both avalanche and field-emission mechanisms. In the former, charge carriers in the collector depletion layer are accelerated by the high field to energies great enough to release bound electrons from the lattice; whilst the latter case occurs when the field strength in the depletion layer is great enough to remove bound electrons without the assistance of any collision mechanisms.

Less frequently met in power transistors is the case where, as a result of a rise in V_{CE}, the collector depletion layer increases in thickness until it reaches the emitter depletion layer. This results in *punch-through* breakdown, which is more often met in the case of high-frequency transistors having very thin base regions.

A mechanism called (somewhat unfortunately) *second breakdown* is also observed. This is characterized by a sudden fall in V_{CE} to a low level from a region of high V_{CE} and I_C. Usually it occurs after "first breakdown" has occurred, when the operating point may shoot to the left from a sustaining region curve, as shown by the lower broken line in figure 9.1(b). It could, however, also occur *before* "first breakdown" has taken place, as shown by the upper broken line. Operating limitations imposed by second breakdown will be considered further in chapter 10 (see figure 10.38).

Second breakdown is thought to be due to the formation of a "hot-spot," or region of high current density, where sufficient localized power is generated to melt a minute volume of the semiconductor material. This hypothesis is supported by the fact that a time lag always occurs before second breakdown takes place, suggesting the existence of a thermal delay. With a thin asymmetric comparatively high resistivity base region, it is not difficult to visualize the resultant transverse fields leading to the "hot-spots" postulated.

In the simple treatment of power amplifiers which follows, the only limiting voltage parameter used will be V_{CEM}, which is the manufacturers' specified maximum collector–emitter voltage mentioned above.

9.2 CLASS A OPERATION

The term "class A" refers to the mode of operation where the collector voltage, V_{CE}, and current, I_C, are allowed to rise and fall about the mean, or quiescent values, V_Q and I_Q (figure 9.2). This is the mode of operation chosen for small-signal amplifiers.

The quiescent voltage and current define the quiescent working point, Q, and since for a large-signal amplifier the voltage and current swings should be as great as possible, the position of Q can be determined from these considerations taking into account the operating limits described previouly.

Firstly, at no time must the working point be allowed to leave the area

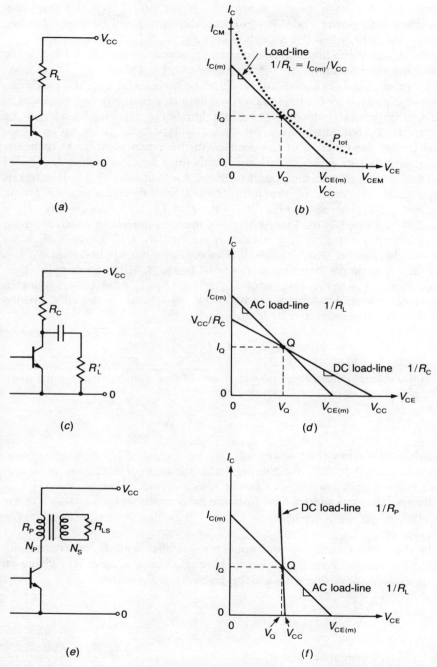

Figure 9.2. Various forms of class A circuit and associated load-line diagrams. (P_{tot}, I_{CM} and V_{CEM} are omitted from diagrams (d) and (f) for clarity.)

below the maximum power dissipation hyperbola. Since the transistor dissipates most power under no-signal conditions (as will be shown below), Q must lie on or below this hyperbola, for $P_{diss(Q)} = V_Q I_Q$. Secondly, $V_{CE(m)}$, the maximum value that V_{CE} is allowed to reach, must not exceed V_{CEM}, the maximum voltage stipulated by the manufacturer. Thirdly, $I_{C(m)}$, the maximum value that I_C is allowed to reach, must not exceed I_{CM}, the maximum current stipulated by the manufacturer. These criteria imply that the locus of the working point (which is a straight line for a resistive load) may be tangential to, but must never cut, the P_{tot} hyperbola as shown in figure 9.2(b); that the junction of this locus with the x axis must lie at or below V_{CEM}; and that its junction with the y axis must lie at or below I_{CM}.

The locus of the working point is called the *load-line*. This is because its slope gives the value of the load applied to the transistor. At $I_C = 0$, $V_{CE} = V_{CE(m)}$, whereas when $V_{CE} = 0$, $I_C = I_{C(m)}$. Consequently $R_L = V_{CE(m)}/I_{C(m)}$, which is the inverse slope of the load-line. If an emitter resistor is present, this slope must, of course, represent $(R_L + R_E)$, where R_L is, as before, the parallel combination of R_C and any other applied load.

If R_L is a simple resistive load—as in figure 9.2(a) for example—then when $I_C = 0$, $V_{CE} = V_{CC}$ and hence $I_{C(m)} = V_{CC}/R_L$. For maximal, symmetrical signal excursions about Q therefore, V_{CC} must be equally divided between the transistor and its load:

$$V_Q = V_{CC}/2 \quad \text{and} \quad I_Q = V_{CC}/2R_L .$$

Thus, under quiescent or no-signal conditions, the transistor must dissipate power given by,

$$P_{diss(Q)} = V_{CC}^2/4R_L \text{ W}$$

and this same amount of power will also be dissipated by the load resistor, the total being $V_{CC}^2/2R_L$ W. This highlights the main disadvantage of class A operation—a major wastage of power under quiescent conditions.

It is possible to isolate a load from the DC component by coupling it to the collector either via a capacitor as in figure 9.2(c) or via a transformer as in diagram (e).

In the former case, R_L' will appear in parallel with R_C at frequencies where the reactance of the capacitor is small compared with R_L', giving an AC or *dynamic load-line* with a slope given by $1/R_L$ where

$$R_L = \frac{R_C R_L'}{R_C + R_L'}$$

which must be smaller than R_C.

Note that if the same Q-point (at half-way along the DC load-line) were retained, there would no longer be symmetrical signal operation. If the bias point were changed to give symmetrical operation, as in diagram (d), and the value of R_L was the same as for the single load-resistor case, then I_Q would have the same value as for that single load-resistor case. However, to achieve this, V_{CC} would have to be increased, because R_C is larger than R_L, as is also shown in diagram (d). Hence, the quiescent power dissipation would be even greater than for the single load-resistor case. Furthermore, some of the AC power would inevitably be wasted in R_C.

If a transformer is used, the value of R_L can be substantially greater than R_C, which is now only the (usually very small) resistance of the primary winding R_P. This is because the resistance R_{LS} is reflected to the primary and multiplied by the square of the transformer turns ratio, when it appears effectively in series with R_P to give,

$$R_L = R_P + \left(\frac{N_P}{N_S}\right)^2 R_{LS} \simeq \left(\frac{N_P}{N_S}\right)^2 R_{LS} \,.$$

This expression assumes that R_P is indeed very small, so that it can be represented on load-line diagram (f) as a near-vertical line. This implies that for symmetrical signal operation, V_{CE} can rise to $2V_{CC}$ as shown, which in turn means that V_{CEM} must be chosen to be greater than $2V_{CC}$.

It is convenient to analyze the class A transformer-coupled output stage in order to compare it with the more efficient class B equivalent and eventually with the modern and much more widely-used transformerless stages. Large sinusoidal signals will be assumed, and average power dissipations taken over any integral number of wavelengths are assigned the following symbols:

P_{tot} = maximum allowed power dissipation within the transistor
P_{diss} = power dissipated within the transistor
$P_{diss(Q)}$ = dissipation under quiescent (no-signal) conditions
$P_{diss(FL)}$ = dissipation under conditions of maximum sinusoidal signal
$P_{diss(m)}$ = maximum power dissipation possible ($P_{diss(m)}$ should be $<P_{tot}$)
P_L = full load power
P_S = power extracted from supply at full load.

For the class A stage, the powers defined above take values as follows. P_{tot} is determined by the manufacturer.

When an output transformer is used, the power in the load will be due only to the alternating, or signal, components of V_{CE} and I_C. If these are $V \sin \omega t$ and $I \sin \omega t$, then the power in the load will be the product of the relevant RMS values, that is $(V/\sqrt{2})(I/\sqrt{2})$ or $\frac{1}{2}VI$. For the maximum excursions shown in figure 9.3 therefore:

$$P_L = \frac{V_{CE(m)}}{2\sqrt{2}} \frac{I_{C(m)}}{2\sqrt{2}} = \frac{V_{CE(m)} I_{C(m)}}{8} \quad \text{or} \quad \frac{V_Q I_Q}{2} \,. \tag{9.1}$$

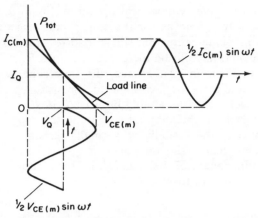

Figure 9.3. Excursions of I_C and V_{CE} for maximum efficiency in class A.

(This assumes that the power dissipated in the very low primary winding resistance is negligible.)

To find the power dissipation within the transistor, it is necessary to involve the absolute magnitudes of V_{CE} and I_C, which are

$$i_C = I_Q + I \sin \omega t$$

and

$$v_{CE} = V_Q - V \sin \omega t$$

where I and V are the peak values of the collector current and voltage sinusoids.

The instantaneous power dissipated is p, where $p = v_{CE} i_C$ or

$$p = V_Q I_Q - VI \sin^2 \omega t + V_Q I \sin \omega t - I_Q V \sin \omega t .$$

The average power dissipated,

$$P_{diss} = \frac{1}{2\pi} \int_0^{2\pi} p \, d(\omega t)$$

$$= \frac{1}{2\pi} \int_0^{2\pi} (V_Q I_Q - VI \sin^2 \omega t) \, d(\omega t)$$

(since both sine terms have an average value of zero)

$$= V_Q I_Q - \frac{VI}{2\pi} \int_0^{2\pi} \tfrac{1}{2}(1 - \cos 2\omega t) \, d(\omega t)$$

or

$$P_{\text{diss}} = V_Q I_Q - \tfrac{1}{2} VI. \tag{9.2}$$

This shows that P_{diss} is greatest at no load, that is, when $I = 0$. Hence,

$$P_{\text{diss(m)}} = P_{\text{diss(Q)}} = V_Q I_Q. \tag{9.3}$$

At full load, $V = V_Q$ and $I = I_Q$ giving:

$$P_{\text{diss(FL)}} = V_Q I_Q - \tfrac{1}{2} V_Q I_Q = \tfrac{1}{2} V_Q I_Q. \tag{9.4}$$

The power extracted from the supply at full load is obviously

$$P_S = P_L + P_{\text{diss(FL)}} = \frac{V_Q I_Q}{2} + \frac{V_Q I_Q}{2} = V_Q I_Q. \tag{9.5}$$

This result, and most of the others, are apparent from figure 9.3, and equations (9.1)–(9.5) should be checked against this diagram. Most power amplifier design problems can be solved by such a graphical approach.

The operating efficiency of the stage is defined as the ratio of maximum load power to power extracted from the supply at full load:

$$\text{operating efficiency} = \frac{P_L}{P_S} = \frac{\tfrac{1}{2} V_Q I_Q}{V_Q I_Q} = 0.50.$$

A more useful criterion of performance is the commercial efficiency, which relates the maximum load power to the maximum dissipation. The maximum dissipation of a power transistor is an important factor in determining its cost, hence the term "commercial efficiency." In the class A case, $P_{\text{diss(m)}} = P_Q$, which gives:

$$\text{commercial efficiency} = \frac{P_L}{P_{\text{diss(m)}}} = \frac{\tfrac{1}{2} V_Q I_Q}{V_Q I_Q} = 0.50.$$

This result means that a power transistor working in class A must be capable of dissipating twice the power that it is to deliver to the load. (Actually, this efficiency can be improved if it can be guaranteed that the drive is present at all times, including switch-on.)

These theoretical values for the efficiencies of a class A stage must now be interpreted in the light of practical considerations.

Firstly, the area near the junction of the load line with the y axis is inside the saturation line (see figure 9.1), which means that, at maximum signal amplitude, one peak of the output signal will be clipped. Further, distortion will occur owing to the fall in h_{FE} at high collector currents.

Secondly, distortion and some clipping will occur at working points near the junction of the load-line and the x axis owing to the reduction of h_{FE} at low collector currents and the imminence of cut-off.

If serious distortion is to be avoided, it is apparent that the magnitude of the signal swings must be reduced, which means that the efficiencies of the stage will fall below 50%.

If the aforementioned limitations on the signal excursions are observed, the remaining distortion can be attributed to the nonlinearity of the current transfer characteristic I_C/I_B, and the input characteristic I_B/V_{BE}. The manner in which distortion arises owing to these causes is shown in figures 9.4 and 9.5, where the nonlinearities are exaggerated for clarity.

In figure 9.4, the effect of the curved transfer characteristic is clearly seen, the sinusoidal input current i_B being amplified in a nonlinear manner to produce the distorted waveform of i_C.

In figure 9.5, the sinusoidal input is a voltage wave v_{BE}, which implies a voltage source of very low (ideally zero) internal impedance. This means that the current, i_B, due to v_{BE} will be a function of the input characteristic, the result being the distorted wave shown. If, however, a current source (of very high internal impedance) is used, the input current is no longer dependent on the relationship between i_B and v_{BE}. In other words, any changes in the input impedance of the transistor (which account for the nonlinear input characteristic) will be swamped by the high internal impedance of the source and distortion will be caused by the transfer characteristic alone.

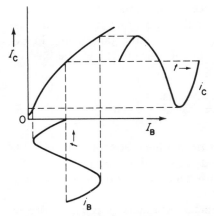

Figure 9.4. Distortion due to transfer characteristic.

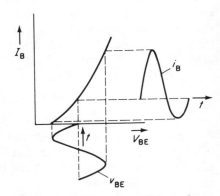

Figure 9.5. Distortion due to input characteristic when voltage source is used.

Neither a near-perfect voltage source nor a near-perfect current source is likely to be encountered in practice and the distortion due to a "practical" source will lie somewhere between the extremes quoted. However, since the *total* distortion is due to the combined effects of the transfer and input characteristics, then, observing from figures 9.4 and 9.5 that the two characteristics have opposite curvature, it is reasonable to suppose that an optimum source impedance exists for which the total distortion is a minimum. This is, in fact, so, but the graphical procedure for determining this optimum value is quite tedious, apart from which the impedance is dictated to some extent by other considerations. It is usually more satisfactory to design a single-frequency power amplifier in such a manner as to minimize distortion by other methods, such as feedback, and concentrate on maximum power gain in the output stage.

For a power amplifier stage, the required load is dictated by the load-line, and is usually quite a low resistance. For example, using a Texas Instruments 2N3903 "Silect" silicon npn transistor in class A, $I_{CM} = 200$ mA, while $V_{CEM} = 40$ V. The load-line joining these two points falls under the maximum power hyperbola if a good heat sink is used, and gives $R_L = 40/0.2 = 200 \, \Omega$. Such a rigidly defined resistance at the output has two implications.

Firstly, it is necessary to employ an output transformer so that whatever load the transistor is required to feed is properly transformed in order that the AC load-line should be in the correct position. As an example, suppose that the 2N3903 were required to feed a $4 \, \Omega$ load. If N_P and N_S are the number of turns on the primary and secondary of an output transformer, then

$$\left(\frac{N_P}{N_S} \right)^2 = \frac{200}{4}$$

or

$$N_P : N_S \simeq 7 : 1 \, .$$

The AC load-line would therefore appear as in figure 9.6(b), while the actual circuit would be as in figure 9.6(a).

Since it would be undesirable to dissipate considerable heat in the primary owing to the high quiescent collector current, the primary resistance R_P should be very low. This gives rise to a DC load-line, also shown in figure 9.6(b), whose slope represents R_P. The DC load-line is actually the locus which would be followed by the operating point if the input signal varied very slowly, so that the output transformer became incapable of reflecting the load to its primary.

Since R_P is very small, the DC load-line has a steep angle, which means that the supply voltage V_{CC} must now be quite close to V_Q as shown. However, when a signal is applied of a frequency which allows the trans-

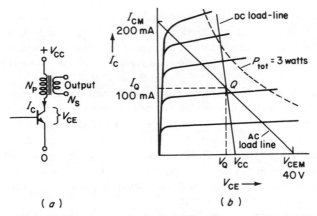

Figure 9.6. (a) Large-signal stage with transformer output; (b) load-lines for Texas 2N3903 connected as above.

former to operate normally, the locus shifts to the AC load-line, and V_{CE} is able to rise to V_{CEM}. This situation sets a limitation for stages with output transformers that $V_{CC} \not> \frac{1}{2} V_{CEM}$.

The second implication of a low load is that it may be considered to approximate to a short-circuit at the output, in which case the input impedance to the transistor becomes h_{ie}. Usually, this small-signal parameter is reasonably accurate when used for large-signal computations since the input characteristic does not deviate too far from a straight line.

Under these circumstances, the ratio of the input transformer (if used) should be such that the source impedance is matched to h_{ie} for maximum power transfer.

Finally, the bias conditions must be considered.

In figure 9.7(a), a series bias system is illustrated. The design of this system is, however, not as simple as was the case for a small-signal amplifier, for if R_E is made large enough to ensure good stability, it will probably dissipate excessive power in the large-signal case. If R_E is reduced, the stability of the circuit becomes poor, and it is usual to find that class A power stages work quite close to the limits of stable operation.

If the maximum junction temperature which the transistor can tolerate is known, then equation (4.17) may be used to estimate the conditions which allow this temperature to be approached from a given ambient. From equation (4.17), the junction temperature T_j is:

$$T_j = T_{amb} + \frac{dT_j}{1 - r}.$$

In this context dT_j is the temperature rise due to switching on the circuit,

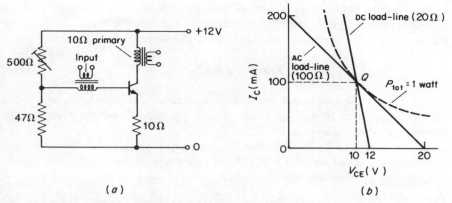

Figure 9.7. (a) Example of class A large-signal amplifier stage; (b) load-lines for amplifier stage.

and if T_j is the maximum safe junction temperature, then T_{amb} must be the maximum ambient temperature at which the transistor is expected to work.

Equation (4.19) gives

$$r = \theta(2V_{CE} - V_{CC})\frac{dI_C}{dT_j}.$$

In the present case it is necessary to calculate dI_C/dT_j and this may be done by using equation (4.10) in conjunction with the manufacturer's data. In that the graphs relating the change in I_C to changes in I_{CEO}, V_{BE} and h_{FE} can be approximated by straight lines over limited ranges, the following expression, derived from equation (4.10), is usually adequate:

$$\Delta I_C = K\,\Delta I_{CEO} - \frac{Kh_{FE}}{R_E + R_B}\,\Delta V_{BE} + KI_B\Delta h_{FE}. \qquad (9.6)$$

Here, each change is taken over the same temperature interval.

The complexity of equation (9.6) means that the extraction of K, knowing the maximum tolerable rise in I_C, becomes a very tedious procedure. Further, no account has been taken of the effect of R_E from the points of view of degeneration, voltage drop, or power dissipation. These facts, coupled with the often incomplete data available from the manufacturer, make a more empirical approach mandatory.

Firstly, R_E should be assigned a value which will not cause excess voltage drop or power dissipation in view of the power supply available. Secondly, its degenerative effect on the signal should be considered. Thirdly, the bias resistor (or resistors) should be assigned values such that K is as small as possible, though again, power dissipation in these resistors must be considered.

If sufficient data are available, K may be inserted into equation (9.6) to give the rise in I_C over and above the chosen quiescent value. Finally, the maximum junction temperature should be calculated:

$$T_j = T_{amb} + \theta P_{diss(m)} \tag{9.7}$$

where T_{amb} is the maximum ambient temperature envisaged, while θ is the thermal resistance (usually given in °C/mW) which relates the temperature rise to the power dissipated.

In the present case, where the low-resistance output transformer maintains the quiescent value of V_{CE} fairly constant and close to the supply voltage V_{CC}, the power dissipated becomes:

$$P_{diss(m)} \simeq V_{CC} I_C$$

where

$$I_C = I_Q + \Delta I_C \,,$$

giving

$$T_j \simeq T_{amb} + \theta V_{CC}(I_Q + \Delta I_C) \,. \tag{9.8}$$

If T_j exceeds the maximum permissible junction temperature, then K must be decreased, or the quiescent power reduced, or a combination of the two.

9.2.1 Simple Design Example

Consider the use of a silicon planar npn medium power transistor used in class A. Suppose that a power supply delivering 12 V is available, as also is a heat sink that will improve the value of θ in free air from 0.2°C/mW to 0.1°C/mW. For a maximum permissible junction temperature of 150°C, this means that if a maximum ambient temperature of 50°C is assumed, the maximum power dissipated may be found by use of equation (9.7):

$$150 = 50 + 0.1 P_{diss(m)}$$

giving $P_{diss(m)} = 1$ W.

The quiescent conditions may now be conveniently set at $(I_Q + \Delta I_C) = 100$ mA and $V_Q = 10$ V as shown in figure 9.7(b). Note that by including ΔI_C in the predetermined quiescent value of I_C, the difficulties inherent in the solution of equation (9.6) have been avoided for the present.

By setting $V_Q = 10$ V, the remaining 2 V of the 12 V power supply is available to provide the voltage drops across the primary resistance of the output transformer and the emitter resistor, R_E. Take these in turn.

Firstly, if 1 V is assumed to be dropped across the primary resistance of the transformer, this resistance, R_P, is fully defined, since the value of the quiescent current is known:

$$R_P = \frac{1}{0.1} = 10\,\Omega.$$

Secondly, if the remaining 1 V is dropped across R_E, then R_E is also 10 Ω. This will involve a power waste of 0.1 W. If the combination of the bias resistors and the input transformer primary shown in figure 9.7(a) gives $R_B = 50\,\Omega$, then $K = 0.22$, which is very satisfactory.

The upper bias resistor R_1 is made variable so that the quiescent conditions can be accurately adjusted when the transistor has reached its working temperature, and readjusted if a transistor of different gain is substituted. Any signal which is applied will then reduce the power dissipated according to the movement of the working point on the AC load-line. In the present case, this AC load-line is the line joining $2V_Q$ and $2I_Q$, giving $R_L = 20/0.2 = 100\,\Omega$. If the actual load is R_{LS}, the ratio of the output transformer is

$$\left(\frac{N_P}{N_S}\right)^2 = \frac{R_L}{R_{LS}}.$$

The efficiency of the circuit has been maximized by allowing the voltage and current excursions to be twice the quiescent values, but if severe distortion is to be avoided, these excursions must be reduced by some 20%. If a sinusoidal input is applied such that V_{CE} swings by 16 V peak-to-peak, while I_C swings by 160 mA peak-to-peak, then the average power output is:

$$P_{av} = \frac{16}{2\sqrt{2}} \times \frac{0.16}{2\sqrt{2}} = 0.32\,\text{W}.$$

The proportion of this power wasted in R_E is quite small in this case:

$$P_{R_E} = \left(\frac{0.16}{2\sqrt{2}}\right)^2 \times 10 = 32\,\text{mW}.$$

Ignoring this small loss, the efficiency of the stage is given by the ratio of the average power output to the maximum dissipation (1 W) in the transistor:

$$\text{commercial efficiency} = \frac{0.32}{1} \quad \text{or} \quad 32\%.$$

9.3 CLASS B OPERATION

From the foregoing discussion, it is apparent that a power transistor working in class A is an inefficient device, owing mainly to the power dissipated as a result of the high quiescent current involved. A further disadvantage is that this quiescent current must flow through the primary of the output transformer, which means that an iron core of large cross-section must be used in order that magnetic saturation should be avoided, especially when the value of I_C approaches its upper limit.

Both of these difficulties may be avoided by using two transistors, each biased near cut-off so that the quiescent current is negligible, one transistor accommodating positive-going signals and the other, negative-going signals.

When a transistor is biased near cut-off as in figure 9.8, it is said to be operating under class B conditions. (Strictly speaking, the transistor is in the *class AB* mode since it is not *quite* biased to cut-off.) Here, under quiescent conditions, I_C is very low, whilst V_{CE} is almost at V_{CC}. This means that the power dissipated during quiescent conditions, $P_{diss(Q)}$, is negligibly small.

Figure 9.8 shows both AC and DC load-lines, and indicates that $V_{CE(m)} \simeq V_{CC}$. The DC load-line, of course, corresponds to the primary resistance of an output transformer.

When a signal is applied at the base, then excursions of one polarity will cut the transistor off, while those of the other polarity will be amplified in the usual manner. This effect is shown in figure 9.9 for a sinusoidal input of the maximum possible amplitude.

P_{tot} will be determined by the manufacturer, and the various other power dissipations in the circuit may be obtained as follows.

Using figure 9.9,

$$P_L = \frac{1}{2} \frac{V_{CE(m)}}{\sqrt{2}} \frac{I_{C(m)}}{\sqrt{2}} = \frac{1}{4} V_{CE(m)} I_{C(m)} . \qquad (9.9)$$

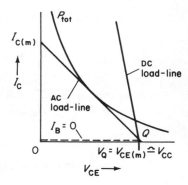

Figure 9.8. Class B operation.

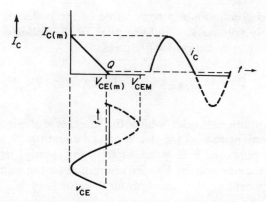

Figure 9.9. Excursions of I_C and V_{CE} under class B conditions: $2V_{CE(m)} \not> V_{CEM}$; $V_{CE(m)} \simeq V_{CC}$.

(The factor $\frac{1}{2}$ arises since the transistor conducts only over alternate half-cycles, while P_L is defined as the maximum load power over an integral number of *full* cycles.)

The power dissipation within the transistor may be obtained in the same manner as for the class A case, except that account must be taken of the non-conducting half-cycle:

$$i_C = I \sin \omega t \atop v_{CE} = V_{CC} - V \sin \omega t \Biggr\} \quad 0 < \omega t < \pi$$

$$i_C = 0 \atop v_{CE} = V_{CC} \Biggr\} \quad \pi < \omega t < 2\pi .$$

Over the first half-cycle the instantaneous power is $p = v_{CE}i_C$, and is zero over the second half-cycle.

Since

$$v_{CE}i_C = V_{CC}I \sin \omega t - VI \sin^2 \omega t$$

$$P_{diss} = \frac{V_{CC}I}{2\pi} \int_0^\pi \sin \omega t \, \mathrm{d}(\omega t) - \frac{VI}{2\pi} \int_0^\pi \sin^2 \omega t \, \mathrm{d}(\omega t)$$

$$P_{diss} = \frac{V_{CC}I}{\pi} - \frac{VI}{4} . \tag{9.10}$$

Under the no-signal condition $I = 0$, equation (9.10) gives:

$$P_{diss(Q)} = 0 . \tag{9.11}$$

Under full-load conditions the peak value of the collector voltage, V, is equal to the supply voltage V_{CC} and the peak of the collector current, I, is equal to $I_{C(m)}$. Hence, equation (9.10) gives:

$$P_{diss(FL)} = V_{CC} I_{C(m)} \left(\frac{1}{\pi} - \frac{1}{4} \right) = 0.068 \, V_{CC} I_{C(m)} \,. \tag{9.12}$$

To find the conditions under which the transistor dissipates maximum power, it is convenient to find the derivative of equation (9.10) and equate to zero. For this purpose, V is given as IR_L, for clearly, in the common-emitter mode, the excursion of V_{CE} will be equal to the voltage excursion across the load, providing that any emitter resistor is included in R_L.

$$P_{diss} = \frac{V_{CC} I}{\pi} - \frac{R_L I^2}{4} \,. \tag{9.13}$$

Therefore

$$\frac{dP_{diss}}{dI} = \frac{V_{CC}}{\pi} - \frac{R_L I}{2} = 0$$

giving

$$I = \frac{2V_{CC}}{\pi R_L} = 0.64 I_{C(m)} \,. \tag{9.14}$$

This means that maximum dissipation occurs when the peak value of the sinusoidal collector current is 64% of its maximum possible value. From equations (9.13) and (9.14)

$$P_{diss(m)} = \frac{V_{CC}^2}{\pi^2 R_L} \,. \tag{9.15}$$

This equation is extremely useful in that it relates the supply voltage and load to the maximum possible dissipation, which must, of course, be equal to or less than P_{tot}. It therefore serves as a fundamental design equation for a class B stage.

The operating efficiency may be obtained from a knowledge of the power extracted from the supply for full load. This will be:

$$P_S = V_{CC} I_{C(average)}$$

$$= \frac{V_{CC}}{2\pi} \int_0^\pi I_{C(m)} \sin \omega t \, d(\omega t)$$

or

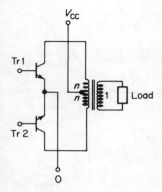

Figure 9.10. Transistors in push-pull.

$$P_S = \frac{V_{CC}I_{C(m)}}{\pi}.$$
(9.16)

Therefore:

$$\text{operating efficiency} = \frac{P_L}{P_S} = \frac{\text{equation (9.9)}}{\text{equation (9.16)}} = \frac{\pi}{4} = 0.78.$$
(9.17)

This is a considerable improvement over the class A case.

$$\text{Commercial efficiency} = \frac{P_L}{P_{\text{diss(m)}}} = \frac{\text{equation (9.9)}}{\text{equation (9.15)}}$$

$$= \frac{V_{CC}I_{C(m)}}{4} \frac{\pi^2 R_L}{V_{CC}^2} = \frac{\pi^2}{4} = 2.47.$$
(9.18)

This means that for a class B stage, the power output can have a value up to 2.47 times the value of P_{tot}, which is nearly five times better than the case for class A.

The methods of connecting transistors in such a manner that both halves of the input signal can be amplified will now be considered. The basic form of connection is called the *push–pull* mode and is shown in figure 9.10.

9.3.1 Push–Pull Operation

It is obviously possible to bias a pair of transistors so that they work in any mode from class A through the range of class AB to class B (and beyond). However, the best compromise between high efficiency and acceptable distortion puts the working point quite near the true class B condition. The form taken by the composite I_C/V_{BE} characteristic is illustrated in figure 9.11, where the optimum bias voltage is clearly defined as that voltage which

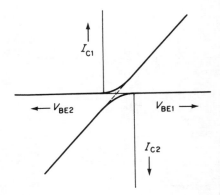

Figure 9.11. Composite I_C/V_{BE} characteristic (with slope exaggerated for clarity).

allows the difference between the characteristics to approximate closely to a straight line near the origin. The curves of figure 9.11 are idealized, since the I_C/V_{BE} characteristic is exponential, not straight, but an approach to this linearity may be made by including an unbypassed emitter resistor. This tends to swamp the very low resistance implied by $\Delta V_{BE}/\Delta I_C$ and so linearize the relevant characteristic. Fortunately the value of R_E in the case of power transistors can be very small ($<1\,\Omega$), which allows both the extra power dissipation and the degeneration to be kept low. Furthermore it helps to provide a much smaller stability factor than would be possible were an emitter resistor excluded altogether.

Strictly speaking, the ordinates in figure 9.11 should overlap slightly, for the class B condition is not quite attained. This overlap is illustrated in figure 9.12 which shows the nature of the composite load-line. The effect of

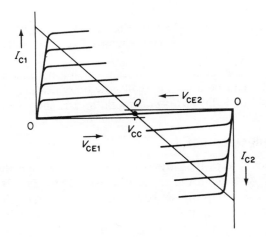

Figure 9.12. Composite load-line for class B push–pull operation.

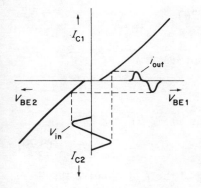

Figure 9.13. Cross-over distortion in an un-biased push–pull stage (with I_C/V_{BE} slopes exaggerated for clarity).

zero bias is shown in figure 9.13, where the nature of the distortion is clearly apparent. It might be thought that, if only sinusoidal signals of a single frequency are involved, as would be the case for a servo-motor driving stage, then such distortion would be of little importance. However, figure 9.13 shows that much of the distortion arises owing to clipping at the lower values of I_C, which means that the sensitivity of the amplifier for low-level signals is considerably reduced. This may result in a "dead spot," which means that the servo-motor delivers no torque over a small angular displacement when nominally stationary. It is therefore advisable to apply bias in all cases of class B push–pull operation and so minimize this *cross-over distortion*.

9.4 DISTORTION

In general, distortion due to the curvature of the I_C/V_{BE} characteristic can be examined by representing this curvature in the form of a power series, so that

$$i_C = I_Q + Av_{BE} + Bv_{BE}^2 + Cv_{BE}^3 + \cdots .$$

If

$$v_{BE} = V \sin \omega t ,$$

then

$$i_C = I_Q + AV \sin \omega t + BV^2 \sin^2 \omega t + CV^3 \sin^3 \omega t + \cdots \qquad (9.19)$$

where A, B, C, etc. are constants. Recalling that

$$\sin^2 \theta = \tfrac{1}{2}(1 - \cos 2\theta)$$

and

$$\sin^3 \theta = \tfrac{1}{4}(3 \sin \theta - \sin 3\theta)$$

equation (9.19) becomes:

$$i_C = (I_Q + \tfrac{1}{2}BV^2) + (AV + \tfrac{3}{4}CV^3) \sin \omega t - \tfrac{1}{2}BV^2 \cos 2\omega t$$
$$- \tfrac{1}{4}CV^3 \sin 3\omega t \cdots. \tag{9.20}$$

This equation shows that the nonlinearity of the I_C/V_{BE} curve results in the introduction of not only a DC or *rectified component*, $\tfrac{1}{2}BV^2$, but also of a series of harmonics. The effect of the rectified component is to shift the operating point, while that of the harmonics is to produce a distorted collector current waveform. Figure 9.14 shows the relationship of the fundamental with the various components of equation (9.20). It will be noticed that the even harmonics (represented by only the second harmonic in figure 9.14) repeat themselves in the second half-cycle of the fundamental *in the same phase* as in the first half-cycle. This point is of importance if two transistors are worked in class A push–pull, for it enables the even harmonics to cancel out. Consider the two collector currents in such a case and let them be represented in the form of equation (9.19). For one transistor, $v_{BE1} = V \sin \omega t$, and for the other, $v_{BE2} = -V \sin \omega t$, giving:

$$i_{C_1} = I_Q + AV \sin \omega t + BV^2 \sin^2 \omega t + CV^3 \sin^3 \omega t + \cdots$$
$$i_{C_2} = I_Q - AV \sin \omega t + BV^2 \sin^2 \omega t - CV^3 \sin^3 \omega t + \cdots.$$

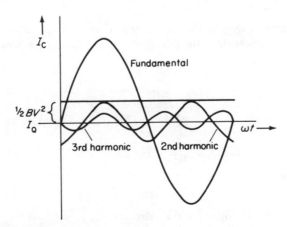

Figure 9.14. Components of distorted collector current wave (0, 1st, 2nd and 3rd harmonics only shown).

From these two equations, the effective current in the primary of the output transformer is:

$$i_{C_1} - i_{C_2} = 2AV \sin \omega t + 2CV^3 \sin^3 \omega t + \cdots.$$

Again using the identity $\sin^3 \theta = \frac{1}{4}(3 \sin \theta - \sin 3\theta)$, this becomes

$$i_{C_1} - i_{C_2} = 2(AV + \tfrac{3}{4}CV^3) \sin \omega t - \tfrac{1}{4}CV^3 \sin 3\omega t + \cdots. \tag{9.21}$$

This equation shows that the second harmonic (and subsequent even harmonics) have cancelled out, and also that the rectified component has disappeared. The result is depicted in figure 9.15(a), where only the components due to the square term have been included.

The situation for the class B case is somewhat different, though the result is similar. Over each odd half-cycle, the full set of harmonics will be present

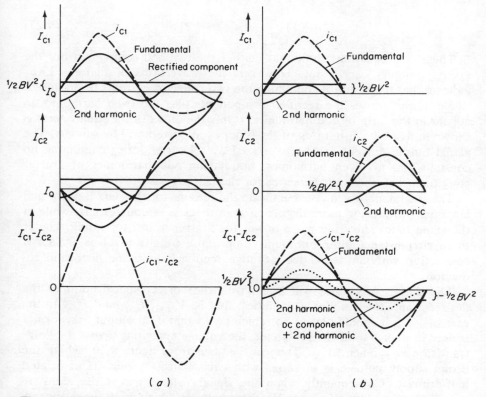

Figure 9.15. Collector and resultant current waveforms in push–pull stages for: (a) class A push–pull; and (b) class B push–pull.

as in equation (9.20), but I_Q will be zero and only one transistor will contribute the effective current:

$$i_{C_1} - i_{C_2} = \tfrac{1}{2}BV^2 + (AV + \tfrac{3}{4}CV^3)\sin \omega t - \tfrac{1}{2}BV^2 \cos 2\omega t - \tfrac{1}{4}CV^3 \sin 3\omega t \tag{9.22}$$

when

$$2n\pi < \omega t < (2n+1)\pi .$$

During even half-cycles, the other transistor will contribute the effective current:

$$i_{C_1} - i_{C_2} = -\tfrac{1}{2}BV^2 + (AV + \tfrac{3}{4}CV^3)\sin \omega t + \tfrac{1}{2}BV^2 \cos 2\omega t - \tfrac{1}{4}CV^3 \sin 3\omega t$$
$$+ \cdots$$

when

$$(2n+1)\pi < \omega t < (2n+2)\pi . \tag{9.23}$$

These two equations show that neither the rectified component nor the even harmonics cancel out. However, the waveforms of figure 9.15(b) indicate that the result of combining the two collector currents to produce a single output causes the rectified components plus the even harmonics to appear in the form of odd harmonics. (Only the effect of the square term is shown in figure 9.15, but that of the other even powers will be similar.) The actual output current can be expressed as a Fourier series containing no constant and no even harmonics, but having odd harmonics of greater amplitude than those of the series for the class A case.

The treatments given above indicate the reasons for the class B push–pull amplifier introducing more distortion than its class A counterpart, while at the same time delivering more power for a given transistor rating. Under these circumstances, the optimum compromise usually involves a class B connection with an associated negative feedback loop for distortion reduction.

The design procedure will begin with the correct specification of the desired transistor handling capabilities, that is, V_{CEM}, I_{CM}, and P_{tot}. In the case of V_{CEM} it should always be remembered that if an output transformer is used in a push–pull class B circuit, the voltage appearing across the "off" transistor may reach $2V_{CC}$. This can be seen from figure 9.10, where the actual supply voltage is in series with each transistor plus its associated half-primary. Consequently when the signal voltage across the alternate transistor approaches $V_{CE(m)}$ ($=V_{CC}$) this voltage, which at all times appears across the alternate primary by normal transformer action, causes the total voltage across the "off" transistor to reach $V_{CC} + V_{CE(m)}$ or $2V_{CC}$. Consequently V_{CEM} for each transistor must be at least $2V_{CC}$. Figure 9.9 shows $V_{CEM} = 2V_{CE(m)}$.

Finally, the temperature and heat dissipation must be considered.

9.5 HEAT DISSIPATION

The heat generated at the collector junction and due to the power dissipated, $V_{CE}I_C$, is conducted away with an efficiency depending on the construction of the transistor. However, it must eventually reach the case having arrived via a thermal resistance θ_{jc}.

The heat flow then continues, partly owing to conduction through the mounting components associated with the transistor and partly as a result of convection. The conduction process is the most important, and each heat conducting component will have an associated thermal resistance, θ (usually measured in °C/W).

This state of affairs is easily susceptible to analysis by analogy, and may be represented by the "thermal circuit" of figure 9.16. Here, P_{diss} watts are dissipated at the collector junction, and this produces a temperature T_j, which is given by:

$$T_j - T_{amb} = P_{diss}\theta_t \qquad (9.24)$$

where θ_t is the total thermal resistance involved and T_{amb} is the ambient temperature.

In the example of figure 9.16, θ_t is the sum of the thermal resistances between junction and case, case and sink, and sink to ambient:

$$\theta_t = \theta_{jc} + \theta_{cs} + \theta_{sa} . \qquad (9.25)$$

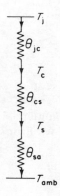

Figure 9.16. Heat flow diagram: j = transistor collector junction; c = transistor case; s = heat sink; and amb = ambient surroundings.

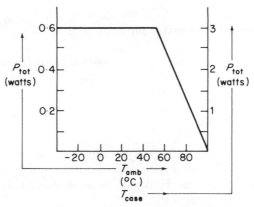

Figure 9.17. Typical de-rating characteristic for a small power transistor.

For a large power transistor, a case having a heavy copper or brass header is used, the collector being bonded to this header not only thermally, but electrically. This means that the header itself must often be electrically insulated from the heat sink upon which the transistor is mounted. This is usually done by interposing a thin mica washer which, referring to equation (9.25), would contribute most of θ_{cs}. (A less common but better alternative is a washer moulded from beryllia, which is a ceramic having a thermal conductivity similar to that of aluminum, but is also an electrical insulator.)

The value of θ_{sa} for the heat sink itself is dependent on its surface area, configuration and, of course, the material used.

Equation (9.24) implies that the permissible power dissipation decreases as the ambient temperature increases, and this leads to a de-rating characteristic of the form shown in figure 9.17.

It is now possible to determine the maximum power which it is permissible to dissipate using any given transistor, providing that *either* a de-rating curve applicable to the requisite conditions is available *or* that the value of θ_t is obtainable. This maximum power must, of course, be less than P_{tot}, the limiting power dissipation at the highest ambient temperature envisaged.

9.5.1   Simple Design Example

As an indication of the order of quantities involved consider the case of a 2N3055 silicon power transistor, for which θ_{jc} is given as 1.5°C/W and, if a mica washer is used, θ_{cs} is about 0.5°C/W. Then suppose that a commercial cooling fin extrusion, finished in matt black, gives $\theta_{sa} = 2.25$°C/W. From these figures,

$$\theta_t = 1.5 + 0.5 + 2.25 = 4.25°C/W .$$

If the maximum anticipated ambient temperature is 50°C and the maximum junction temperature is given as 200°C then the maximum allowable power dissipation is given using equation (9.24):

$$P_{\text{tot}} = \frac{T_j - T_{\text{amb}}}{\theta_t} = \frac{200 - 50}{4.25} = 35 \text{ W}.$$

If the transistor were working in class A, then the value of $P_{\text{diss(m)}}$ would be given by the quiescent conditions:

$$P_{\text{diss(m)}} = P_{\text{diss(Q)}} = V_Q I_Q.$$

This means that the maximum power output (neglecting distortion) would be $35 \times \frac{1}{2} = 17.5$ W for a single-ended class A stage or 35 W for a push–pull class A stage.

For a single-ended class B stage, the maximum output power would be 35×2.47. For two transistors in class B push–pull this would be $35 \times 2.47 \times 2$, nearly 173 W.

The "off" transistor would experience $2V_{CC}$ under conditions of maximum signal in the "on" transistor and since the value of V_{CEM} for the 2N3055 is given as 60 V, then,

$$V_{CC} \leq 60/2 = 30 \text{ V}.$$

9.5.2 Transient Operation

A final concept suggested by the electrical analogy is that of thermal time constant. So far, all heat dissipation considerations have been essentially those relating to the steady state. However, it would be theoretically possible to exceed the specified maximum power dissipation of a transistor for a short period, providing that this period were so short that the junction temperature did not reach T_j. This implies a thermal equivalent of a capacitor connected across the "thermal resistor," θ_{jc}, resulting in a time constant given by the product of the two. This concept, though sometimes useful, will not be considered further here since transient operation is of marginal interest in servo-drive amplifier design.

9.6 THE OUTPUT TRANSFORMER

In figure 9.10, the turns ratio of the output transformer for a push–pull stage is given as $2n:1$. If the actual load at the secondary has a value R_{LS}, the transformer ratio may be derived simply for both the class A and class B cases.

Figure 9.18(a) represents the class A push–pull case, where both transis-

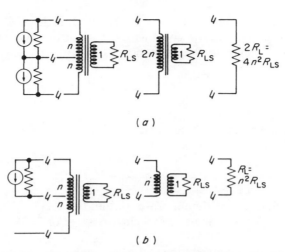

Figure 9.18. Development of equivalent circuits for push–pull output stage loads: (a) class A, both transistors operating concurrently; (b) class B, transistors operating alternately.

tors are operating concurrently. If these transistors are properly matched, then the net signal current in the connection from the emitters to the transformer center-tap will be zero. The two optimum loads are therefore in series and the following relationship will hold:

$$2n = \left(\frac{2R_L}{R_{LS}}\right)^{1/2} \quad \text{or} \quad n = \left(\frac{R_L}{2R_{LS}}\right)^{1/2}. \tag{9.26a}$$

In the class B case, only one transistor is operative at any given time, which leads to the equivalent circuit of figure 9.18(b). The relationship giving the transformer ratio must therefore involve one half of the transformer primary only:

$$n = \left(\frac{R_L}{R_{LS}}\right)^{1/2} \quad \text{or} \quad 2n = \left(\frac{4R_L}{R_{LS}}\right)^{1/2}. \tag{9.26b}$$

The relevant values of R_L must be inserted into equations (9.26a) and (9.26b) if a consistent comparison between the two cases is to be made, and the basis for this comparison will be that of equal output powers using transistors having identical values of V_{CEM}. Also $V_{CE(m)}$ will be assumed to take its maximum possible value.

Taking the class A case, equation (9.1) gives $P_L = \frac{1}{8}V_{CE(m)}I_{C(m)}$ and since $V_{CE(m)} = V_{CEM}$, this becomes:

$$P_L = \frac{1}{8}V_{CEM}I_{C(m)}. \tag{9.27a}$$

For the class B case, equation (9.9) gives $P_L = \frac{1}{4}V_{CE(m)}I_{C(m)}$ and since $V_{CEM} = 2V_{CE(m)}$, this becomes:

$$P_L = \tfrac{1}{8}V_{CEM}I_{C(m)} . \tag{9.27b}$$

These two expressions are identical, which means that, for equal output powers, the two values of $I_{C(m)}$ must also be identical.

Using figures 9.3 and 9.9, the two values for R_L are:

$$R_L(\text{class A}) = V_{CEM}/I_{C(m)} \tag{9.28a}$$

$$R_L(\text{class B}) = V_{CEM}/2I_{C(m)} . \tag{9.28b}$$

Putting these expressions into equations (9.26a) and (9.26b) gives:

$$\text{for class A} \quad 2n = \left(\frac{2V_{CEM}}{I_{C(m)}R_{LS}}\right)^{1/2}$$

$$\text{for class B} \quad 2n = \left(\frac{2V_{CEM}}{I_{C(m)}R_{LS}}\right)^{1/2}$$

These two expressions prove that the output transformer turns ratios for class A and class B push–pull pairs are identical when the power output is the same and transistors having identical values of V_{CEM} are used at maximum efficiency.

However, equations (9.28) indicate that the load seen by a class B transistor is only half that seen by its class A counterpart. This means that the power gain of a class B pair is only half that of a class A pair, which implies that the drive power for this class B stage must be twice that for the class A stage. This is actually apparent as a result of examining figures 9.3 and 9.9, for whereas the peak-to-peak voltage swings are the same in both cases ($=V_{CEM}$), the peak-to-peak current swing in the class B case is twice that for the class A case.

The frequency response of the push–pull stage will depend on the resistances and inductances of the transformer windings. The equivalent circuit for a transformer is shown in figure 9.19(a). R_1 and L_1 are the winding resistance and leakage inductance referred to the primary. R_m and L_m are the components of the magnetizing impedance. An h-parameter output circuit is shown feeding the transformer.

For low frequencies the effect of L_1 will be negligible, but the magnetizing inductance, L_m, will seriously shunt the primary winding as shown in figure 9.19(b). Assuming that $1/h_{oe}$ and R_m are very large, the current I_C from the transistor will divide equally between L_m and the primary winding (i.e., the reflected load R_L) when $R_L = \omega L_m$. This represents the −3 dB point:

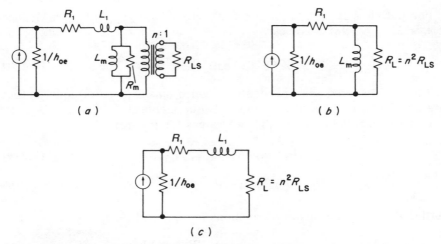

Figure 9.19. Equivalent circuits for a transformer: (*a*) full equivalent; (*b*) low-frequency equivalent; (*c*) high-frequency equivalent.

$$f_{(-3\,\mathrm{dB})} = \frac{R_L}{2\pi L_m} \quad \text{for low frequencies .}$$

This shows that for a good low-frequency response the magnetizing inductance must be specified by taking the value of R_L into account during the design procedure, and maximizing L_m.

At high frequencies the magnetizing inductance becomes unimportant, but the leakage inductance severely reduces the current flow to the primary. In this case the -3 dB point is found by determining the frequency at which ωL_1 is equal in magnitude to $1/h_{oe}$. R_1 and R_L are assumed to be small in comparison with ωL_1.

$$f_{(-3\mathrm{dB})} = \frac{1}{2\pi h_{oe} L_1} \quad \text{for high frequencies .}$$

Under conditions of very low L_1, it is usual for the capacity between turns to become significant enough to restrict the high-frequency response.

9.7   THE DRIVER STAGE

Three distinct functions are performed by the driving stage. They are:

  (i) provision of adequate power to the output stage;
 (ii) provision of bias voltage and current to the output stage;
(iii) phase splitting if the output is a push–pull pair.

The simplest drive system that will accomplish these three functions is illustrated in figure 9.20 for a class AB push–pull stage.

Firstly, the peak-to-peak signal current needed by the output transistor bases will obviously be $2I_{C(m)}/h_{FE(min)}$, and this must be provided by the transformer secondary. Hence, knowing the peak-to-peak current swing available from the driver transistor, the transformer ratio can be chosen accordingly.

Secondly, the diode D is held in conduction by a current defined by R, which must be greater than the peak value of output transistor base current, otherwise the diode will cut off. When operating properly, the value of R can be empirically adjusted so that the diode voltage drop supplies a V_{BE} base bias voltage to the output transistors which will hold them in the class AB condition, that is, just beginning to conduct, to minimize cross-over distortion. Also, if the diode is mounted on the same heat sink as the output transistors, its voltage drop will track the temperature-induced variation in V_{BE} to some extent, so offering a degree of compensation.

Finally, the anti-phase drive is automatically accomplished by the center-tapped secondary.

Should class A operation be required, more than one diode can be used, or a resistor may be substituted; and in the latter case, an appropriate resistor-thermistor combination can be designed to provide thermal compensation.

Even though it is very easy to apply negative feedback, as is also shown in figure 9.20, this does not solve any of the basic problems associated with

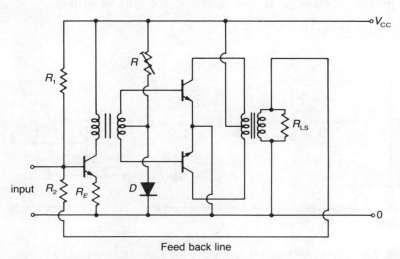

Figure 9.20. A push–pull class AB power amplifier circuit using transformer output and drive, and including feedback.

transformers, such as cost and bulk, though it does improve the poor frequency response. In principle, it is better to use electronic phase splitting, and figures 9.21(*a*) and (*b*) show two common circuits, both of which have already been met. However, because their outputs do not inherently provide proper biasing for the power transistors, they are normally AC coupled, which makes extra biasing circuitry necessary. The simplest way to accomplish this is shown in diagram (*c*) where a divider chain includes resistors R_A and R_B, which each provide an appropriate biasing voltage for its associated power transistor.

In diagram (*a*) the resistor R_O has been included to equalise the two drive impedances, for obviously that from the emitter is much lower than that from the collector. The long-tailed pair of diagram (*b*), however, offers equally high drive impedances. Unfortunately, the output transistors constitute highly asymmetrical loads when biased into the class B mode, because only one is operative at a time. This again makes some extra circuitry necessary such as that also shown in figure 9.21(*c*), which provides diode/resistor conducting paths for negative-going signals.

It should now be clear that in the rather primitive form of power amplifier so far presented, there exist many disadvantages, including the necessity for at least one transformer. It does, however, serve to illustrate the use of the derived ideal equations and leads naturally to the use of power amplifiers to drive motors, which is the next topic. However, modern power amplifiers are transformerless and though this implies the acceptance of gross load mismatching, excellent performance can be obtained via overall feedback, which can also be used to enhance bias stability, since a DC path exists through the network. Further, the technique lends itself to integration, albeit at fairly low powers at the time of writing.

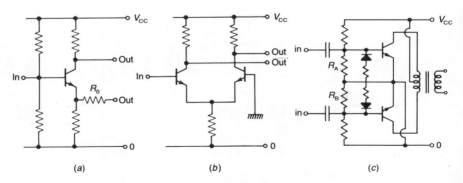

Figure 9.21. Phase splitters (*a*) and (*b*); and a simple biasing circuit with class B drive impedance equalisation (*c*).

9.8 SERVO-MOTOR DRIVES

Many small servo-motors are of the AC two-phase type. These require two fixed frequency supplies at a phase angle of 90°. One of these supplies has fixed amplitude and serves only to make possible the production of a magnetic field which rotates in space, thus turning the rotor, to which there are no connections. Commonly, this fixed supply comes from the mains or from a 400 Hz supply as is usual in aircraft. It may also be derived from some part of a servo-system, such as the chopper wheel in a double-beam electro-optical instrument. The symbol for such a motor is shown in figure 9.22(a).

The second supply must remain accurately in quadrature with the first, but must have an amplitude dependent on the drive signal. This means that the actual driving signal is usually provided by a power amplifier, and in order to make possible the transformerless operation of a conventional push–pull amplifier, the control winding of the servo-motor is usually provided with a center-tap. The form of drive circuit is then as shown in figure 9.22(b).

Since the signal and reference windings of the servo-motor are highly inductive, it is usual to connect a capacitor in parallel with such a winding if it is driven from an amplifier, for if the capacitance is chosen with care the resultant impedance reverts to a resistance only. This is known as "power factor correction."

It should be noted that the optimum load to be presented by a servo-motor winding will be dependent on the mode of operation of the output transistors, for it has been pointed out that the apparent load "seen" by a

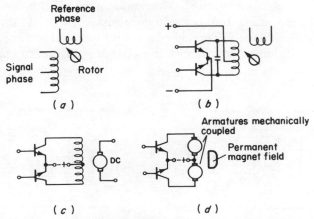

Figure 9.22. (a) Two-phase servo-motor; (b) push–pull transistor stage for driving (a); (c) split-field DC servo-motor drive; (d) twin armature DC servo-motor drive.

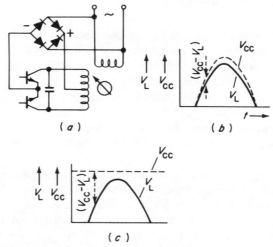

Figure 9.23. Servo-motor operation: (*a*) using unsmoothed DC, the servo-mode; (*b*) waveforms in the servo-mode; (*c*) waveforms in the normal class B mode.

class B stage is half that of a class A stage, leading to current drive which is twice that for the class A.

It is often possible to dispense with a smoothed DC supply for the output transistors and use instead a rectified signal as shown in figure 9.23(*a*). This implies that each output transistor is operating in class B but with a variable value of V_{CC}, and this state of affairs may be illustrated by comparing the collector voltage waveforms with V_{CC}, as has been done in figure 9.23(*b*).

The equivalent comparison for the class B amplifier having a smooth DC supply is given in figure 9.23(*c*). The major point to note is that in the servo-mode, the voltage drop across the transistor, $(V_{CC} - V_L)$, is negligible since the supply voltage wave is in phase with the load voltage wave. This implies high efficiency, and in fact the operating efficiency can approach 100%.

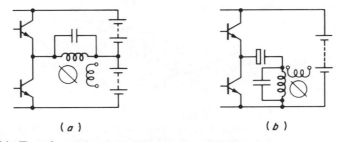

Figure 9.24. Transformerless output stages for: (*a*) two-rail supply; and (*b*) one-rail supply.

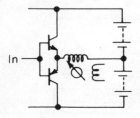

Figure 9.25. Complementary pair output stage.

In the event of a DC servo-motor or actuator being used, a power amplifier employing transformers is obviously useless but, given an appropriate DC-coupled driver stage, there is no reason why the output transistors themselves should not be coupled directly to the field windings of a split-field DC servo-motor, as shown in figure 9.22(c).

The design of such a stage must take into account the static load-line, for this will represent the locus of the transistor operating points for slow signal changes. The dynamic load-line will come into operation at higher frequencies. This, of course, is equivalent to considering the split field as a resistive load for very low frequencies and an inductive load at higher frequencies.

Clearly the motor ceases to rotate when the net field is zero, and until a small but finite out-of-balance current flows the holding or restoring torque will be zero. Sometimes this is an undesirable situation, and a twin-armature motor having a permanent magnet field may be used to ensure that the holding torque is large. This implies a class A mode so that the quiescent currents through the windings are equal at null. Such a motor would be connected as in figure 9.22(d).

If a center-tapped field winding for either an AC split-phase or DC split-field winding is not available, the two-rail supply system shown in figure 9.24(a) may be used. In the case of the AC motor only, a capacitively coupled load may be applied as in figure 9.24(b)[1].

Figure 9.25 shows an example of *complementary* symmetry. Here, it will be noticed that because the two output transistors are of opposite polarity in such a stage, they may be driven from a single-ended driver, and that no phase-splitting stage is necessary. A brief treatment of this form of power amplifier will now be given.

9.9 TRANSFORMERLESS COMPLEMENTARY POWER AMPLIFIERS

The basic form of the complementary power amplifier is given in figure 9.26(a). Here, a simple CE stage is used to drive a pair of complementary power transistors Tr2 and Tr3. If low distortion is required, the characteristics of Tr2 and Tr3 should be well matched (which is not an easy require-

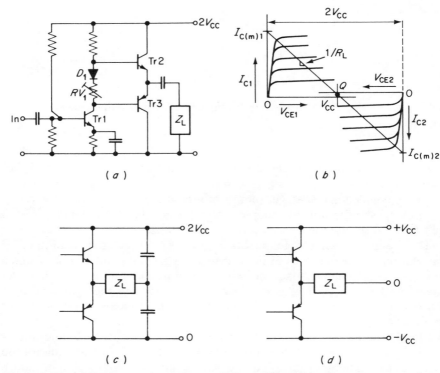

Figure 9.26. (*a*) Basic complementary power amplifier; (*b*) composite output characteristic for a complementary pair; (*c*) load isolation using dual capacitors; (*d*) two-rail method of making load isolating capacitor unnecessary.

ment to fulfil), and they should be biased so that the two output characteristics complement each other as in figure 9.26(*b*).

This, it will be noticed, is similar to the normal class B composite characteristic of figure 9.12, which means that all the relevant equations will also be similar. However, as there is no transformer, then (for a resistive load) when either output transistor is saturated, the full supply voltage will appear across the other, so that it becomes convenient to call this full supply voltage $2V_{CC}$. Hence, $2V_{CC} \le V_{CEM}$, and under quiescent conditions this voltage will be shared equally between Tr2 and Tr3, and the common point will be at V_{CC} as for the class B stage with an output transformer.

To achieve this quiescent condition (that is, the slight overlapping of the output characteristics), there must be a small bias voltage applied between the two bases, This may be provided by one or more diodes, and is sometimes individually adjusted by a pre-set resistor. D_1 and RV_1 perform this function in figure 9.26(*a*).

Notice that the load is supplied via a capacitor so that the quiescent conditions are not invalidated by the existence of a DC path through the load. This capacitor often has a very large value—for example, when the load has a low resistance, as does a loudspeaker—and this may prove an embarrassment due to its large charging transient at switch-on. Two alternative connections which solve this problem are shown in figure 9.26(c) and (d). In diagram (c) two capacitors are in parallel from the signal point-of-view, so that each may have half the capacity of the original unit. From a charging transient point-of-view, however, they are in series, and are connected across the supply rails, so that the charging current no longer passes through the load. An alternative solution is given in diagram (d) where a positive and a negative rail are used, $+V_{CC}$ and $-V_{CC}$. Here, the two emitters are at $0\,V$ in the quiescent condition, so that a capacitive connection to the load is unnecessary. This technique is widely used in monolithic amplifiers (see chapter 7) not only for power amplification purposes but also as a general level-shifting technique to bring the output quiescent potential to the same level as the input.

It being no longer necessary to provide an anti-phase drive because of the complementary signal requirements of the output transistors, numerous circuits may be designed to provide biasing and cross-over distortion minimization in addition to that of figure 9.26(a). One which has already been met in the case of operational amplifier complementary output stages, is the V_{BE}-multiplier, and figure 9.27 gives a basic circuit using this technique.

Here, the driver transistor has a collector load R_C, and the signal is level-shifted by Tr2. Because $V_{BE(2)}$ exists across R_2, then if $RV_1 = R_2$, then

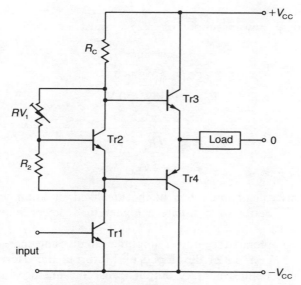

Figure 9.27. A complementary pair output stage with V_{BE}-multiplier biasing.

$2V_{BE(2)}$ must exist across the series combination. This is applied across the bases of the Tr3/Tr4 combination, and again RV_1 may be empirically adjusted to give minimal cross-over distortion. If Tr2 is similar to Tr3 and Tr4, and is mounted on the same heat sink, then good temperature compensation may be achieved using this circuit. Also, it is an obvious candidate for integration, given good enough heat removal facilities, and indeed several small-power integrated circuits are available using this method.

9.9.1 Simple Design Example

A complementary pair like that in figure 9.27 is required to deliver 24 W of power to a resistive load of $5 \, \Omega$. Determine the power supply voltages needed to achieve this, and specify P_{tot}, V_{CEM} and I_{CM} for the transistors used.

Each of the two power transistors will be called upon to deliver 12 W to the load, so that using the idealized commercial efficiency of 2.47, the maximum power dissipation of each transistor will be about $12/2.5 \approx 5 \, W$.

Using the basic class B design equation (9.15),

$$P_{\text{diss(max)}} = \frac{V_{CC^2}}{\pi^2 R_L} \,,$$

$$V_{CC} = \pi (P_{\text{diss(max)}} R_L)^{1/2}$$

$$= \pi (12 \times 5)^{1/2} = 24.3 \, V \,.$$

Hence, a dual power supply of $\pm 25 \, V$ must be used, and for the transistor,

$$P_{tot} > 5 \, W \text{ each}$$

$$V_{CEM} > 2V_{CC} \approx 50 \, V \,,$$

and

$$I_{CM} > V_{CE(m)} / R_L = V_{CC} / R_L$$

$$= 25/5 = 5 \, A \,.$$

Note that this calculation refers to the idealized situation: in reality, it would have been better to postulate a higher load power requirement of (say) 30 W at least.

A practical complementary power amplifier circuit is given in figure 9.28, which shows a 5 W amplifier developed by Motorola Inc. Here, it will be seen that the two complementary output transistors are driven by a com-

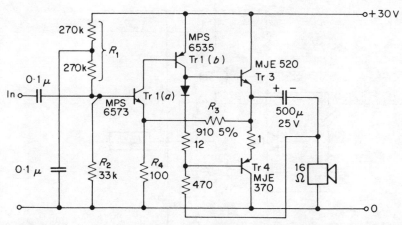

Figure 9.28. 5 W complementary power amplifier. (Reproduced by permission of Motorola Inc.)

pound driver stage consisting of Tr1(a) and Tr1(b). From a DC point-of-view, the base of Tr1(a) is held essentially at a constant voltage by R_1 and R_2, while its emitter voltage depends on the voltage drop in R_4. The current passing from the output point into R_4 via R_3 determines this voltage drop in R_4 and hence defines the emitter voltage of Tr1(a), resulting in good bias stability as a result of the overall negative feedback. To demonstrate the operation of the loop, assume that the output point drifts positive. This causes an increased current in R_3 and R_4 so sending the Tr1(a) emitter also positive with respect to the base. Hence, the collector current decreases, as does that in Tr1(b) and so the drive to Tr2 and Tr3 goes negative, taking the output point with it.

The second transistor of the compound driver stage feeds the two output transistors in the usual way, but its load resistor is bootstrapped by applying a positive feedback voltage from the load itself. This is so that the impedance to the bases of the two output transistors is high, resulting in an approach to a current drive and making for lower distortion. Note that although the collector current of the MPS 6535 passes through the (loud-speaker) load, it is small enough to be neglected in comparison with the drive current from the output point itself.

Note also that when the output is taken from the emitters, as in the present circuit, the drive voltage must swing by the same amount as the output voltage. A smaller drive swing can be used if the complementary transistors are inverted and their collectors commoned[2] but this, of course, results in a high impedance or current generator output as is used in the transconductance operational amplifiers mentioned in chapter 7.

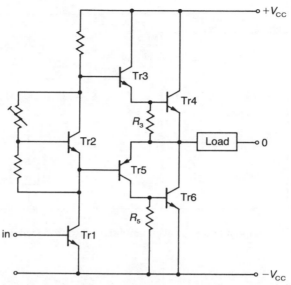

Figure 9.29. A quasi-complementary pair output stage with V_{BE}-multiplier biasing.

9.9.2 Quasi-Complementary Power Amplifiers

In practice, it is difficult to match npn and pnp transistors closely, particularly in the case of high-power discrete devices, and a very common way of overcoming this problem is to use matched power transistors of the same polarity, and drive them using a lower-power complementary pair. The basic circuit for the resulting *quasi-complementary power amplifier* is shown in figure 9.29 where the actual output stage consists of a matched pair of npn power transistors. Tr4 is part of an npn Darlington pair like that of figure 5.17, and is driven via the emitter of Tr3. Conversely, Tr5/Tr6 is a quasi-Darlington pair, the base of Tr6 being driven from the collector of the pnp transistor Tr5. Resistors R_3 and R_5 ensure that Tr4 and Tr6 can be turned OFF more quickly than Tr3 and Tr5 would otherwise allow. Also, when either Tr4 or Tr6 is supposed to be OFF, the resistor R_3 or R_5 prevents the leakage current I_{CEO} of its Darlington driver from holding it in a slightly conducting state.

Note that the V_{BE}-multiplier must supply three V_{BE} voltage drops in this case, to accomodate Tr3, Tr4 and Tr5.

9.10 POWER MICROCIRCUITS

Both hybrid[3,4] and monolithic[5,6] integrated circuits are available for power application, and figure 9.30 gives the functional circuit for a typical hybrid module, the HC2500, which will deliver up to 100 W. Here, it will be

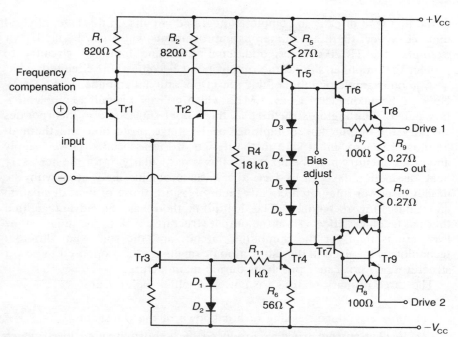

Figure 9.30. Functional circuit for the HC2500 high-power operational amplifier. (Reproduced by permission of Harris Semiconductor Inc.)

seen that the output stage is quasi-complementary as in figure 9.29, with the addition of small degeneration resistances R_9 and R_{10}, which improve linearity. The input is via a long-tailed pair Tr1/Tr2, and it will be noted that the output is taken to a pnp transistor from one collector only, as in the simple system of figure 5.34. The defined bias current flows down R_4 to produce a voltage drop across structures represented in this functional circuit by D_1 and D_2. This is applied to mirror transistor Tr3, which provides the long-tail current to the Tr1/Tr2 difference pair; and also to Tr4, which is the active load for Tr5. The structures D_3–D_6 provide biasing for proper class AB operation, and these are shunted by an external $1\,k\Omega$ variable resistor for cross-over distortion minimization. Normally, the "drive 2" pin is connected to the negative line and the output taken from the usual point, but the provision of alternate drive outputs is useful in switching applications.

Frequency compensation is achieved by connecting a capacitor from the collector of Tr5 (via the bias adjust pin) back to the collector of Tr1. Given a sufficiently large capacitance here (about 500–1000 pF), the resultant dominant lag will result in a 6 dB roll-off, giving full operational amplifier performance down to unity gain.

This circuit is an excellent example of how techniques already developed in the text lead directly to complete integrated circuits, in this case a hybrid unit. However, there are numerous improvements which can be made; for example, the HC2000H has additional "load-line limiting" circuitry to protect the amplifier should the output point become short-circuited.

A good example of a monolithic power operational amplifier is the 150 W National Semiconductor Corp. LM12, which utilizes a genuine complementary pair of Darlington structures as the output stage. It also incorporates circuitry for output current-limiting, over-voltage protection and thermal-limiting. Even so, great care in the design of the associated power supply and other circuitry must be taken and this is particularly so for the necessary heat dissipation layout. In fact, the basic problems associated with the design of power integrated circuits themselves are those of heat dissipation and fabricating techniques. The hybrid method has the advantage that thermal feedback effects from the output structures back to the input can be less serious than for the monolithic circuit; and the somewhat different technologies involved in the fabrication of small-signal, as opposed to power structures, do not cause problems during manufacture.

The three primary reasons for power amplifier failures are:

(1) an over-voltage, leading to breakdown of the transistors;
(2) an over-current (such as produced by a short-circuited load) which can lead to various breakdown modes from second breakdown of a transistor to fusing of the interconnections or bonding leads;
(3) an excessive temperature rise, leading to destruction of the transistors.

The reverse over-voltage problem can be solved by the inclusion of protective diodes across the output devices, provided that such an over-voltage is transient only, otherwise the diodes themselves could be destroyed. A forward over-voltage is usually damaging only if combined with an over-current, so taking the transistor outside the safe working area, and figure 9.31 shows a basic method of sensing this condition and cutting off the bias current to the transistor at risk.

In this circuit, Tr6 is normally OFF, because the base bias current supplied via R_6 is extracted via D_1. However, if a load current becomes excessive the drop in R_{SC} becomes large enough to reverse-bias D_1, when the bias current is able to turn Tr6 ON, so that its collector diverts the bias current away from the Tr2–Tr4 combination thus turning it OFF. Notice that R_6 is in such a position that an increase in the voltage across the Tr2–Tr4 combination also tends to turn Tr6 ON by virtue of increasing the current through R_6. (A particularly ingenious realization of this technique is where the internal bonding lead from the integrated circuit to the output pin of the header is made from resistance wire and is used as R_{SC}.)

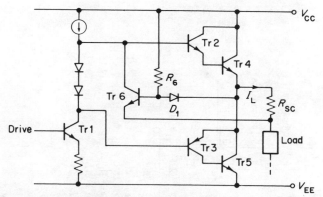

Figure 9.31. Basic method of over-current/over-voltage protection.

Protection for the quasi-pnp pair can be provided in a similar manner.

Should an excessive temperature occur, it is also necessary to shut down the amplifier, and this again can be accomplished by removing bias current at an appropriate point. Various methods of temperature sensing exist, and the most appropriate in the context of an integrated circuit is to use the temperature dependence of a transistor itself[5].

9.11 SUMMARY OF POWER AMPLIFIERS

The foregoing chapter has shown how transistors may be used, within well-defined limits, for power amplification, and how distortion is introduced when these limits are approached. It has also been seen that further distortion occurs due to the curvature of the "straight" parts of the characteristics. Means of minimizing this distortion have been discussed and series of expressions have been derived *which apply only to perfect transistors operating at maximum efficiency*. These are reproduced in table 9.1 and should be used with care, for the performance obtainable from real transistors is always poorer than the calculations indicate. Furthermore, the equations are valid only for a sine wave, this being useful in the present context since AC servo-motors will require such a sinusoidal output. For other forms of signal some of the expressions will contain a constant which is easily determined. Such constants for the case of a triangular and square wave are given in reference 1.

The chapter concluded with a brief introduction to integrated circuit power amplifiers, and to methods of circuit protection against overvoltage, over-current and excessive temperatures.

TABLE 9.1. Design equations for an ideal power transistor (or *each* of a pair) using transformer coupling and working with a sinusoidal signal. (Valid also for transformerless pair with power supply $2V_{CC}$)

	Class A	Class B
	In both cases $R_L = V_{CE(m)}/I_{C(m)}$	

Class A: ac load-line, dc load-line, $I_{C(m)}(=2I_Q)$, I_Q, Q, V_Q $V_{CE(m)}$ $(=V_{CC})$ $(=2V_Q)$ — $V_{CE(m)} \leqslant V_{CEM}$

Class B: ac load-line, dc load-line, $I_{C(m)}$, I_Q, Q, $V_{CE(m)}$ $2V_{CC}$ $(=V_{CC})(=V_Q)$ — $2\,V_{CE(m)} \leqslant V_{CEM}$

	Class A		Class B	
Full output power P_L	$\frac{1}{2}V_Q I_Q$	(9.1)	$\frac{1}{4}V_{CE(m)} I_{C(m)}$	(9.9)
Quiescent dissipation $P_{diss(Q)}$	$V_Q I_Q$	(9.3)	almost zero	
Maximum possible dissipation $P_{diss(m)}$	$V_Q I_Q$	(9.3)	$V_{CC}^2/\pi^2 R_L$	(9.15)
Dissipation for maximum signal $P_{diss(FL)}$	$\frac{1}{4}V_Q I_Q$	(9.4)	$0.068\,V_{CC}\cdot I_{C(m)}$	(9.12)
Power extracted from supply at full load P_S	$V_Q I_Q$	(9.5)	$V_{CC} I_{C(m)}/\pi$	(9.16)
Operating efficiency $\dfrac{P_L}{P_S} \times 100$	50 %		78.5 %	
Commercial efficiency $\dfrac{P_L}{P_{diss(m)}} \times 100$	50 %		247 %	

Heat equation: $P_{diss(m)}$ must be less than P_{tot}.

$T_j - T_{amb} = \theta_t P_{diss}$ (9.24) and $\theta_t = \theta_{jc} + \theta_{cs} + \theta_{sa}$ (9.25).

REFERENCES

1. *Fairchild Semiconductor Ltd. Application Report No. AR-48*, Push-pull class AB transformerless power amplifiers.
2. Freyling, N., Complementary solid state audio amplifiers, *Motorola Application Note AN-230*.
3. Peterson, W. R., General application considerations for the RCA-HC1000 hybrid linear power amplifier, *RCA Application Note AN-4483*.
4. Nappe, J., General application considerations for the RCA-HC2000 power hybrid operational amplifier, *RCA Application Note AN-4782*.
5. Long, E. L. and Frederickson, T. M., 1971, High-gain 15-W monolithic power amplifier with internal fault protection, *IEEE J. Solid St. Circuits*, **SC-6**, 35–44.
6. Gray, P. R., 1972, A 15-W monolithic power operational amplifier, *IEEE J. Solid St. Circuits*, **SC-7**, 474–480.

QUESTIONS

1. Why can the hybrid-π or h-parameter models not be used for most power amplifier stage calculations?

2. Most modern power transistors are silicon types. However, germanium power transistors do offer certain advantages. What are they?

3. Under what conditions does a transistor working in class A dissipate the most power, and why?

4. What are the advantages of coupling a load to a transistor collector via a transformer? What are the disadvantages?

5. What is meant by the term "thermal resistance?"

6. Why is it not normally possible to operate a single transistor in the class B mode?

7. Equation (9.15) is the basic design expression for the class B stage. Why?

8. How does cross-over distortion arise?

9. What are the functions of the driver stage to a power transistor output stage?

10. What is the difference between a complementary and a quasi-complementary output stage?

11. For what purpose does Tr6 exist in the circuit of figure 9.31, and how does it work?

PROBLEMS

9.1. A power transistor works in the class A mode with a resistive load $R_L = 48\,\Omega$ in its collector. If the signal excursions are symmetrical about a quiescent point at $I_Q = 0.25\,A$, what value of supply voltage V_{CC} is needed? Also, what minimum values of I_{CM} and P_{tot} must the transistor type exhibit?

9.2. A power transistor works under class A conditions with $I_Q = 1.5\,A$ and $V_Q = 10\,V$. If it is properly matched to a load via an output transformer with a very low primary winding resistance, what is the maximum power which can be delivered to this load in theory and neglecting any transformer losses? Estimate what it would be in practice.

The transistor header has a thermal resistance of 2.5°C/W and it is mounted on to a heat sink via a mica washer of thermal resistance 0.5°C/W. If the collector junction must not exceed 100°C and the ambient temperature can rise to 35°C, specify the maximum allowed thermal resistance of the heat sink.

9.3. A germanium power transistor is biased for the class A condition, and is mounted on a heat sink via a mica washer. The relevant thermal resistances are $\theta_{jc} = 3$°C/W, $\theta_{cs} = 0.5$°C/W and $\theta_{sa} = 1.9$°C/W. If the collector junction temperature is not to exceed 90°C and the ambient temperature may reach 50°C, what is the theoretical maximum power which the transistor can deliver to the load? What would be a more realistic value for P_L? If a pair of transistors working in class B push–pull conditions were to supply the same power, what would be the minimum value of P_{tot} chosen for each transistor?

9.4. A single power transistor working in class A supplies 10 W of sinusoidal power to a servo-motor via an 85% efficient transformer. What would be the theoretical reduction in transistor power dissipation under this full-load condition if a pair of transistors working in class B (via a transformer of similar efficiency) were substituted? Assume that the full-load condition implies maximal signal excursions.

9.5. A class B push–pull power amplifier is to be designed for which the output transformer is 85% efficient. It must work at an ambient temperature of 40°C and the maximum allowable junction temperature is 90°C. If the transistors are fixed to heat sinks so that $\theta_{total} = 5$°C/W, what is the (theoretical) maximum power which can be delivered to the load?

If the supply available is 30 V DC and the (resistive) load is 3 Ω, what is the ratio of the output transformer, and what must be the minimal value of V_{CEM} for the transistors chosen?

9.6. A class B push–pull power amplifier drives a power-factor-corrected servo-motor which presents a $40\,\Omega$ resistive load and which, at full power, takes $\frac{1}{2}$ A RMS. If a 24 V DC power supply operates the amplifier, what is the turns ratio of the output transformer?

Specify P_{tot}, V_{CEM}, and I_{CM} for the output transistors. Assume that the output transformer is 80% efficient.

9.7. Show that when a transistor is working under class B conditions with a sinusoidal input, it dissipates maximum power when the peak value of the collector current is 64% of the maximum possible peak value, $I_{\text{C(m)}}$.

The theoretical maximum power output to the $5\,\Omega$ load of a class B transformerless push–pull power amplifier is 50 W. What supply voltage will be needed if the two transistors share this voltage equally under quiescent conditions?

9.8. A class B push–pull output stage is to drive a $15\,\Omega$ vibrator unit at a power level of up to 10 W, and must work in an ambient temperature of 50°C. A 12 V power supply is used along with an output transformer of 85% efficiency. What is the (theoretical) maximum power which could be dissipated by each transistor if a sinusoidal drive were used? Could a pair of germanium transistors be used if they were mounted on heat sinks having thermal resistances of 2.25°C/W? For these transistors, $\theta_{\text{jc}} = 3°\text{C/W}$ and the maximum permissible temperature is 90°C.

9.9. Draw a circuit for a transformerless power amplifier consisting of a complementary pair of transistors driven by a CE stage along with a V_{BE}-multiplier. Use a ±15 V power supply and a $4\,\Omega$ load.

Calculate the minimum acceptable rating of the power transistors in terms of P_{tot}, V_{CEM}, and I_{CM}. What is the maximum theoretical power which can be delivered to the load, and what is a more practical value for this power?

9.10. A quasi-complementary output pair of transistors drives an AC servo-motor which is power-factor-corrected to present a resistive load at the frequency of operation. At full power this motor takes 2 A RMS at 50 V RMS.

Draw the basic circuit for the stage including a V_{BE}-multiplier, and quote the ratio of the V_{BE}-multiplier resistors.

9.11. A complementary pair of power transistors working in class B push–pull drives a $10\,\Omega$ resistive load and uses ±24 V power supplies. The transistors each have thermal resistances of 1.2°C/W from junction to case, and their insulating washers have thermal resistances of 0.4°C/W. The two transistors are mounted side-by-side on a heat sink and must work in ambient temperatures up to 40°C.

From the following list, choose the cheapest heat sink which will keep the transistor collector junctions below their maximum allowed temperature of 120°C, knowing that the lower the thermal resistance, the higher the cost.

$$\text{Type A } \theta_{sa} = 4.5°\text{C/W} \qquad \text{Type B } \theta_{sa} = 5.5°\text{C/W}$$
$$\text{Type C } \theta_{sa} = 6.5°\text{C/W} \qquad \text{Type D } \theta_{sa} = 7.5°\text{C/W}$$

9.12. A 400 Hz AC motor is properly power-factor-corrected to present a resistive load. It requires 25 V RMS at 1 A RMS for full power operation.

Design the output stage of a complementary class B power amplifier which will drive this motor, and choose an appropriate pair of transistors from the list given below.

Also, specify the voltage and current of a dual-voltage ($\pm V_{CC}$) power supply suitable for this circuit.

Assume that the maximum ambient temperature which will be encountered is 50°C, and that heat sinks of $\theta_{sa} = 3.5°$C/W are available, along with mica washers of $\theta_{cs} = 0.5°$C/W.

The following specifications apply to each transistor in the pair. For all transistors, the maximum allowable junction temperature is 80°C.

	V_{CEM}	P_{tot}	I_{CM}	θ_{jc}
Pair A	60 V	10 W	2 A	2°C/W
Pair B	80 V	8 W	2 A	2.5°C/W
Pair C	100 V	6 W	2 A	1.5°C/W

10

The Bipolar Transistor as a Switch

An ideal switch may be defined as a device having zero impedance when ON and an infinite impedance when OFF. Such a definition clearly precludes the manufacture of an ideal switch, which means that some general expression must be obtained which will serve as a figure of merit for a real switch. The usual method of quoting such a figure of merit is to divide the OFF impedance by the ON impedance. For mechanical switches and relays both impedances are largely resistive, and the figure of merit may easily reach magnitudes of 10^{10} and above.

This factor, however, does not by any means completely specify a switch. Other parameters include the maximum ON current, the maximum current and voltage which can be broken, and the maximum OFF voltage. Further, these factors must be related to the life of a switch in terms of the number of operations possible before burnt contacts, weakened springs etc. cause the device to fall below its initial performance.

The latter consideration leads to the postulation of purely electronic switches having no moving parts, for providing such switches are not subject to destructive voltages, currents or dissipations, their life is theoretically infinite. Moreover, the actual time taken to complete a switching operation is very much less than that for a mechanical switch, owing to the absence of moving parts.

In recent decades, the use of solid-state switches has led to the discipline of *digital electronics*. The vast area encompassed by this term ranges from the design of logic systems using "hard-wired" monolithic logic gates, registers etc., through the application of microprocessors, which are essentially general purpose programmable logic systems fabricated on one chip, to the technology of digital computers themselves. The increase in importance of digital electronics depends upon continued progress in the development and manufacture of solid-state digital monolithic circuits, which again is a vast topic in itself.

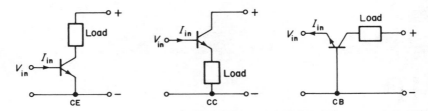

Figure 10.1. Switching modes.

Because many hitherto "analog-type" functions can now be realized using digital techniques (such as digital and switched capacitor filters), and because analog and digital circuits are often in close juxtaposition (as in A–D and D–A converters, for instance), it is necessary to study both technologies in parallel. The present book concentrates on analog design, however, because any attempt to introduce digital electronics within the confines of a volume of reasonable size would necessarily be so superficial as to be virtually useless. Therefore, the present chapter will present only those aspects of solid-state switching which are *not* concerned with logic applications.

When a transistor is used as a switch it may take any of the three basic modes of connection, as illustrated in figure 10.1. However, in the common-collector mode the operating voltage V_{in} must be slightly larger than the load voltage. Conversely, in the common-base mode, the operating current must be slightly larger than the load current. This leaves the common-emitter mode, and figure 10.2 shows a load-line representing this mode for an arbitrary transistor.

When the transistor is in the OFF state I_B is zero or has a small reverse value, and I_C therefore lies between I_{CBO} and I_{CEO}. The case when $I_B = 0$ is depicted in figure 10.2, and the dissipation under these conditions is clearly

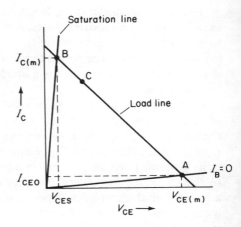

Figure 10.2. Common-emitter collector characteristic.

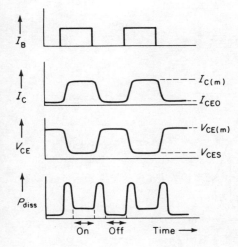

Figure 10.3. Switching profiles.

$V_{CE(m)}I_{CEO}$, which is small. Furthermore, the OFF resistance R_{OFF} is $V_{CE(m)}/I_{CEO}$, which is clearly a large value. (Note that R_{OFF} is *not* an incremental resistance, since the leakage current I_{CEO} is not a direct function of applied voltage.)

When the transistor is ON it may or may not be saturated, depending on the base current. If saturated, the working point will move to B, when the power dissipated will become $V_{CES}I_{C(m)}$. This is again quite a low value, as is the corresponding ON resistance, $R_{CES} = V_{CES}I_{C(m)}$.

Although the power dissipated at the OFF and (saturated) ON points is small, the working point of the transistor must pass through the "normal" active region, where the dissipation is significant. Figure 10.3 shows the variation of I_C, V_{CE}, and P_{diss} for a transistor switched into and out of saturation by a square wave input applied to the base. From it, the average power dissipated is seen to be a function of (*a*) energy dissipated per operation and (*b*) number of operations per unit time. Clearly, the faster a transistor switches from OFF to (saturated) ON, the less heat will be generated within the device. This leads to a consideration of switching speed.

10.1 SWITCHING SPEED

Figure 10.4(*a*) illustrates the way in which I_C changes for a square wave change in I_B, and shows (in more detail than figure 10.3) the rise and fall profile. The nature of the various components of the switching time may be explained as follows.

When the base current rises, this is equivalent to lowering the potential barrier across the emitter–base depletion layer, thus allowing charge carriers to enter the base region from the emitter (see chapter 2). In the base region,

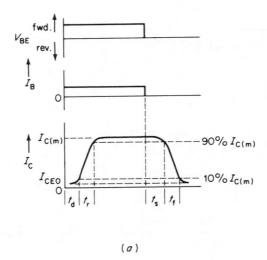

(a)

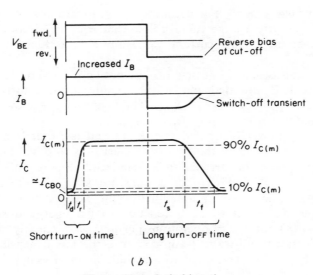

(b)

Figure 10.4. Switching times.

most of these minority carriers diffuse across the field-free base to the collector, and it is the time taken for this diffusion process which contributes most of the *delay time*, t_d.

Clearly, since diffusion is a random process, not all of the initial batch of minority carriers will arrive at the collector together, which means that the collector current will slowly build up. The effect is accentuated by the low value of h_{FE} which obtains at low values of I_C. The definition of t_d is therefore given as the time taken for the collector current to rise to 10% of its final value after the application of base current having a very fast (ideally infinite) rise. I_B is assumed large enough to saturate the transistor.

It is worth recalling here that those minority carriers in the base region that do not reach the collector but recombine within the base region are the cause of base current, since further carriers must flow in to maintain electrical equilibrium within the crystal as a whole. In other words, if a given number of charges enter the emitter from the external circuit and a smaller number flow out of the collector, then the difference must be made up from the base circuit. In fact, it is physically valid to say that the base current does not actually *control* the collector current, but that the emitter–base potential difference, V_{BE}, simply makes possible the surmounting of the emitter–base potential barrier by charge carriers. The relationship between base and collector currents is therefore only a ratio which has been assigned the name of current gain, h_{FE}.

If the emitter–base junction has been reverse biased during its OFF state, it can be thought of as a charged capacitor, and the discharge time of this capacitor also contributes to t_d.

The collector current rise time, t_r, is really the "straight" part of the profile, and is due to the finite time taken for the charge carriers to cross the base region in sufficient numbers to bring I_C up to 90% of its final value. The turn-ON time, $t_d + t_r$, can be significantly reduced by overdriving the base as shown in figure 10.4(*b*) but this also has the effect of increasing the turn-OFF time as will be seen later.

The transistor is now saturated and the collector voltage falls below the base voltage, so that both junctions are forward biased. This means that the total current passing through the transistor is limited only by the external circuit, and that this total current is the difference between the current injected into the base from the emitter and that returned to the base from the collector, less the external base current itself. This mechanism accounts for the storage of minority carriers within the base region, and when the base current is switched rapidly OFF, these stored charges must be removed before the collector current can fall. The *storage time*, t_s, is the time taken for most of these carriers to diffuse away, and it is defined as the time taken for the collector current to fall by 10% after the removal of base current. Storage time is lengthened if the base is overdriven, so any attempt to improve the turn-ON time by overdriving will also lengthen the turn-OFF time.

The fall time, t_f, is a function of the time taken for minority carriers to diffuse out of the base, and is defined as the time taken for I_C to fall from 90% to 10% of its ON value. Both t_r and t_f are dependent on the constructional details of the transistor and are related to the cut-off frequency f_α (mentioned in chapter 5). The switch-OFF time, $t_s + t_f$, may be reduced by applying a reverse bias to the emitter–base junction. This will result in the flow of a *reverse* base current whilst the stored carriers are being removed, but this current will disappear and only I_{EBO} will flow when the transistor is in its normal OFF condition. This is also shown in figure 10.4(*b*). If the reverse bias is large enough, I_C will approach I_{CBO}.

Storage time is also a function of h_{FE} and the value of t_s given by a manufacturer's data sheet will be associated with a (usually quoted) range of h_{FE}, which may well be much smaller than the production spread of h_{FE} for that transistor.

One way of reducing t_s to a very small value is to maintain the transistor in a nonsaturated state when ON. This implies that the collector—base junction will never become forward biased, and so no charges will be injected into the base from the collector.

Unfortunately, the ON dissipation of the transistor becomes comparatively large, for a point such as C on the load-line of figure 10.2 now represents the ON condition. For discrete transistor work, there is no point in using nonsaturating switches where power switching is concerned, since apart from the fact that a saturating switch is very much faster than any mechanical device, it is also much more efficient than the nonsaturating switch owing to its negligible ON dissipation.

It is possible to find the energy dissipated during a switching operation by using a linear approximation of the I_C and V_{CE} profiles shown in figure 10.5. Here, four regions are delineated and the energy dissipated during each is found as follows:

(i) Region 1—switch ON

$$I_C = I_{C(m)} \frac{t}{t_r} \quad \text{and} \quad V_{CE} = V_{CC}\left(1 - \frac{t}{t_r}\right)$$

giving

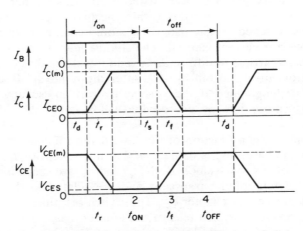

Figure 10.5. Linear approximation of switching profiles: region $1 \simeq t_r$; region $2 = t_{ON} \simeq t_{on} + t_s^{\smile} - t_d - t_r$; region $3 \simeq t_f$; region $4 = t_{OFF} \simeq t_{off} + t_d - t_s - t_f$.

$$\frac{R_{C1}}{R_E} > \frac{R_1}{R_2} > \frac{R_{B1}R_{C1}}{R_E(R_{B1} + R_{C1})} . \tag{10.21}$$

When this equation is satisfied the circuit will operate.

The order of dwell time required determines the value of C and fine control may be obtained by varying R_{B1}, but within the limits of the design equations.

The trigger sensitivity is very difficult to determine analytically, but in order to ensure that Tr1 is desaturated, it is necessary for the incoming charge (i.e., the area of the input pulse) to be greater than the charge control parameter Q_{BS}, and ideally greater than Q_{OFF}.

In the preceding discussion, only the case where the ON transistor is saturated has been taken. This, however, is not the only mode of operation and other analyses will apply when nonsaturated switching is used.

10.5   THE TOTEM-POLE POWER BUFFER

When power is to be extracted from the collector point of a regenerative switching transistor, that transistor must be in the OFF state, so that its collector voltage is high. This means that the load current must be extracted from V_{CC} via R_C, and this could cause a significant voltage drop in R_C. One effect of this could be a change in the timing waveforms and another could be a reduction in the voltage needed by the load. Further, any decrease in the value of R_C to accommodate a comparatively high load current could again result in a change in the timing waveforms; and would certainly cause a marked increase in the total power consumed when the transistor switched to the saturated, or ON state.

The best solution to this problem is to include an output buffer, and several of these were presented in chapter 9 for linear operation. However, buffers for switching circuits are simpler by virtue of the fact that a pair of output transistors of the same polarity can be used, because only the two states, OFF or (saturated) ON need be considered.

Figure 10.21 shows a transistor Tr1 (which may be one of a regenerative pair) driving a *totem-pole power buffer stage*, so-called because one transistor sits upon the other.

When Tr1 switches ON, the voltage drop in the emitter resistor R_{E1} provides V_{BE2}, so turning Tr2 ON also. In this state, Tr2 could act as a current sink if the load were connected from V_{CC} instead of from ground; but given the connection shown, the load is unenergized. Because both Tr1 and Tr2 are ON, the voltage at the collector of Tr1 will be:

$$V_{C1} = V_{BE2} + V_{CES(1)} \simeq 0.7 + 0.3 = 1 \text{ V}$$

This is also the voltage applied to the base of Tr3, which will nevertheless remain OFF because in the ON condition, its emitter voltage would be:

where V_{C_1} and V_{C_2} are the initial and final collector–emitter voltages and U is the emitter–base voltage (about 0.3 V for Ge or 0.7 V for Si).

The total switch-ON charge is Q_{ON} where

$$Q_{ON} = \Delta Q_B + Q_V$$
$$= \Delta I_C \tau_C + Q_V .$$

To this charge must be added a further component if the base–emitter junction was initially reverse biased. This may be found in the same way as Q_V, for when the base–emitter junction is reverse biased, it too may be considered to be a capacitor, but of value C_{te}.

If the transistor is driven to saturation, the stored charge Q_{BS} must be extracted before switch-OFF can occur. This is due to any base current in excess of $I_C/h_{FE(O)}$, where $h_{FE(O)}$ is measured almost at saturation. If this excess current is I_{BS}, then $Q_{BS}/I_{BS} = \tau_S$, the *saturation time factor*.

To switch the transistor OFF, a charge equal to the sum of the three previously discussed components must be extracted from the base. That is, Q_{OFF} is the sum of the stored charge and the charge needed to reduce I_C by ΔI_C and that needed to change the collector voltage by $(V_{C_1} - V_{C_2})$:

$$Q_{OFF} = I_{BS} \tau_S + \Delta I_C \tau_{CO} + Q_V$$

(here, τ_{CO} is τ_C measured at $V_{CB} = 0$ since it varies somewhat with V_{CB}).

To this charge must be added the extra component made necessary if the base–emitter junction is to be reverse biased.

In order to make the necessary calculations the charge control parameters C_{tc}, C_{te}, τ_C, and τ_S must be known, and some manufacturers do give these data. Further, $h_{FE(O)}$ must also be known, and sometimes it is necessary to use $h_{FE(S)}$ ($< h_{FE(O)}$) which is the current gain *almost immediately available* when the transistor switches ON.

The actual calculation of Q_B depends on the nature of the junction, and is a simple expression only for an abrupt junction, which is often not a valid assumption[4].

The real advantage of the system is that the transient times relating to a transistor in a real circuit can be found in terms of the charge control parameters, whereas if the manufacturer's data included a list of values for t_d, t_r, t_s, and t_f, these would apply only to the performance of the transistor in a particular test circuit. This is the reason why most manufacturers define the test circuit used, which is usually that which allows the transistor to perform at its optimum rating.

As a case in point, consider the Texas Instruments BFR40, which is an npn epitaxial planar transistor in the "Super Silect" range. Here, the characteristic switching times are measured in a circuit which produces an $I_{C(m)}$ of 250 mA using a base current of 25 mA to drive the transistor into saturation and also using a reversed base current of 25 mA as a switch-off transient to improve the value of t_s. The resulting "typical" values are:

$$t_d = 25 \text{ ns} \quad t_r = 30 \text{ ns} \quad t_s = 140 \text{ ns} \quad t_f = 35 \text{ ns} .$$

These characteristic times are very short, and insofar as low-power switching is concerned can be regarded as instantaneous in comparison with the capacitive and/or inductive time constants found in most such circuits. Even for much higher power transistors, inherent switching times are still short for modern devices. For example, in the case of the Motorola 2N6546, when switching 10 A at 250 V, the maximum switching times are given as:

$$t_d = 0.05 \ \mu\text{s} \quad t_r = 1 \ \mu\text{s} \quad t_s = 4 \ \mu\text{s} \quad t_f = 0.7 \ \mu\text{s} .$$

Switching transistors are available not only as discrete devices but also in integrated circuit arrays, as, for example, where it is desired to drive a seven-segment filament or LED display. Such an application may require several tens of milliamps per segment, so counting as a small-power application and one in which the heat dissipation capability of a dual-in-line encapsulation (DIL-pac) is entirely adequate.

Some examples of switching applications now follow, and most have been treated from a biasing rather than a speed point-of-view, for the latter approach is more relevant to much lower-power digital technology. However, some "nondigital" applications, such as choppers for DC amplifiers and class D amplifiers do depend upon high switching speeds and are also included.

10.2 TRANSISTOR CHOPPERS FOR DC AMPLIFIERS

Chapter 7 included a discussion of DC amplifiers, and showed how an electromechanical chopper could be used to modulate a direct current signal (which, in this context, means "convert to AC") after which it could be amplified by a normal AC coupled amplifier and demodulated by the same chopper mechanism. The chopping frequency is clearly limited by mechanical considerations, which consequently limit the upper frequency of the incoming signal.

The use of transistors in place of electromechanical choppers has two basic advantages. Firstly, they can be operated at much higher frequencies, thus enabling the complete amplifier to accept higher input frequencies (a common chopper frequency is 4000 Hz, though units using chopping speeds up to 100 kHz have been constructed). Secondly, having no moving parts, a chopper transistor has a theoretically infinite life.

There are several ways in which transistors may be connected to chop a DC signal. Basically, the modulator transistor may either present successive

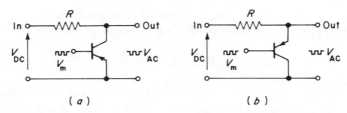

Figure 10.6. Shunt choppers: (*a*) normal connection; (*b*) inverse connection.

low and high resistances in shunt with an incoming signal, as in figure 10.6, or in series with it as in figure 10.7.

Figure 10.6(*a*) shows that the proportion of the input signal arriving at the right of R is short-circuited to the common line each time the base of the transistor is driven negative by the modulating signal v_m. The resistance R may represent the internal resistance of the signal source R_g, in which case the full signal voltage appears across the transistor when it is in its OFF state. Conversely, the nature of the source may be such as to prohibit its being short-circuited, in which case R may include an external series resistance R_{ex}. These facts indicate that the ON and OFF resistances of the transistor itself can be of considerable importance. Consider the output voltage when the transistor is in the OFF state and the source resistance is R_g, the external resistance at the input is R_{ex} and the OFF resistance of the transistor is R_{OFF}:

$$V_{AC(max)} = \frac{V_{DC} R_{OFF}}{R_g + R_{ex} + R_{OFF}}.$$

Clearly, if $R_{OFF} \gg R_g$ or R_{ex}, then $V_{AC(max)} \to V_{DC}$.

Thus the first requirement of a modulator transistor is a high OFF resistance.

Unfortunately, this expression is not accurate at low current levels, for when the transistor is OFF a leakage current still flows, and this will lie between I_{CEO} and I_{CBO} depending on whether the base is open or reverse biased.

Now consider the case when the transistor is ON and has a resistance R_{ON}.

$$V_{AC(min)} = \frac{V_{DC} R_{ON}}{R_g + R_{ex} + R_{ON}}.$$

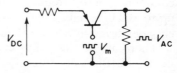

Figure 10.7. Series chopper.

Clearly, if R_{ON} is very small, $V_{AC(min)} \to 0$, which indicates that R_{ON} should in fact be as low as possible.

Again, this form of expression does not tell the whole story. Even though the applied V_{CE} may be very small there will still be a collector–emitter voltage due to the base current and this means that the output voltage can never fall below this value. In other words this *offset voltage* forms a pedestal upon which the output wave is superimposed.

Finally, carrier storage and capacitance effects combine to produce a switching transient, and both this and the offset voltage are shown in figure 10.8. Here, the signal voltage is assumed to be zero and the waveshape is due entirely to the chopping voltage. The switching-ON spike is normally smaller than the switching-OFF spike since it is immediately damped by the resistive circuit formed by the source in series with the conducting transistor. The decay time of the switch-OFF spike determines the maximum operating frequency of the chopper, since the transistor must not be switched ON again until this spike has decayed completely.

One method of reducing the height of the offset voltage is to invert the transistor as shown in figure 10.6(b). This reduction occurs because the expression for the offset voltage in either mode is in part an inverse function of the current gain in the alternate mode, and this current gain is greater for normal operation than for inverted. Thus $V_{EC(offset)} < V_{CE(offset)}$. Since both the magnitude and temperature dependence of $V_{EC(offset)}$ are functions of base current, I_B, then the $V_{CE(offset)}/I_B$ characteristic is the first design curve which should be consulted if a chopper stage is to be optimized. Sometimes the manufacturer's data include a maximum figure for $V_{EC(offset)}$ when the transistor is worked under given conditions, which simply means that such an optimization has already been performed, and production transistors are selected to fall within the desired limits.

Typical chopper transistors exhibit offset voltages of less than 5 mV, but for many low-level applications even this figure is much too high. It is clear that the offset voltage will appear as a square wave at the input of the main AC amplifier and, if this has a gain of only a hundred, a 5 mV offset will lead to an output voltage of 0.5 V. Many methods have been devised to counteract this offset[5], but solutions involving extra circuitry are now giving place to simpler techniques using field-effect transistors or photoresistive cells, both of which will be considered in later chapters.

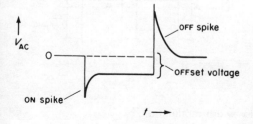

Figure 10.8. Chopping characteristics for zero input voltage (polarities for pnp transistor).

10.3 THE PULSE-WIDTH MODULATED POWER AMPLIFIER

During the discussion on power amplifiers in chapter 9 it was found that some considerable proportion of the total power output was dissipated within the transistors themselves. This was shown to be at least 50% for a transistor biased to class A conditions, at least 31.5% for a class B transistor, and a smaller amount for the servo-mode amplifier. Conversely, the present chapter has shown that the dissipation in a saturated switching-mode transistor is very low indeed, and this leads to the question of whether a power amplifier—and particularly a fairly wide-band power amplifier—can be so designed that the output transistors operate only as switches.

Such a design is in fact feasible, and depends on the principle of *pulse-width modulation*. This concept is illustrated in figure 10.9. Diagram (*a*) shows the output from a square-wave generator being applied to an inductive load, for example, a motor or loudspeaker. The repetition frequency of the square wave is assumed to be so high that little current at this frequency can pass through the inductance, which means that a voltage nearly equal to the generator voltage appears across the inductive part of the load.

If the applied wave is symmetrical in amplitude and time as shown in figure 10.9(*b*), then the *average* voltage appearing across the inductance is zero. However, if the wave is asymmetrical in time as shown in figures 10.9(*c*) and (*d*), then the average voltage is *not* zero, but can take either a positive or a negative value given by:

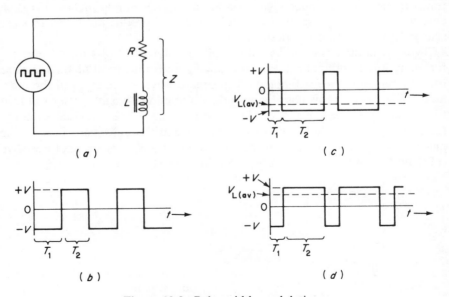

Figure 10.9. Pulse-width modulation.

$$V_{L(av)} = \frac{VT_1 - VT_2}{T_1 + T_2} = \frac{V}{T}(T_1 - T_2).$$

This means that a net current of $V_{L(av)}/R$ would flow and that, if the load were a permanent-magnet-field DC motor, it would rotate one way or the other, or, if it were a loudspeaker, the cone would move in or out.

If it were possible to construct a device which produced a square wave of constant frequency and amplitude, but whose dwell time difference ($T_1 - T_2$) varied linearly with an input signal amplitude, then clearly the average voltage applied to the load would also be a linear function of this input signal. If the frequency of the square wave is called the *pulse repetition frequency* (PRF) then it is clear that this must be greater than the maximum signal frequency which is to be encountered. In fact the PRF should be several times greater than the maximum signal frequency, otherwise the inductive load (i.e., the low-pass filter) will not be capable of regenerating the original signal. Thus, for the case of a loudspeaker driven by an audio signal, the PRF should approach 100 kHz if audio frequencies up to 20 kHz are to be accommodated at reasonable distortion levels. Keeping in mind that a 100 kHz square wave contains a series of (odd) harmonics of the fundamental, it is immediately apparent that the switching speed of the output transistor(s) becomes important. Also, radio-frequency interference (RFI) can be generated by such circuits.

For these and other reasons, pulse-width-modulated switching (or class D) audio amplifiers have not become widespread compared with class B feedback amplifiers, but the technique has become common in *switch-mode power supplies* (SMPS)[6] which are introduced in chapter 14; and in motor control[7].

Figure 10.10 shows one method of pulse-width modulating a signal. In the

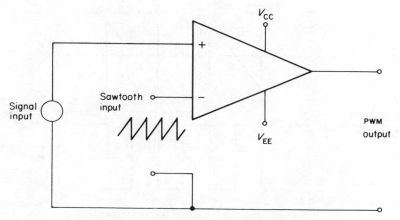

Figure 10.10. A pulse-width modulation method.

example a sawtooth wave is applied to one input of an operational amplifier and the signal is applied to the other. If the op. amp. has a very high gain, then only a very small voltage difference between the inputs will be needed to swing the output either maximally positive or negative. Therefore, using the op. amp. input polarities shown in figure 10.10 the output will be a square wave for zero signal input, as in figure 10.11. (Here, the op. amp. is being used as a *comparator*, a function which will be described in detail later in the chapter.) This implies that the output polarity reverses as the input sawtooth crosses the (zero) voltage applied to the noninverting input terminal. If now a finite signal is applied to this terminal, as in figure 10.12, the output will reverse when the sawtooth crosses whatever value this signal takes at the instant of coincidence. Hence, the output square wave will be width-modulated, which is the desired result.

The pulse-width modulated waveform may now be converted to a power level using a switching transistor and applied directly to an inductive load such as a motor, which acts as its own low-pass filter, or via a separate low-pass filter if the inductance of the driven device is low, as in the case of a loudspeaker. Also, such a circuit can be used for driving a filament lamp in which the thermal delay inherent in the filament is sufficient to prevent flicker at the PRF used.[8]

For a loudspeaker or a bidirectional motor the output stage should be a push–pull pair of switching transistors, used along with dual power supplies so that a bipolar output signal can be provided (see also figure 11.55(*b*)).

The method described is but one of many—for example, a falling as opposed to a rising sawtooth may be used, as may a full triangular wave; all that is required is a linear rise and/or fall. It does, however, serve as a basic introduction to a general concept.

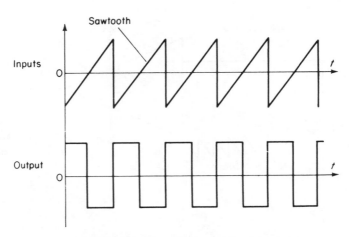

Figure 10.11. PWM with zero signal input.

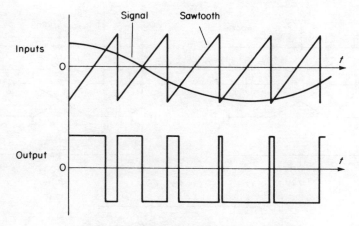

Figure 10.12. PWM with finite input signal.

10.4   REGENERATIVE SWITCHING

It is possible to couple two transistors together in such a manner that they exist inside a loop having high positive feedback. This means that if one of them begins to conduct, the feedback signal will augment the procedure and the transistor will quickly switch hard ON. Normally, the second transistor is forced quickly OFF by the same process.

Such an arrangement can, by relatively minor circuit changes, be caused to exhibit three different but allied properties. Firstly, if the ON condition is made unstable, the switching back and forth of the transistors becomes a self-maintaining operation and the *multivibrator* or *astable* circuit results. This device is often the basis of square-wave generation and, as has been stated, one of its uses is the provision of a modulation signal for transistor choppers.

Secondly, the ON condition can be made stable, but the circuit can be arranged so that an incoming pulse will trigger the transistors and cause them to reverse their states. This is the *bistable*, *binary* or *Eccles–Jordan* circuit. One of its main uses is as a memory element and as such it is employed in digital electronics. This property is also valuable in small power work—for example, a lamp may be lit, or a relay closed by a signal pulse, and will remain so until the arrival of a further pulse.

Thirdly, one of the transistors may be arranged to have a stable ON state and the other an unstable ON state. This constitutes the *monostable* circuit or *one-shot multivibrator*. Here, an incoming pulse causes the states of the transistors to reverse, and they remain in these reversed states for a well-defined time which depends on the circuit constants, then revert to their original states. One of the primary uses of the monostable circuit is to obtain a pulse well-defined in amplitude and duration when the incoming

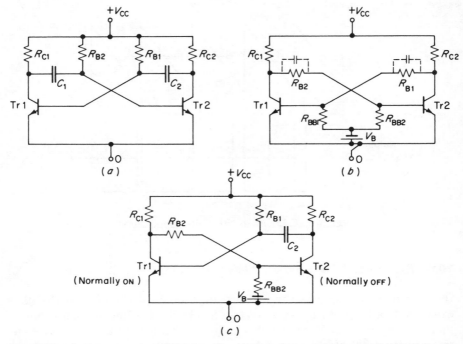

Figure 10.13. Regenerative switching circuits: (*a*) the multivibrator; (*b*) the bistable circuit; (*c*) the monostable circuit.

pulse is small or ill-defined, as is the case, for example, with pulses arriving from several forms of radioactivity detector. Another application is the provision of a second pulse at some fixed time after the arrival of a signal pulse. This secondary pulse is extracted as the circuit reverts to its stable state.

The two-transistor regenerative switch may take several different forms—for example, the transistors may be arranged to switch ON at the same time and OFF at the same time or one of several complementary pair circuits may be employed. However, before describing any of the more exotic arrangements, the design discussion will be confined to the simple npn regenerative pairs shown in figure 10.13.

10.4.1 The Multivibrator

Consider the circuit of figure 10.13(*a*) and its associated wave profiles given in figure 10.14. Let Tr1 be ON and Tr2 be OFF.

C_1 will charge via Tr1 and R_{B_2} until it allows the base of Tr2 to become slightly positive, and Tr2 switches ON. When this happens the collector of Tr2 changes potential from V_{CC} to a small saturation voltage V_{CES}. Hence, a

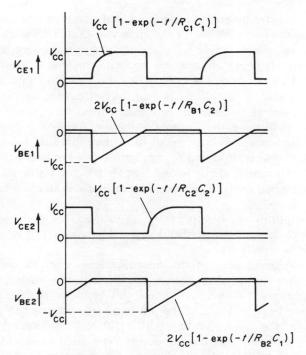

Figure 10.14. Multivibrator wave profiles.

voltage of $-V_{CC}$ is suddenly applied to the base of Tr1, causing it to switch hard OFF. This is because the left-hand side of C_2 has previously been held near zero voltage (actually at $V_{BE(sat)}$) by the base of Tr1, while the right-hand side had been nearly at V_{CC} since Tr2 was OFF.

Since Tr1 is now OFF, C_1 must charge through R_{C1} and the base–emitter junction of Tr2. This implies that, since the base–emitter junction presents a very low resistance compared with R_{C1}, then the collector voltage of Tr1 must change with a time constant $R_{C1}C_1$. This is shown in figure 10.14 where it is seen to reach V_{CC}.

Also, because Tr1 is OFF, its base–emitter junction must be reverse biased and so present a very high resistance. Consequently, C_2 must discharge through R_{B1} and Tr2, which implies a time constant $R_{B1}C_2$. This is the time constant which controls the pulse repetition frequency (PRF) of the circuit, for when C_2 has fully discharged and *just* begun to charge positively on the left-hand side, then Tr1 will again be brought into conduction and the cycle will repeat. The switching-ON time is very short since base current is provided via a pure resistance, as is collector current, and the positive feedback signal from each collector to each base arrives via a loop which includes the gain of both transistors. The sharpness of the collector voltage ON profile may be seen in figure 10.14.

This qualitative treatment has assumed that the transistors are able to switch very rapidly compared with the time constants of the circuit. Were this not so the discussion would have had to take into account the effect of the continuously changing impedances presented by the transistors, which would have led to considerable complexity. For all moderate frequencies this is a reasonable assumption, but at high frequencies the transistor parameters themselves become the limiting factors.

In order to determine the limiting values of the components and the frequency of operation, it is convenient to assume that the transistors act as perfect switches; that is, R_{CE} and R_{in} are zero when the transistor is ON and infinitely high when OFF. This means that in the ON state, the base and collector are clamped to the common line, but present an open circuit when the transistor is OFF.

These assumptions are by no means unjustified, for they simply imply that V_{CC} is much greater than either V_{CES} or V_{BE} and if V_{CC} is greater than 5 V or so, this is clearly true.

Using the foregoing assumptions, an equivalent circuit or model can be drawn to elucidate the switching action. Let this action be such that Tr1 goes OFF and Tr2 goes ON. Immediately before this happens, the right-hand side of C_2 must have been at V_{CC} and the left-hand side clamped to zero by the base of Tr1. In the model of figure 10.15(a) this corresponds to switch position A. When the switch-over occurs, the right-hand side of C_2 is suddenly clamped to zero as Tr2 turns ON, which means that the left-hand side of C_2 follows it down and finds itself at $-V_{CC}$. It must therefore try to recharge to $+V_{CC}$ via R_{B1}; that is, through a voltage of $2V_{CC}$. This corresponds to switch position B.

Of course, in the real circuit, as soon as the left-hand side of C_2 reaches a voltage just above zero, Tr1 switches ON again and forces Tr2 OFF. Nevertheless, the time T_2 at which it does so must be derived from the exponential curve associated with the true rate of discharge, which is:

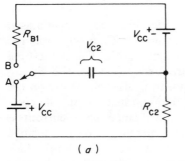

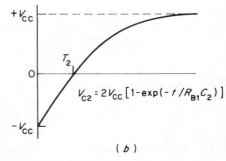

$$V_{C2} = 2V_{CC}[1 - \exp(-t/R_{B1}C_2)]$$

(a) (b)

Figure 10.15. (a) Switching model; (b) charging exponential.

$$v_{C_2} = 2V_{CC}\left[1 - \exp\left(-\frac{t}{R_{B1}C_2}\right)\right] \tag{10.6}$$

when $t = T_2$, $v_{C_2} = V_{CC}$, giving:

$$V_{CC} = 2V_{CC}\left[1 - \exp\left(-\frac{T_2}{R_{B1}C_2}\right)\right]$$

or

$$\frac{1}{2} = \exp\left(-\frac{T_2}{R_{B1}C_2}\right)$$

giving

$$T_2 = 0.69 R_{B1} C_2 . \tag{10.7a}$$

Here T_2 is the *dwell time* during which Tr2 is ON. Similarly

$$T_1 = 0.69 R_{B2} C_1 . \tag{10.7b}$$

If an output signal is taken, the *mark-space ratio*—that is, the ratio of the ON interval to the OFF interval—is clearly the ratio of the two dwell times. Further, the output PRF will be $f = 1/(T_1 + T_2)$. If the circuit is symmetrical, so that $T_1 = T_2 = T$, the PRF is simply $1/2T$.

Having obtained an expression for the dwell time, it is now possible to use it in a determination of the minimum value of R_B. When Tr1 switches OFF, its collector voltage changes to V_{CC} as C_1 discharges through R_{C1} according to the law:

$$v_{C_1} = V_{CC}\left[1 - \exp\left(-\frac{t}{R_{C1}C_1}\right)\right].$$

It is reasonable to assume that T_2 should be at least long enough to allow v_{C_1} to reach 0.98 of its final value before switch-over occurs, and this being so,

$$0.98V_{CC} \leq V_{CC}\left[1 - \exp\left(-\frac{T_2}{R_{C1}C_1}\right)\right]$$

giving

$$T_2 \geq 4R_{C1}C_1 . \tag{10.8a}$$

Similarly

$$T_1 \geq 4R_{C2}C_2 . \tag{10.8b}$$

If $C_1 = C_2 = C$, then equations (10.7a) and (10.8a) can be compared:

$$0.69R_{B1}C \geq 4R_{C1}C$$

or

$$R_{B1} \geq 5.8R_{C1} . \tag{10.9a}$$

Similarly

$$R_{B2} \geq 5.8R_{C2} . \tag{10.9b}$$

Having determined a lower limit for R_B, the upper limit will complete the specification. This is found simply by ensuring that R_B is not so large that it prevents its associated transistor from saturating fully. If $h_{FE(O)}$ is the current gain almost at saturation,

$$\frac{V_{CC}}{R_{B1}} \geq \frac{I_{CS1}}{h_{FE(O)}} = \frac{V_{CC} - V_{CES}}{h_{FE(O)}R_{C1}} \simeq \frac{V_{CC}}{h_{FE(O)}R_{C1}}$$

giving

$$R_{B1} \leq R_{C1}h_{FE(O)} . \tag{10.10a}$$

Similarly

$$R_{B2} \leq R_{C2}h_{FE(O)} . \tag{10.10b}$$

Combining equations (10.9) and (10.10) gives

$$h_{FE(O)}R_{C1} > R_{B1} > 5.8R_{C1} \tag{10.11a}$$

and

$$h_{FE(O)}R_{C2} > R_{B2} > 5.8R_{C2} . \tag{10.11b}$$

Equations (10.7) and (10.11) are the basic design equations for the multivibrator. The values of V_{CC} and $h_{FE(O)}$ will be fixed, and I_{CS} will depend on the application in question. Consequently, R_C is easily determined and this value gives the permissible range of values for R_B by equation (10.11). This, in conjunction with the desired PRF, will give C by use of equation (10.7).

Should it be desired to control the PRF, either C_1 and C_2 or R_{B1} and R_{B2}

may be made variable. It is usually convenient to switch the capacitors as a coarse control and use ganged potentiometers forming R_{B1} and R_{B2} as a fine control.

Occasionally, it is necessary to make the mark-space ratio variable, and this may be done by controlling C_1 with respect to C_2 or R_{B2}.

10.4.2 Simple Design Example

Consider the design of a symmetrical 1.2 kHz multivibrator using a pair of Texas "Silect" transistors type 2N3709 at ON collector currents of 1 mA. If a 12 V power supply is used,

$$R_C = \frac{V_{CC} - V_{CES}}{I_{CS}} \simeq 12 \, k\Omega \, .$$

Assuming that $h_{FE(0)} \simeq h_{FE(min)}$, then from equations (10.11)

$$45 \times 12 > R_B > 5.8 \times 12$$

that is, R_B lies between 540 and 70 kΩ.

For a 1.2 kHz PRF, $T_1 = T_2 = 0.42$ ms and from equations (10.7),

$$0.42 \times 10^{-3} = 0.69 R_B C$$

giving $R_B C = 6.1 \times 10^{-4}$.

Both relationships involving R_B are approximately satisfied if

$$R_B = 270 \, k\Omega \quad \text{and} \quad C = 0.0022 \, \mu F \, .$$

10.4.3 The Bistable Circuit

The bistable circuit operates in a slightly different manner from the astable circuit. The ON state must be stable for either transistor, and this is achieved by the use of a voltage source V_B in figure 10.13(b) which maintains the alternate transistor in the OFF state. A reversal of states may come about only by the application of a pulse at some suitable point—for example, a negative pulse at the base of the ON transistor, or a positive pulse at the base of the OFF transistor.

To establish the conditions under which the ON and OFF states are stable, consider figure 10.13(b) and let Tr1 be ON and Tr2 be OFF.

For Tr1 to remain ON, the base current must be great enough to ensure saturation:

$$I_{B1} \geq \frac{I_{CS1}}{h_{FE(O)}} = \frac{V_{CC} - V_{CES}}{h_{FE(O)} R_{C1}} \, .$$

But

$$I_{B1} = \frac{V_{CC} - V_{BE}}{R_{C2} + R_{B1}} - \frac{V_B + V_{BE}}{R_{BB1}}.$$

Therefore

$$\frac{V_{CC} - V_{BE}}{R_{C2} + R_{B1}} - \frac{V_B + V_{BE}}{R_{BB1}} \geq \frac{V_{CC} - V_{CES}}{h_{FE(O)} R_{C1}}. \tag{10.12}$$

In this equation, V_{CES} and V_{BE} have been retained in case they are significant compared with V_B. However, both R_{CES} and the base–emitter resistance have been omitted since they are small compared with the values of the circuit resistors.

The reverse bias at the base of Tr2 must be sufficient to hold it hard OFF; that is, this base must not be allowed to go positive with respect to the common line.

$$V_{EB2} = V_B - \frac{V_B + V_{CES}}{R_{BB2} + R_{B2}} R_{BB2}. \tag{10.13}$$

Equations (10.12) and (10.13) are the basic design equations for the bistable circuit. The simplest design procedure is to establish a value for R_C knowing $h_{FE(O)}$, V_{CC}, and the collector current required, then insert trial values of R_B and R_{BB} into these equations.

The capacitors shown by broken lines in figure 10.13(b) are to enable the transistors to switch from one state to the other more quickly. They are usually termed *speed-up capacitors*.

The charge control concept is of use in the determination of the value of the speed-up capacitor. If the stored charge at saturation is Q_{BS}, then the speed-up capacitance C will be not less than the value required to extract this charge:

$$C \geq \frac{Q_{BS}}{V_{CC}}.$$

10.4.4 The Monostable Circuit

In figure 10.13(c), Tr1 is normally ON and Tr2 is normally OFF. To hold Tr1 ON, adequate base current must be provided:

$$I_{B1} \geq I_{CS1}/h_{FE(O)}.$$

That is,

$$\frac{V_{CC} - V_{BE}}{R_{B1}} \geq \frac{V_{CC} - V_{CES}}{R_{C1} h_{FE(O)}}.$$

Neglecting V_{BE} and V_{CES}, this becomes

$$R_{B1} \le R_{C1} h_{FE(O)} . \tag{10.14}$$

Also, the emitter junction of Tr2 must be reverse-biased by an amount V_{EB2}, which must not exceed V_{EBM}:

$$V_{EB2} = V_B - \frac{V_B + V_{CES}}{R_{BB2} + R_{B2}} R_{BB2} .$$

Rearranging this equation gives:

$$\frac{R_{BB2}}{R_{BB2} + R_{B2}} = \frac{V_B - V_{EB2}}{V_B + V_{CES}} . \tag{10.15a}$$

Trial values of R_{BB2} may be inserted and checked for validity by ensuring that sufficient base current is available to saturate Tr2 during the metastable state; that is, $I_{B2} > I_{CS2}/h_{FE(O)}$:

$$\frac{V_{CC} - V_{BE}}{R_{C1} + R_{B2}} - \frac{V_B + V_{BE}}{R_{BB2}} \ge \frac{V_{CC} - V_{CES}}{R_{C2} h_{FE(O)}} . \tag{10.15b}$$

If a negative pulse is applied to the base of Tr1, or a positive pulse to the base of Tr2, the states will change over. As in the multivibrator, the right-hand side of C_2 will be clamped to the common line and the left-hand side will go down to $-V_{CC}$. The capacitor will then recharge via R_{B1}, and when the left-hand side rises just above zero, Tr1 will again turn ON. The dwell time for the metastable state is derived in the same way as equations (10.7):

$$T = 0.69 R_{B1} C_2 . \tag{10.16}$$

Note that a speed-up capacitor may also be incorporated in the circuit, and this should be connected across R_{B2}.

10.4.5 Practical Aspects of Regenerative Switches

All three circuits of figure 10.13 involve the minimum number of components necessary to demonstrate their respective switching actions. Any practical circuit will, however, have to perform certain other functions. Paramount among these is the necessity of protecting the base–emitter junction of the OFF transistor against over-voltage. It will be recalled that V_{EBM} for most transistors is quite small (usually a few volts) and anything over the stipulated value will destroy the junction. In both the multivibrator and the monostable circuit, the OFF transistor can be subjected to a V_{EB} of up to V_{CC} (as can the OFF transistor in the bistable circuit if speed-up capacitors are used).

This is not an easy problem to solve since any additional components included in the base circuit will have an effect upon the dwell time. Possibly the best solution is that shown in figure 10.16(*a*) where a silicon diode of very high reverse resistance, D_1, is inserted into the base lead so that it blocks the incoming negative signal. Clearly, what it actually does is to pass a very small reverse saturation current which causes voltage drops to appear across the diode and across the base–emitter junction of the transistor. If the reverse resistance of the diode is much greater than that of the base–emitter junction then most of the applied voltage will appear across the diode and the transistor will thereby be protected. If the success of this scheme is in doubt because a good silicon transistor of high reverse resistance is being used, a second diode, D_2, can be included. Here, however, it must be established that the drop across D_2 is sufficient to hold Tr1 OFF, which is why two diodes in series have been shown.

A second method is to insert a high reverse resistance diode into the emitter lead as in figure 10.16(*b*). This enables the emitter to rise in potential along with the base, again protecting the transistor. In the few cases where the base–emitter forward resistance has to be taken into account, the forward resistance of this diode must be included.

In figure 10.13(*b*) and (*c*) a battery of voltage V_B is shown. This is used to establish the stability of the OFF transistor, and can readily be replaced by a Zener diode as depicted in figure 10.17. It is useful to ensure that the Zener diode is permanently in a conducting state by including a bleed resistor R, shown dotted.

The triggering of bistable and monostable circuits is a topic which has been the subject of a great deal of analysis. In principle, there are four points in the circuits to which triggering pulses may be applied—the two bases and the two collectors. The ON base and collector will offer low impedances to incoming pulses and the OFF base and collector, high impedances. In that triggering pulses are usually extracted from the collector of a trigger transistor switching ON, this implies a low impedance, and the pulses can be either positive (for a pnp trigger) or negative (for a npn trigger).

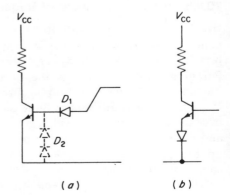

Figure 10.16. Base–emitter junction protection.

(a) (b)

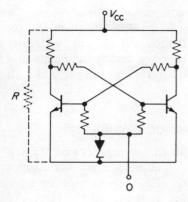

Figure 10.17. Use of Zener diode in a bistable circuit.

An interesting trigger circuit is given for a bistable circuit in figure 10.18. Here, a pair of pulse *steering diodes* is connected so that incoming negative trigger pulses are routed to the correct points. If Tr1 is ON, then D_1 will be biased OFF and the trigger pulse will be blocked. However, Tr2 will be OFF, so the negative trigger pulse will easily pass D_2. It will then have the dual effect of sending the collector of Tr2 negative, tending to switch it ON and of sending the base of Tr1 negative tending to switch it OFF.

The dwell times of both multivibrators and monostable circuits can be altered somewhat by changes in temperature. This is due in part to the change in V_{BE} for this defines the voltage which it is necessary to apply to the base to switch the transistor over. Also, the leakage current I_{CBO} is

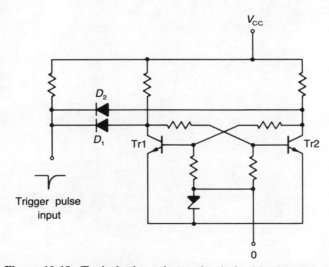

Figure 10.18. Typical trigger input circuit for bistable switch.

additive to the timing current, which again affects dwell time. Finally, a drop in h_{FE} will occur at low temperatures, as a result of which the ON transistor could be unsaturated. However, all these effects can be greatly minimized by choosing silicon transistors having low leakage currents and ensuring that the circuit components allow full saturation to occur under all the environmental conditions anticipated.

10.4.6   Other Forms of Regenerative Switching Circuits

1. The Complementary Pair. The three forms of regenerative switch discussed above can all be constructed using a complementary pair so that both transistors are ON or both OFF concurrently. The basic connection from which the three circuits are derived is shown in figure 10.19. Here the collector current of Tr1 is also the base current of Tr2, and it can be seen that if this current is rising, then the voltage at the second collector is going negative. This constitutes positive feedback to the base of Tr1 and so results in a rapid switch-ON for both transistors. When either transistor is subjected to a perturbation tending to switch it OFF, the converse occurs and the feedback loop causes both transistors to switch hard OFF. When OFF only leakage currents flow, but at elevated temperatures, these may be great enough to switch the circuit ON again. This is most easily obviated by the correct choice of the resistor R.

A considerable number of papers have appeared in which are devised ingenious methods of associating the complementary pair with external timing or triggering circuitry. However, such circuitry will not be considered here since such complementary pairs are analogous to the four-layer semiconductor device which will be discussed in chapter 13.

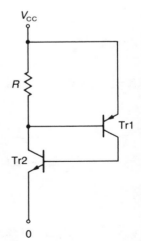

Figure 10.19. Basic complementary bistable circuit.   0

2. The Emitter-Coupled Switching Circuit. The monostable form of this circuit (figure 10.20(*a*)) is very common. Here Tr1 is normally ON and Tr2 is normally OFF, and under these conditions the steering diode is unbiased. Consequently, a negative trigger pulse will be allowed to pass to the base of Tr1 and the collector of Tr2, so initiating a change of state. When Tr1 switches OFF its emitter voltage goes negative, and this also being the emitter voltage of Tr2 a forward bias is applied to the base–emitter junction of that transistor, so turning it ON. This is because the base voltage of Tr2 is held fairly constant by resistors R_1 and R_2. Consequently, Tr2 switches ON and the negative-going transient at its collector is applied to the base of Tr1 via the capacitor C, so forcing Tr1 hard OFF.

Because Tr2 is now saturated, its base will clamp the left-hand side of C very near to the emitter voltage V_2. The right-hand side of C will charge

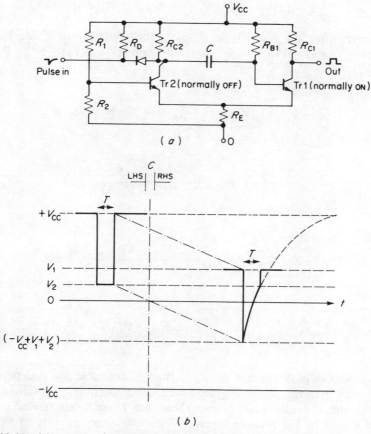

Figure 10.20. (*a*) Emitter-coupled monostable circuit. (*b*) Voltage excursions at each side of C for emitter-coupled monostable circuit.

towards $+V_{CC}$ via R_{B1} and R_E along with the saturated Tr2 in series. If $R_{B1} \gg R_E$, the time constant will be approximately $R_{B1}C$.

The emitter voltage is now V_2, or $I_{E2}R_E$, and when the right-hand side of C reaches a voltage a little higher than this, switch-over will again take place.

Referring to figure 10.20(b), the left-hand side of C is seen to fall by $(V_{CC} - V_2)$ when Tr2 switches ON. Hence, the right-hand side of C also falls by this amount, reaching a voltage $V_1 - (V_{CC} - V_2)$, or $(-V_{CC} + V_1 + V_2)$.

Using this voltage as a datum line, it is clear that the right-hand side of C will try to charge up to $+V_{CC}$, that is, through a voltage

$$V_{CC} + (V_{CC} - V_1 - V_2) = 2V_{CC} - V_1 - V_2 .$$

The relevant law is therefore

$$v_c = (2V_{CC} - V_1 - V_2)[1 - \exp(-t/R_{B1}C)] .$$

However, after the dwell time T, the right-hand side of C reaches the switch-over voltage V_2, having risen through $(V_{CC} - V_2) - (V_1 - V_2)$, or $(V_{CC} - V_1)$.

Hence,

$$(V_{CC} - V_1) = (2V_{CC} - V_1 - V_2)[1 - \exp(-T/R_{B1}C)]$$

giving

$$\exp(-T/R_{B1}C) = \frac{V_{CC} - V_2}{2V_{CC} - V_1 - V_2}$$

whence

$$-T = R_{B1}C \ln\left(\frac{V_{CC} - V_2}{(V_{CC} - V_2) + (V_{CC} - V_1)}\right)$$

or

$$T = R_{B1}C \ln\left(1 + \frac{V_{CC} - V_1}{V_{CC} - V_2}\right) . \tag{10.17}$$

This equation is rather less accurate than is desirable for it depends on many approximations, the most important being the infinitely rapid switching of Tr1 and Tr2. However, it does show that T can be controlled by both R_{B1} and C and by the relative levels of V_1 and V_2, the stability of V_2 being determined by the combination R_1 and R_2.

The recovery of the circuit after an excursion is not immediate as the foregoing treatment suggests, for C will recover via R_{C2} and the parallel combination of R_{C1} and R_E. This time constant is normally small, but it should be ensured that a further trigger pulse does not appear until two or three recovery time constants have elapsed.

Finally, the bias situation must be considered. In order to maintain Tr1 ON and Tr2 OFF during standby, then V_1 must be more positive then V_2, as is shown in figure 10.20(b); that is,

$$\frac{V_{CC}R_E}{R_E + [R_{B1}R_{C1}/(R_{B1} + R_{C1})]} > \frac{V_{CC}R_2}{R_1 + R_2}$$

$$\frac{1}{1 + \{R_{B1}R_{C1}/[R_E(R_{B1} + R_{C1})]\}} > \frac{1}{1 + (R_1/R_2)}$$

$$\frac{R_1}{R_2} > \frac{R_{B1}R_{C1}}{R_E(R_{B1} + R_{C1})} . \qquad (10.18)$$

This assumes that Tr1 is fully saturated when in the standby condition, which may be assured by making the base current at saturation large enough; that is,

$$I_B > I_{CS}/h_{FE(O)}$$

or

$$\left(\frac{V_{CC} - V_1}{R_{R1}}\right) > \frac{V_{CC} - V_1}{h_{FE(O)}R_{C1}}$$

giving

$$R_{C1}h_{FE(O)} > R_{B1} . \qquad (10.19)$$

During the dwell time, Tr2 must be saturated, which means that by similarity with equation (10.19), $h_{FE(O)}R_{C2} > R_1$. R_2 may now be found by noting that the voltage at the junction of R_1 and R_2 should be positive with respect to the base of Tr1 if the base lead were open-circuited and Tr1 continued to conduct; that is

$$\frac{R_1}{R_2} < \frac{R_{C1}}{R_E} . \qquad (10.20)$$

The design equations have now been established. The current through the transistors is first decided, giving R_{C1} and R_{C2}. Saturation is assured by making R_{B1} and R_1 small enough, and equations (10.18) and (10.20) are combined to suggest values of R_E and R_2:

$$\text{energy dissipated} = J_r = V_{CC}I_{C(m)}\int_0^{t_r} \frac{t}{t_r}\left(1 - \frac{t}{t_r}\right) dt$$

$$= \tfrac{1}{6}V_{CC}I_{C(m)}t_r \tag{10.1}$$

(ii) Region 3—switch OFF

$$I_C = I_{C(m)}\left(1 - \frac{t}{t_f}\right) \quad \text{and} \quad V_{CE} = V_{CC}\frac{t}{t_f}$$

giving

$$\text{energy dissipated} = J_f = \tfrac{1}{6}V_{CC}I_{C(m)}t_f \tag{10.2}$$

(iii) Region 2—ON period

$$J_{ON} = V_{CES}I_{C(m)}t_{ON} \tag{10.3}$$

(iv) Region 4—OFF period

$$J_{OFF} = V_{CC}I_{CEO}t_{OFF} . \tag{10.4}$$

If the transistor is to switch repetitively, and the various times involved are known, then clearly the power dissipated in watts (i.e., joules per second) must be the sum of equations (10.1)–(10.4) divided by T where

$$T = t_r + t_f + t_{ON} + t_{OFF} \tag{10.5}$$

and t_{ON} and t_{OFF} are defined as in figure 10.5.

Before leaving the discussion of switching transients, an alternative method of describing switching behavior must be mentioned. This is the charge control parameter system of Beaufoy and Sparkes[1,2,3]. Here, the actual charges which must flow into, or out of, the base region to initiate or stop conduction are considered.

Firstly, a change in collector current ΔI_C is assumed to be proportional to a change in base charge ΔQ_B, the constant of proportionality having the dimensions of time:

$$\frac{\Delta Q_B}{\Delta I_C} = \tau_C$$

where τ_C is called the collector time factor.

Secondly, if this change in collector current is accompanied by a change in collector *voltage*, then a further change in base charge Q_V must occur to change the potential across the base–collector depletion layer, which can be considered as a capacitance C_{tc}.

From the basic relationship $q = CV$, the charge Q_V may be found:

$$Q_V = \int_{V_{C_2}+U}^{V_{C_1}+U} C_{tc}\, dV$$

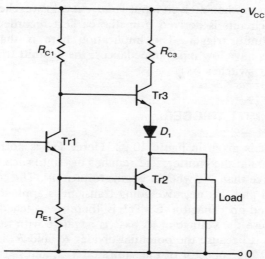

Figure 10.21. The totem-pole output buffer.

$$V_{E3} = V_{CES(2)} + V_{D1}$$

which is too close to 1 V to allow sufficient $V_{BE(3)}$ to effect turn-ON, and which explains the function of the diode.

When Tr1 turns OFF, the base of Tr2 is effectively grounded through R_{E1}, so it also turns OFF. Under these circumstances, Tr3 can obtain base current via R_{C1}, which turns it ON. The load (connected as shown) is now energized via R_{C3}, Tr3 and D_1, which contribute only small voltage drops. The purpose of R_{C3} is simply to limit the current during transients, and particularly because (due to charge storage) Tr2 usually turns OFF more slowly than Tr3 turns ON, when for a very brief period, current can flow down the Tr3–D_1–Tr2 chain. The value of R_{C3} is therefore that which will limit the transient current to the maximum allowed by the smaller of the two transistors, I_{CM}, and its value may therefore be very low.

Note that had the load been taken to V_{CC}, it would have been un-energized when Tr3 was ON, of course.

From the above explanation, it is apparent that Tr2 and Tr3 can be small power transistors capable of delivering useful load currents without significantly altering the time constants in the preceding circuitry. They can also be capable of quickly charging or discharging any capacitance associated with the load, making for fast transient operation. However, the presence of R_{E1} is not entirely convenient in the context of the regenerative circuits so far presented, so Tr1 is often associated with an extra transistor which is itself operated from the collector of one of the regenerative pair.

So far, the discussion of regenerative circuits using external inputs has been confined to those requiring input *current pulses*. A useful variant is that which changes state when a given input *voltage* level is reached. The generic name for these circuits is derived from the earliest recorded vacuum tube version, the Schmitt trigger. The implication here is that whereas the foregoing circuits have low input impedances, the Schmitt trigger presents a high one (until it switches ON).

10.6 THE SCHMITT TRIGGER

The basic circuit is given in figure 10.22. Here, Tr1 is normally ON and a voltage V_1 appears at the emitter. Tr2 remains OFF until such time as its base goes more positive than V_1, when it begins to turn ON. The collector voltage of Tr2 then falls and this negative-going transient is applied to the base of Tr1 via the speed-up capacitor C. Tr1 is thereby switched hard OFF and remains so because the voltage at its base is negative with respect to that at its emitter. This is because the potential divider R_1 and R_2 is now fed from the voltage V_2 at the collector of the saturated Tr2, and not from the much more positive V_{CC}.

The reversed state will hold until the input voltage falls below V_2, when the circuit will revert to its standby condition.

Figure 10.23 gives the idealized waveforms. From it, an important principle becomes apparent; that V_1 must be greater than V_2. This is because, if the converse were true, the circuit would revert to its standby state immediately after triggering, and the sequence would repeat continuously while the input voltage lay between V_1 and V_2.

If a trigger circuit is required which has a very small backlash $(V_1 - V_2)$, then V_{BE} and V_{CES} become important, for their values can be comparable

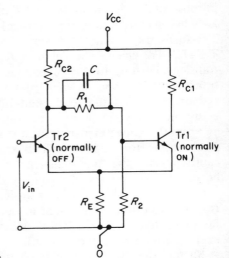

Figure 10.22. Schmitt trigger circuit.

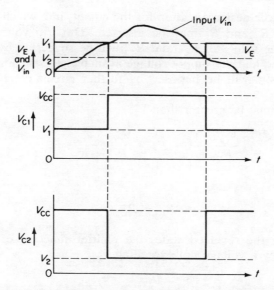

Figure 10.23. Idealized wave profiles for Schmitt trigger.

with the backlash voltage. Remembering that V_{BE} is a function of temperature, it is possible that the circuit may operate satisfactorily over a very limited temperature range only. Further, the resistor tolerances become closer as backlash is decreased, and the stability of the supply voltage becomes important. The most convenient solution to this problem is for a trigger circuit with a fairly high backlash to be preceded by a suitable pre-amplifier to raise the operating voltage levels. This system has two further advantages: firstly, it enables the overall backlash to be varied without altering any component in the actual trigger circuit; and secondly, it can be designed to act as a buffer stage, for the trigger circuit presents a low resistance at the input when ON.

The design procedure will simply be a calculation of the various bias resistor values so as to assure saturation of Tr1 in the standby condition and saturation of Tr2 in the ON condition. The following derivation will include V_{BE} and V_{CES} in order to show their effect on the backlash voltage.

In the stable condition,

$$I_{B1} > I_{CS1}/h_{FE(O)}$$

that is

$$\frac{V_{CC} - (V_1 + V_{BE})}{R_{C2} + R_1} - \frac{V_1 + V_{BE}}{R_2} > \frac{V_{CC} - (V_1 + V_{CES})}{R_{C1}h_{FE(O)}}. \quad (10.22)$$

This is the basic design equation for the circuit, into which trial values of R_{C1}, R_{C2}, and R_1 and R_2 may be inserted. The trip voltage V_1 will be specified and the values of R_{C1} and R_{C2} will be suggested by the collector currents required and the supply voltage available. Restrictions on collector and base resistors will be imposed by further circuit conditions detailed below.

Because V_1 must be greater than V_2,

$$\frac{V_{CC} - V_{CES}}{R_{C1} + R_E} R_E > \frac{V_{CC} - V_{CES}}{R_{C2} + R_E} R_E$$

or

$$R_{C1} < R_{C2} . \tag{10.23}$$

Also, during the reversed state, the emitter junction of Tr1 must be reverse-biased by a voltage not exceeding V_{EBM}:

$$V_2 - \frac{V_2 + V_{CES}}{R_1 + R_2} R_2 \ngtr V_{EBM} .$$

Neglecting V_{CES}, this becomes

$$\frac{R_1}{R_1 + R_2} \ngtr \frac{V_{EBM}}{V_2} . \tag{10.24}$$

All the switching circuit functions mentioned above, plus many others, can be (and usually are) performed by integrated circuits. In particular, the basic three functions, astable, bistable, and monostable, can easily be realized using integrated circuits, but only for low power logic applications. At higher voltages and currents, design using discrete transistors remains necessary unless the logic chips are followed by interfacing circuitry to higher levels.

For the more complex functions, integrated circuits which are comparable to the complete operational amplifiers described in chapter 7 become useful. For example, the Schmitt trigger is simply a comparator insofar as its function is concerned, and the use of an operational amplifier as a comparator was described with reference to figure 7.27. This and some other switching applications of integrated circuits are described below.

10.6.1 Comparators

One of the fundamental disadvantages of the simple Schmitt trigger is that the input impedance to Tr2 falls when the circuit changes state, which usually means that a buffer stage must be provided. The operational amplifier connected as a comparator overcomes this problem inherently.

Furthermore, by appropriate choice of ancillary circuitry, it can compare either voltages or currents; can change state as an incoming signal changes sign (a zero-crossing detector); and can be used in such a way as to provide controllable backlash, or *hysteresis*.

Figure 10.24(a) shows a very simple zero-crossing detector. Here, the input is referred to zero by connecting the noninverting input of an operational amplifier to ground.

If e_{in} goes positive then the output goes negative, but the output voltage level is limited by the Zener diode Z_d^- in the feedback loop. Conversely, if e_{in} goes negative, the output goes positive, v_{out} now being limited by the Zener diode Z_d^+. (The functions of diodes D_1 and D_2 are to prevent Z_d^+ and Z_d^- conducting when forward-biased.) This usage of Zener diodes is an example of a *limit circuit*, which enables an output voltage to be defined irrespective of the amplifier or power supply parameters.

The circuit of figure 10.24(a) has a very low backlash voltage defined by v_{in} (between the inputs) and R_G. Using the first-order assumption that v_{in} is zero, then the state change occurs when the feedback current becomes zero; that is, when,

$$i_f = i_g - i_b^- = \frac{e_{in}}{R_G} - i_b^- = 0 . \tag{10.25}$$

From this equation, it is seen that the state change occurs when $e_{in} = i_b^- R_G$, and since for a good operational amplifier i_b is very small, clearly a zero-crossing voltage comparator results. If a current comparator is needed, a number of inputs may be used in the manner of a summing amplifier (figure 7.25(b)). When these currents add algebraically to zero, the circuit will change state. Taking the small bias current i_b^- into account, this may be expressed as:

$$i_g - i_b^- = 0$$

or

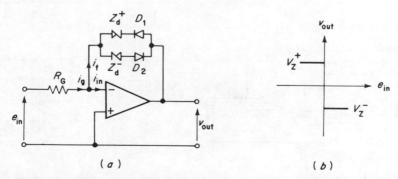

$$(a) \qquad\qquad (b)$$

Figure 10.24. (a) A simple zero-crossing detector and (b) its transfer characteristic.

$$i_g = i_b^- .$$ \hfill (10.26)

The circuit as described is a zero-crossing detector only because the noninverting input was referred to ground. If this input is referred to a voltage other than ground, then it becomes a *level detector*, as shown in figure 10.25(a), and its transfer performance is as shown in figure 10.25(b). (Should either of these circuits be required to produce an output voltage near zero for one of the states, then the appropriate Zener diode can be omitted.)

The circuits so far presented have no backlash, which means that the output voltage will "judder" up and down when the input signal is near zero. In fact, noise peaks on a slowly changing, near-zero, input signal are often sufficient to cause the circuit to change state many times. To avoid this, it is convenient to allow the output signal to modify the reference voltage, so introducing *hysteresis*. Figure 10.26(a) shows how this may be accomplished, and it operates as follows.

If $v_{out} = V_Z^+$, then V_{ref} is given by:

$$V_{ref}^+ = V_Z^+ \frac{R_{F2}}{R_{F1} + R_{F2}} .$$

If e_{in} goes more positive than this, then the circuit state changes and the output goes negative to V_Z^-. Hence, the reference voltage becomes:

$$V_{ref}^- = V_Z^- \frac{R_{F2}}{R_{F1} + R_{F2}} .$$

If e_{in} continues to go positive, the output will remain at V_Z^-.

This output will also remain at V_Z^- even if e_{in} falls below V_{ref}^+. It must now go negative with respect to V_{ref}^- before a state change again occurs. This prevents "juddering" of the output and also produces a fast "snap action" state change as soon as e_{in} reaches the appropriate level.

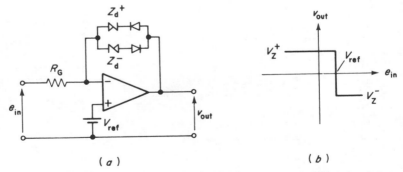

(a) \hfill (b)

Figure 10.25. (a) A level detector and (b) its transfer characteristic.

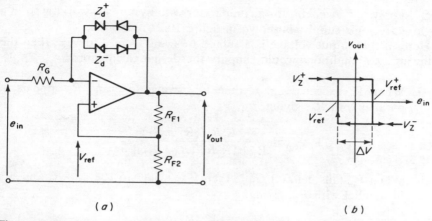

Figure 10.26. (*a*) A level detector with hysteresis and (*b*) its transfer characteristic.

The transfer characteristic is shown in figure 10.26(*b*), and it should be carefully studied to ensure a proper appreciation of the operation of this circuit. Notice that the width of the hysteresis loop is actually analogous to the backlash voltage of the Schmitt trigger, and is given by:

$$\Delta V = V_Z^+ \left(\frac{R_{F2}}{R_{F1} + R_{F2}} \right) - V_Z^- \left(\frac{R_{F2}}{R_{F1} + R_{F2}} \right)$$

$$= (V_Z^+ - V_Z^-) \frac{R_{F2}}{R_{F1} + R_{F2}} \qquad (10.26)$$

(keeping in mind that V_Z^- is a *negative* voltage).

This hysteresis voltage width ΔV can be made controllable by substituting a potentiometer for R_{F1} and R_{F2}. The effects of offset voltage and drift can be minimized in the standard ways described in chapter 7; but since minimizing the effects of the operational amplifier bias currents involves making $R_{F1} \| R_{F2} = R_G$, the substitution of a potentiometer for fixed values of R_{F1} and R_{F2} should be approached with care.

Although an operational amplifier can be (and often is) used as a comparator, there are advantages in using purpose-designed monolithic comparator modules. These are essentially high-gain operational amplifiers designed to work in the open-loop mode, and which exhibit very fast slewing rates, usually at the expense of linearity, which is not important for a comparator. They are also designed to have no latch-up problems, so that the output may swing between the supply voltages, omitting the Zener diodes. Further, the design is often such that the common-mode voltage swing can exceed the power supply voltages by considerable margins without damage to the device. Finally, like the more recent operational amplifiers, many comparators can work from low-voltage supplies, and often from a single supply rail. A circuit for a comparator with hysteresis, working from a single $+5\,V$ logic supply, is shown in figure 10.27(a).

Here, if the output voltage is assumed to swing from 0 to V_{CC}, then the following two conditions would simplify the design calculations:

(i) When V_{in} exceeds $V_{ref(hi)}$, then V_{out} swings to 0 and R_3 appears in parallel with R_2, giving,

$$V_{ref(lo)} = \frac{V_{CC}R_2R_3/(R_2+R_3)}{R_1+[R_2R_3/(R_2+R_3)]}.$$

(ii) When V_{in} falls below $V_{ref(lo)}$, then V_{out} swings to V_{CC} and R_3 appears in parallel with R_1, giving,

$$V_{ref(hi)} = \frac{V_{CC}R_2}{[R_1R_3/(R_1+R_3)]+R_2}.$$

This results in the hysteresis characteristic of figure 10.27(b).

In practice, the input voltages of this LM 339 comparator can fall to zero, and it will still operate. However, they can only rise to within about 1.3 V below V_{CC}, otherwise operation will cease. The reason for this can be seen from the basic circuit shown in figure 10.28, which shows that the input stage is a pnp Darlington-connected difference configuration. Hence, as the common-mode input voltage falls towards zero the input stage continues to conduct. However, if the input voltage rises, then when less than $2V_{BE}$ is available at the input, the Darlington pair can no longer conduct and operation ceases. This $2V_{BE}$ is referred to the V_{CC} supply, so that the input voltage must not be allowed to exceed $(V_{CC}-2V_{BE})$, that is V_{in} and V_{ref} must not rise to more than about 1.3 V below V_{CC}.

In addition to basic, operational-amplifier derived comparators, various special purpose monolithic integrated circuits for switching purposes exist, the best known of which is the type 555 timer module, originated by Signetics Inc.

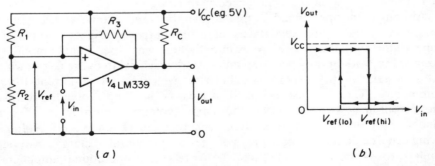

Figure 10.27. (a) A single-supply comparator and (b) its transfer function.

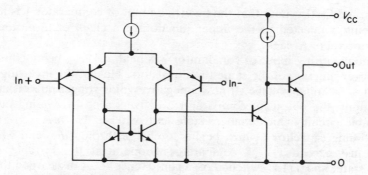

Figure 10.28. Basic circuit for the LM 339 comparator.

10.7   THE TYPE 555 MONOLITHIC TIMER MODULE

This module contains, among other active elements, two complete comparators and a bistable unit (or flip-flop). Figure 10.29 is a partial block diagram of the microcircuit, and minimal external components have been shown in order to describe the basic operation of the circuit.

Assume that the bistable circuit is in the ON state, which is defined as that state which will turn Tr1 ON. This means that the timing capacitor C_t is

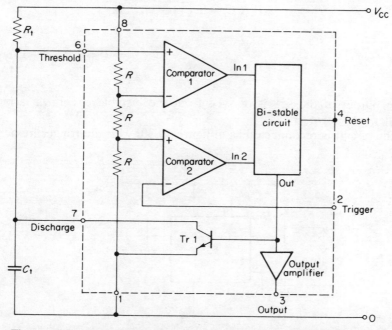

Figure 10.29. Basic block diagram for the 555 timer integrated circuit.

short-circuited. Also note that the inverting input of comparator 1 is held at $\frac{2}{3}V_{CC}$, being connected to the upper junction of a chain of three identical resistors of value R each.

The noninverting input of comparator 2 is held at $\frac{1}{3}V_{CC}$, being connected to the lower junction of the same three resistors. Hence, the inverting input will float at a similar voltage. If now a negative-going trigger pulse is applied to this input, the output of comparator 2 delivers a positive-going pulse to the bistable circuit, which changes state and switches Tr1 OFF.

The timing capacitor C_t now begins to charge through R_t, and when its voltage just exceeds $\frac{2}{3}V_{CC}$, comparator 1 operates, the bistable circuit changes state, and Tr1 is switched ON again. Hence, C_t is discharged through Tr1 and the 555 is ready to accept another pulse.

This sequence actually describes monostable operation, in which the dwell time is the time required for C_t to charge to $\frac{2}{3}V_{CC}$ via R_t:

$$v_{C_t} = \tfrac{2}{3}V_{CC} = V_{CC}[1 - \exp(-T/R_tC_t)]$$

giving

$$\exp(-T/R_tC_t) = \tfrac{1}{3}.$$

That is,

$$-T/R_tC_t = \ln\left(\tfrac{1}{3}\right) = -1.1$$

or

$$T = 1.1R_tC_t. \tag{10.27}$$

The output of the 555 is derived from the bistable circuit via a buffer amplifier, as shown.

If the circuit is reconnected as in figure 10.30, with the trigger input (pin

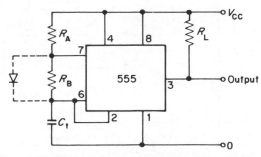

Figure 10.30. The 555 timer connected as an astable circuit.

2) connected to the threshold input (pin 6); and where R_t is replaced by two resistors, R_A and R_B, then astable operation results, as follows.

When C_t charges up via $(R_A + R_B)$ to a voltage just above $\frac{2}{3}V_{CC}$, then comparator 1 causes the bistable circuit to change state so that Tr1 switches ON. This means that C_t must now discharge via R_B only, and when its voltage falls just below $\frac{1}{3}V_{CC}$, comparator 2 causes the bistable circuit to change state, switching Tr1 OFF. Hence, the voltage on C_t oscillates between $\frac{1}{3}$ and $\frac{2}{3}$ of V_{CC}, with a charging time constant of $(R_A + R_B)C_t$ and a discharging time constant of $R_B C_t$.

This results in a square-wave output having "mark" and "space" times T_M and T_S as follows.

From the foregoing derivation, the time T taken to charge from zero to $\frac{2}{3}V_{CC}$ is $1.1 (R_A + R_B)C_t$.

Also the time T_1 taken to charge from zero to $\frac{1}{3}V_{CC}$ would be given by:

$$T_1 = -(R_A + R_B)C_t \ln (1 - \tfrac{1}{3})$$
$$= 0.405(R_A + R_B)C_t$$

so that

$$T_M = T - T_1 \simeq 0.69(R_A + R_B)C_t . \qquad (10.28)$$

Also the time to discharge from $\frac{2}{3}V_{CC}$ to $\frac{1}{3}V_{CC}$ via R_B would be:

$$T_S \simeq 0.69 R_B C_t . \qquad (10.29)$$

The PRF is now:

$$f \simeq \frac{1}{T_M + T_S} = \frac{1}{0.69(R_A + 2R_B)C_t} = \frac{1.45}{(R_A + 2R_B)C_t} . \qquad (10.30)$$

Notice that for this circuit, the mark/space ratio is:

$$\frac{T_M}{T_S} = \frac{R_A + R_B}{R_B} = 1 + \frac{R_A}{R_B}$$

which is always greater than unity.

To produce a mark/space ratio which is equal to, or less than unity, a diode may be connected across R_B as shown dashed in figure 10.30. Now, when charging, C_t receives current via R_A and the diode only, so that when charging from $\frac{1}{3}V_{CC}$ to $\frac{2}{3}V_{CC}$:

$$T_M \simeq 0.69 R_A C_t$$

(from 10.28).

Then, C_t discharges from $\frac{2}{3}V_{CC}$ to $\frac{1}{3}V_{CC}$ via R_B as before because the diode is now nonconducting. Hence,

$$T_S \simeq 0.69 R_B C_t \qquad (10.29)$$

so

$$\frac{T_M}{T_S} = \frac{R_A}{R_B}$$

which can take any value including unity. (In fact, if a single potentiometer is used, R_A being on one side of the wiper and R_B on the other, the mark/space ratio can be manually adjusted through unity.)

Also,

$$f = \frac{1}{T_M + T_S} = \frac{1}{0.69(R_A + R_B)C_t} = \frac{1.45}{(R_A + R_B)C_t} .$$

Numerous other functions can be accomplished by this circuit, and it must be remembered that it is only the progenitor of very much more sophisticated timer microcircuits.

10.8 INDUCTIVELY-COUPLED REGENERATIVE CIRCUITS

Each of the regenerative circuits so far discussed has been resistively or capacitively coupled with the capacitors determining, for the multivibrator and monostable circuit, the dwell time. This coupling can equally well be performed inductively, using a transformer, and all three forms of regenerative circuit can be realized in this manner. These variants will not be discussed in detail since their applications have become rather specialized, but the general action of the basic astable circuit will be considered briefly as an introduction to the transistor inverter.

The basic astable *blocking oscillator* circuit is shown in figure 10.31. It is representational only and much simpler than a practical circuit; it does, however, serve to explain blocking action more lucidly than would a more complex diagram. When the circuit is energized, the bias supply V_{BB} causes the transistor to conduct slightly. The rise in collector current through the primary winding of the transformer causes a voltage to appear across the secondary in the sense shown by the dot. This augments the base voltage, causing a greater current to flow in the base circuit which in turn produces a greater collector current. This cumulative action quickly saturates the transistor so that practically the whole of V_{CC} appears across N_P. When this happens the current in N_P increases linearly, assuming that the inductance of the transformer remains constant. Basically

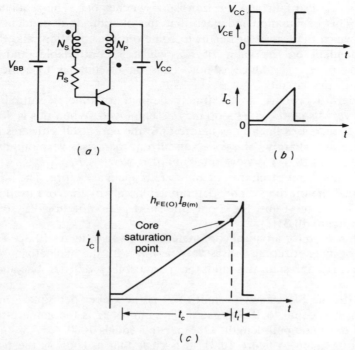

Figure 10.31. Blocking oscillator: (*a*) circuit (simplified); (*b*) wave profiles; (*c*) collector current profile for blocking oscillator using saturable square-loop core. t_c is part of ON time defined by core and t_t part of ON time defined by transistor.

$$e = N \frac{\mathrm{d}\Phi}{\mathrm{d}t} = N \frac{\mathrm{d}\Phi}{\mathrm{d}i} \frac{\mathrm{d}i}{\mathrm{d}t}$$

or

$$e = L \frac{\mathrm{d}i}{\mathrm{d}t}$$

giving

$$I_C = \frac{V_{CC}}{L} t . \tag{10.31}$$

This situation is shown in the wave profiles of figure 10.31(*b*). Eventually, the collector current reaches a maximum, for the rate of change of current in N_P is constant, leading to a constant base voltage. This in turn produces a constant base current $I_{B(m)}$, and when I_C reaches $h_{FE(O)}I_{B(m)}$ the transistor begins to come out of saturation. Thus V_{CE} begins to rise, causing the primary voltage $(V_{CC} - V_{CE})$ to fall. This results in a fall in secondary voltage, so I_B also falls and the transistor switches OFF. The sequence then repeats. This explanation indicates that the ON time is dependent on the point at which the transistor begins to come out of saturation, and this is in turn dependent on the value of R_S, which is controllable, and on the magnitude of $h_{FE(O)}$, which is not. Thus on ON time is to some extent arbitrary in this circuit.

In order to produce an accurately defined ON time, a saturable core having a very "square" B/H loop may be employed. When this is done, the value of I_C increases linearly as before until the core itself saturates and the inductance falls sharply. The collector current then rises very rapidly, since L in equation (10.31) is now small, until it reaches $h_{FE(O)}I_{B(m)}$ when the transistor comes out of saturation and I_C falls again as before. The important point is that the ON time is now determined largely by the core itself and the contribution of the transistor is only the short period occupied by the spike shown in figure 10.31(c).

The B/H loop for a typical core material is given in figure 10.32. The field strength of magnetization B rises very rapidly with the magnetizing force H until the material saturates and the permeability dB/dH becomes very small.

(N.B. In the SI system the unit of B is the weber per square metre or tesla, which is equal to 10 000 gauss. The unit of H is the ampere-turn per metre of magnetic path-length. One oersted equals $1000/4\pi$ A m^{-1}.)

Referring again to figure 10.31, it is clear that as soon as the transistor has switched OFF, the sequence will begin again, and practical forms of

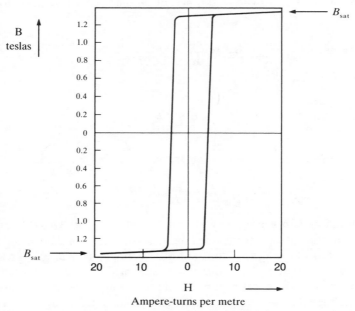

Figure 10.32. DC hysteresis loop for typical 0.5 mm square-loop material.

blocking oscillator include a capacitor so that a predetermined OFF time will also appear. Thus the collector voltage waveform will be square.

Another method of producing a square wave is by using two transistors in the well-known inverter configuration, which will now be described.

10.8.1 Transistor Inverters

The basic function of the inverter is the conversion of a direct current to an alternating current, usually, though not necessarily, having a square wave-form. This output may be rectified and smoothed, in which case the complete apparatus may be regarded as a DC transformer, the output voltage depending on the turns ratio of the transformer involved. Such inverters have been constructed to deal with powers of many kilowatts, but utilize *thyristors* rather than transistors. (Thyristors are considered in a later chapter.) Smaller inverters using normal transistors have applications as low-power, high-voltage supplies, as accurate square-wave generators, and as voltage-level-to-frequency converters in some measuring systems. These small transistor inverters form the subject matter of this section, and the basic design principles are formulated.

The commonest form taken by the inverter uses two transistors and a square loop transformer coupled as shown in figure 10.33(*a*), and its operation is conveniently explained with the help of the relevant wave profiles given in figure 10.33(*b*).

Suppose that Tr1 begins to conduct. The collector current in N_{1P} rises, and by transformer action produces a voltage in N_{1S} in the sense indicated by the dot so that a forward base current flows in Tr1 and turns it hard ON. This means that the collector voltage reaches V_{CES} very quickly as shown in the wave profile. The magnetizing current, however, increases more slowly, according to equation (10.31). This current forms only part of the collector current, and the remainder, which supplies the load via normal transformer action, increases rapidly as shown. The result is a square current wave topped by a slowly rising "cap." Eventually the magnetization of the core reaches the saturation flux level B_{sat}, and as the inductance falls so the magnetizing current rises sharply until it reaches $h_{FE(O)}I_{B(m)}$. The transistor then comes out of saturation, the primary voltage $(V_{CC} - V_{CE})$ falls and the regenerative switch-OFF occurs.

At the same time, the voltage induced in N_{2S} switches Tr2 ON, and the effective current in the primary reverses. This current increases until saturation occurs in the opposite direction and another reversal is initiated. This sequence of events is repetitive and results in a square-wave output, one transistor supplying the load during the positive swing, the other during the negative swing.

The voltage rating of the transistors is a prime design parameter, and in specifying it two important points must be taken into account. Firstly, when a transistor is OFF, a voltage of magnitude V_{CC} is induced in the relevant half

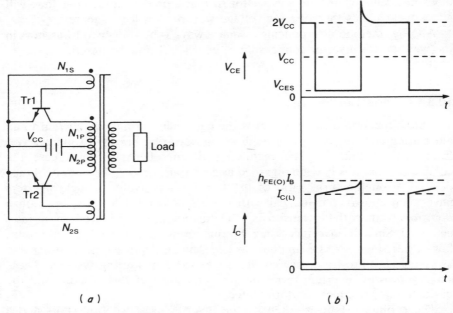

Figure 10.33. (*a*) Basic inverter circuit and (*b*) inverter wave profiles for one transistor.

of the primary in a sense *additive* to the supply voltage. Thus, a voltage of $2V_{CC}$ appears across the OFF transistor.

Secondly, when a transistor switches OFF, this is equivalent to opening an inductive circuit, for the transformer will inevitably include some leakage inductance. Since $e = -L \, di/dt$, a high reverse voltage can be induced, and appears as the *overshoot* spike shown in the wave profile for V_{CE}.

The transistor chosen must be capable of withstanding the transient overshoot and the longer period at $2V_{CC}$. In practice, the overshoot is minimized by designing the transformer to have a low leakage reactance. Further protection may be provided by connecting a Zener diode across the transistor so that any excessive voltage is obviated. This precaution also takes care of the case where the load itself contributes sufficient inductance to generate an excessive overshoot.

The current specification of the transistors depends on the design of the transformer, for the maximum collector current will be that current needed to saturate the core fully plus a little extra to accommodate the current peak at saturation, all this being in addition to the current requirement of the load referred to the primary.

The value of saturation flux, Φ_{sat}, can be related to the supply voltage and operating frequency quite simply by applying the basic relation $e = N \, d\Phi/dt$.

When a change of state occurs the ON transistor saturates and almost the whole of the supply voltage appears across the associated primary winding, which has N_P turns. The magnetic flux changes from Φ_{sat} in one direction to Φ_{sat} in the other, during one half-period, $\frac{1}{2}T$. This rate of change must produce an EMF equal and opposite to V_{CC}, neglecting the small resistive voltage drop in the winding. Putting these facts into the basic equation gives:

$$V_{CC} = N_P \frac{2\Phi_{sat}}{T/2}$$

and since the frequency $f = 1/T$, this becomes:

$$V_{CC} = 4N_P\Phi_{sat}f . \tag{10.32}$$

This equation shows that the frequency is proportional to the applied voltage, which explains how the inverter can be used as a voltage-to-frequency converter.

Before proceeding with the discussion on the design technique itself, some notes on the nature of square-loop material will be in order. The most commonly used materials are nickel–iron alloys which have been mechanically treated so that the individual metal crystals are all aligned in the same direction. This is achieved mainly by cold rolling and though the magnetic properties of the material after such treatment are optimum, they are subject to modification by shock or excessive heat. This means that the toroids or laminations should be treated very gently; they should not be dropped, bent, or even clamped down heavily after winding. (A toroid is a closely wound spiral of magnetic material in ribbon form.)

In use, the material must be magnetized along, or perpendicular to, the direction of rolling, for a square-cornered loop will be observed only when the individual crystals are all magnetized in an optimum direction. This may be any one of three mutually perpendicular directions, the [100], [010] or [001]. When either a toroid or transformer laminations are used, the winding space automatically ensures that magnetization takes place in the correct direction. The toroid is superior to transformer laminations in this respect, and can in general be expected to lead to a more sharply defined loop with a smaller area than that resulting from the use of a stack of laminations. Unfortunately, it is more difficult to wind a toroid than it is to wind an E or U section, for the wire must be threaded through a toroid, whereas it can be wound on to a bobbin and stampings inserted later in the case of a laminated core.

The area of the B/H loop represents the work done in reversing the magnetization. This *hysteresis loss* is part of the total electrical power loss involved in operating the transformer, and so should be as small as possible. Here again the toroidal core proves somewhat superior to the laminated E or U section core.

Some *eddy current loss* also occurs in the material of the transformer owing to heating by the eddy currents set up by the alternating magnetic field. This is minimized by insulating one side of the strip from which a toroid is wound or one side of each stamping from which a stack of laminations is built up. Such insulation prevents the eddy currents from flowing transversely across the core.

A further power loss occurs owing to the heating effect of the current in the transformer windings. This is termed the *copper loss*.

The final major power loss occurs in the transistor. The early part of this chapter showed that the power dissipated during the ON and OFF conditions is quite small, but that the instantaneous dissipation during the actual switching process can be high. It is necessary to calculate the average dissipation only in order to specify a suitable heat sink, for a transistor which will handle the peak currents and voltages involved is normally large enough to cope with the inherent power dissipation quite easily.

The efficiency of an inverter may be calculated by summing the losses described above, but in most cases this efficiency will be found to lie in the region of 75–85%, and in the preliminary design procedure it is safe to assume an efficiency of 80%.

The actual design of an inverter is in part an art, for the size of the transformer is related not only to electrical parameters but also to the purely mechanical process of determining whether the requisite number of turns can be wound on the core with ease but without too much spare winding area.

The simplest procedure is as follows:

(1) Determine the output power required, $P_L = V_L I_L$. This assumes that the output wave is square and that V_L and I_L are the amplitudes of the relevant waves.

(2) Knowing V_{CC}, find the part of the collector current that forms the reflected load current, $I_{C(L)}$. Assume an efficiency of 80%:

$$I_{C(L)} = \frac{P_L}{0.8 V_{CC}}.$$

(3) Assume a core dimension and check the electromagnetic performance. It is here that a pragmatic approach is necessary, for although two equations can be written (as soon follows), the actual core dimensions cannot be extracted directly. Also, in that cores of certain preferred dimensions are easily obtainable, the simplest procedure is to establish that the requisite number of turns will fit into the winding space available. If not, it will be necessary to redesign for a larger size.

For a given core, the manufacturer's data will indicate the number of ampere-turns per meter (H_0) necessary to produce a saturation flux density B_{sat}. Then, the following design equation may be used:

$$H_0 = \frac{N_P I_{C(mag)}}{l} \tag{10.33}$$

where $I_{C(mag)}$ is the magnetizing component of the collector current and l is the magnetic path length.

Also, equation (10.32) can be used:

$$V_{CC} = 4 N_P \Phi_{sat} f = 4 N_P B_{sat} A_{eff} f$$

where A_{eff} is the *effective core area*. This is the actual core area multiplied by the *stacking factor*, which takes account of that part of the overall area occupied by insulation and minute air gaps. Typically, the stacking factor is about 0.9 for 0.1 mm strip.

The base windings are found either by arranging for their generated voltage to be a little higher than V_{BE} at saturation, or by making this generated voltage much higher than V_{BE}—say 5–6 V—and limiting the base current by using a resistor. This latter method is the better, since replacement of transistors by units having different values of V_{BE} will have little effect.

10.8.2 Simple Design Example

Suppose that a voltage of 1000 V is required at 5 mA, as the basis of a photomultiplier power supply, and that this is to be derived from a 12 V line:

$$P_L = 1000 \times 0.005 = 5\,\text{W}.$$

Therefore, if the efficiency of the requisite inverter is 80%,

$$I_{C(L)} = \frac{5}{12 \times 0.8} = 0.52\,\text{A}.$$

From equation (10.32),

$$N_P = \frac{V_{CC}}{4 B_{sat} A_{eff} f}.$$

Choosing a commercial toroid with $B_{sat} = 1.5$ tesla, a magnetic field length of 10 cm. and a cross-sectional area of 0.4 cm^2, along with a stacking factor of 0.9 gives,

$$N_P = \frac{12}{4 \times 1.5 \times [(0.4 \times 0.9)/10\,000] \times 1000}$$

$$= 56 \text{ turns for 1000 Hz operation.}$$

The magnetizing current may now be obtained from equation (10.33):

$$I_{C(mag)} = \frac{H_0 l}{N_P}.$$

If the manufacturers' data show that the magnetizing force to fully saturate the core at 1 kHz is about 240 ampere–turns per metre, this becomes

$$I_{C(mag)} = \frac{240 \times 10}{56 \times 100} = 0.43 \text{ A}.$$

Sufficient information is now available to define the transistor:

$$V_{CEM} > 2.5 \times 12 = 30 \text{ V} \quad \text{and} \quad I_{CM} > 0.52 + 0.43 \simeq 1 \text{ A}.$$

The collector winding must carry up to 1 A and the base winding must carry up to $1/h_{FE(O)}$ A. If it is to develop about 6 V and use a limiting resistor, $N_S = (6/12) \times 56 \simeq 28$ turns. The load windings will have $(1000/12) \times 56 = 4667$ turns.

Wire tables must now be consulted to determine what gages to use and the collector windings should be wound together (bifilar winding) in order to ensure maximum coupling and hence minimal overshoot.

10.8.3 Further Notes on the Transistor Inverter

The discussion so far has centered on the common-emitter configuration; that is, the base windings have connected the commoned emitters to the two bases. This is not the only possible configuration[9,10] and the common collector circuit may be recommended where high power dissipations are to be expected in the transistors, for they can be mounted directly on the same heat sink in this configuration. This means that the collectors will be automatically commoned since most power transistors have their collectors mechanically and electrically bonded to the header for heat dissipation reasons.

The two-transistor inverter sometimes fails to start when the supply is switched on, especially when the load is already connected. This is due to a failure by either transistor to switch ON and if reliability is to be obtained, a starting circuit must be incorporated[11].

Starting is most easily achieved by introducing a little forward bias into the base circuit as shown in figure 10.34. A small silicon diode may be fed via the bleed resistor R, and its forward drop applied to the transistor bases as shown. This will ensure reliable starting.

Finally, it must be pointed out that the very simple design equations

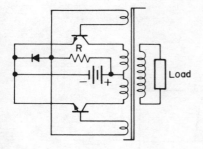

Figure 10.34. Incorporation of starting diode.

utilized have ignored the reduction in the available supply voltage due to resistive drops in the windings. If silicon transistors are used, V_{BE} may easily be of the order of 1 V and with the diode and resistive drops the sum total may approach 2 or 3 V. This is significant if, for example, a 12 V power supply is used, and accurate calculations must clearly take account of it.

10.8.4   The Two-Transformer Inverter

As its name suggests, this form of the transistor inverter, due to Jensen[12], utilizes two separate transformers, and in doing so avoids several of the problems posed by the single-transformer circuit.

The two-transformer circuit is useful primarily in the higher power ranges, for its main advantage is that it economizes in square-loop core material by using a small saturating transformer in the base circuit and a large non-saturating conventional load transformer. This proves economical since square-loop core material is much more expensive than standard transformer iron. Furthermore, in the case of large single-transformer inverters, the heat generated as a result of the iron and copper losses becomes difficult to dissipate.

A representative two-transformer circuit is given in figure 10.35(a) and it should be noted that this is only one of many variations[13,14]. The operation is very similar to the single-transformer circuit. Suppose that Tr1 begins to conduct. The base winding of the load transformer, N_B, applies a voltage to the primary of the saturating transformer N_1 in such a sense that the secondary voltage switches Tr1 hard ON and Tr2 hard OFF. The voltage across N_B is now constant at $V_{CC}(N_B/N_C)$ and the resulting current is limited largely by R. This current is comprised of the base drive current plus the magnetizing current, and the latter increases linearly until the transformer saturates. When this happens the magnetizing current increases rapidly, which it can do only at the expense of the base drive current. Tr1 therefore begins to switch OFF, and the regenerative action assists this process while switching Tr2 ON. The sequence repeats, so producing a square-wave output at the load transformer which, however, never saturates.

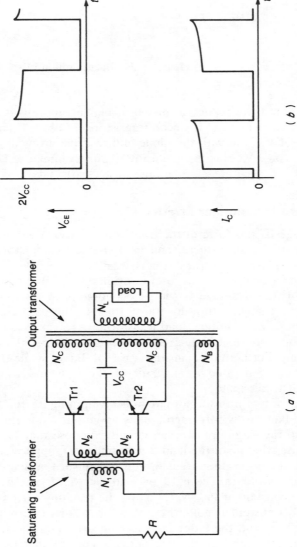

Figure 10.35. (a) Basic two-transformer inverter; (b) wave profiles for two-transformer inverter.

Several advantages accrue from this mode of operation. Since the base current is responsible for saturation, it extracts quite small current pulses from the supply as saturation in each direction is reached. Further, the actual magnitude of the magnetizing current itself is much smaller for the small base-current transformer than was the case for the large collector-current transformer. (It will be recalled that, in the design example, $I_{C(mag)}$ turned out to be of the same order as $I_{C(L)}$.) In the case of the higher power inverters, these are real advantages, for the requirement of high-level bursts of current can embarrass many power supplies.

The fact that the peak current can be limited to little more than $I_{C(L)}/h_{FE(O)} + I_{mag}$ means that, when that, when interrupted by the transistor switching OFF, only a small voltage spike is induced. The result is the very satisfactory set of wave profiles shown in figure 10.35(b).

Looked at another way, R can be said to swamp the impedance changes presented by the saturating transformer to the output transformer and supply, thereby reducing the resultant transients.

The calculation of the correct number of turns for N_2 and hence N_1 is complicated by the existence of R and the self-inductance of the load transformer, because the basic design equation (10.32) assumed a constant applied voltage which is no longer a valid assumption. The accurate expression is quite complex and contains factors not usually known, so it is recommended that a crude approximation be made by inserting a factor of 8 instead of 4. This arises if it is assumed that the impedance in the circuit due to the extra items is equal to the reactance of the saturating transformer.

The output transformer is designed in the manner described for normal audio-frequency output transformers. If a sharp-cornered square wave is to be maintained, this transformer must have a good frequency response.

The effect of loads other than pure resistances is of importance from the point of view of power dissipation, but since the topic is of general interest in power switching it will be considered under a separate heading.

10.9   POWER SWITCHING

So far, the discussions on various types of transistor switching circuits have tacitly assumed the presence of resistive loads. In the case of the noninductive regenerative circuits this load has taken the form of a collector resistor only. It is therefore necessary to consider the effect of loading a switching transistor, not only resistively, but inductively, which presents a more complex, and also a more common problem.

If a load is to be applied to one or both of the transistors in a regenerative pair, care must be taken that this load does not adversely affect the dwell time, or even the regeneration capability; that is, the load current must either be very small compared with the collector current or the collector current must actually *be* the load current. The former case usually involves the addition of a buffer stage which applies only a low loading to the output transistor while still providing a high working current to the load itself.

Most loads for power switching transistors take the form of relays or other electromechanical devices having predominantly inductive impedances. It has already been mentioned that a high voltage spike appears when the current in an inductive circuit is broken, and that this spike can lead to the breakdown of a transistor.

If the wave profile of such a spike is examined, it is found to consist of a high-frequency exponentially damped sinusoid. This sinusoid is caused by the resonance of the *LCR* circuit formed by the inductance of the coil with its stray, or winding, capacitance and its inherent resistance. The exponential damping is due to the gradual dissipation in the resistance of the stored energy $(\frac{1}{2}LI_C^2)$. The first few half-cycles of the voltage wave may have very high amplitudes and these must be restricted to levels safe for any associated transistor.

The best way of protecting the transistor is to connect a diode across the inductive load as shown in figure 10.36. Under normal ON conditions, the diode is reverse biased by the voltage drop across the load and so is nonconducting. When the transistor switches OFF, the voltage spike rises in such a sense as to bring the diode into conduction, which restricts its magnitude to something less than a volt and allows the stored energy to dissipate rapidly. (The sense of the spike can be determined very easily by invoking Lenz's law, which states that inductance tends to *prevent* changes in current. In the case of figure 10.36 this acts to maintain the current through the load when the collector becomes an open-circuit, and this leads to a circulating current in the direction shown by the arrow, which obviously produces a *forward* drop across the diode.) This so-called *fly-back* diode should be chosen so that it is capable of carrying the current pulses involved. Conversely, the diode itself can be protected by connecting a current limiting resistor in series with it.

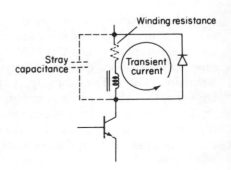

Figure 10.36. Inductive load with spike-limiting diode.

The tendency of an inductance to prevent changes in current produces a distortion of the load-line on an output characteristic as shown in figure 10.37(a). Here, the inductance acts to prevent the current from rising rapidly during switch-ON; or falling rapidly during switch-OFF. The switch-OFF path will be seen to take the transistor into areas of high instantaneous power dissipation, which can be dangerous, especially at high repetition rates. This situation has been met in the case of inverters, and a typical operating path is shown in figure 10.37(b). It will be recalled that the OFF transistor is subjected to a voltage of $2V_{CC}$, and this is represented by point A. When the voltage falls the working point moves to B, after which the transistor begins to conduct. This is represented by a movement up the DC load-line to C. (Strictly speaking, the path would be slightly concave below the load-line.) At this point the transistor is saturated, but the current continues to rise until the transformer core saturates and the base drive is eventually cut off. This happens at point D, and as the transistor switches OFF, the inductance of the transformer takes the operating path through the region of high dissipation and back to point A via the maximum point of the voltage spike E.

It will now be apparent from both the material above and that relating to the power amplifiers of Chapter 9, that a power transistor must never be operated outside a "safe" region on the output characteristic family of curves. Such a region is called the *safe operating area* or SOA, and is depicted in figure 10.38. This extends the concept of the working area of figure 9.1(a) to include the second breakdown limitation depicted in figure 9.1(b); and also takes account of permissible fast transients across the $>P_{tot}$ region.

In the case of some high-power transistors, the maximum current may be limited by the current-carrying capacity of the lead wire to the emitter even before the transistor itself becomes liable to damage, but either mechanism is taken into account by limiting line A in figure 10.38.

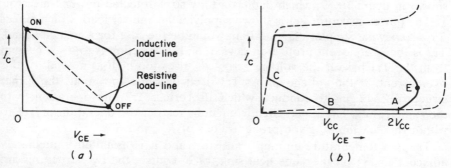

Figure 10.37. (a) Effect of inductive load; (b) load-line for inverter circuit.

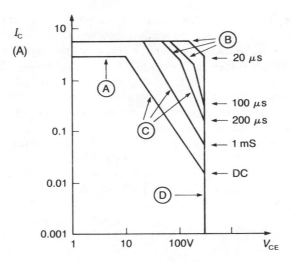

Figure 10.38. The safe operating area, or SOA.

A locus defining the onset of second breakdown can be drawn, and this will often reduce the area of operation below that defined by heat dissipation problems. In figure 10.38, this is shown as limiting line C, whereas B encompasses simply the P_{tot} limit. As has been mentioned, fast transients may exceed these limits, and this is taken into account in the diagram by the limiting lines over and above the DC case, wherein the speed of such transients is also limited.

The usual maximum voltage above which avalanche breakdown may occur, is given by limiting line D. (This is the so-called "first breakdown" mechanism which gives second breakdown its name).

A transistor may be constrained to operate within its prescribed limits by appropriate circuits techniques, and this is called "SOA protection." Over-current (sometimes called overload) protection has already been met in operational amplifier output stages (figure 7.11), and this technique is also shown in figure 10.39, which depicts a fully SOA-protected power transistor.

The power structure Tr1 is shown supplying an emitter load, which makes for a somewhat simpler diagram than would be the case for a collector load. Tr2 is the overcurrent protection device, and its function is to draw current from the Tr1 base if the voltage across R_E rises too high as a result of this overcurrent. If the voltage across Tr1 itself becomes too great, the Zener structure Z_1 begins to conduct, which also brings Tr2 into operation, with the same result. These two circuit functions restrict the operation of Tr1 to within the SOA limits, given proper values of R_{Z_1} and R_E.

Tr3 is a temperature-limiting transistor, and it is mounted immediately adjacent to Tr1 on the same heat sink (or of course, the two structures are oriented next to each other in the monolithic case). Should the temperature of Tr1 rise to an unacceptable level for any reason—either internal power

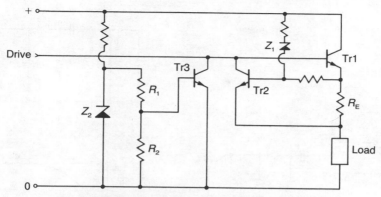

Figure 10.39. An SOA-protected power transistor.

dissipation, ambient temperature rise or both—then Tr3 begins to extract current from the Tr1 base. This is accomplished by choosing the value of the voltage reference structure Z_2 and its potential divider R_1 and R_2 so that Tr3 is normally in a nonconducting state; that is, V_{BE3} is held to 0.4 V or so. However, because V_{BE} falls with increasing temperature (at about -2 mV/°C), then at a sufficiently high temperature, Tr3 will conduct and so remove base current from Tr1.

10.10 CAPACITIVE LOAD

Although rare compared with inductive loads, the capacitive load does occur in a few circuits, notably where a capacitor is connected directly across the output from a power supply source for smoothing purposes. This particular application will be considered in detail in a later chapter and the present section will be confined to comparatively small signal work.

An example of how a capacitive load may arise is given by the circuit of figure 10.40(a). This is the well-known *diode pump* or *cup and bucket* circuit. Another reason for its inclusion here is that it is essentially a frequency-to-voltage convertor, which is the antonym of the transistor inverter circuit. This characteristic has many uses, including frequency measurement, pulse rate measurement (e.g., in nucleonics where pulses from Geiger, proportional, or scintillation counters are accepted) and the demodulation of FM signals in communications.

Figure 10.40(b) shows a suitable model for the analysis of the circuit. When the switch is at position a, this corresponds to the ON state of the transistor, and the height of the input pulse is nearly V_{CC}. Capacitors C_P and C_R receive equal charges providing that $V_{CC} > v_{out}$ so that D_2 conducts. At the end of the pulse the input voltage drops to zero, this condition being represented by switch position b. The voltage across C_P is now in such a direction as to bring D_1 into conduction, and a circulating current flows until

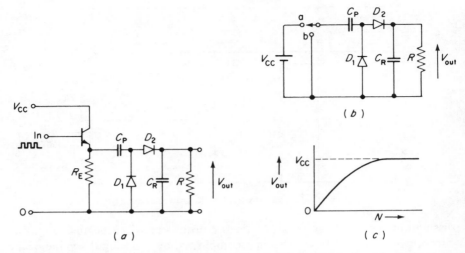

Figure 10.40. (*a*) Diode pump circuit; (*b*) model for analysis; (*c*) diode pump characteristic.

C_P is discharged. The diode D_2 is nonconducting during this time, so C_R discharges via R.

If the pulse rate is great enough a continuous current flows in R, producing a voltage drop which is a function of this pulse rate:

charge on C_P (and C_R) due to each pulse $= q = C_P(V_{CC} - v_{out})$.

If the pulse rate is N pulses per second, then the current input equals the current output or

$$NC_P(V_{CC} - v_{out}) = I_R = v_{out}/R$$

giving

$$v_{out} = V_{CC}\left(\frac{NC_P R}{1 + NC_P R}\right) = V_{CC}\left(\frac{1}{(1/NC_P R) + 1}\right). \qquad (10.34)$$

This equation is plotted in figure 10.40(*c*). When $N = 0$, $v_{out} = 0$, and when N is very large $v_{out} = V_{CC}$.

It is apparent that v_{out} is directly proportional to N only when N is very small. This limited range can be extended by making C_R very much larger than C_P, for this technique will ensure that the voltage across C_R will increase very little with each pulse and so the incoming charges will differ very little. It is also possible to change the range of the circuit by adjusting the value of R.

The nonlinearity of the circuit is clearly due to the fact that, as v_{out} builds up, each incoming pulse charges C_P and C_R to a lesser degree, since $q = C_P(V_{CC} - v_{out})$.

Linearity may be achieved if the voltage at the diode side of the reservoir capacitor is kept to a negligibly low value. This can be accomplished if C_R is replaced by an operational amplifier having a feedback capacitor. This *Miller integrator* has been mentioned in the discussion relevant to figure 7.25(c). In this application, which is depicted in figure 10.41, it may be assumed that a virtual earth point exists at VE if the voltage gain of the amplifier is so great that v_{in} is very small when v_{out} is large. This means that from the point of view of the input circuit the apparent value of C_R is in fact $(1 - A)C_F$, which can be very much larger than C_R. This "apparent" capacitance charges up linearly, but slowly. The voltage across it is amplified by an amount A, and so has the same value as before but varies linearly with the PRF.

To show that the "apparent" reservoir capacitor has a value $C_F(A - 1)$ it is necessary only to draw the equivalent circuit for the Miller integrator (figure 10.42). The input voltage, v_{in}, is amplified by an amount A (where A has a negative numerical value). Consequently, the voltage across the capacitor is:

$$v_{in} - v_{out} = \frac{1}{C_F} \int i \, dt .$$

But

$$v_{out} = Av_{in}$$

so

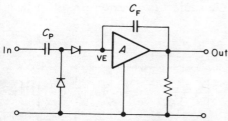

Figure 10.41. Linearized diode pump.

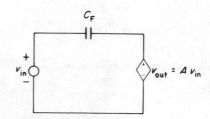

Figure 10.42. Equivalent circuit for Miller integrator.

$$v_{in}(1 - A) = \frac{1}{C_F} \int i \, dt$$

or

$$v_{in} = \frac{1}{C_F(1 - A)} \int i \, dt.$$

Comparing this with $v_{in} = 1/C_F \int i \, dt$ shows that the effective capacitance is $C_F(1 - A)$, which is much greater than C_F because A has a very large negative value.

The operational amplifier can take a very simple form if extreme linearity is not required, and one transistor will often suffice.

Note that for both the nonlinear and linear versions of the diode pump circuit the load resistor may be omitted, in which case the voltage across the reservoir capacitor will build up in a series of steps. This could form the basis of a pulse counter, for a trigger circuit could be operated at some output voltage corresponding to N pulses. The pulses themselves must, of course, be of equal heights and widths, and this can be ensured by triggering a monostable circuit with nonuniform input pulses, then feeding the output to the diode pump circuit.

This is the sort of function which can be more easily accomplished using digital integrated circuits, and so it will not be pursued in this volume. However, in conclusion, it is worth noting that if in figure 10.41 the load resistor is connected across the shunt diode and the op. amp. section is omitted altogether, then the simple *diode clamp* of figure 10.43(*a*) results. Here, the positive-going portion of a bipolar signal is always short-circuited by the diode, so that the capacitor eventually charges to the mean of the input voltage and the output appears as a negative-going signal only. If the diode is inverted, so is the output signal, as is shown in figure 10.43(*b*). Note that this circuit is useful only for signals much greater than a diode volt drop, for the output waveform will never reach zero but will always appear to sit upon a pedestal of this magnitude. To achieve a genuine reference to zero voltage, it is necessary to use an operational amplifier with a diode in the feedback loop. Such circuits can also be used to produce logarithmic amplifiers, and are an introduction to the subject of *function circuits*, which is a large topic in its own right[15].

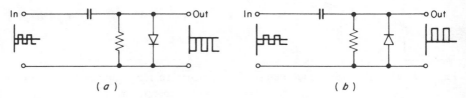

Figure 10.43. Diode clamp circuits: (*a*) negative; (*b*) positive.

REFERENCES

1. Sparkes, J. J., 1959, The measurement of transistor transient switching parameters, *Proc. IEE, B*, **106**, *Suppl.* 15, 562–567.
2. Beaufoy, R., 1959, Transistor switching circuit design using charge control parameters, *Proc. IEE, B*, **106**, *Suppl.* 17, 1085–1091.
3. Sparkes, J. J., 1960, A study of the charge control parameters, *Proc. IRE*, October.
4. den Brinker, C. S. and Fairbairn, D., 1963, An analysis of the switching behaviour of graded base transistors, *Electron. Eng.*, August.
5. Graeme, J. G., Tobey, G. E. and Huelsman, L. P., 1971, *Operational Amplifiers* (New York: McGraw-Hill).
6. Hnatek, E. R., 1980, *Design of Solid State Power Supplies* (New York: McGraw Hill).
7. Pearman, R. A., *Power Electronics*, 1980 (Reston, VA: Reston).
8. Watson, J., 1968, The design of stabilized power supplies for incandescent light sources, *IEEE Trans. Instrum. Meas.*, **IM-17**, No. 2 March.
9. Bell, J. S. and Wright, P. G., 1961, The choice and design of d.c. convertors, *Electron. Eng.*, April.
10. Fleming, G. C., 1959, Transistors, and saturable-core transformers as square wave oscillators, *Electron. Eng.*, September.
11. Stephenson, W. I., Morgan, L. P. and Brown, T. H., 1959, The design of transistor push–pull d.c. convertors, *Electron. Eng.*, October.
12. Jensen, 1947, An improved square-wave oscillator circuit, *IRE Trans. Circuit Theory*, September.
13. Nowicki, J. R., 1961, Improved high power d.c. convertors, *Electron. Eng.*, October.
14. Pelc, K. and Tove, P. A., 1963, Two-transformer d.c. to a.c. power convertors, *Electron. Eng.*, November.
15. Wong, Y. L. and Ott, W. E., 1976, *Function Circuits* (New York: McGraw-Hill).

QUESTIONS

1. What are the advantages and disadvantages of a solid-state switch compared with an electromechanical device?

2. Why does a fast solid-state switch produce less internal heat than a slower one in the same application?

3. Explain the meaning of the terms "delay time" and "storage time" in relation to a switching transistor.

4. Why is the class D power amplifier stage more efficient then either the class A or class B for similar load power?

5. Briefly describe the three basic two-transistor regenerative switch circuits.

6. With the aid of a diagram, explain how a totem-pole power buffer circuit operates.

7. What is a Schmitt trigger circuit?

8. Describe both a zero-crossing and a hysteresis comparator, and draw circuits for both using a purpose-designed IC which requires no limiting circuitry and which operates from ±15 V supplies.

9. What is a "blocking oscillator?"

10. Why is square-loop, easily saturable magnetic core material useful in inverter designs?

11. For what is a "fly-black" (or "free-wheeling") diode used?

12. What is meant by the "SOA" for a power transistor?

13. Describe the diode pump circuit.

14. What is a "diode clamp" circuit?

PROBLEMS

10.1. A power transistor is connected as a CE switch with a resistive load of $3.3\,\Omega$ and operates from a 12 V supply. A 5 kHz square wave of $1:3$ mark/space ratio is applied to the base and has sufficient magnitude to just ensure saturation, under which conditions the following parameters apply:

$$V_{CES} = 0.5\,\text{V} \quad I_{CEO} = 0.1\,\text{mA} \quad t_d = 0.05\,\mu\text{s} \quad t_r = 1\,\mu\text{s}$$
$$t_s = 4\,\mu\text{s} \quad t_f = 0.7\,\mu\text{s}.$$

What is the average power dissipated in (i) the load, and (ii) the transistor?

10.2. The comparator of a pulse-width-modulated amplifier receives a 100 kHz sawtooth wave of magnitude ±5 V; and a 100 Hz sine wave of magnitude ±1 V peak-to-peak (see figure 10.12). Neglecting any delays in the comparator, what are the shortest and longest widths of the output pulses?

10.3. A multivibrator circuit like that of figure 10.13(*a*) is to be designed for a PRF of 100 Hz. If a 12 V power supply is available, along with transistors having values of $h_{FE(o)} = 50$, calculate the values of collector and base resistors for saturation currents of 5 mA. What are the timing capacitor values? (Assume a $1:1$ mark/space ratio.)

10.4. One unit of an LM339 is operated from a single $+5\,\mathrm{V}$ supply. The noninverting input is taken to the junction of two $47\,\mathrm{k\Omega}$ resistors which are connected across the same supply. What is the value of the feedback resistor which must be connected from the output to this noninverting input if the comparator is to switch over when a voltage applied to the input rises to $+3\,\mathrm{V}$ or falls to $+2\,\mathrm{V}$? (Assume the existence of an appropriate load resistor.)

10.5. A 555 timer is connected as an astable circuit to deliver a square wave of $1:1$ mark/space ratio at $483\,\mathrm{Hz}$. Draw the appropriate circuit, using a $15\,\mathrm{V}$ supply, and calculate values for the timing components R_A, R_B and C_t.

10.6. In the circuit shown in figure 10.44, pin 7 of the 555 timer module normally presents a short circuit to ground. When the push-button is operated pin 7 presents an open circuit until the voltage at pin 6 rises above $\frac{2}{3}V_{CC}$, when it reverts to a short circuit again.

  (i) Draw the shape of the pulse which appears at the output when the push-button is operated, and calculate its height and duration.

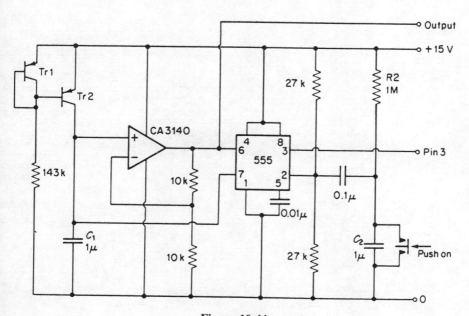

Figure 10.44.

(ii) How could the duration of the output pulse be doubled by adding one extra component? Calculate the value of this component. How could it be doubled by removing one component?

(iii) What is the polarity of the trigger pulse which appears at pin 2 when the push-button is operated, and why does it appear at all? Take $V_{BE} = 0.7$ V and $kT/q = 26$ mV at 25°C.

(iv) What pulse height and length appears at pin 3?

11

The Field-Effect Transistor— Discrete and Integrated

It has been seen in earlier chapters that the CE-connected bipolar transistor is a low input resistance device when operating at and above moderate collector currents; that is, currents of the order of 1 mA upwards. When $|I_C|$ enters the microamp region, however, the input resistance can increase to the order of a megohm. Whether the bipolar transistor is thought of as fundamentally a current- or a voltage-driven device therefore depends on how, under the prevailing operating conditions, the input generator resistance compares with R_{in}. The field-effect transistor, however, always presents a very high value of R_{in}—10^{10} Ω upwards, in fact—when connected in the analogous common-source mode. For this reason it may be thought of as a true voltage amplifier for all values of input generator resistance normally encountered.

The field-effect transistor consists essentially of a *channel* of semiconductor material having ohmic contacts at each end called the *source* and *drain*. Formed along this channel is either a junction, or a metal film insulated from the channel itself. The former is variously called a junction FET, a unipolar transistor, or simply a FET. The latter is termed an insulated gate FET, or IGFET; and often, more specifically, a metal oxide-semiconductor transistor, or MOST. In this book, the terms FET and IGFET will be used.

11.1 THE JUNCTION FET

Figure 11.1 shows an idealized cross-section through an n-channel FET. Here, the n-type channel has been formed in a p-type substrate, and a further p-type diffusion has been made, so forming the gate region. (Actually, an n-layer is usually deposited epitaxially, and the acceptor impurity is diffused in to form not only the gate, but also the p-type isolating regions.)

421

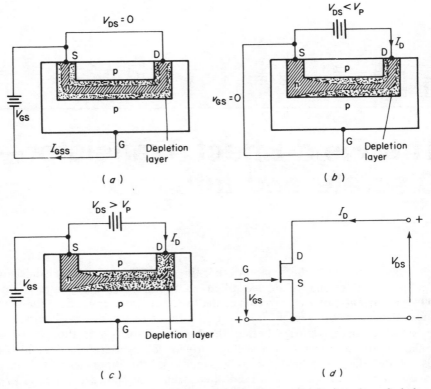

Figure 11.1. The n-channel junction FET (idealized cross-section) and symbol show-ing polarities.

If the gate is biased negatively with respect to the channel, a reverse-biased junction results, and a depletion layer forms at this junction. Figure 11.1(a) shows this depletion layer penetrating into the n-channel but has been omitted in the p-substrate and gate for reasons of clarity. The ends of the channel, the source and drain, are shown short-circuited, and under these conditions only the usual temperature-dependent reverse leakage current I_{GSS} flows. For a low-leakage junction FET, I_{GSS} can be of the order of 1 pA (10^{-12} A) at 25°C.

If the gate voltage, V_{GS}, increases (negatively), the depletion layer increases in thickness until the channel is almost completely depleted. This effect may also be achieved by short-circuiting the source to the gate and applying a drain-source voltage, V_{DS}, as in figure 11.1(b). A *drain current* now flows along the channel, and a voltage drop is produced between the n-channel and the p-gate/substrate which rises from zero near the source to a maximum near the drain, so that the depletion layer thickness increases as shown. If V_{DS} is increased, the drain current also rises until the depletion layer so constricts or *pinches-off* the channel that very little further current rise can occur. When this state of equilibrium has been reached, $V_{DS} = V_P$, the *pinch-off voltage*, and $I_D = I_{DSS}$, the *saturation current*.

When the FET is working in the *saturation region*, I_D cannot be significantly increased, but it *can* be decreased by applying a gate voltage as in figure 11.1(c). When V_{GS} is of the same magnitude as V_P (but of opposite sign), I_D will be reduced to zero. Hence, when $|V_{GS}| = |V_P|$, it is called the *gate pinch-off voltage*.

The foregoing discussion explains the shapes of both the output and transconductance characteristics of figures 11.2 and 11.3. These curves have

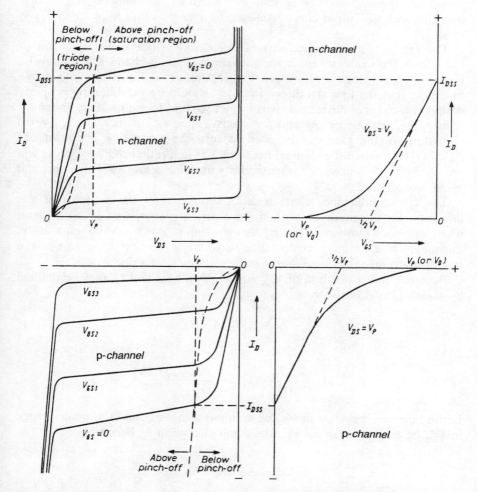

Figure 11.2. Output characteristics for junction FETs.

Figure 11.3. Transconductance characteristics for junction FETs.

been sketched for both n-channel and p-channel devices, and it will be observed that all polarities are reversed for the p-channel FET compared with those for the n-channel.

The incremental channel resistance, r_{ds}, is the inverse slope of an output curve at any point and it will be seen that in the saturation region ($V_{DS} > V_P$) such curves approach the horizontal, so that r_{ds} has a very high value.

Below pinch-off, however, the output characteristics exhibit differing slopes, depending on the value of V_{GS}, and in particular near the origin, the incremental resistance becomes identical with the chord resistance. That is, when $V_{DS} \to 0$, then $r_{ds} = r_{DS}$. This means that below pinch-off, the channel becomes a *voltage-controlled resistor* (VCR), and in fact is used as such in numerous applications. (Note that this attribute is absent in the bipolar transistor, whose output curves coalesce to a single value R_{CES} when V_{CE} is small.)

The FET is basically a majority-carrier device, since either electrons *or* holes form the main current in the n-channel or the p-channel, respectively. Hence, because electrons have greater mobilities than holes, it is reasonable to expect that the lowest values of $r_{DS(ON)}$ should be exhibited by properly designed n-channel devices. This is in fact so, and for some large geometry units $r_{DS(ON)}$ can be as low as a few ohms. (Here $r_{DS(ON)}$ is r_{DS} at $V_{GS} = 0$.)

If desired, the gate can be used to turn the FET ON ($V_{GS} = 0$) or OFF ($|V_{GS}| > |V_P|$), when the channel becomes a switch rather than a VCR. As will be seen later, the FET switch is particularly useful as a low-level chopper, and as an analog gate.

The transconductance characteristics show that as the modulus of V_{GS} increases, so the modulus of I_D falls. Note, however, that in both cases, when V_{GS} goes more positive, I_D also goes more positive. Mathematically, this means that the transconductance curve has a positive slope, so that the transconductance itself, g_{fs}, has a positive numerical value.

Because I_D is a function of V_{GS} and V_{DS}, both g_{fs} and r_{ds} may be defined by taking partial derivatives:

$$I_D = f(V_{GS}, V_{DS})$$

giving

$$\delta I_D = \frac{\partial I_D}{\partial V_{GS}} \delta V_{GS} + \frac{\partial I_D}{\partial V_{DS}} \delta V_{DS}.$$

Letting the small change in V_{GS} be an input signal v_{gs}, and the small change in V_{DS} be an output signal v_{ds}, the small change in I_D becomes

$$i_d = g_{fs} v_{gs} + \frac{1}{r_{ds}} v_{ds}$$

where

$$g_{fs} = \frac{i_d}{v_{gs}} \; (v_{ds} = 0) \quad \text{(i.e., output short-circuited)}$$

and

$$r_{ds} = \frac{v_{ds}}{i_d} \; (v_{gs} = 0) \quad \text{(i.e., input short-circuited)} .$$

(Note that g_{fs} is sometimes written g_m by analogy with the mutual conductance of a bipolar transistor).

For the junction FET, the following expression[1] closely describes the transconductance curve:

$$I_D = I_{DSS} \left(1 - \frac{V_{GS}}{V_P} \right)^2 . \tag{11.1}$$

The index of the bracketed term actually lies between about 1.9 and 2.1.

Figure 11.3 indicates that V_P is rather an arbitrary voltage because it occurs where the true transconductance curve becomes asymptotic to the V_{GS} axis. For this reason, manufacturers often quote V_P as that gate-source voltage which reduces I_D to some small value, typically 10 nA. However, a more satisfactory definition is afforded by equation (11.1), from which g_{fs} may be concurrently derived:

$$g_{fs} = \frac{\partial I_D}{\partial V_{GS}} = -\frac{2I_{DSS}}{V_P} \left(1 - \frac{V_{GS}}{V_P} \right) . \tag{11.2}$$

If $V_{GS} = 0$, that is, $I_D = I_{DSS}$, g_{fs} becomes

$$g_{fso} = \frac{-2I_{DSS}}{V_P} \quad \text{or} \quad \frac{-I_{DSS}}{\frac{1}{2}V_P} . \tag{11.3}$$

This is the tangent to the curve at $I_D = I_{DSS}$ and, if produced, this tangent will obviously cut the abscissa at $\frac{1}{2}V_P$ so defining the position of V_P. This is shown in figure 11.3.

When the applied channel voltage, V_{DS}, becomes sufficiently high, breakdown occurs and the output curves rise very sharply at a characteristic value of V_{DS}. For $V_{GS} = 0$, the breakdown voltage is termed BV_{DSS}, which implies that the source is short-circuited to the gate (by the usual convention). Actually, the mechanism involved is that of junction breakdown, because for this short-circuited source-gate condition $BV_{DSS} = BV_{DGS}$. Often BV_{GSS} is quoted, which is a more direct way of defining the gate-to-channel junction breakdown voltage.

11.1.1   The Equivalent Circuit

It has been tacitly assumed in the discussion above that the FET has been connected in the common-source (CS) configuration; that is, with the gate acting as the signal input point and the drain as the output point. This, along with the common-drain (CD) connection, is the most useful configuration in low- and medium-frequency work, while the common-gate mode (like the common-base) has applications mainly at the higher frequencies.

The small-signal equivalent circuit of figure 11.4 reflects these premises and is relevant to the pinched-off region only. The transconductance term, g_{fs}, appears in the output current generator, and the direction of this dependent current is such as to conform with the usual "positive-in, negative-out" convention. The channel resistance appears in shunt with the current generator, as does the associated channel capacitance, C_{ds}. (In fact, C_{ds} is largely the stray header capacitance and is therefore very small, typically less than 0.4 pF.)

Because the input resistance r_{gs} is that associated with a reverse-biased junction, it has an extremely high value. The input capacitance C_{gs} is therefore the dominant parameter, and this is usually of the order of a few picofarads.

The FET is not a unilateral device, even at moderate frequencies, and this fact is demonstrated by the existence of the gate-drain capacitance C_{gd}, which is also usually a few picofarads. Clearly, because both C_{gs} and C_{gd} are the depletion capacitances of a reverse-biased junction, they are both highly dependent on the junction bias voltage. As this voltage increases, causing the depletion layer to become thicker, so the values of C_{gs} and C_{gd} decrease.

For convenience in measurement, manufacturers usually specify the capacitance at the input with the output short-circuited, C_{iss}, and that at the output with the input short-circuited, C_{oss}. An inspection of figure 11.4

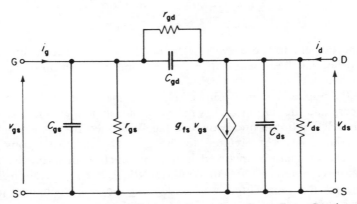

Figure 11.4. FET equivalent circuit where $C_{gd} = C_{rss}$, $C_{ds} + C_{gd} = C_{oss}$ (or C_{dss}) and $C_{gs} + C_{gd} \simeq C_{iss}$ (or C_{gss}).

shows that $C_{iss} = C_{gs} + C_{gd}$ and $C_{oss} = C_{ds} + C_{gd} \simeq C_{gd}$ because C_{ds} is very small.

From these two relationships it is clear that

$$C_{gs} \simeq C_{iss} - C_{oss} \, .$$

(Note that the capacitance at the gate with the source short-circuited to the drain, C_{iss}, may also be written C_{gss}, and that the reverse transfer capacitance with the gate short-circuited to the source may be written C_{rss}.)

The gate-drain resistance r_{gd}, being associated with a reverse-biased junction, is very high, like r_{gs}, and may also often be ignored for this reason.

11.1.2 Typical FET Parameters and their Variations

Table 11.1 gives a list of parameters typical of a small-signal n-channel FET. Notice that I_{DSS} is not referred to the condition where $V_{DS} = V_P$ according to the definition implied by figures 11.2 and 11.3, but instead quotes a V_{DS} of 20 V. Clearly, this value of I_{DSS} will be somewhat higher than that when $V_{DS} = V_P$, because the output characteristics slope upwards slightly in the saturation or pinched-off region. This is, however, a common practice which does give a realistic value for I_{DSS} relevant to the actual working conditions of the device.

The pinch-off voltage is defined in table 11.1 for the E102 as that gate voltage which brings I_D down to 10 nA when $V_{DS} = 20$ V.

The value of g_{fs}, 1 mA/V, is quoted for the upper end of the transconductance curve where $V_{GS} = 0$. The value for any other working point may be obtained from equation (11.2).

Like all semiconductor devices, the FET has parameters which vary with temperature. The effect on the transconductance curve is shown in figure 11.5, from which it will be seen that the change in I_D is negative for low values of V_{GS} and becomes positive for high values of V_{GS}. This means that it passes through a stable point, and this has led to the concept of zero drift biasing.

The reasons for the change-over in drain-current temperature drift polari-

TABLE 11.1. Parameters for the Siliconix E102 Epoxy Encapsulated n-Channel FET (reproduced by permission).

I_{DSS} (mA)	V_P (V)	g_{fso} (μA/V)	r_{ds} (Ω)	C_{rss} (pF)	C_{iss} (pF)
$V_{DS} = 20$ V	$V_{DS} = 20$ V $I_{D(off)} = 10$ nA	$V_{DS} = 20$ V $f = 1$ kHz	$V_{DS} = V_{GS} = 0$ $f = 1$ kHz	$V_{DS} = 20$ V $f = 1$ MHz	$V_{DS} = 20$ V $f = 1$ MHz
min max	min max	min max	min max	min max	min max
0.9 4.5	-0.8 -4.0	1000	1200	3	8

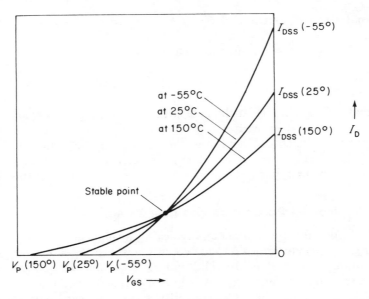

Figure 11.5. Temperature variation of transconductance curve for a high pinch-off FET.

ty are twofold. Firstly, the channel resistivity exhibits a positive temperature coefficient because of the decrease in mobility of the charge carriers as the temperature rises; and secondly the potential drop across the junction depletion layer exhibits the usual negative temperature coefficient of about $-2.2 \, \text{mV}/°\text{C}$. This latter effect results in a rise in I_D with temperature, and the stable point exists where the two phenomena cancel out. (The FET in question must, however, exhibit a V_P of more than a volt or so for this explanation to be valid.)

Temperature variations in the capacitances associated with the junction are quite marked, as are the variations resulting from changes in reverse voltage. For these reasons, manufacturers' data usually incorporate graphs delineating these variations.

The relationship between the geometry of a FET and its parameters is such that a high I_{DSS} will be associated with a unit of high V_P, and vice versa. This situation simplifies the biasing problem as explained below.

11.1.3 Biasing the FET

Figure 11.6 shows a very simple and common method of biasing the FET for either CS or CD operation. Here, the gate bias voltage V_B is held constant by R_1 and R_2 (I_G being very small) so that the voltage drop in R_S due to I_D acts as a DC negative feedback input. Figure 11.7 (which has been drawn in the

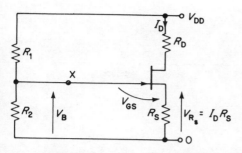

Figure 11.6. The series DC feedback bias method.

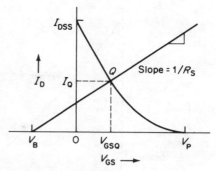

Figure 11.7. The bias line.

first quadrant to represent the general case for either p- or n-channel devices) describes this situation.

Summing bias voltages round the input loop gives

$$V_{GS} = R_S I_D - V_B . \tag{11.4}$$

This straight-line equation is plotted on the transconductance curve of figure 11.7 and is known as the *bias line*. The inverse slope of the bias line gives R_S, and it crosses the abscissa at V_B. The operating or quiescent point for the FET is where the bias line crosses the transconductance curve. Here, $I_D = I_Q$ and $V_{GS} = V_{GSQ}$.

If figure 11.7 is extended to cover the two transconductance curves representing the limiting parameters $I_{DSS(min)}$, $I_{DSS(max)}$, and $V_{P(min)}$, $V_{P(max)}$, then the resultant diagram becomes that of figure 11.8. Here, the bias line defines the upper and lower limits of I_Q and V_{GSQ}, and from the geometry of the diagram, the following relationships may be simply derived[2]:

$$R_S = \frac{V_{GSQ(max)} - V_{GSQ(min)}}{I_{Q(max)} - I_{Q(min)}} = \frac{\Delta V_{GSQ}}{\Delta I_Q} \tag{11.5}$$

and

$$V_B = \frac{V_{GSQ(min)} I_{Q(max)} - V_{GSQ(max)} I_{Q(min)}}{I_{Q(max)} - I_{Q(min)}} . \tag{11.6}$$

These equations mean that if $I_{Q(max)}$ and $I_{Q(min)}$ are known, a complete bias design may be performed. This is because equation (11.1) may be transformed to give V_{GSQ} in terms of I_Q, knowing I_{DSS} and V_P:

$$V_{GSQ(min)} = V_{P(min)} \left[1 - \left(\frac{I_{Q(min)}}{I_{DSS(min)}} \right)^{1/2} \right] \tag{11.7a}$$

and

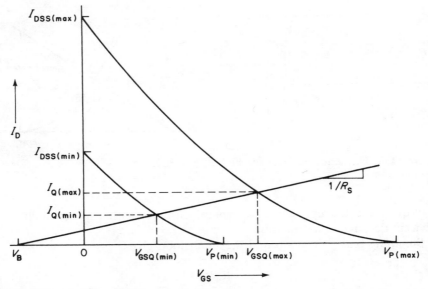

Figure 11.8. Bias line superimposed on limiting tolerance transconductance curves.

$$V_{\mathrm{GSQ(max)}} = V_{\mathrm{P(max)}}\left[1 - \left(\frac{I_{\mathrm{Q(max)}}}{I_{\mathrm{DSS(max)}}}\right)^{1/2}\right]. \qquad (11.7b)$$

Normally, ΔI_{Q} is in fact known, for the permissible variation in V_{Q} (the quiescent value of V_{DS}) clearly limits the maximum output signal swing of the stage. A simple design example will illustrate this.

11.1.4   Simple Bias Design Example

Consider the case of a Siliconix E102 used in a cs stage with an 18 V supply. Assume that the maximum permissible variation in V_{Q} is 2.0 V, centered at about 14 V. Because the E102 has an $I_{\mathrm{DSS(min)}}$ of 0.9 mA, a reasonable value for I_{Q} is 0.5 mA.

Hence,

$$R_{\mathrm{D}} = \frac{V_{\mathrm{DD}} - 14}{I_{\mathrm{Q(nominal)}}} = \frac{18 - 14}{0.5} = 8\,\mathrm{k\Omega}.$$

The nearest preferred value resistor is

$$R_{\mathrm{D}} = 8.2\,\mathrm{k\Omega}$$

which leads to

$$\Delta I_Q = \Delta V_Q / R_D = 2.0/8.2 \simeq 0.24 \text{ mA} \,.$$

This means that I_Q may take a value between 0.38 and 0.62 mA, so that equations (11.7) give

$$V_{\text{GSQ(min)}} = 0.8\left[1 - \left(\frac{0.38}{0.9}\right)^{1/2}\right] = 0.28 \text{ V}$$

and

$$V_{\text{GSQ(max)}} = 4.0\left[1 - \left(\frac{0.62}{4.5}\right)^{1/2}\right] = 2.51 \text{ V}$$

giving

$$\Delta V_{\text{GSQ}} = 2.51 - 0.28 = 2.23 \text{ V} \,.$$

Equation (11.5) now gives R_S:

$$R_S = \Delta V_{\text{GSQ}} / \Delta I_Q = 2.23/0.24 \simeq 9.1 \text{ k}\Omega$$

and equation (11.6) gives V_B:

$$V_B = \frac{0.28 \times 0.62 - 2.51 \times 0.38}{0.24} = 3.25 \text{ V} \,.$$

To obtain a bias voltage of 3.25 V, R_1 may conveniently be 150 kΩ and R_2 may be 33 kΩ.

Note that R_1 and R_2 define the input resistance; if a large R_{in} is required, a suitable resistor may be inserted at point X. This modification will be discussed later.

The accuracy of this method of biasing calculation is dependent on the fit of the theoretical square-law curve to the actual transconductance characteristic of the device. In point of fact, the crux of this matter is the value of drain current at which V_P is measured. It is unlikely that the manufacturer takes this point into account, but if his specified I_D is lower than the optimum (that is, much less than $I_{\text{DSS}}/100$), the calculation will still be valid, albeit rather less accurate than would be otherwise possible.

11.1.5 The Common-Source Amplifier Stage

Having biased the FET into an operating condition, consider the small-signal performance. In figure 11.9, the full equivalent circuit has been simplified for very low frequencies by ignoring the various capacitances. Also, the composite generator resistance at the gate has been called R_G, and the load resistance R_L, which includes R_D.

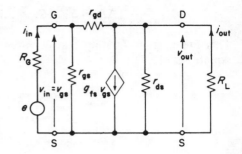

Figure 11.9. Low-frequency equivalent circuit.

By definition, the voltage gain $A_v = v_{out}/v_{in}$ or v_{ds}/v_{gs}, or if r_{gs} and r_{gd} are very large

$$A_v = -g_{fs}v_{gs}\left(\frac{r_{ds}R_L}{r_{ds} + R_L}\right)\frac{1}{v_{gs}}$$

$$= -\frac{g_{fs}R_L}{1 + (R_L/r_{ds})} \ . \tag{11.8}$$

If $r_{ds} \gg R_L$, which is usual in the pinched-off region,

$$A_v \simeq -g_{fs}R_L \ . \tag{11.9}$$

Because g_{fs} has a positive numerical value, A_v is negative; that is, a phase reversal has occurred.

When higher frequencies are involved, the equivalent circuit must retain its capacitances as shown in figure 11.10, which means that the Miller effect comes into operation. The high resistances r_{gs} and r_{gd} have been omitted because their effects would be swamped by the reactances of C_{gs} and C_{gd}.

If sinusoidal quantities are used,

$$I_{in} = V_{gs}j\omega C_{gs} + (V_{gs} - V_{ds})j\omega C_{gd}$$

whence

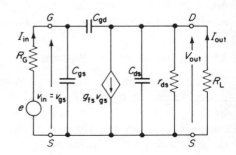

Figure 11.10. High-frequency equivalent circuit.

$$Y_{in} = \frac{I_{in}}{V_{gs}} = j\omega[C_{gs} + (1 - A_v)C_{gd}] \tag{11.10}$$

which means that an input capacitance C_{in} appears, where

$$C_{in} = C_{gs} + (1 - A_v)C_{gd} . \tag{11.11}$$

If C_{ds} is small so that $A_v \simeq -g_{fs}R_L$, C_{in} becomes

$$C_{in} = C_{gs} + (1 + g_{fs}R_L)C_{gd} . \tag{11.12}$$

Notice that if the output is short-circuited to AC ($R_L = 0$), then

$$C_{in} = C_{iss} \simeq C_{gs} + C_{gd} .$$

Similarly, if the input is short-circuited to AC, then $C_{out} = C_{oss} = C_{ds} + C_{gd}$. Finally, because $C_{rss} = C_{gd}$, then $C_{gs} = C_{iss} - C_{rss}$.

Normally, the output is not short-circuited to AC which means that C_{in} is much larger than C_{iss}, and it is this capacitance which limits the upper frequency response of the CS stage. If the composite resistance at the input is R_G, then the high-frequency cut-off point will be given by

$$f_H = \frac{1}{2\pi R_G C_{in}} . \tag{11.13}$$

This result may also be obtained by inspection of the simplified circuit of figure 11.11.

In a real CS circuit it is usual to take advantage of the high input resistance of the FET, and to accomplish this an input resistor R_3 is inserted between the bias point and the gate so as to diminish the shunting effect of R_1 and R_2. This technique also means that R_1 and R_2 may take quite low values.

This input resistor has been incorporated into the CS stage of figure 11.12, which also shows a signal generator capacitively coupled to the input. Also

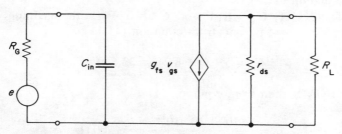

Figure 11.11. High-frequency equivalent circuit (simplified).

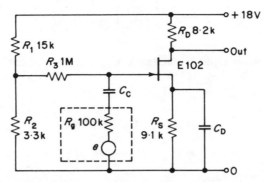

Figure 11.12. A practical common-source amplifier.

shown is the source decoupling capacitor C_D, which has the same bypassing function as in the bipolar CE stage and which is largely responsible for defining the low-frequency cut-off point.

Figure 11.12 clearly shows that because $R_1 \| R_2 \ll R_3$ the internal resistance of the signal generator R_g is effectively in parallel with R_3, so that

$$f_H \simeq \frac{R_3 + R_g}{2\pi R_3 R_g C_{in}} . \tag{11.14}$$

11.1.6 Design Example (Continued)

Figure 11.12 is a CS stage using the biasing components previously calculated, along with an input resistor R_3 (1 MΩ), a coupling capacitor C_c and a bypass capacitor C_D.

For this circuit, the voltage gain is

$$A_v \simeq -g_{fs}R_L = -1 \times 8.2 = -8.2 .$$

(Notice that this is a very approximate calculation, because g_{fs} should, strictly speaking, have been calculated for the actual working point of the FET from equation (11.2).)

From Table 11.1, $C_{iss} = 8\,\text{pF}$ and $C_{rss} = 3\,\text{pF}$ (both maximum values). Hence, $C_{gs} = 8 - 3 = 5\,\text{pF}$ and from equation (11.11):

$$C_{in} = 5 + (1 + 8.2)3 = 32.6\,\text{pF} .$$

If $R_g = 100\,\text{k}\Omega$, then from equation (11.14)

$$f_H = \frac{(10^6 + 10^5)10^{12}}{2\pi(10^6 \times 10^5)32.6} = 53\,700\,\text{Hz} .$$

This means that no cs stage designed as above would have a high-frequency cut-off point below about 54 kHz. (An experimental design using an arbitrary E102 had a voltage gain of -8.5 and an f_H of 72 kHz.)

11.1.7 The Common-Drain or Source-Follower Stage

The basic CD configuration is shown in figure 11.13 and, if the FET equivalent circuit (neglecting r_{gs} and r_{gd}) is incorporated, a model for analysis results as shown in figure 11.14. To determine the voltage gain, it is convenient to sum currents at the junction of the source and the load resistor R_F. If sinusoidal quantities are used

$$V_{out}\left(\frac{1}{R_F} + \frac{1}{r_{ds}}\right) - V_{gs}j\omega C_{gs} = g_{fs}V_{gs}$$

and because $V_{gs} = (V_{in} - V_{out})$, this becomes

$$V_{out}\left(\frac{1}{R_F} + g_{ds}\right) + (V_{out} - V_{in})(j\omega C_{gs} + g_{fs}) = 0$$

or

$$V_{out}\left(\frac{1}{R_F} + g_{ds} + g_{fs} + j\omega C_{gs}\right) = V_{in}(j\omega C_{gs} + g_{fs})$$

giving

$$A_v = \frac{V_{out}}{V_{in}} = \frac{g_{fs} + j\omega C_{gs}}{(1/R_F) + g_{ds} + g_{fs} + j\omega C_{gs}}.\qquad(11.15)$$

Here there is no phase reversal because A_v obviously has a positive value. At low frequencies, where $\omega C_{gs} \to 0$,

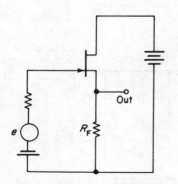

Figure 11.13. Basic common-drain stage (source-follower).

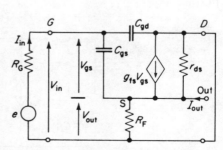

Figure 11.14. Equivalent circuit for source-follower.

$$A_{v(lf)} \simeq \frac{g_{fs}}{(1/R_F) + g_{ds} + g_{fs}}$$

and because the FET is working in the pinched-off region, where $r_{ds} \gg R_F$, this further reduces to

$$A_{v(lf)} \simeq \frac{g_{fs}R_F}{1 + g_{fs}R_F} \qquad (11.16)$$

and if $g_{fs}R_F \gg 1$, then $A_{v(lf)} \to 1$.

The input impedance to the stage is again largely capacitive, and may be determined by summing currents at the gate in figure 11.14:

$$I_{in} = V_{in}j\omega C_{gd} + (V_{in} - V_{out})j\omega C_{gs}$$

giving

$$Y_{in} = \frac{I_{in}}{V_{in}} = j\omega[C_{gd} + (1 - A_v)C_{gs}] .$$

That is,

$$C_{in} = C_{gd} + (1 - A_v)C_{gs} . \qquad (11.17)$$

Here again the existence of a Miller effect is seen, but because $A_v \simeq +1$, C_{in} for the CD stage is very much less than for the CS stage. Hence, the high-frequency performance of the CD stage is good.

The output impedance may be obtained by redrawing figure 11.14 so that the input generator is suppressed, and a voltage V_{out} is supplied externally at the output (figure 11.15).

Letting the parallel combination of R_G and $X_{C_{gd}}$ be the complex impedance $\bar{Z}_{gd}$,

$$I_{out} = V_{out}\left(\frac{1}{R_F} + \frac{1}{r_{ds}} + \frac{1}{\bar{Z}_{gd} - jX_{C_{gs}}}\right) + g_{fs}V_{gs} .$$

Putting

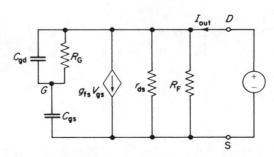

Figure 11.15. Equivalent circuit for determination of Y_{out}.

$$V_{gs} = V_{out}\left(\frac{-jX_{C_{gs}}}{\bar{Z}_{gd} - jX_{C_{gs}}}\right)$$

gives

$$I_{out} = V_{out}\left(\frac{1}{R_F} + g_{ds} + \frac{1 - jX_{C_{gs}}g_{fs}}{\bar{Z}_{gd} - jX_{C_{gs}}}\right)$$

so that

$$Y_{out} = \frac{I_{out}}{V_{out}} = \frac{1}{R_F} + \dot{g}_{ds} + \frac{g_{fs} + j/X_{C_{gs}}}{1 + j\bar{Z}_{gd}/X_{C_{gs}}} . \tag{11.18}$$

At low frequencies where $X_{C_{gs}}$ and $X_{C_{gd}}$ are large, this becomes

$$G_{out} = \frac{1}{R_F} + g_{ds} + g_{fs} .$$

Usually $1/R_F$, $g_{ds} \ll g_{fs}$ so that

$$G_{out} \simeq g_{fs} \quad \text{or} \quad R_{out} \simeq \frac{1}{g_{fs}} . \tag{11.19}$$

The prime use of the CD stage is as a buffer or impedance changer. At low frequencies where $X_{C_{in}}$ is large, the input impedance is clearly very high, for $R_{in} \simeq r_{gs} \| r_{gd}$. Conversely, the output resistance is low, typically of the order of 1000 Ω.

11.1.8 The Degenerate CS Stage

Figure 11.16 shows a CS stage with part of its source resistance unbypassed. The gain of this stage from gate to drain is

$$A_{v(CS)} = -\frac{v_{out}}{v_{in}} = -\frac{i_d R_D}{v_{in}} = -\frac{i_s R_D}{v_{in}} . \tag{11.20}$$

Now $i_s R_F$ is the signal voltage developed across R_F, which is by definition the input signal multiplied by the CD voltage gain; that is, from equation (11.16),

$$i_s R_F = A_{v(CD)}v_{in} = \frac{g_{fs}R_F v_{in}}{1 + g_{fs}R_F} . \tag{11.21}$$

Rewriting equation (11.20) as $A_{v(CS)} = -i_s R_F R_D / v_{in} R_F$ and putting equation (11.21) into this gives

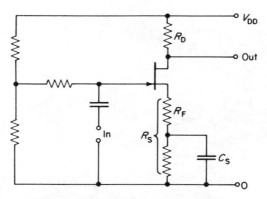

Figure 11.16. A degenerate cs stage.

$$A_{v(CS)} = -\frac{g_{fs}R_D}{1 + g_{fs}R_F}. \tag{11.22}$$

If $g_{fs}R_F \gg 1$ (which is *not* necessarily true), (11.22) reduces to

$$A_{v(CS)} \simeq -\frac{R_D}{R_F}.$$

Notice the resemblance between equation (11.22) and the ideal feedback equation $A_{v(FB)} = A/(1 - AB)$. The denominator $(1 + g_{fs}R_F)$ is clearly analogous to the feedback factor, and by introducing degeneration into the cs stage, both internal distortion and noise along with the gain are reduced by this factor.

11.1.9 The Source Bypass Capacitor

It has been shown above that, for the cd stage, the output resistance is approximately $1/g_{fs}$. This means that, in the cs stage, if R_S is bypassed by a capacitor C_S, then a low-frequency cut-off point will be defined when $|X_{C_S}| = 1/g_{fs} \| R_S$, and if $R_S \gg 1/g_{fs}$, then

$$f_L = \frac{g_{fs}}{2\pi C_S} \tag{11.23}$$

which is the design equation for C_S. (Note that often R_S will be comparable to $1/g_{fs}$ and so should be included in this equation.)

11.1.10 The Bootstrap Follower

The input resistance to a practical junction FET amplifier stage is determined largely by the biasing components. A useful technique for raising this input resistance and also providing automatic biasing, is illustrated in figure 11.17, where the bootstrap follower is depicted. Here use is made of the fact that the gate must be held at opposite polarity to the drain so that it can be supplied from a resistor chain in the source connection as shown, or indeed from the common line directly (when $R_{F2} = 0$).

In this configuration, the gate bias is provided by the voltage drop in R_{F1}:

$$-V_{GSQ} = I_Q R_{F1}$$

so that the usual square-law equation becomes

$$I_Q = I_{DSS}\left(1 + \frac{I_Q R_{F1}}{V_P}\right)^2$$

giving the value of R_{F1} for a chosen quiescent drain current:

$$R_{F1} = \frac{-V_P}{I_Q}\left[1 - \left(\frac{I_Q}{I_{DSS}}\right)^{1/2}\right]. \tag{11.24}$$

The limits of V_P and I_{DSS} may be inserted into this equation to give a fully toleranced bias design and R_{F2} chosen on the basis of the signal swing desired.

The bootstrap-follower is a useful direct-coupled buffer stage, but it exhibits an offset voltage from input to output equal to V_{GSQ}. To minimize this a FET having a low V_P should be chosen, and such a choice also results in a maximal value of R_{in}, which may be shown as follows:

$$i_{in} = \frac{v_{in} - v_{out(R_{F2})}}{R_3} = \frac{v_{in}(1 - A_{v(R_{F2})})}{R_3}$$

or

$$R_{in} = \frac{v_{in}}{i_{in}} = \frac{R_3}{(1 - A_{v(R_{F2})})}$$

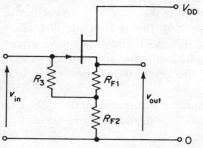

Figure 11.17. The bootstrap follower.

where $v_{\text{out}(R_{F2})}$ is the signal voltage developed across R_{F2} and $A_{v(R_{F2})}$ is the corresponding voltage gain from the gate to the top of R_{F2}. Knowing the source-follower gain from equation (11.16), this is:

$$A_{v(R_{F2})} = \frac{g_{fs}(R_{F1} + R_{F2})}{1 + g_{fs}(R_{F1} + R_{F2})} \frac{R_{F2}}{(R_{F1} + R_{F2})}$$

$$= \frac{g_{fs}R_{F2}}{1 + g_{fs}(R_{F1} + R_{F2})}$$

which is maximized when R_{F1} is minimized; that is, when V_{GSQ} is small, as with a FET of low V_P.

Putting this into the expression for R_{in} gives:

$$R_{in} = \frac{R_3}{1 - A_{v(R_{F2})}} = R_3\left(1 - \frac{g_{fs}R_{F2}}{1 + g_{fs}(R_{F1} + R_{F2})}\right)^{-1}$$

$$= R_3\left(\frac{1 + g_{fs}(R_{F1} + R_{F2})}{1 + g_{fs}R_{F1}}\right) \tag{11.25}$$

which is also maximized when R_{F1} is minimized.

If R_3 is connected to ground then R_{F2} disappears and equation (11.25) becomes simply R_3. Conversely, if R_{F1} is short-circuited to the signal by a capacitor then equation (11.25) becomes:

$$R_{in} = R_3(1 + g_{fs}R_{F2}) = \frac{R_3}{1 - A_{v(CD)}}$$

which is larger than R_3 because the common-drain (or source-follower) gain is less than unity.

11.2 DISTORTION IN FET STAGES

The prime cause of both harmonic and intermodulation distortion in saturated FET stages is the square law transconductance curve. A secondary cause is the small variation in r_{ds} as a function of V_{DS} and V_{GS}. Considering only the transconductance curve, it is obvious that minimum distortion occurs if the quiescent point is placed in the most linear portion, where $I_Q \rightarrow I_{DSS}$. This, however, would lead to an unstable bias situation[2], so that a better solution is to bias elsewhere on the transconductance curve and apply feedback if necessary.

To demonstrate harmonic distortion, assume a sinusoidal input voltage so that $v_{GS} = [V_{GSQ} + V \sin(\omega t)]$ and determine the resultant drain current i_D from

$$i_D = I_{DSS}\left(1 - \frac{v_{GS}}{V_P}\right)^2$$

$$i_D = I_{DSS}\left(1 - \frac{V_{GSQ} + V\sin(\omega t)}{V_P}\right)^2$$

giving

$$i_D = \frac{I_{DSS}}{V_P^2}\left[(V_{GSQ} - V_P)^2 + V^2\sin^2(\omega t) + 2V(V_{GSQ} - V_P)\sin(\omega t)\right].$$

This reduces to

$$i_D = \left[I_Q + \frac{I_{DSS}}{2}\left(\frac{V}{V_P}\right)^2\right] - \left[\frac{I_{DSS}}{2}\left(\frac{V}{V_P}\right)^2\cos(2\omega t)\right] + \left[g_{fs}V\sin(\omega t)\right].$$

Here, the first term shows that the quiescent current has increased slightly; the second term is a second harmonic; and the final term is the fundamental. (A more complete analysis would include the effects of both departures from the ideal square law transconductance curve, and the existence of some bulk resistance in the channel[3].)

A harmonic distortion factor D_h may be defined as the ratio of the RMS value of the distortion component to the RMS value of the fundamental:

$$D_h = \frac{I_{DSS}V^2}{(2\sqrt{2})V_P^2}\frac{\sqrt{2}}{g_{fs}V}$$

which, using equation (11.2), becomes

$$D_h = -\frac{V}{4(V_P - V_{GSQ})} \quad \text{or} \quad -\frac{25V}{V_P - V_{GSQ}} \text{ per cent}. \tag{11.26}$$

This expression demonstrates the expected: that the harmonic distortion is proportional to the signal amplitude, and is minimal at $V_{GSQ} \to 0$; that is, near $I_Q \to I_{DSS}$.

Intermodulation distortion arises as a result of the interaction between any two sinusoids producing sum and difference components. Knowing that all cyclic signals can be expressed as a series of sinusoids, it is reasonable to suppose that intermodulation distortion will arise as a result of interaction between each pair of these sinusoids. Many of the resultant sum and difference components will be above or below the pass-band of the amplifier, but others will contribute to the output.

To demonstrate intermodulation distortion, it is convenient to take an input signal of the form $v_{GS} = V_{GSQ} + V_A\sin(\omega_A t) + V_B\sin(\omega_B t)$ and insert it into the usual square-law equation as was done in the previous section.

Upon squaring out this equation[2], it will be found that the quiescent current has increased to

$$I_Q + \frac{I_{DSS}}{2V_P^2} (V_A^2 + V_B^2)$$

which is of the same form as that for a single sine-wave input but includes the sum of the squares of the coefficients of all the input sinusoids. This clearly suggests a generality.

The intermodulation term will be found to be

$$\frac{I_{DSS}}{V_P^2} V_A V_B \{\cos [(\omega_A - \omega_B)t] - \cos [(\omega_A + \omega_B)t]\} . \qquad (11.27)$$

An intermodulation distortion factor D_i may now be defined as the ratio of the RMS value of the intermodulation component to that of the fundamental. Knowing that the RMS value of a series of sinusoids is the root of the sum of the squares of those sinusoids, and that the fundamental is simply

$$g_{fs}[V_A \sin (\omega_A t) + V_B \sin (\omega_B t)]$$

it will be found that

$$D_i = \frac{V_A V_B}{\sqrt{2}(V_{GSQ} - V_P)(V_A^2 + V_B^2)^{1/2}}$$

or

$$\frac{70.7 V_A V_B}{(V_{GSQ} - V_P)(V_A^2 + V_B^2)^{1/2}} \text{ per cent} . \qquad (11.28)$$

Although both the harmonic and intermodulation distortion factors show, as expected, that distortion is minimal when $V_{GSQ} \to 0$, they fail to show the real extent of distortion when $V_{GSQ} \to V_P$. This is because the square-law transconductance curve becomes invalid at low values of I_Q, and the distortion is in fact much greater than the simple theory predicts[3].

Secondary effects, involving changes in r_{ds} with working point, are dealt with by Sherwin[4], and will not be repeated here because they are subsidiary to the main argument.

Distortion may be reduced by applying feedback, and if this is done by leaving part of R_S unbypassed, the values of D_h and D_i will be divided by the feedback factor for this particular case. From equation (11.22), this is $(1 + g_{fs}R_F)$.

11.3   NOISE IN JUNCTION FIELD-EFFECT TRANSISTORS

Three main noise sources exist in the FET: Johnson noise in the channel, shot noise due to gate leakage current, and flicker or $1/f$ noise. The shot noise is very small, because, for a FET, the gate leakage current is extremely low.

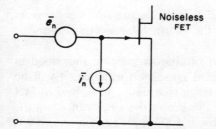

Figure 11.18. Two-noise-generator model.

Figure 11.18 shows the "two-noise-generator" equivalent circuit discussed in chapter 6, and figure 11.19 gives a plot of spot noise-generator values for a Siliconix U 168 p-channel FET. An inspection of this latter diagram will reveal two important points. Firstly, flicker noise is represented by a rise in $\bar{e}_n$ as the frequency falls, whereas (unlike the case for the bipolar transistor) $\bar{i}_n$ remains virtually constant. Secondly, the values of $\bar{e}_n$ and $\bar{i}_n$ are such that unless R_G is very large indeed, $\bar{e}_n^2/R_G \gg \bar{i}_n^2 R_G$. This fact is important, because it can be used to simplify the noise factor equation, (6.18). If the correlation coefficient is zero, this simplification is as follows:

$$\mathrm{NF} = 1 + \frac{1}{4kT}\left(\frac{\bar{e}_n^2}{R_G} + \bar{i}_n^2 R_G\right)$$

$$\simeq 1 + \frac{\bar{e}_n^2}{4kTR_G}. \tag{11.29}$$

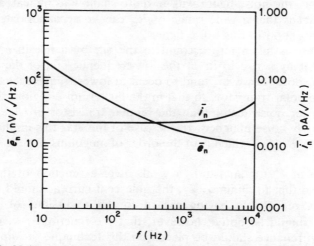

Figure 11.19. Noise generator characteristics for Siliconix U 168. (Reproduced by permission.)

(For a FET it is easy to accept the noncorrelation of the two noise generators at low frequencies, because $\bar{e}_n$ represents the channel Johnson noise and $\bar{i}_n$ represents the gate-current shot noise.)

The noise-equivalent resistance method of characterization described in chapter 6 is also useful for FETs. For a good noise performance, it will be recalled that the ratio R_{ni}/R_{nv} should be large (see figure 6.8), and in fact this is invariably so for FETs. For example, inserting the values of $\bar{e}_n$ and $\bar{i}_n$ given by figure 11.19 for the U 168 working at 100 Hz into equations (6.38) and (6.39) gives

$$R_{nv} \simeq \frac{\bar{e}_n^2}{16.5} \simeq \frac{45^2}{16.5} \simeq 123 \text{ k}\Omega \quad \text{at } 25°\text{C}$$

$$R_{ni} \simeq \frac{16.5}{\bar{i}_n^2} \simeq \frac{16.5}{0.022^2} \simeq 35 \text{ M}\Omega \quad \text{at } 25°\text{C}$$

so that

$$\frac{R_{ni}}{R_{nv}} \simeq \frac{34\,000}{123} = 270 .$$

The value of R_{nv} in the region above f_F has been shown by van der Ziel[5] to approach $0.6/g_m$ for a FET wherein the complete channel length is modulated by the gate. Some FETs do in fact approach this ideal, but for more common types $R_{nv} \simeq 0.75/g_m$.

The optimum value of R_G for a minimal noise figure is $\bar{e}_n/\bar{i}_n$ (or $\sqrt{(R_{nv}R_{ni})}$) according to equations (6.19) and (6.34). For the FET, this implies that $R_{G(opt)}$ is likely to be very large. However, because R_{ni}/R_{nv} is also very large, the noise figure will be quite shallow as indicated by figure 6.8, which means that a wide range of R_G can be accommodated without serious deterioration in this noise figure.

Flicker noise is taken into account in the FET by a rise in $\bar{e}_n$, and this manifests itself by a rise in NF at the low-frequency end of the spectrum. This rise in $\bar{e}_n$ does, however, tend to occur at lower frequencies than is the case for the bipolar transistor, so that in the flicker noise region, the FET can provide a better noise factor than the bipolar transistor *provided* that the stage is properly noise matched. In the case of the FET, this implies that R_G should be near $R_{G(opt)}$, which is of the order of megohms since it is given by $\bar{e}_n/\bar{i}_n$.

The value of $\bar{e}_n$ for an FET is a weak inverse function of the channel current, I_D, so that to minimize $\bar{e}_n$, the quiescent current should be as near to I_{DSS} as is reasonable having regard to the excursions of the signal voltage[5]. It should be noted, however, that the function *is* weak, so that not too much reliance should be placed on this technique of improving the noise factor.

Figure 11.20 is an idealized plot of $\bar{e}_n$ against frequency, the numerical values being appropriate for a typical low noise FET. The idealized slope for $1/f$ noise of 3 dB/octave for $\bar{e}_n$ has been assumed (i.e., 6 dB/octave for $\bar{e}_n^2$); and after a sharp break-point at f_F, the magnitude of $\bar{e}_n$ has been taken as constant, to represent the white noise region.

In FET data sheets, it is usual to quote $\bar{e}_n$ at two frequencies, typically 10 Hz and 10 kHz; that is, values within both the $1/f$ region and the white noise region are given. If a true $1/f$ slope is assumed, then the value of $\bar{e}_n$ anywhere in the flicker noise spectrum can be determined, as follows.

If the given value of spot-noise at a frequency f_1 is provided, then the value at f_2 is given by,

$$\bar{e}_{n(f_1)}^2 f_1 = \bar{e}_{n(f_2)}^2 f_2$$

or

$$\bar{e}_{n(f_2)}^2 = \bar{e}_{n(f_1)}^2 \frac{f_1}{f_2} .$$

The break-point, f_F is defined as occurring when the magnitude of the pink noise equals that of the white; that is,

$$\bar{e}_{n(f_F)}^2 = \bar{e}_{n(f_w)}^2$$

giving

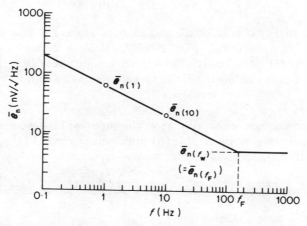

Figure 11.20. Typical pink noise spectrum (idealized) for a junction FET.

$$\bar{e}^2_{n(f_p)} \frac{f_p}{f_F} = \bar{e}^2_{n(f_w)}$$

or

$$f_F = f_p \frac{\bar{e}^2_{n(f_p)}}{\bar{e}^2_{n(f_w)}} \tag{11.29}$$

where $\bar{e}^2_{n(f_p)}$ is the value of the spot-noise voltage generator given at some frequency f_p in the $1/f$ region.

The contribution of pink noise to the mean square noise voltage at the input over any frequency band may now be determined. Assuming a band f_1 to f_2, then,

$$\overline{v^2_{ni(p)}} = \int_{f_1}^{f_2} \bar{e}^2_n \, df$$

$$= \bar{e}^2_{n(f_1)} f_1 \int_{f_1}^{f_2} \frac{1}{f} \, df$$

$$= \bar{e}^2_{n(f_1)} f_1 \ln\left(\frac{f_2}{f_1}\right). \tag{11.30}$$

(This equation has already been met: it is equation (6.44).)

An example will serve to illustrate the use of the foregoing material.

Example

A junction FET exhibits the following spot-noise voltage generator values:

$$\bar{e}_n = \begin{cases} 20\ \text{nV}/\sqrt{\text{Hz}} & \text{at 10 Hz} \\ 5\ \text{nV}/\sqrt{\text{Hz}} & \text{at 10 kHz} . \end{cases}$$

Assuming that $\bar{i}_n$ is very small, and that the FET is the first stage of a DC amplifier having a single roll-off at 20 kHz, calculate the mean square noise voltage referred to the input. The effective input generator resistance is 5 kΩ, and the noise below 0.1 Hz is to be counted as drift.

Solution

The total mean square noise voltage at the input will have components due to: (a) white thermal noise of R_G; (b) pink noise from 0.1 Hz to f_F; and (c) white noise from f_F to 20 kHz; that is,

$$\overline{v^2_{ni}} \simeq 4kTR_G\Delta f + \bar{e}^2_{n(10)}[10 \ln(f_F/f_1)] + \bar{e}^2_{n(10\,000)}\tfrac{1}{2}\pi f_H .$$

(Note: the last term assumes that $f_H \gg f_F$.)

The numerical values of the three terms are as follows:

(i) $\qquad 4kTR_G\Delta f = 8 \times 10^{-17}\Delta f \quad$ (from figure 6.5(b))

$$= 8 \times 10^{-17} \times 20\,000 \times \tfrac{1}{2}\pi$$

$$= 2.513 \times 10^{-12} \text{ V}^2 .$$

(ii) Here, $f_1 = 0.1$ Hz and f_F is given by equation (11.29)

$$f_F = 10(20/5)^2 = 160 \text{ Hz}$$

giving

$$\bar{e}^2_{n(10)}10 \ln\left(\frac{f_F}{f}\right) = (20 \times 10^{-9})^2 10 \ln\left(\frac{160}{0.1}\right)$$

$$= 0.030 \times 10^{-12} \text{ V}^2 .$$

(iii) $\qquad \bar{e}^2_{n(10\,000)} \times \tfrac{1}{2}\pi f_H = (5 \times 10^{-9})^2 \tfrac{1}{2}\pi 20\,000$

$$= 0.785 \times 10^{-12} \text{ V}^2 .$$

The sum of these components is:

$$\overline{v^2_{ni}} = (2.513 + 0.030 + 0.785) \times 10^{-12} \text{ V}^2$$

$$= 3.328 \times 10^{-12} \text{ V}^2$$

or

$$\sqrt{(v^2_{ni})} = 1.82 \ \mu\text{V} .$$

This example clearly demonstrates the overriding contribution of R_G to the total noise of the amplifier stage.

11.4   THE JUNCTION FET DIFFERENCE AMPLIFIER

A long-tailed pair of matched junction FETs is shown in Figure 11.21, and may be analyzed by using small-signal input voltages v_{g_1} and v_{g_2}, and letting any small-signal fluctuation at the sources be v_s.

The signal voltages at the drains will be:

$$v_{d1} = i_{d1}R_D = g_{fs}v_{gs1}R_D = g_{fs}R_D[v_{g1} - v_s]$$

and

$$v_{d_2} = i_{d2}R_D = g_{fs}v_{gs2}R_D = g_{fs}R_D[v_{g2} - v_s]$$

giving

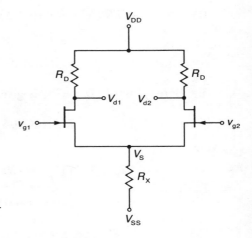

Figure 11.21. The junction FET difference stage.

$$(v_{d1} - v_{d2}) = g_{fs}R_{D2}[v_{g_1} - v_{g2}]$$

so that

$$A_{v(\text{diff})} = \frac{v_{d1} - v_{d2}}{v_{g1} - v_{g2}} = g_{fs}R_D \;.$$

This unsurprising result shows that the difference-in-to-difference-out voltage gain is the same as for a simple CS stage, which it will be recalled, is low because of the small value of g_{fs} presented by the junction FET when compared with typical values of g_m for the bipolar transistor.

The common-mode gain is also analogous to that of the bipolar pair, and may be found as follows:

$$
\begin{aligned}
(v_{d1} + v_{d2}) &= g_{fs}R_D[v_{g1} + v_{g2} - 2v_s] \\
&= g_{fs}R_D[v_{g1} + v_{g2} - 2(i_{d1} + i_{d2})R_x] \\
&= g_{fs}R_D\left[v_{g1} + v_{g2} - \frac{2R_x}{R_D}(v_{d1} + v_{d2})\right]
\end{aligned}
$$

or

$$(v_{d1} + v_{d2})(1 + 2g_{fs}R_x) = g_{fs}R_D(v_{g1} + v_{g2})$$

giving

$$
\begin{aligned}
A_{v(\text{CM})} &= \frac{v_{d1} + v_{d2}}{v_{g1} + v_{g2}} = \frac{g_{fs}R_D}{1 + 2g_{fs}R_x} \\
&\approx \frac{R_D}{2R_x} \quad \text{when } R_x \text{ is large} \;.
\end{aligned}
$$

This is, of course, the gain of a degenerate stage, and in practice, R_x will usually be the dynamic resistance of some form of current mirror.

11.4.1 The BiFET Operational Amplifier

Although the junction FET presents a high input impedance, it does not offer a high voltage gain, as has been seen. Therefore, in the case of complete amplifiers, it is advisable to follow FET input stages with bipolar transistors. However, the combination of the two types of device in a monolithic IC has presented some problems, as has the adequate matching of two input FETs to form a difference stage[6].

One method of overcoming such problems is to use ion implantation in which atomic species are ionized, then accelerated to high velocities in a vacuum using electrostatic fields. The high-velocity ions are fired at the silicon substrate through masks, and both their depth of penetration and density are subject to a high degree of control. Hence, accurate doping levels can be achieved at very much lower temperatures than are needed for diffusion techniques[7]. This is the method used by both Motorola and Texas Instruments in their BiFET process, which has led to a large range of operational amplifiers and other integrated circuits.

Figure 11.22 shows the basic circuit for the Texas Instruments TL080 op.amp. series. Here the FETs Tf1 and Tf2 form a long-tailed input pair, and are loaded by a group of bipolar transistors in a current-mirror

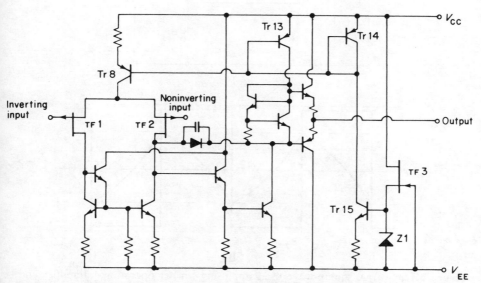

Figure 11.22. Basic circuit for the Texas Instruments TL080-series BiFET operational amplifier (reproduced by permission).

configuration similar to that of the 741 op. amp. of figure 7.12. The second stage and complementary output pair are also similar to the 741.

However the main biasing network is derived from a 5.2 V Zener structure Z_1, driven by Tf3, which acts as a constant-current source or *current-limiter*, the operation of which is described later in the chapter. This 5.2 V is applied to the base of a mirror transistor Tr15 having an emitter resistor, and it in turn drives the diode-connected transistor Tr14. The base drive for the long-tail transistor Tr8 and the dynamic load transistor Tr13 is provided from this diode.

Operational amplifiers using FET input stages can be extremely useful in applications where both low bias currents and high gains are required, such as integrators, though here, the offset voltage may be a problem.

An example of a circuit which takes advantage of low bias current is afforded by the square-wave oscillator circuit shown in figure 11.23(a),

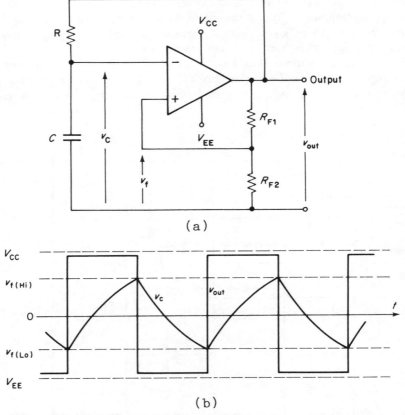

(a)

(b)

Figure 11.23. (a) A square-wave oscillator using a BiFET operational amplifier and (b) the waveforms produced.

which can be used at very low frequencies if this input bias current is small enough not to contribute significantly to the charging and discharging of the capacitor C. This circuit is similar to the hysteresis comparator in that two different reference voltages, $v_{f(hi)}$ and $v_{f(lo)}$ are produced at the junction of R_{F1} and R_{F2} depending upon whether the output is near the positive or the negative supply voltage; that is,

$$v_{f(hi)} \simeq \frac{V_{CC}R_{F2}}{R_{F1} + R_{F2}} \quad \text{and} \quad v_{f(lo)} \simeq \frac{V_{EE}R_{F2}}{R_{F1} + R_{F2}}.$$

Suppose that v_{out} is initially near V_{CC}. The capacitor charges to a value of v_C just above $v_{f(hi)}$, when comparator action causes the output to fall rapidly to nearly V_{EE}, as shown in the waveform diagram of figure 11.23(b). The capacitor then discharges to just below $v_{f(lo)}$, the output rises to nearly V_{CC} and the sequence continues. Hence, the output is a square wave as shown, provided that the magnitudes of V_{CC} and V_{EE} are equal.

The period of the wave may be calculated by observing that at any output changeover time, an instantaneous voltage $V_{CC} - v_{f(lo)}$ (or $V_{EE} + v_{f(hi)}$) appears across the capacitor and resistor. Hence,

$$v_C = V_{CC}\left(1 + \frac{R_{F2}}{R_{F1} + R_{F2}}\right)[1 - \exp(-t/CR)].$$

Charging from $v_{f(lo)}$ to $v_{f(hi)}$ is complete after a half-period, so

$$\frac{2V_{CC}R_{F2}}{R_{F1} + R_{F2}} = V_{CC}\left(1 + \frac{R_{F2}}{R_{F1} + R_{f2}}\right)\left[1 - \exp\left(-\frac{1}{2}\frac{T}{CR}\right)\right]$$

or

$$\frac{2R_{F2}}{R_{F1} + 2R_{F2}} = 1 - \exp\left(-\frac{1}{2}\frac{T}{CR}\right)$$

giving

$$T = 2CR \ln\left(\frac{R_{F1} + 2R_{F2}}{R_{F1}}\right). \tag{11.31}$$

In addition to the square-wave output, an exponentially rising and falling "triangular" wave appears at the inverting input. This could be linearized by replacing R with parallelled current sources, and buffered using another BiFET operational amplifier. By this means, a triangle-and-square-wave signal generator may be designed, which is one implementation of the block diagram of figure 8.21.

11.5 THE VOLTAGE-CONTROLLED RESISTOR

Below pinch-off, the channel of an FET behaves as a resistor whose value is a function of V_{GS}. Very near the origin, this channel resistor becomes almost linear, as is demonstrated by figure 11.24(a), which shows a set of output

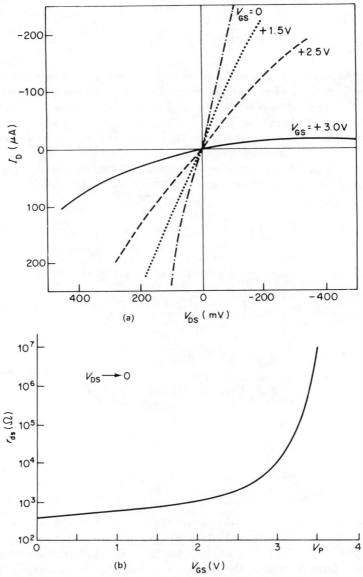

Figure 11.24. (a) Low level drain characteristics for Siliconix p-channel VCR (reproduced by permission). (b) Channel resistance versus gate-source voltage for Siliconix VCR6P p-channel FET (reproduced by permission).

characteristics for a p-channel VCR FET at values of V_{DS} below about 0.4 V. Notice that the curves pass through the origin, and into the third quadrant. Here, operation as a VCR is still possible providing that the gate-junction does not become forward-biased.

In this region near the origin, the incremental value of the channel resistance r_{ds} is similar to the DC or chord resistance, r_{DS}, and well below $V_{GS} = V_P$, the relevant law is approximately exponential:

$$r_{DS} \simeq r_0 \, e^{b|V_{GS}|}$$

where b is a parameter of the device (related to the gate contact potential and hence temperature-dependent) and r_0 is r_{DS} at $V_{GS} = 0$ and some datum temperature. Above $V_{GS} = V_P$, the value of r_{ds} rises very rapidly indeed as shown by figure 11.24(b). Because of the increasing difficulty in controlling r_{ds}, as $|V_{GS}|$ rises above the exponential region, the development of remote pinch-off VCRs which achieve a wider exponential range would be desirable.

Figure 11.25 illustrates two typical VCR applications. In diagram (a), the VCR forms the lower section of a potentiometer, so that

$$v_{out(max)} = v_{in}\left(\frac{r_{DS(max)}}{R + r_{DS(max)}}\right)$$

and

$$v_{out(min)} = v_{in}\left(\frac{r_{DS(min)}}{R + r_{DS(min)}}\right).$$

The use of a VCR as an emitter resistor for a bipolar transistor is shown in figure 11.25(b). Here, the voltage gain of the stage is approximately R_C/r_{DS} over the range where r_{DS} permits normal operation of the bipolar transistor.

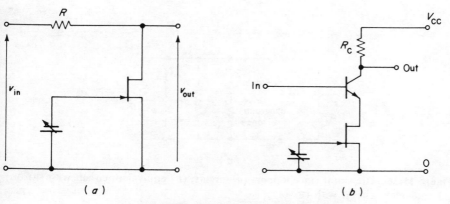

Figure 11.25. Typical VCR applications.

11.6   THE FET AS A CHOPPER

From the discussion given in the foregoing section, it will be recalled that near the origin, the channel resistance is essentially linear and nonpolarized. This suggests that the FET would be useful as a chopper of low-level DC (i.e., very slowly fluctuating) signals for use in DC amplifiers. In such an application, an important advantage of the FET is that no offset voltage is observed, which means that the techniques necessary for offset cancellation when bipolar transistors are used are no longer necessary. (There is, however, a small offset due to transients, which will be explained below.)

Figure 11.26(*a*) shows the circuit for a simple shunt FET chopper. From the equivalent circuit representing the FET in the ON condition (figure 11.26(*b*)) the output voltage—which should, ideally, be zero—is given by

$$V_{\text{out}} = \frac{ER}{R_{\text{g}} + R_1 + R} \qquad (11.32a)$$

where R is the parallel combination of $r_{\text{DS(on)}}$ and R_{L}.

Because R_{L} is usually the input resistance to a following amplifier, it is reasonable to assume that it will be large compared with $r_{\text{DS(on)}}$, so that equation (11.32*a*) becomes

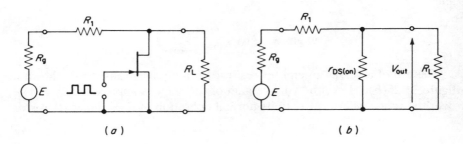

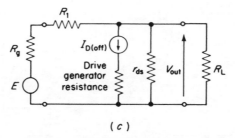

Figure 11.26. The shunt FET chopper: (*a*) circuit; (*b*) equivalent circuit with FET ON; (*c*) equivalent circuit with FET OFF.

$$V_{\text{out}} = \frac{Er_{\text{DS(on)}}}{R_g + R_1 + r_{\text{DS(on)}}} \, . \tag{11.32b}$$

This suggests that to make V_{out} negligibly small, R_1 should be as large as possible. Unfortunately, this conclusion conflicts with the requirements for the OFF conditions, shown in figure 11.26(c). Here, even if both r_{DS} and R_{L} are very large indeed, a leakage current $I_{\text{D(off)}}$ will produce a voltage drop across $(R_g + R_1)$, so reducing V_{out} below E:

$$V_{\text{out}} \simeq E - I_{\text{D(off)}}(R_g + R_1) \, . \tag{11.32c}$$

For V_{out} to be almost equal to E, clearly $(R_g + R_1)$ should be as small as possible, which is the converse of the previous requirement. The volt drop in $(R_g + R_1)$ will, of course, be greater if R_{L} or r_{ds} is not extremely large.

Techniques exist for leakage compensation in shunt chopper circuits, but a major improvement is observed by employing the shunt-series technique shown in figure 11.27. Here, a series and a shunt FET are operated in anti-phase, so that when one is ON the other is OFF; that is, when Tf$_1$ is OFF, so presenting a high channel resistance, Tf$_2$ is ON, which clamps the output point to the common line. Hence, equation (11.32b) applies, where R_1 becomes the Tf$_1$ channel resistance.

Conversely, when Tf$_1$ is ON, and Tf$_2$ is OFF, equation (11.32c) is relevant, and R_1 is now small.

Thus, both optimum conditions for the shunt chopper are fulfilled by adding a series FET.

The major disadvantage of the FET chopper is that spikes from the drive waveform are able to pass into the signal route via interelectrode capacitances, and the average value of these spikes produces an offset voltage which can be significant when very small signals are to be chopped. Here again the shunt-series configuration is helpful, particularly when small-geometry FETs having low capacitances are utilized. Unfortunately, a small geometry construction also implies a comparatively high channel resistance.

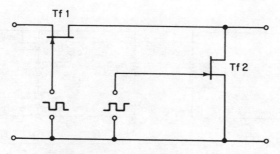

Figure 11.27. The shunt-series FET chopper.

11.7   THE FET ANALOG GATE

An analog gate may be defined as a switch which allows a signal to pass undistorted when in the ON position. Because mechanical switches are slow, and have limited lives, the use of semiconductor switches is a desirable design aim, particularly in commutating or multiplexing applications. Here, the requirement is that signals from a number of different channels are applied sequentially to a single measuring and recording system. Currently, the most satisfactory means of achieving essentially distortionless commutation is by the use of mechanical reed switches, but these are giving way to FET (or IGFET) commutators as the characteristics of these latter devices improve[8].

Unlike the chopper FET, which usually feeds into the very high input resistance of a capacitively-coupled AC amplifier, the analog gate FET must often feed a comparatively low resistance. Figure 11.28(a) illustrates the relevant situation, from which it will be seen that when the FET is ON

$$\frac{V_L}{E} = \frac{R_L}{R_g + r_{DS(on)} + R_L}. \tag{11.33a}$$

This expression will approach unity if R_L is large. For the OFF condition,

$$V_L \simeq I_{D(off)}R_L \tag{11.33b}$$

which will approach zero if $I_{D(off)}$ is very small.

One problem with the analog gate is that of ensuring that the FET is switched hard ON and OFF by the drive voltage. In figure 11.28(a), if E goes negative, then the magnitude of the effective drive or control voltage V_C will be $(|V_C| - |E|)$, and if this is less than $|V_P|$, the FET will not turn fully OFF. Consequently, it must be established that $(|V_C| - |E|) > V_P$, which means that the excursion of V_C may be quite large, leading inevitably to greater spike feedthrough.

If $V_C = 0$ for the ON condition, and E remains negative, it will forward-

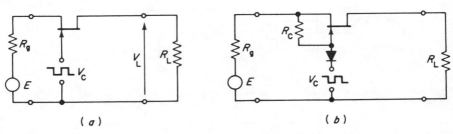

(a)   *(b)*

Figure 11.28. (a) Analog gate circuit; (b) analog gate circuit with floating drive.

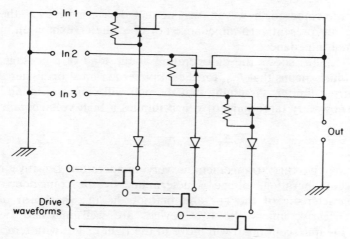

Figure 11.29. Junction FET commutator.

bias the gate–source junction and gate current may flow. Conversely, if E goes positive, its direction is such that the FET will tend to be turned OFF.

These disadvantages may be overcome by using IGFET analog gates, as will be described later, but sometimes it is worth while increasing the complexity of the circuit in order to retain the lower $r_{DS(on)}$ of the junction FET. For example, figure 11.28(b) shows how a diode in series with the gate will block forward current when E goes negative and $V_C = 0$. The high value resistor R_C is then able to maintain the gate at the same potential as the source, so that the FET remains fully ON. Further, it is possible to shunt the diode with a capacitor for speed-up purposes.

Figure 11.29 shows three elements of a junction FET commutator, and is self-explanatory.

11.8 THE CURRENT LIMITER

A final example of the application potential of the junction FET is the constant current generator (see also figure 7.26). For a fixed gate-source voltage, the drain current will be constant, and the channel will present a high value of r_{ds} if $V_{DS} > V_P$. Under these circumstances only temperature variations will affect the value of I_D. An extension of this concept is to connect the gate and source internally, so producing a two-terminal device which will pass a constant current I_{CL} equal to I_{DSS}. This device is usually termed a current limiter (CL) and its only operating criterion is that the voltage across it should at all times be greater than V_P. It will then provide an electrical function which is the dual of the Zener diode, so that the designer has at hand both constant current and constant voltage elements. It should be remembered, however, that neither device is perfect—the Zener diode does not present zero impedance nor does the CL exhibit an infinitely high internal impedance.

Figure 11.30(a) shows the CL operating as the load of a cs stage. Here, $I_Q = I_{CL}$, which means that I_{DSS} for the amplifier FET must be higher than I_{CL} for the current limiter. Assuming that the pair will operate, the AC load on the FET is now very large indeed, which implies a high voltage gain:

$$A_v \simeq -g_{fs}r_{ds(CL)} .$$

Unfortunately, this arrangement is very temperature sensitive because the quiescent operating voltage is determined by the coincidence of the output characteristic of the FET and that of the CL, as shown in figure 11.30(b). Clearly, unless the two devices are well matched (which is unlikely), the quiescent point will move to the right or left with temperature until clipping of the signal occurs as one or the other approaches desaturation. For this reason, even if I_Q is pre-set by means of an adjustable source resistor, the circuit is still inherently highly temperature sensitive.

Some further examples of CL applications are given in figure 11.31. Here, circuit (a) shows the CL used as a load in the essentially self-biasing CD stage; while circuit (b) illustrates its use as the constant-current element of a difference amplifier. Here, either field-effect or bipolar transistors may be used depending on the input impedance requirements.

Figure 11.31(c) shows an FET with a source resistor and with its gate connected to ground. This can be analyzed in the same way as was the bootstrap-follower, that is by noting that $V_{GS} = -I_Q R_S$. This gives:

$$R_S = -\frac{V_P}{I_Q}\left[1 - \left(\frac{I_Q}{I_{DSS}}\right)^{1/2}\right] \tag{11.24}$$

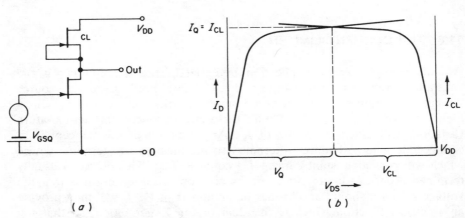

Figure 11.30. Circuit and voltage division diagram for a cs amplifier using a CL load.

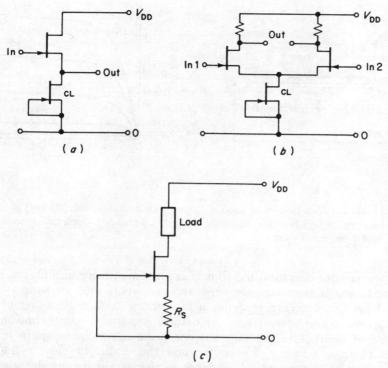

Figure 11.31. Further CL applications: (*a*) as load for CD stage; (*b*) as constant-current source for difference amplifier; (*c*) as a programmed current sink.

which shows that any current I_Q which is less than I_{DSS} can be defined by this method. This constitutes a "programmed" CL.

A Zener diode may be substituted for R_S, as is shown in the BiFET op. amp. circuit of figure 11.22 and this results in an almost constant value for V_{GS}. However, it does not provide linear feedback compensation as does the resistor.

11.9 THE INSULATED GATE FET OR IGFET

As its name suggests, the IGFET has no pn gate junction, but instead, a gate electrode is insulated from the channel by a very thin layer of (usually) silicon dioxide. (The use of SiO_2 is very common, and has led to the alternative terms MOST or MOSFET, meaning metal oxide silicon transistor. However, other insulating layers including silicon nitride are also used, so that IGFET is a more generic acryonym.)

An idealized cross-section of an n-channel depletion-type IGFET is given in figure 11.32(*a*). Here, two regions of donor impurity have been diffused into

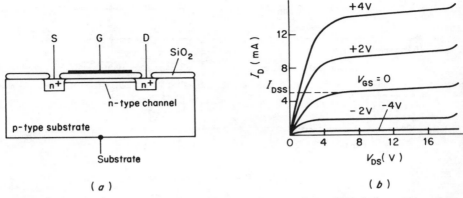

Figure 11.32. (a) Idealized cross section through an n-channel depletion-type IGFET; (b) output characteristics for Siliconix 2N3631 n-channel depletion-type IGFET. (Reproduced by permission.)

the p-type substrate to form the source and drain. A layer of SiO_2 has been deposited, and a metallic gate superimposed upon this. Because of the physical processes occurring at an oxide–silicon surface (bending of the Fermi levels), an inversion layer, or natural n-channel, occurs, and in the structure of figure 11.32(a) this will constitute a conductive path in the p-type substrate between the source and the drain. In most n-channel depletion IGFETs, however, this channel is augmented by the diffusion of donor impurity.

The application of a negative voltage to the gate will tend to deplete this channel of electrons, so reducing any channel current; and the application of a positive voltage will enhance the number of electrons in the channel, so increasing I_D. Hence, the transconductance curve for the device actually crosses the ordinate as shown in figure 11.33(a); that is, the so-called depletion-type IGFET (also called a type B FET according to a JEDEC convention) can operate in *both* the depletion *and* enhancement modes. This capability is also made apparent by the family of output characteristics shown in figure 11.32(b), where I_{DSS} is clearly exceeded when V_{GS} reverses. However, because a conducting channel does exist at $V_{GS} = 0$, the depletion IGFET is (like the junction, or type A FET) a *normally-ON* device.

An IGFET working purely in the enhancement mode results if acceptor impurity is diffused into an n-type substrate to form a p-type source and drain as shown in figure 11.34(a). Clearly, two back-to-back diodes are formed by the source–channel and channel–drain junctions, so that there is essentially an open-circuit between source and drain in the absence of gate voltage. When a negative voltage of sufficient magnitude is applied to the gate, a p-type channel is induced, which enhances the conductivity so that I_D rises. Figure 11.33(d) shows that the enhancement IGFET (or type C FET) has a transconductance characteristic which is shifted entirely past the y-axis, so

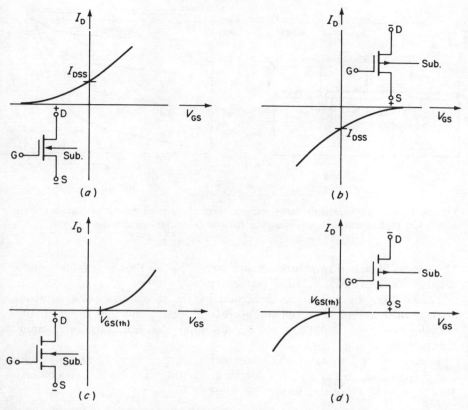

Figure 11.33. Symbols and transconductance curves for the family of insulated-gate FETs: (*a*) n-channel depletion (type B) IGFET; (*b*) p-channel depletion (type B) IGFET; (*c*) n-channel enhancement (type C) IGFET; (*d*) p-channel enhancement (type C) IGFET.

that it can never operate in the depletion region. It is therefore a *normally-off* device.

The p-channel enhancement IGFET has been the most common of the family, and was in fact the first active element used in large-scale integrated MOSFET circuits. This was mainly because it is the simplest of the family to fabricate, but also because it has drain and gate voltages of the same (negative) polarity, which makes DC coupling very convenient. Output characteristics for a typical p-channel enhancement IGFET are given in figure 11.34(*b*).

It is also possible to produce n-channel enhancement and p-channel depletion types, but the natural formation of an n-type layer at the oxide–silicon interface makes the relevant processes more difficult. Figure

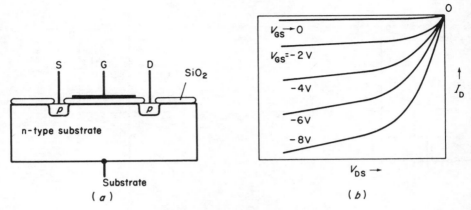

Figure 11.34. (*a*) Idealized cross-section through a p-channel enhancement-type IGFET; (*b*) typical output characteristics for the p-channel enhancement FET.

11.33 gives sketched transconductance curves for all four types, and relates each to the appropriate (IEEE recommended) symbol.

Notice that the polarity is indicated by the direction of the substrate arrow, which refers to the channel-substrate diode; and that the normally-OFF devices have broken lines representing the normally open-circuit channels.

In each of the four cases, the transconductance curve will be seen to have a positive slope, so that like the junction FET, the IGFET has a transconductance g_{fs} having a positive numerical value.

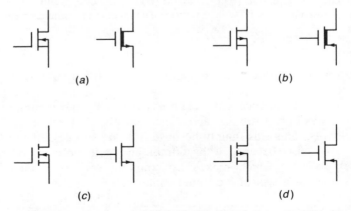

Figure 11.35. Alternative IGFET symbols for devices having substrates connected to sources. (*a*) n-channel depletion; (*b*) p-channel depletion; (*c*) n-channel enhancement; (*d*) p-channel enhancement types.

Figure 11.33 should be carefully studied, because it effectively relates the IGFET symbols to their transconductance characteristics, and a thorough understanding of the current and voltage directions and polarities as indicated by these symbols and characteristics will allow each to be rapidly deduced from the other.

However, the symbols shown are not unique and figure 11.35 repeats them along with a set of equivalents which are often used when each substrate is assumed to be connected to the source, which is a very common mode of connection, as will be seen later. Note that in each case, the arrow defines the direction of current in the source lead, and this convention should not be confused with the "pn" arrow used in the approved symbols. Also, note that a "thickened" channel implies a depletion-type IGFET.

11.9.1 IGFET Models and Parameters

A large-signal model for the IGFET, relevant to the saturated condition (and to the SPICE 2 parameters used in many computer-aided-design procedures) is given in figure 11.36. This illustrates the presence of diode junctions from the channel to the source and drain regions which must, of course, always be reverse-biased, and will exhibit the usual leakage currents. Also, they will have associated capacitances to the channel, depicted here as C_{BS} and C_{BD}.

The bulk or parasitic resistances r_S and r_D represent those portions of the channel which are not under the influence of the gate voltage (but are small and have little effect); and C_{GB} is the capacitance from the gate to the

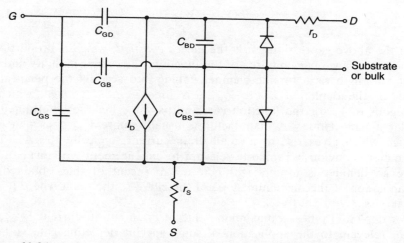

Figure 11.36. A large-signal n-channel IGFET equivalent circuit (for p-channel, reverse diode polarities).

substrate, or bulk. C_{GS} and C_{GD} are as for the junction FET (and have the same values as their small-signal equivalents C_{gs} and C_{gd}).

It will be noticed that a drain-to-source resistance has not been included. This is because the expression for the drain current—that is, the value of the dependent current generator—takes this into account, as follows:

$$I_D = \frac{b}{2}(V_{GS} - V_{GS(th)})^2(1 + \lambda V_{DS}) \qquad (11.34)$$

where λ is the *channel length modulation parameter* which describes the spread of the pinch-off region into the channel, so reducing its effective length as V_{DS} increases. This parameter, when multiplied by V_{DS}, gives the rise in the output characteristics otherwise described by r_{ds} in a small-signal model (λ has typical values of below 0.02 for integrated circuit IGFETs[9]).

Should r_{ds} be needed for small-signal calculation purposes, it may be obtained simply by differentiating I_D with respect to V_{DS} to obtain the associated conductance as follows:

$$g_{ds} = \frac{dI_D}{dV_{DS}} = \frac{b(V_{GS} - V_{GS(th)})^2}{2} \cdot \lambda$$

$$= \frac{\lambda I_D}{1 + \lambda V_{DS}} \simeq \lambda I_D . \qquad (11.35)$$

The transconductance may be similarly obtained:

$$g_{fs} = \frac{dI_D}{dV_{GS}} = b(V_{GS} - V_{GS(th)})(1 + \lambda V_{DS})$$

$$\simeq b(V_{GS} - V_{GS(th)}) . \qquad (11.36)$$

In the above expressions, b is the *device constant*, which is proportional to the capacitance per unit area of the junction, the carrier mobility and the width to length ratio for the channel, taking into account the penetration depth of the depletion layer. $V_{GS(th)}$ is the *threshold voltage*, which is analogous to V_p for the junction FET, and its value may be determined by similar means. However, manufacturers usually quote $V_{GS(th)}$ as that gate voltage which gives rise to a small drain current, typically 10 μA. (This parameter is obviously primarily relevant to enhancement-type IGFETs.) The value so defined is usually within a few per cent of that obtained by extrapolation of the maximum slope of the curve to the point where it cuts the abscissa.

For depletion types, a maximum value of drain cut-off current $I_{D(OFF)}$ is given, relevant to the application of some specific depleting gate voltage. $I_{D(ON)}$ is the drain current guaranteed to flow when V_{GS} takes some specific enhancing value and is relevant to both depletion and enhancement IGFETs.

TABLE 11.2. Siliconix 2N3631 n-channel depletion IGFET.

At $V_{DS} = 10$, and $V_{GS} = 0$,	$I_{DSS} = 2\text{–}10$ mA
At $V_{DS} = 5$ V, and $V_{GS} = -10$ V,	$I_{D(OFF)} = 100$ pA max
At $V_{DS} = 10$ V, and $I_D = 1$ μA.	$|V_{GS(th)}| \ngtr 6$ V

TABLE 11.3. Siliconix M511 p-channel enhancement IGFET.

At $V_{DS} = -20$ V, and $V_{GS} = 0$,	$I_{DSS} = -10$ nA max
At $V_{DS} = -10$ V, and $V_{GS} = -10$ V,	$I_{D(ON)} = -3$ mA min
At $V_{GS} = V_{DS}$, and $I_D = -10$ μA	$V_{GS(th)} = -3$ to -6 V

In the case of depletion IGFETs, a current I_{DSS} will flow when $V_{GS} = 0$. (This is precisely the same definition as for the junction FET.) For enhancement IGFETs, the drain current is minute at $V_{GS} = 0$, but it is sometimes quoted to demonstrate the switching capabilities of the device.

Because the IGFET has a genuine but extremely large ohmic resistance between its gate and channel, it is entirely valid to quote this resistance. This numerical value may easily be 10^{16} Ω, and this order of magnitude means that the gate capacitance, if charged, will remain so for long periods. An important practical point is that if even a small charge (often acquired due to friction in handling) is applied to the gate capacitance, a high enough gate voltage will result to effect breakdown of the gate insulation. For this reason, some IGFETs contain internal protective Zener diodes; in other cases, great care should be exercised to connect all leads into circuit before removing the short-circuiting device applied by the manufacturer.

Tables 11.2 and 11.3 given above contain some parameters relevant to a typical n-channel depletion IGFET and a p-channel enhancement type. Breakdown voltage parameters are similar to their junction FET or bipolar equivalents.

11.9.2 Application of the IGFET

The versatility of the IGFET family has led to its exploitation in applications ranging from all-IGFET operational amplifiers to logic systems including very large scale integrated (VLSI) circuits, and also including the voltage-driven power devices to be described later. Much of this has been made possible by new chip geometries and fabrication methods, which have progressively improved IGFET performance in many areas, but notably in the reduction of $V_{GS(th)}$ to levels which are compatible with standard supply voltages, including 5 V logic supplies.

Amongst these techniques is the use of phosphorus-doped polycrystalline silicon ("polysilicon") as a gate material. This has the added advantage that the polysilicon can be deposited as part of the normal processing sequence, then the phosphorus can be diffused in to form both the conductive polysilicon gate and the source and drain regions. This implies self-alignment of the gate with respect to the source and drain, so that very small structures with optimum geometries can be realized. Such geometries can exhibit extremely low values of C_{gs} and C_{gd}, with concomitant good frequency responses and switching speeds.

Another technique is to grow silicon nitride over the silicon dioxide insulation. The Si_3N_4 has a much higher dielectric constant than SiO_2, which again reduces the gate threshold voltage.

Ion implantation (which has already been mentioned in the context of BiFET operational amplifiers) may also be used to reduce the threshold voltage by modifying the doping level of the substrate under the gate. However, of even more importance is the fact that this technique can also make possible the fabrication of both n-channel and p-channel IGFETs on the same chip, leading to the extremely important complementary (CMOS) structures to be introduced later. Furthermore, ion implanation is a process which is also compatible with self-alignment of the gate with respect to the source and drain.

Because the discrete IGFET has a comparatively low value of g_{fs}, and also exhibits a higher noise figure than the junction FET[10], it is used in simple amplifier configurations only in special cases, such as electrometry. Here, DC measurements of extremely low currents are involved, so that the enormous gate resistance of the IGFET becomes invaluable. However, should a simple IGFET stage become necessary, it may be designed using an approximate equivalent circuit similar to that used for the junction FET, and in fact, manufacturers often quote small-signal parameters in the same form. Care should be taken when biasing such stages, however, having regard to the range of gate–source polarities made possible by the family of IGFETs.

In analog applications, IGFET structures are found in many IC operational amplifiers, and often as the only active elements therein. They are also found in combination with bipolar devices, and in analog blocks (sometimes along with logic cells) in uncommitted arrays and application-specific integrated circuits (ASICs). In such IC applications, IGFET structures are used in difference amplifier pairs, as current mirrors, and as active resistors or loads, to name but a few! It is therefore appropriate to consider these configurations, and to note that the various "active-load" circuits are applicable also to logic inverters.

11.9.3 The IGFET Current Mirror and Difference Amplifier

IGFET long-tailed pairs and current mirrors may be combined to form amplifier cells in similar ways to bipolar structures, and again IC fabrication is necessary to obtain the mandatory close matching of parameters. Figure 11.37(a) shows a current mirror using n-channel enhancement IGFETs as an example. Here, the defined current I forms the channel current of Tf1,

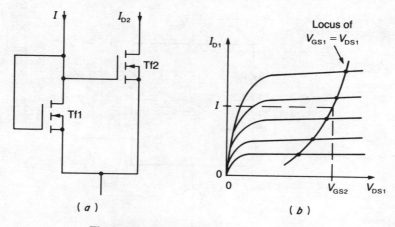

Figure 11.37. An IGFET current mirror.

which then produces a value of $V_{GS(1)}$ lying on the locus shown in diagram (*b*). This voltage is applied to Tf2, so that $I_{D(2)} = I$. Given identical devices, these currents are also nearly identical because there are no errors due to base currents as in the use of bipolar transistor mirrors.

Knowing that $V_{GS(1)}$ is always great enough to hold Tf1 in saturation, equation (11.34) may be used to determine the relative values of I and $I_{D(2)}$:

$$\frac{I_{D(2)}}{I} = \frac{b_2(V_{GS(2)} - V_{GS(th)(2)})^2}{b_1(V_{GS(1)} - V_{GS(th)(1)})^2} \frac{(1 + \lambda_2 V_{DS(2)})}{(1 + \lambda_1 V_{DS(1)})}.$$

Neglecting the effect of channel length modulation, and assuming the threshold voltages to be identical, this becomes simply b_2/b_1. The implication here is $I_{D(2)}$ can be made different from I by making b_1 different from b_2, and this is actually accomplished by designing different length-to-width (or *aspect*) ratios for the two channels.

Finally, it must be recognized that approximations inherent in the above treatment can cause significant errors. However, as with the bipolar equivalents, more sophisticated versions of the circuit (including the IGFET version of the Wilson current mirror) will lead to improvements[9].

An IGFET difference amplifier stage (also using n-channel enhancement structures) is shown in figure 11.38. A small-signal analysis may be carried out in a similar manner to that previously employed in the use of the JFET pair; but large-signal analysis for either the JFET or IGFET, leading to a graph of drain current versus gate difference voltage, is more complex than for the bipolar pair owing to the square-law form of the transconductance characteristic. Basically, it may be performed as follows.

Using equation 11.34 for saturated, identical IGFETs, and extracting the gate–source voltage gives:

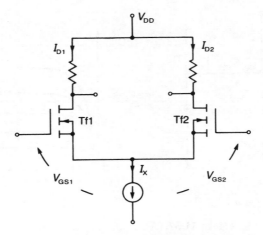

Figure 11.38. An IGFET difference stage.

$$V_{GS(1)} - V_{GS(2)} = V_{in} = \left[\frac{2I_{D(1)}}{b}\right]^{1/2} - \left[\frac{2I_{D(2)}}{b}\right]^{1/2}.$$

Now $(I_{D(1)} + I_{D(2)}) = I_x$ or $I_{D(2)} = (I_x - I_{D(1)})$ and inserting this gives:

$$V_{in} = \left[\frac{2I_{D(1)}}{b}\right]^{1/2} - \left[\frac{2[I_x - I_{D_1}]}{b}\right]^{1/2}.$$

Squaring this twice leads to a quadratic equation, the solution of which is:

$$I_{D(1)} = \frac{I_x}{2} \pm \frac{I_x}{2}\left[\frac{bV_{in}^2}{I_x} - \frac{b^2V_{in}^4}{4I_x^2}\right]^{1/2}$$

The second term in the bracket will be much smaller than the first for small input increments, so that,

$$I_{D(1)} \simeq \frac{I_x}{2} \pm \left[\frac{bI_x}{4}\right]^{1/2} V_{in} \tag{11.37}$$

which indicates a region of linear operation for small signals. (Clearly, $I_{D(2)}$ will be given by a similar expression.)

Equation (11.37) may be differentiated to obtain the (single-ended-output) transconductance;

$$g_{fs} = \frac{dI_{D(1)}}{dV_{in}} = \left[\frac{bI_x}{4}\right]^{1/2}. \tag{11.38}$$

This will have a maximal value where $I_x = 2I_{D(1)} = 2I_{D(2)} = 2I_Q$ so that

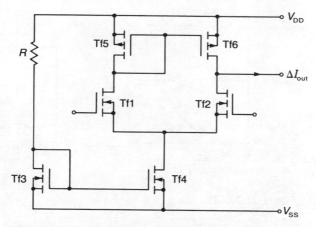

Figure 11.39. A biased and loaded IGFET difference stage.

$$g_{fs(max)} = \left[\frac{bI_Q}{2}\right]^{1/2}$$

and substituting for I_D from equation (11.34) will give:

$$g_{fs(max)} = \frac{b}{2}\left(V_{GSQ} - V_{GS(th)}\right).$$

The difference-output transconductance will obviously be twice this, which is the same as the transconductance for a simple CS stage (equation (11.36)) which is an expected result by analogy with the bipolar long-tailed pair.

A difference amplifier circuit using an n-channel IGFET long-tailed pair both biased and loaded by current mirrors is shown in figure 11.39. Here, the long-tail current is defined by the resistor R (or a current source) and Tf3 determines the value of V_{GS} for the mirror IGFET Tf4. Similarly, the current through Tf5 is the same as that through Tf1 and is mirrored by Tf6, which is the active load for Tf2. The output current ΔI_{out} is therefore the difference between the drain currents of Tf1 and Tf2 (compare figure 7.3).

The concept of loading an IGFET with an active element (as in the case of the Tf2/Tf6 combination above) is a very powerful one, having regard to the many different polarity combinations made possible by the family of IGFETs, and as such, merits further consideration.

11.9.4 Active Loads and Complementarity

Consider first the circuit of figure 11.40(*a*). Here, an n-channel enhancement IGFET has its drain and gate connected so that $V_{GS} = V_{DS}$. The locus of this is included on the set of output characteristics of figure 11.37(*b*), because the

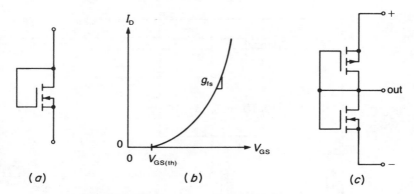

Figure 11.40. The IGFET as a nonlinear resistor.

connection is actually the same as for Tf1 in the current mirror configuration. Because $V_{GS} = V_{DS}$, then saturation is assured provided the available V_{DS} is always greater than $V_{GS(th)}$.

Figure 11.40(*b*) repeats the transconductance curve to make the point that the device may be thought of as a nonlinear resistor. Clearly, the same reasoning also applies to p-channel IGFETs, and diagram (*c*) shows how such a *complementary pair* can be used as a voltage divider. This is also an example of how the IGFET structure can replace comparatively space-consuming diffused resistors in IC design.

The incremental resistance of the *active resistor* is obviously $1/g_{fs}$, and this will be presented at either terminal, it now being a simple two-terminal device. At the upper end of the curve, an approach to linearity appears, so it can also reasonably be used as a load for small signals. Figures 11.41(*a*)

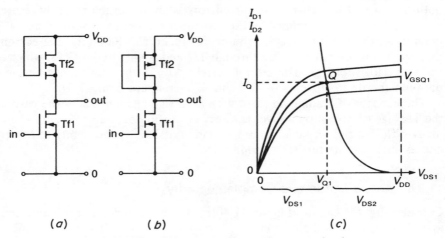

Figure 11.41. The IGFET active resistor as a load.

and (b) show active resistor loads of both similar and complementary form, respectively, and diagram (c) shows how the working point is defined by an appropriate Tf1 output curve and a Tf2 transconductance characteristic. (Note that here, the active resistor transconductance characteristic must be plotted from right to left because $V_{DS(2)} = V_{DD} - V_{DS(1)}$.)

Unfortunately, the voltage gain of such a stage is poor, being given as usual by the transconductance multiplied by the load, which in this case is $1/g_{fs(2)}$:

$$A_v = -\frac{g_{fs(1)}}{g_{fs(2)}} \tag{11.39}$$

so that the combination serves best as an inverting buffer.

If the gate of Tf2 in diagram (b) is held at a fixed voltage V_{GG}, as in figure 11.42(a), then it acts as a *current source load*, and its drain presents a high incremental resistance to Tf1. This makes for a high voltage gain, but does so at the expense of some bias stability because the working point Q is now defined by the confluence of two output characteristics, as shown in figure 11.42(b). Although this arrangment does provide good small-signal gain, the output voltage swing is limited. This is because, when Tf1 is OFF, the output voltage will rise to nearly V_{DD}; but when Tf1 is ON, the output voltage fall is limited by the threshold voltage. Hence, the transfer characteristic has the form shown in figure 11.42(c).

In switching applications, a major disadvantage of $V_{out(min)}$ being non-zero, is that power is consumed whilst Tf1 is in the OFF condition. Furthermore, power is also consumed during the ON to OFF excursion, for though the gain is high, it does not offer an adequately fast transient because Tf2 is essentially a static load.

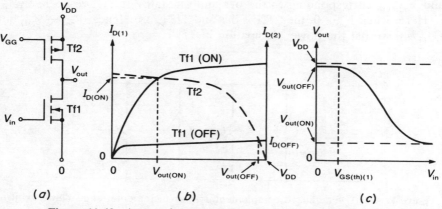

Figure 11.42. A CMOS inverter stage with a current-source load.

The voltage gain may be determined using the simple equivalent circuit shown in figure 11.43, which is valid for all forms of complementary pair.

Because the sources of the devices are connected to power supplies, they are effectively coupled together insofar as the signal is concerned, so that the two IGFETs appear in parallel, which makes the relevant analysis quite simple. For the present case, $v_{gs(2)}$ is zero, so that using expressions (11.34)–(11.36),

$$A_v = \frac{-g_{fs(1)}}{g_{ds(1)} + g_{ds(2)}} = \frac{-b_1(V_{GS(1)} - V_{GS(th)(1)})}{I_D(\lambda_1 + \lambda_2)}$$

and since $(V_{GS} - V_{GS(th)}) = \left(\dfrac{2I_D}{b}\right)^{1/2}$

$$A_v = \frac{-1}{\lambda_1 + \lambda_2}\left(\frac{2b_1}{I_D}\right)^{1/2}. \tag{11.40}$$

This shows that the voltage gain is inversely proportional to the root of the channel current, which is convenient for the design of low-current IC structures. Also, because $(\lambda_1 + \lambda_2)$ is small, the voltage gain must be high, which is to be expected having regard to the high incremental load resistance presented by the drain of Tf2.

Note that by inspection of figure 11.43, the output conductance will be simply $(g_{ds(1)} + g_{ds(2)})$ or $1/I_D(\lambda_1 + \lambda_2)$.

Note also that the signal and V_{GG} inputs could have been interchanged, in which case a *current sink load* would have resulted.

The dynamic range and the voltage gain may be further improved by driving *both* gates of the complementary pair, as shown in figure 11.44(*a*); and the ON and OFF conditions have again been defined by drawing the extreme output curves for both devices in diagram (*b*). Here, because both devices are active, the limits of the output voltage have been labelled $V_{out(Hi)}$ and $V_{out(Lo)}$ corresponding to the OFF and ON states of Tf1, respectively.

Here, when $V_{in} \to 0$, then Tf1 is OFF and Tf2 is ON. Hence, $V_{out} \to V_{DD}$ but $I_{D(1)} = 0$ so that the power dissipation in Tf1 is zero.

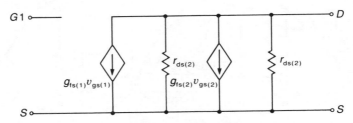

Figure 11.43. A generalized complementary-pair incremental equivalent circuit.

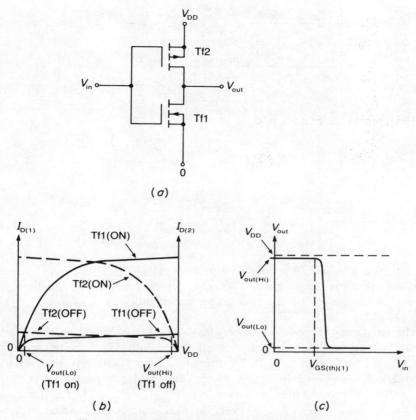

Figure 11.44. The push-pull CMOS inverter stage.

Conversely, when $V_{in} \rightarrow V_{DD}$, then Tf1 is ON and Tf2 is OFF. Hence $V_{out} \rightarrow 0$, but now $I_{D(2)} = 0$ so that the power dissipation in Tf2 is zero.

Hence, power is dissipated only during the transient between ON and OFF (unless a resistive external load is present, in which case the ON device will dissipate a little power); and this transient is rapid because both devices are being driven by the common gate signal. The transfer characteristic of figure 11.44(c) illustrates this, and shows how a very close approximation to an ideal switching action is achieved by the configuration.

The foregoing properties have led to the push-pull CMOS pair becoming of major importance in IC logic applications including microprocessors. However, in analog amplifier stages, the working point will be defined by the confluence of two intermediate output characteristics, in which case the voltage gain may again be determined using figure 11.43. Here, because the gates are connected together, $v_{gs(1)} = v_{gs(2)} = v_{in}$ so that,

$$A_v = \frac{-(g_{gs(1)} + g_{fs(2)})}{g_{ds(1)} + g_{ds(2)}}$$

$$= \frac{-[b_1(V_{GS(1)} + V_{GS(th)(1)}) + b_2(V_{GS(2)} + V_{GS(th)(2)})]}{I_D(\lambda_1 + \lambda_2)}.$$

Again using $(V_{GS} + V_{GS(th)}) = \left(\dfrac{2I_D}{b}\right)^{1/2}$

$$A_v = \frac{-\left[b_1\left(\dfrac{2I_D}{b_1}\right)^{1/2} + b_2\left(\dfrac{2I_D}{b^2}\right)^{1/2}\right]}{I_D(\lambda_1 + \lambda_2)}$$

$$= \frac{-1}{\lambda_1 + \lambda_2}(\sqrt{b_1} + \sqrt{b_2})\left(\frac{2}{I_D}\right)^{1/2}. \tag{11.41}$$

Note that this would reduce to equation (11.40) for the current source/ sink-loaded CMOS stage given identical device parameters. Further, the output conductances of the two stages are identical at $(g_{ds(1)} + g_{ds(2)})$ or $1/I_D(\lambda_1 + \lambda_2)$.

The frequency response of either the CMOS or current source/sink-loaded inverters may be determined by including the internal capacitances in the small-signal equivalent circuit of figure 11.43 (along with any external load impedance), as shown in figure 11.45. The Miller capacitance $(C_{gd(1)} + C_{gd(2)})(1 - A_v)$ would then be added to to $(C_{gs(1)} + C_{gs(2)})$ in the usual way, and the network analyzed. However, the maximum possible value of f_H will occur when a voltage generator is applied to the input, when $(C_{gd(1)} + C_{gd(2)})$ will be effectively grounded at their left-hand ends, and so appear in

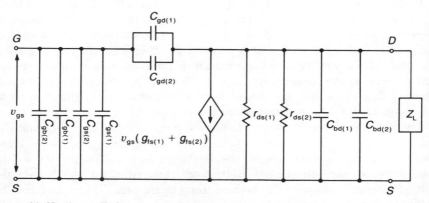

Figure 11.45. A small-signal equivalent circuit for CMOS frequency-response calculations.

parallel with all the other capacitances. For no external load, this value of f_H is also F_T and is given by $1/(2\pi CR)$ where

$$C = C_{gd(1)} + C_{gd(2)} + C_{bd(1)} + C_{bd(2)} \quad \text{and} \quad R = 1/(g_{ds(1)} + g_{ds(2)}) .$$

11.9.5 An IGFET/Bipolar Operational Amplifier

The fabrication of bipolar transistors and IGFETs of both polarities on the same silicon chip has led to some advanced integrated circuits[11,12,13] of which the CA3130 series operational amplifier is a good example. The functional circuit for this unit is shown in figure 11.46, and a description of its operation follows.

Tf6 and Tf7 are a pair of interdigitated p-channel enhancement structures, connected as a long-tailed pair, which presents a difference input resistance of over 1000 MΩ, with gate currents better than 10 pA at room temperature. Two bipolar npn structures, Tr9 and Tr10, are connected as a current mirror active load (see figure 7.3) which also accomplishes difference-to-single-ended conversion.

The second stage is a bipolar common-emitter amplifier Tr11 which has an IGFET active load comprised of Tf3 and Tf5. This is where the bulk of the amplification takes place: Tr11 has a gain of some 6000 compared with only about 5 in the IGFET input stage.

The output stage, Tf8 and Tf12, consists of a complementary IGFET pair, which has the major advantage that the output voltage swings can closely approach the supply voltages. The gain of this stage is very dependent upon the applied load, however, as has been pointed out, and is not likely to exceed about 30.

Because three distinct stages of voltage gain are involved, the provision of Miller compensation to effect a full 6 dB/octave roll-off characteristic can reduce the available gain (of some 110 dB) significantly. Therefore, the designer has the choice of what value of compensating capacitor (or other roll-off circuit) to insert, effectively across Tr11. (Other versions such as the CA3160 are internally compensated.)

The Zener diode, Z1, defines a voltage of about 8.3 V (provided that the total supply voltage is over about 10 V, below which Z1 cannot operate at all), and this is divided down by R_1, diodes D1–D4, and the p-channel enhancement IGFET Tf1. This latter structure, Tf1, is diode-connected and drives the mirror IGFET's Tf2 and Tf3. These structures are cascoded with Tf4 and Tf5, respectively, which are connected in the common-gate configuration. The cascode current-source load technique has several advantages, one of which is to increase the incremental resistance presented at the collector of the amplifing device. Consider, for example, Tf3 and Tf5. The equivalent circuit for this cascode pair is shown in Figure 11.46(*b*), where the dependent current generator of Tf3 has been omitted because $v_{gs(3)}$ is

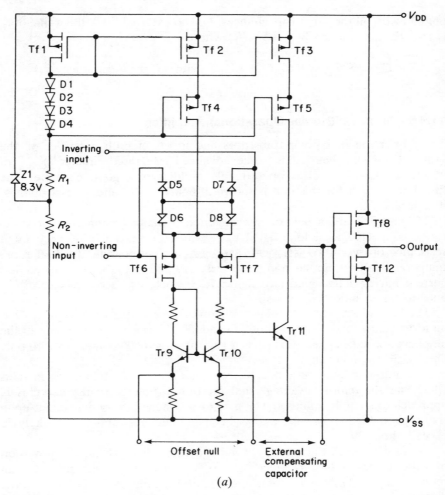

Figure 11.46. (*a*) Basic circuit of the CA3130 operational amplifier (reproduced by permission of Harris Semiconductor).

zero, the gate being connected to a power supply. Conversely, the Tf5 current generator is given by:

$$g_{fs(5)}v_{gs(5)} = -g_{fs(5)}v_{ds(3)} \cdot$$

This is because, although the gate of Tf5 is also taken to a power supply, its source is driven by the drain of Tf3, so making it a true common-gate configuration.

This identity has been included in the Thevenin equivalent of diagram (*c*), which is convenient for analysis. To determine the incremental resis-

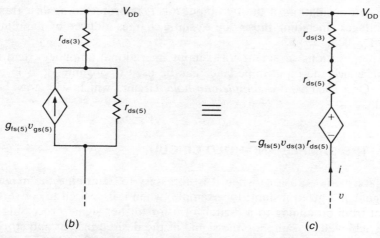

Figure 11.46. (b) and (c) Equivalent circuits for a cascode current source.

tance at the drain of Tf5, apply a voltage v and let the resulting current be i. Then,

$$i = \frac{v - g_{fs(5)}v_{ds(3)}r_{ds(5)}}{r_{ds(5)} + r_{ds(3)}} .$$

But $v_{ds(3)} = ir_{ds(3)}$,

so

$$i = \frac{v - g_{fs(5)}ir_{ds(3)}r_{ds(5)}}{r_{ds(5)} + r_{ds(3)}}$$

giving

$$i\left[1 + \frac{g_{fs(5)}r_{ds(3)}r_{ds(5)}}{r_{ds(5)} + r_{ds(3)}}\right] = \frac{v}{r_{ds(5)} + r_{ds(3)}} .$$

Hence, if the second term in the square brackets is much greater than unity, the incremental resistance at the drain of Tf5 is:

$$r_{(5)} = \frac{v}{i} \simeq \{g_{fs(5)}r_{ds(5)}\}r_{ds(3)} \qquad (11.42)$$

which shows that the channel resistance of Tf3 is significantly increased by cascoding Tf5. This very high incremental load resistance, presented to the collector of Tr11, explains why the voltage gain of this stage is so high.

The Tf2–Tf4 combination performs a similar function and provides the long-tail current for the input stage.

The gate insulation of the two input IGFETs is protected against transient high voltages (including those due to static charges induced by handling) by diodes D5–D8.

The applications of FET or IGFET-input operational amplifiers tend to be those in which the input impedance can be used to advantage, as in buffer stages. One such is the *sample-and-hold* circuit, which will now be described.

11.10 THE SAMPLE-AND-HOLD CIRCUIT

There are many occasions when it is necessary to determine the magnitude of a signal at a given instant; for example, when it is desired to convert such a signal from an analog to a digital form for further signal processing. This can be achieved by sampling the signal at the desired instant and storing or holding that magnitude until the appropriate operations have been performed on it.

In principle, this can be achieved by the simple circuit of figure 11.47(a),

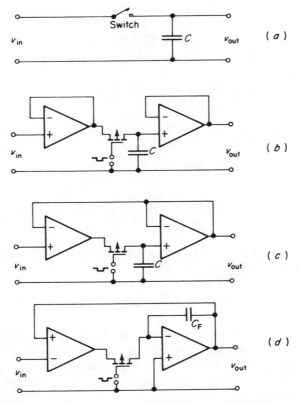

Figure 11.47. Development of the sample-and-hold circuit.

in which a capacitor is driven from the signal source until the switch is opened, when it holds the signal magnitude for a time dependent upon its own load. Clearly, the switch can be a FET or IGFET or a more complex combination of either. Even so, the circuit as shown has several disadvantages, and some of these, along with some possible solutions, are as follows.

(i) The storage capacitor presents a load to the signal source: this may be negated by using a FET or IGFET operational amplifier connected in the noninverting mode (figure 7.26(b)) as a voltage-follower, or buffer.

(ii) The load on the capacitor may tend to discharge it during the time it should be holding a sample: here again a FET or IGFET buffer may be used to advantage. The resulting circuit now appears as in figure 11.47(b), in which a p-channel IGFET has been included as the switch. This circuit highlights a third problem, namely:

(iii) The offsets of the two buffer amplifiers detract from the accuracy of the sample magnitude. This situation can be improved by the provision of overall feedback, as in figure 11.47(c).

(iv) For large signals, it is difficult to keep the IGFET switch hard ON or hard OFF during the "sample" and "hold" periods, respectively (as explained in section 11.11 on analog switches). This problem can be avoided by using a Miller integrator in place of the second buffer amplifier, so that the switch now works into a virtual earth, or summing point (see figure 7.25(d)). Overall feedback establishes unity gain, so that the input is now applied to the inverting terminal of the input amplifier.

An integrated circuit realization of this system has been described[14] which uses an operational transconductance amplifier (OTA) at the input, so that the Miller capacitance is driven by a current source, which improves the slew rate (and hence sampling speed) of the module. Slew rate has been considered in relation to operational amplifiers in chapter 7, but here, terminology concerned with the sample-and-hold function is more relevant, as can be explained with reference to figure 11.48.

Here, the broken curve represents an arbitrary signal waveform, whilst the full curve represents the waveform at the output of the sample-and-hold module. When the switch turns ON, a *switching spike* (negative-going in the case of a p-channel IGFET) is transmitted via the gate-channel capacitance to the output. (This spike will be very small if an IGFET having a low C_{gd} is used.) After this, the voltage on the sampling capacitor builds up to the level of the signal waveform. The time taken for the capacitor voltage to change to within a specified percentage of the signal level is called the *acquisition time*. This is usually specified for the worst-case condition, where the capacitor voltage must change over a full-scale excursion. Most of the acquisition time is due to the CR time constant of the circuit, but there is also a portion representing the delay (including that in the switching logic)

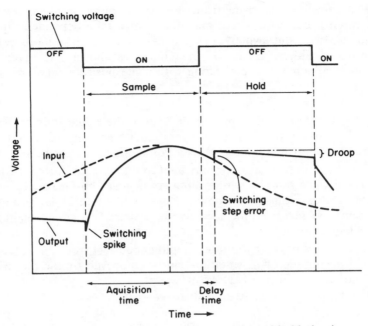

Figure 11.48. Performance of a sample-and-hold circuit.

before the IGFET switch closes; and a *settling time* which accounts for any overshoot as the capacitor voltage reaches the signal level. Clearly, the "sample" time must exceed the worst-case acquisition time.

When the switching waveform reverts to the "hold" condition, a small delay occurs, representing the finite *aperture time* it takes for the switch to open plus a *settling time* during which the transients decay. There is also a switching step error as shown in the diagram, when the IGFET is driven OFF.

After this, the charge on the capacitor begins to leak away, albeit very slowly because of the very high IGFET switch OFF resistance and input resistance of the following buffer amplifier. The amount by which the capacitor voltage changes during a "hold" period is called the *droop*, and this again is usually specified for worst-case conditions.

All the various sources of error must be taken into account when critical sample-and-hold functions are necessary, and it should be remembered that the speed of operation is limited by these errors. For example, the acquisition time limits the shortness of the "sample" period, whilst the droop limits the length of the "hold" period.

A comprehensive survey of the use of sample-and-hold modules in data acquisition and processing, and a more detailed treatment of the errors implicit in them, is given by Sheingold[15].

11.11   FURTHER IGFET APPLICATIONS

The IGFET has applications both as a chopper and as an analog gate, and in both cases the p-channel enhancement type is particularly suitable because of its unique biasing requirements and the comparative ease with which multi-device integrated circuits may be fabricated. Figure 11.49 shows the circuit for an integrated shunt-series IGFET chopper, complete with driver stage. Notice that because the gate currents are essentially zero, and because the driver gate and drain voltages rise and fall in opposition, these driver voltages may be applied directly to the two chopper gates. Notice also that the substrates have been connected to a positive voltage V_{DD2}. This is so that the input signal may go positive without forward-biasing the junctions between sources, drains, and substrates.

Figure 11.50 shows part of an IGFET multiplexer array. This is also highly suitable for monolithic integrated circuit fabrication, and numerous multi-element units are readily obtainable.

The complementary IGFET configuration is particularly useful as an analog gate, which is often referred to as the *CMOS transmission gate*. In this application, depicted in figure 11.51, the two channels are connected in parallel; and the gates are switched in antiphase so that the IGFETs are both ON or both OFF.

When the IGFETs are ON, an input signal v_{in}—which may take either polarity—will be transmitted to the output with little distortion. It must, of course, not exceed V_{DD} or V_{SS} in magnitude, otherwise the channel–substrate junction will become forward-biased, which can be serious, as will be explained in the next section. When the IGFETs are OFF, the signal is effectively blocked.

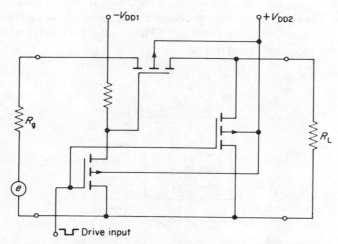

Figure 11.49. Series-shunt integrated IGFET chopper.

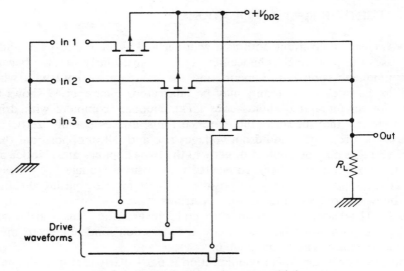

Figure 11.50. Integrated IGFET multiplexer.

In the ON condition the value of r_{ds} will be a function of the actual voltage applied between the gate and the source for each IGFET. This voltage, v_{gs}, must be given by the difference between the applied gate voltage and the signal voltage, v_{in}:

$$v_{gs} = v_g - v_{in} \,.$$

Notice that for the n-channel IGFET, v_g must be positive for the ON condition, so that when v_{in} is also positive, v_{gs} will be less than v_g. This means that $r_{ds(n)}$ will be increased, especially if v_{in} approaches v_g in magnitude. However, for the p-channel IGFET, v_g is negative, so that $(v_{g(p)} - v_{in})$ is *more* negative than $v_{g(p)}$. Hence, the p-channel IGFET is held hard ON, and $r_{ds(p)}$ will be minimal.

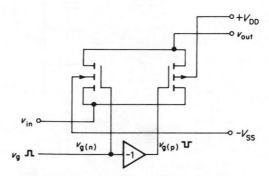

Figure 11.51. The CMOS transmission gate.

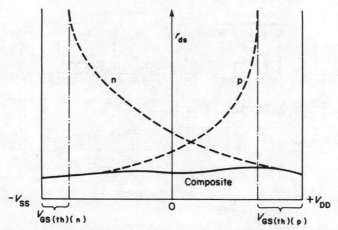

Figure 11.52. The composite ON characteristic of the CMOS transmission gate.

This situation can be visualized by plotting the two curves of r_{ds} versus v_{gs} as in figure 11.52 and combining them in parallel to give the ON resistance characteristic shown. At the time of writing, this rather oddly shaped characteristic takes values of (typically) a few hundred ohms for contemporary CMOS transmission gates.

From a dynamic point of view, the interelectrode capacitances of the CMOS structures are important because they limit the operating speed and also transmit pulses at the leading and trailing edges of the gate drive waveforms. These limitations are not serious, however, and CMOS transmission gates form the bulk of available analog gates and multiplexers.

11.12 INTEGRATED CIRCUITS

FET and IGFET structures are widely used in integrated circuits, notably of course in the form of logic modules, but also in uncommitted arrays, operational amplifiers, and analog switches and multiplexers.

In the particular case of transmission gate modules, the circuit of figure 11.51 is commonly realized as an integrated circuit in which the gate drive inverter also takes the form of a CMOS pair. Further, multi-channel modules of this form are integrated along with decoding logic so that a digital input may be used for sequentially switching ON the set of transmission gates. Such an integrated circuit forms a complete CMOS multiplexer.

Figure 11.53(a) shows an idealized cross-section through a CMOS pair diffused into an n-type substrate. Here it will be seen that the n-channel source and drain have been diffused into a "well" (or "tub") of p-type material which is isolated from the substrate only by a (reverse-biased) pn junction.

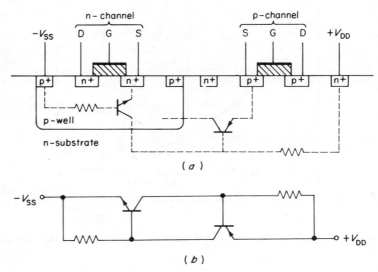

Figure 11.53. (*a*) Idealized cross-section of CMOS pair showing parasitic bipolar structures; and (*b*) circuit equivalent for bipolar structures (compare with figure 12.9).

This form of construction not only results in a complementary pair of IGFETs, but also in a pair of parasitic bipolar transistors (shown dotted). Taken together, these two transistors form a positive-feedback pair as in figure 11.53(*b*), and in chapter 12 it will be shown that if the product of the current gains of such a pair exceeds unity, then a high current flows which can only be removed by an external circuit interruption. Although the current gains of the parasitic transistors are normally very low they can be made high if a large current is injected into the loop, even transiently. Such a condition can occur if, for example, the channel–substrate junction of the n-channel IGFET becomes forward-biased; that is, if v_{in} goes more negative than $-V_{SS}$, as has been mentioned. (Note that this can quite easily happen if a negative v_{in} is present *before* $-V_{SS}$ is switched on.) This condition implies that a forward current flows between the p-well and the n⁺ source or drain of the n-channel IGFET. This initiates the condition known as *latch-up*[16].

Several manufacturers have evolved different methods of countering latch-up: for example, the Harris Semiconductor Corporation completely isolates the two IGFET structures using a glassy insulating layer. This "dielectric isolation" prevents any form of parasitic interaction between the structures. Some other manufacturers endeavor to reduce the h_{FE} product of the parasitic transistors by more conventional techniques: Siliconix uses a heavily doped p-layer buried under the p-well, and Intersil isolates the p-well from the negative line but connects it via a reverse-biased diode structure to the positive line.

All of these methods have advantages and disadvantages, and in particular all affect the degree of over-voltage protection which can be included in the integrated circuit. Such considerations serve to illustrate the level at which the manufacturer and the circuit designer must have an understanding of each other's technology. The need for the electronic systems designer to appreciate the problems presented by the physical nature of the devices employed is apparent when it is seen that a typical 16-channel CMOS multiplexer contains not only the 16 IGFET pairs, but also an inverter for each pair, plus a set of logic gates for operating each channel according to an applied 4-bit digital signal. In addition, structures for over-voltage protection and facilities for "make-before-break" operation are included.

As in the case of the integrated operational amplifier, the entire circuit is represented by a block diagram with a set of input, output, and drive pins. It is the function of the designer to recognize all the limitations presented by the various components, such as those involving signal magnitude and frequency, supply line transients, latch-up (if any), and output offsets due to the switching voltage.

The next section deals with power IGFETs, but at this point it must be mentioned that both they, and bipolar power devices, may be integrated along with low-level logic on the same chip using modified versions of the isolation techniques described above. These so-called "smartpower" ICs are able to perform logic functions which determine when their power switching devices should change state. A very wide range of smartpower chips is available for applications ranging from the operation of stepper motors or gas discharge displays to the control of the switch-mode power supplies described in chapter 14.

11.13 THE POWER IGFET

Two of the problems posed by the power bipolar transistor are firstly that the base requires significant drive current, and secondly that minority carrier storage limits the speed of such devices. Both problems can be overcome by the power IGFET, because its gate demands only a voltage drive, and being a majority carrier device it can be much faster than its bipolar equivalent. However, the input capacitance to a power IGFET, C_{iss}, can be several thousand picofarads, so that unless the drive circuit can transiently charge this capacitance the full speed capability of the device cannot be attained.

The first really practical power IGFETs were introduced by Siliconix Inc., who used a vertical, rather than the usual lateral structure to achieve a short channel length. An idealized cross-section of an n-channel power IGFET using this "V-MOS" structure is shown in figure 11.54(a). Here, an n^+-type substrate with an n^- epitaxial layer is used as the drain and a p-type channel region is diffused into it. An n^+ region is diffused in to form the source, and finally a V-groove is etched using potassium hydroxide. Silicon dioxide is

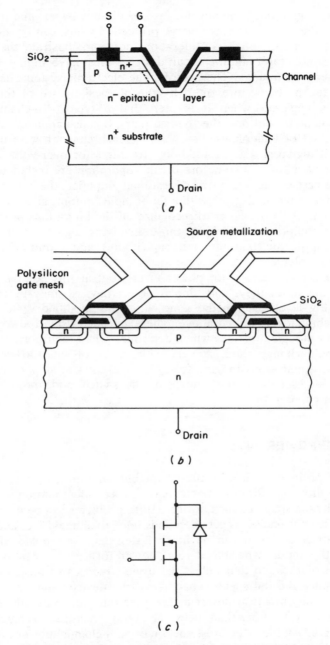

Figure 11.54. (*a*) Idealized cross-section through V-groove IGFET structure. (*b*) Idealized cross-section and upper surface of International Rectifier HEXFET® structure (reproduced by permission). (*c*) A power IGFET symbol including the parasitic reverse diode.

grown over the surface, including into the V-groove, and aluminum metallization is deposited through windows to form the electrodes, as shown.

This is now similar to a vertical bipolar transistor structure, with the addition of the V-groove, but the operation is quite different. The actual n-channel is formed at the interface of the p-region and its SiO_2 layer, and controlled by a positive voltage applied to the metallization within the groove, which is the gate.

This vertical structure results in a very short channel length, making for a low value of $V_{gs(th)}$, and its width can be increased to accept high currents, but with an inevitable rise in C_{iss}; small power devices exhibit values in the tens of picofarads. The flattening of the V-groove apex is a development which results in a lower value of $r_{DS(on)}$ and a higher breakdown voltage than for the earlier "sharp apex" format.

Power IGFETs do not experience second breakdown as do bipolar devices, nor do they suffer thermal runaway. This latter attribute is because the channel exhibits the usual positive temperature coefficient of resistivity, so that any concentration of current into a small region would heat that region, so increasing its resistance and reducing the local current density. In fact, this phenomenon makes possible two very important developments. Firstly, it means that power IGFETs can be connected in parallel to increase the current-carrying capability of a circuit, and second, it allows many individual small IGFET structures on a monolithic substrate to be paralleled using metallization, resulting in a single device with major power-handling characteristics. This latter realization is best exemplified by the International Rectifier Inc. HEXFET® structure, shown in figure 11.54(b). Here, the n-substrate forms a common drain for a matrix of some 77 500 hexagonal IGFET cells per square centimeter.

Each cell is fabricated by a double diffusion (DMOS) process, during which the p-regions and n-regions are formed as shown. The channels are again located at the interfaces of the p-regions and the overlying SiO_2 layers. Notice that the gate is also common to all the cells and is a mesh of polysilicon which is completely enclosed and insulated by the SiO_2. This configuration makes possible the metallization of the entire surface of the structure, so that all the sources are effectively paralleled to form a single HEXFET.

This is yet another technique of forming a short-channel IGFET, and one which can lead to some highly desirable power-handling characteristics. For example, consider the IRF 450:

$$V_{DS(max)} = 500 \text{ V}$$

$$I_{D(max)} \text{ (at } T_c = 25°C) = 13 \text{ A}$$

$$r_{DS(max)} \text{ (at } T_c = 100°C; I_D = 8 \text{ A)} = 0.4 \text{ }\Omega$$

$$C_{iss(max)} = 3000 \text{ pF} .$$

These parameters show that although loads of several kilowatts can be switched, the transient drive currents needed to charge the high input capacitance must be large, otherwise the inherent very fast switching speeds of the device cannot be used to advantage. For a load current of 5 A at 250 V, these speeds are (maximally):

$$t_d = 35 \text{ ns} \quad t_r = 50 \text{ ns} \quad t_s = 150 \text{ ns} \quad t_f = 70 \text{ ns}$$

which may be compared with those for the 2N6546 bipolar power transistor quoted in the last chapter.

The DMOS structure of most modern power IGFETs implies that the channel is always shunted by a pn junction from source to drain, as can be seen in figure 11.54(b), or more clearly in the magnified version of figure 11.57(a). To stress this point, the symbol often includes a diode, as in figure 11.54(c), for it is obviously of importance when there is any possibility of a reverse voltage being applied to the device, even transiently, and so should always be taken into account at the design stage.

Power IGFETs exhibit large values of g_{fs} (10 A V^{-1} for the IRF 450, for example) and after the threshold voltage has been reached, the transconductance curve approaches a straight line rather than a square-law characteristic. Hence, the output characteristics become fairly evenly spaced within the saturation region.

Like all IGFETs, the gate insulation can be broken down by over-voltage (including frictionally-induced), so that care must be taken when handling and installing unless protective diodes are incorporated between the gate and the source.

Applications of power IGFETs encompass most of those which can be accomplished using power bipolar transistors, such as power amplification and switching, but it is possible to extend the frequency range or switching speed in such circuits provided that the drive networks are properly designed. Also, in some cases, the very high input resistance of the IGFET can be put to good use, an example being the filament lamp driver shown in figure 11.55(a). If the IGFET were turned ON quickly, there would be a large current inrush via the cold—and hence low resistance—lamp filament, and this could be higher than the maximum allowable channel current level. However, if a long RC time constant is included as shown, the IGFET is turned ON slowly, so that the filament current and resistance also rise slowly.

Power IGFETs are available in both polarities, making possible complementary output stages. This is of interest not only insofar as "linear" amplifiers are concerned, but as fast switching output stages for Class D pulse-width-modulated circuits. The basic design problem to be faced here is ensuring that the gate drive has adequate transient current capability to quickly charge and discharge the high input capacitances presented by the power IGFETs. If the comparator of the pulse-width-modulator circuit of figure 10.10 can supply sufficient output current, then it may be used to

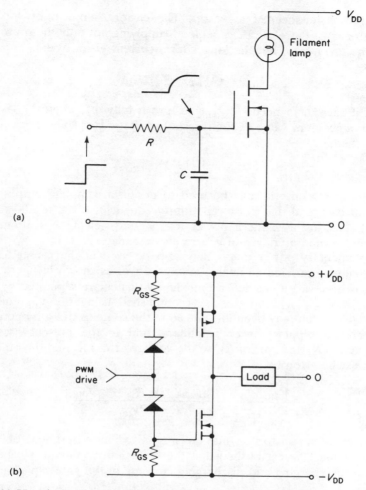

Figure 11.55. (*a*) Slow turn-on technique for filament lamp operation. (*b*) A complementary IGFET Class D amplifier output stage.

drive the complementary IGFET output stage of figure 11.55(*b*) to result in a complete Class D power amplifier. Here, a matched pair of Zener diodes is used to establish that the IGFETs cannot both be ON concurrently, and if the two supply voltages are equal ($\pm V_{DD}$), then in order to prevent a continuous current flow through these Zener diodes V_Z must be greater than V_{DD}.

The drive voltage switches from nearly $+V_{DD}$ to nearly $-V_{DD}$, and when passing though zero volts both of the Zener diodes will be nonconducting, so that each IGFET is held OFF by its gate–source resistor R_{GS}. When the drive voltage increases in either direction, one of the Zener diodes will eventually conduct, so that the voltage across one of the resistors R_{GS} will rise to turn its associated IGFET ON. The current supplied by the drive comparator must therefore be capable of supplying not only the transient to charge the IGFET gate, but the longer term current given by:

$$I_{\text{drive}} = (2V_{\text{DD}} - V_{\text{Z}})/R_{\text{GS}} \, .$$

Also, if each IGFET needs $V_{\text{GS(ON)}}$ to turn it fully ON (where V_{GS} is made up by the maximum value of the threshold voltage plus some extra defined voltage) then,

$$(2V_{\text{DD}} - V_{\text{Z}}) \not< V_{\text{GS(ON)}} \, .$$

The foregoing criteria may be used to design a complete output stage, and it will be found that pulse-repetition-frequencies of up to 200 kHz are easily attained using modern power IGFETs. Fast-slewing operational amplifiers and comparators are of course also mandatory.

Complementary power IGFETs may also be used in the source-follower configuration to provide an analog voltage drive, as shown in figure 11.56. Here, it is necessary to establish that the two gates are separated by about $\{V_{\text{GS(th)(1)}} + V_{\text{GS(th)(2)}}\}$ and this is accomplished using the V_{BE}-multiplier shown. This has already been introduced in the contexts of both operational amplifiers and bipolar power amplifiers, and in the present case, V_{BE} appears across R_1 and part of RV so the voltage across R_2 and the other part of RV must be given by

$$\frac{V_{\text{BE}}}{R_1 + RV_{(1)}} = \frac{V_{\text{GS(th)(1)}} + V_{\text{GS(th)(2)}} - V_{\text{BE}}}{R_2 + RV_{(2)}} \, .$$

Here, crossover distortion may be minimized by adjustment of RV. In addition, it must be established that the bipolar drive circuit is capable of providing the charging and discharging currents to the gate capacitances of

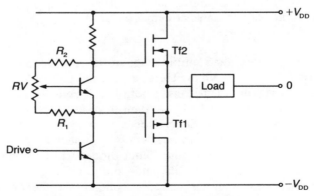

Figure 11.56. A CMOS pair output stage.

Tf1 and Tf2, otherwise the high-frequency performance of the circuit will be severely curtailed.

When making a choice between power bipolar transistors or IGFETs, particularly in switching circuit applications, the relative power dissipations should be carefully considered. This is because the IGFET channels can dissipate large amounts of power in the ON condition at high currents compared with their bipolar equivalents. To minimize this problem, the (already low) channel resistances may be further reduced by choosing *conductivity modulation* devices. Figure 11.57(a) gives an idealized cross-section for the basic DMOS structure, such as is used in the multi-cell power IGFETs already described; and diagram (b) shows how the n⁻-type epitaxial

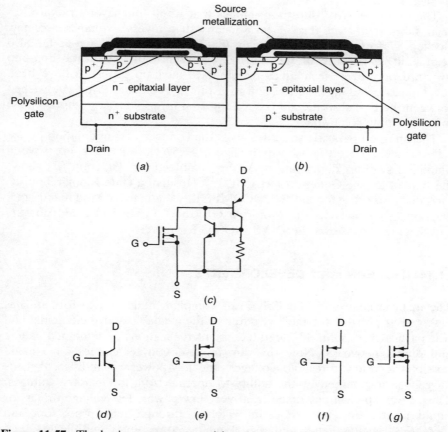

Figure 11.57. The basic DMOS structure (a) and its conductivity modulation modification (b) with an equivalent circuit (c). Symbols for the General Electric IGT™ (d), the RCA COMFET (e), the Motorola GEMFET (f) and the IXYS MOSIGT (g). (Note: RCA and General Electric semiconductors are part of the Harris Semiconductor group).

layer may be grown on a p$^+$-type substrate in the conductivity modulation device. The high resistivity of the n$^-$-epitaxial layer limits the ON-state current density of the normal DMOS device, but this may be markedly increased by the injection of minority charge carriers (holes in this case) from the p$^+$-substrate in the conductivity modulation structure.

The current-carrying capacity of the conductivity modulation device is much higher than for a comparable power IGFET[17], but this is bought at the price of a slower turn-OFF time, which in some cases can be several microseconds, though the turn-ON time remains at a fraction of a microsecond. This is because the minority carriers in the epitaxial region must recombine with the majority carriers before conduction ceases; whereas of course, the normal IGFET is simply a majority-carrier device which requires no such mechanism.

The presence of p$^+$-substrate implies that a full four-layer structure has been fabricated, and this in turn means that reverse voltages are now blocked. Unfortunately, it also means that the potential for latch-up has also been built in because of the inevitable presence of a parasitic npn-pnp structure. However, techniques to minimize latch-up problems have been developed, including the provision of a thin n$^+$-epitaxial layer sandwiched between the p$^+$-substrate and the n$^-$-epitaxial layer. This also improves the switch-off time somewhat.

The npn-pnp parasitic structure gives the conductivity-modulation device an equivalent circuit as shown in figure 11.57(c), but there are several simplified symbols depending upon the manufacturer. Diagram (d) shows the symbol for the General Electric IGT^{TM} (Insulated Gate Bipolar Transistor); diagram (e) is for the RCA COMFET (Conductivity Modulated FET); diagram (f) represents the Motorola GEMFET (Gain Enhanced MOSFET) and diagram (g) shows the IXYS MOSIGT symbol.

11.14 MODERN IGFET DEVELOPMENTS

The IGFET structure—particularly in its complementary (CMOS) format—has emerged as possibly the most versatile of the available active elements, and it is particularly useful in integrated circuit realizations of combined analog and digital networks. Now that ion implantation techniques have made possible very low threshold voltages, the low-power capabilities of CMOS have been augmented by the ability to operate from low battery voltages. Thus, CMOS applications range from very low-power, low-voltage drives for liquid crystal displays (LCDs) in watches, through microprocessors and memories, analog-digital and digital-analog converters, multiplexers, and operational amplifiers, to large output devices.

Further, IGFET switches can be used actually within analog circuitry, and an example has been met in chapter 7, where the very low drift Intersil CAZ amps and flying capacitor operational amplifiers were described. Such circuits depend upon the capability of IGFET switches to act as bidirectional analog gates for small signals, and this attribute has also made possible *switched-capacitor filters*[18].

As has been previously mentioned, it is difficult to produce accurate on-chip resistors, particularly in high values, and such structures also consume considerable chip area. Therefore: if smaller area on-chip capacitors can be substituted, complete single-chip filters become possible. In fact, this can be accomplished using the basic switching method shown in figure 11.58(*a*). Here, S is a two-way switch which moves back and forth between contacts "a" and "b" at a rate of f_c changeovers per second. When it is at position "a," the capacitor C charges to v_1 and when it moves to position "b," the voltage on C charges to v_2. Hence, the amount of charge transferred between "a" and "b" is Q, where:

$$Q = C(v_1 - v_2).$$

Because the switch moves forward and back every period, the mean current flow will be:

$$I = \frac{Q}{T} = \frac{C(v_1 - v_2)}{T} = f_c C(v_1 - v_2)$$

which implies that the circuit is equivalent to a resistance R, where,

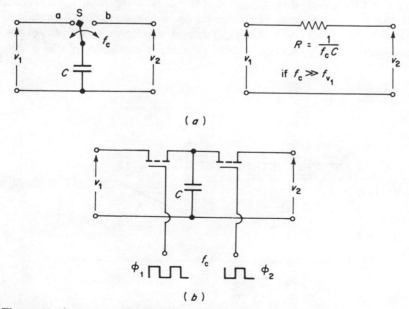

(*a*)

(*b*)

Figure 11.58. The switched-capacitor concept (*a*) and its realization (b).

$$R = \frac{v_1 - v_2}{I} = \frac{1}{f_c C} . \tag{11.43}$$

Notice, however, that this can only be realistic if the switching frequency f_c is much greater than the highest frequency present in the voltage v_1. If this is so, this *sampling* circuit does substitute directly for a resistance. Figure 11.58(*b*) shows how a pair of IGFET structures can produce the switching action needed, and these can be incorporated on the chip along with the capacitor.

Complex filters of high order have been produced by this method, but the integrating module of figure 11.59 will here suffice as an illustration of a single chip realization of IGFET switches, on-chip capacitors and an operational amplifier.

Using equation 7.40 for a normal op. amp. integrator, and incorporating equation 11.43 gives:

$$v_{out} = \frac{1}{RC_F} \int e \, dt = \frac{f_c C_G}{C_F} \int v_{in} \, dt . \tag{11.44}$$

Notice here that it is the *ratio* of the two capacitors which partly defines the gain, and that it is ratios as opposed to absolute values which can be well defined in monolithic ICs. Further, the value of this gain can be changed externally by altering the frequency f_c of the two-phase gate drive.

When a pair of IGFET structures is used to pass a charge from one capacitor to another, it is obvious that the drive waveforms must be so timed that the two switches do not close or open together. This situation is not difficult to achieve using logic gates, and appropriate two-phase clocking techniques are well known. If such a system is extended to a long string of

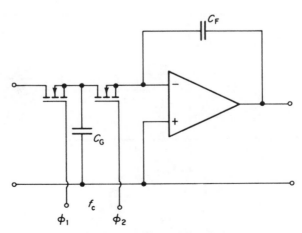

Figure 11.59. A switched capacitor integrator.

IGFET/capacitor structures, then charges can be moved along the string in a time given by the cycling period $(1/f_c)$ multiplied by the number of elements in the string. This is the principle of the *bucket-brigade* device[19] which has numerous applications mainly involving the delaying of a (sampled) analog input signal. Delay times can be of the order of milliseconds and devices can be cascaded, so that in audio engineering, multiple loudspeaker systems can be properly phased, and reverberation and chorus effects can be obtained in electronic organs.

This is one of the general family of *charge transfer* devices, which utilize the transfer of "packets" of charge rather than continuous currents. One of the family, which holds considerable promise in electronic engineering, is the *charge-coupled device* or CCD in which packets of minority carriers are moved along within a substrate under the influence of a series of IGFET-type gates.

11.14.1 Charge Control Devices (CCD)

These components can accept either digital or analog signals, and can act as stores, shift registers or delay lines. They can also accept optical signals and are used in solid state television cameras.

Both n- and p-channel CCD exist, and can be fabricated so that the charge carriers move along a $Si-SiO_2$ interface, or within a buried channel. The simplest configuration to understand is the n-channel surface type, and this is represented in figure 11.60 which will now be used to explain the principle of operation.

The idealized cross-section of a piece of the CCD (diagram (a)) shows that two n-type diffusions into a p-type silicon substrate have been made. These can be represented as diodes in the equivalent circuit of diagram (b), and termed the input and output diodes. A layer of SiO_2 is grown over the substrate, and (in this example) a set of six parallel metal stripes are deposited between the input and output metallizations. The stripe nearest to the input diode is called the input gate, and it can function in the same way as the gate of an n-channel IGFET.

The next three stripes are connected to three lines carrying the voltage waveforms v_1, v_2, and v_3 shown in diagram (c). These are followed by a further set of three stripes carrying the same sequence of waveforms.

Finally, an output gate (similar to the input gate) and an output diode (similar to the input diode) are shown.

At the input end of the structure, charges (electrons in this case) can be injected by the input diode, the applied voltage determining the *amount* of charge so injected. If the input gate is biased positively (or receives a positive pulse) the resulting positive surface potential causes a charge packet to move along the $Si-SiO_2$ interface, to the right.

Now consider the waveforms of diagram (c).

At time t_a, voltage v_1 (on the first stripe) produces a positive surface

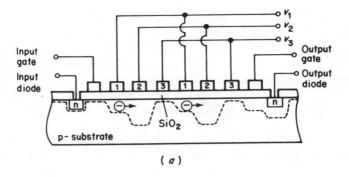

(a)

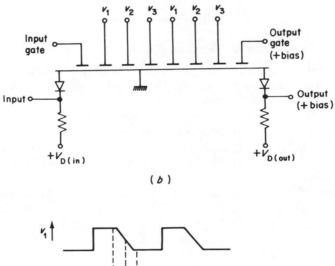

(b)

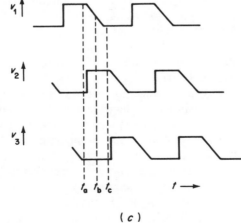

(c)

Figure 11.60. The three-phase surface channel CCD: (a) idealized cross-section showing depletion layer (dotted) for voltages v_1, v_2, and v_3 at time t_b; (b) equivalent circuit; (c) idealized timing waveforms.

potential which causes the charge packet to move under this stripe; and establishes a deep depletion layer. Voltages v_2 and v_3 are zero, however, which prevents the charge packet moving further to the right.

At time t_b, voltage v_2 on the second stripe has risen, whilst v_1 is in the process of falling. The resulting depletion layer depth is represented in diagram (a) by a dotted line; and the (negative) charge packet can be seen moving under the second stripe as its associated surface potential becomes more positive.

At time t_c, only v_2 is positive, and so the charge lies under the second stripe; that is, the charge has beeen transferred by the voltage waveforms applied to the stripes. These stripes are therefore called *transfer electrodes*.

After time t_c, the third transfer electrode is energized, and the charge packet again moves to the right.

Finally, the first (and fourth) transfer electrodes are again energized, and a new charge packet is acquired if permitted by the input gate *and* the input diode. At the same time, the original charge packet moves to the second *element* of three transfer electrodes.

A CCD may have any number of elements, but at the end of the series the emerging charge packets are collected by the output diode, which is biased positively for this purpose. The output gate must also be positive when it allows the output diode to perform the collection; hence, this gate can turn the output ON or OFF, irrespective of the input condition.

The sequence of three voltage waveforms applied to each element of three transfer electrodes consists of three identical waveforms having the phase relationships shown in diagram (c). The sequence is therefore referred to as *three-phase operation*.

If the frequency of the drive waveforms is f_0, then the time taken for a charge packet to cross one element must be $1/f_0$. Therefore, the *delay time*, T_D, for a charge packet to move across the total number of elements N between input and output must be:

$$T_D = N/f_0 . \tag{11.45}$$

If an analog signal is applied to the input diode, then when the input gate is pulsed positively, a charge packet will be transferred under it which is proportional to the magnitude of the applied signal at the moment of sampling. (In this sense, the input of the CCD performs a sample-and-hold function.) This charge packet may then be transferred along the CCD by the transfer electrodes, and will reach the output after the elapse of the delay time, T_D.

Given a CCD having some hundreds or thousands of elements, then a "section" of an analog signal can be stored in terms of a sequence of discrete charge packets, which can be retrieved at will. One obvious application is that of time expansion or compression; the stored signal can be read out at a different rate from which it is read in.

Clearly, the above sequence is equally valid for either analog *or* digital signals, the latter implying that the CCD can be used as a delay line, shift register, or dynamic memory. Insofar as analog signals are concerned, the output signal appears as a series of discrete spikes, so that a low-pass filter must be included to smooth out these spikes into a replica of the continuous input signal.

Tompsett and Zimany[20] point out that this filter should cut off at $f_0/2$ so that pick-up from the drive voltage pulses (at f_0) should be avoided; and that a similar filter should appear in the input circuit so that "beating" resulting from signal components over $f_0/2$ should also be avoided.

In practical CCDs, there is always a small fraction, ϵ of each charge packet which is left behind at each transfer. This leads to some degree of "signal spreading," since most of the small amount of charge left behind is not lost, but is collected during later transfers. This *transfer inefficiency* must increase with f_0 because as the operating frequency increases, progressively less time is available for each transfer. Except at the highest frequencies, the major cause of transfer inefficiency is the effect of interface states in temporarily trapping some electrons. This effect can be reduced by permanently passing a small standing charge through the CCD by appropriate biasing of the input diode and gate. This graphically named "fat zero" technique tends to keep the interface states permanently filled, so minimizing their effect on the signal charge packets. By such techniques, the value of ϵ can be brought below 10^{-4}, and the maximum value of f_0 above 10 MHz (at the time of writing).

A further improvement is possible by diffusing an n-layer on to the p-substrate immediately under the SiO_2 layer. This produces the *buried-layer* CCD, in which the charge packets do not come into contact with the $Si-SiO_2$ interface and its surface traps.

The maximum possible value of T_D is limited by the thermal generation of hole–electron pairs which eventually fill (static) depletion regions with electrons. This is the so-called "dark-current" phenomenon (see chapter 13) and it implies that the CCD is essentially a dynamic or "short-term" memory device. Short-term in this context implies maximum delays of about a tenth of a second for surface types, and up to several seconds for buried-layer types, especially when cooled.

The reason for the term "dark-current" is that the phenomenon deals with the generation of hole–electron pairs in the absence of incident radiant energy, including visible light. Hence, the mechanism involves only thermal energy, as does Johnson noise. However, if an optical system is arranged to produce an image on the surface of the CCD, then a corresponding charge pattern appears under the (transparent) SiO_2 layer, the light quanta having imparted sufficient energy to generate charge carriers. This charge pattern may be collected in the same way as if the charge packets had been injected as usual and transferred along the length of the device. Thus, a linear strip of a two-dimensional image can be "self-scanned" and converted into a series of pulses whose heights represent the shades of grey across that strip. If a number of CCDs lying parallel are fabricated, then the entire two-dimensional image can be converted to a sequence of discrete spikes. This is the principle of the CCD camera[21].

The CCD so far described is but one of a family; the alternate polarity (n-substrate) CCD is an obvious sibling. However, there are also other means of realizing charge transfer than the basic three-phase system presented here. For example, both two- and four-phase CCDs have been fabricated[22], and the former have become the more common. In these devices the surface potential is arranged to form a "downstairs-to-the-right" configuration, so that the charge packets always move one way. This structure is convenient in that the two sets of transfer electrodes can be taken to busbars at each side of the CDD, whereas the three-phase structure needs cross-under facilities.

Such simplification is very desirable in terms of cost, for fewer processing steps are required. However, it also hightlights a factor of very considerable portent for the future; that is, the CDD is actually the forerunner of a family of devices which operate *directly* on charge packets, which move in a homogeneous solid, as opposed to through a plethora of junctions. Hence, the CCD (unlike the bucket-brigade device) cannot be regarded as the monolithic equivalent of an array of discrete devices.

This concept has also reached fruition in the *magnetic-bubble device*[23], in which domains of oriented magnetism are caused to move in defined ways, as are the charge packets of CCD. At the time of writing, magnetic bubble memories having megabit storage capabilities are available in production form, whereas CCD memories have apparently not lived up to their earlier promise. They have, however, been extensively developed in other ways[24], notably in the realization of one- and two-dimensional imaging devices, transversal filters and signal conditioning modules[22].

REFERENCES

1. Gosling, W., Townsend, W. G. and Watson, J., 1971, *Field-Effect Electronics* (London: Butterworths).
2. Watson, J., 1968, Biasing considerations in FET amplifier stages, *Electron. Eng.*, **40**, 489.
3. Vogel, J. S., 1967, Non-linear distortion and mixing processes in field-effect transistors, *Proc. IEEE*, **55** (December).
4. Sherwin, J. S., 1966, Knowing the cause helps cure distortion in FET amplifiers, *Electronics*, **39**, 25 (December).
5. van der Ziel, 1962, Thermal noise in field-effect transistors, *Proc. IRE*, **50**, 1808–1812.
6. Zapf, H. L., 1978, DC transfer characteristic, offset voltage sensitivities, and CMRR of FET differential stages, *IEEE J. Solid St. Circuits*, **SC-13**, 262–265.

7. Cave, D. L. and Davies, W. R., 1977, A quad JFET wide-band operational amplifier integrated circuit featuring temperature-compensated bandwidth, *IEEE J. Solid St. Circuits*, **SC-12**, 382–388.

8. Hamade, A. R., 1979, Temperature compensated quad analog switch, *IEEE J. Solid St. Circuits*, **SC-14**, 944–952.

9. Allen, P. E. and Holberg, D. R. (1987), CMOS Analog Circuit Design (New York: Holt, Rinehart and Winston).

10. Bertails, J.-C., 1979, Low frequency noise considerations for MOS amplifier design, *IEEE J. Solid St. Circuits*, **SC-14,** 773–776.

11. Schade, O. H., 1978, BiMOS micropower ICs, *IEEE J. Solid St. Circuits*, **SC-13**, 791–798.

12. Fullager, D., 1980, CMOS comes of age, *IEEE Spectrum*, **17**, 24–27 (December).

13. Schade, O. H. and Kramer, E. J., 1981, A low-voltage BiMOS op. amp., *IEEE J. Solid St. Circuits*, **SC-16**, 661–668.

14. Stafford, K. R., Gray, P. R. and Blanchard, R. A., 1974, A complete monolithic sample/hold amplifier, *IEEE J. Solid St. Circuits*, **SC-9**, 381–387.

15. Sheingold, D. H. (ed), 1972, *Analog-Digital Conversion Handbook* (Norwood, Mass: Analog Devices Inc.).

16. Troutman, R. R., 1986, Latchup in CMOS Technology (The Netherlands: Kluwer).

17. Russel, J. P., Goodman, A. M. and Goodman, L. A., 1983, The COMFET—a new high conductance MOS-gated device, *IEEE Elec. Dev. Lett.*, **EDL-4**, 63–65.

18. Broderson, R. W., Gray, P. R. and Hodges, D. A., 1979, MOS switched-capacitor filters, *Proc. IEEE*, **67**, 61–75.

19. Butler, W. J., Barron, M. S. and Puckette, C. McD., 1973, Practical considerations for analog operation of bucket-brigade circuits, *IEEE J. Solid St. Circuits*, **SC-8**, 157–168.

20. Tompsett, M. E. and Zimany, E. J., 1973, Use of charge-coupled devices for delaying analog signals, *IEEE J. Solid St. Circuits*, **SC-8**, 151–157.

21. Melen, R. D. and Meindl, J. D., 1974, A transparent electrode CCD image sensor for a reading aid for the blind, *IEEE J. Solid St. Circuits*, **SC-9**, 41–48.

22. Beynon, J. D. E. and Lamb, D. R., 1980, *Charge-Coupled Devices and their Applications* (Maidenhead: McGraw-Hill (UK)).

23. Bernhard, R., 1980, Bubbles take on discs, *IEEE Spectrum*, **17**, 30–33 (May).

24. General C.C.D. papers, 1976, *IEEE J. Solid St. Circuits*, **SC-11** (February).

QUESTIONS

1. The junction FET is modeled only as a voltage-driven device. Why?

2. Why is the junction FET regarded as a "normally-ON" device?

3. For a junction FET, I_{DSS} can never be exceeded. Why?

4. In a junction FET common-source stage, C_{gd} has more effect on limiting the high-frequency response than C_{gs}. Why is this, even though $C_{gs} > C_{gd}$?

5. What is the approximate output resistance of a junction FET source-follower if g_{fs} is much greater than either g_{ds} or the load conductance?

6. In the two-generator noise model for a junction FET, the noise current generator may usually be neglected. Why?

7. The junction FET voltage-controlled-resistor (VCR) can only accept very small signals. Why is this?

8. The two types of bipolar transistor, the two types of junction FET, and the four types of IGFET all exhibit positive transconductances. Sketch the transconductance curve for each to demonstrate this.

9. What effect does the channel length modulation parameter λ have on the family of output characteristics of an IGFET?

10. Draw a circuit for a simple IGFET current mirror in which the current directions and voltage polarities are the same as those for a simple pnp transistor current mirror.

11. How can a complementary pair of IGFETs be used as a voltage divider?

12. Why is the CMOS inverter of major importance in switching applications, including logic functions?

13. Compare the bipolar and IGFET cascode circuits and list their attributes.

14. List and describe the major errors in a sample-and-hold circuit.

15. Sketch the circuit and characteristics of a CMOS transmission gate.

16. What is "latch-up"?

17. Why is it possible, in principle, to connect power IGFETs in parallel, but not bipolar power transistors (without extra components)?

18. What are the differences between a DMOS IGFET and a conductivity-modulated device in terms of both structure and characteristics?

19. Explain the switched-capacitor principle.

20. What is a charge-control device?

PROBLEMS

11.1. A FET is connected into the circuit of figure 11.61. If the quiescent drain current is 0.5 mA and the output voltage swings about a nominal 12 V with respect to the common line, determine the values of the following circuit parameters:

(a) R_D (b) R_S (c) $A_{v(fm)}$

(d) C_{in} (e) f_H (f) f_L

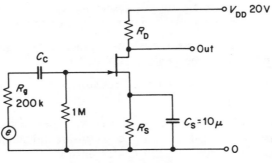

Figure 11.61.

Assume that C_C is very large and that the FET has the following parameters: $I_{DSS} = 1\,\text{mA}$, $V_P = -3\,\text{V}$, $C_{iss} = 8\,\text{pF}$, and $C_{rss} = 3\,\text{pF}$.

11.2. For the FET amplifier stage shown in figure 11.62, calculate the quiescent drain current I_Q if the mid-band voltage gain is -8. Also, determine the value of V_B and suggest values of R_1 and R_2 which will produce it. What is the high-frequency cut-off point if $C_{gs} = 5\,\text{pF}$ and $C_{gd} = 3\,\text{pF}$? For the FET used, $I_{DSS} = 4\,\text{mA}$ and $V_P = -3\,\text{V}$.

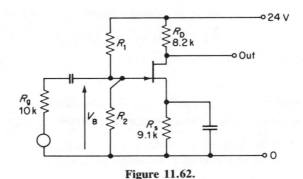

Figure 11.62.

11.3. The (unbypassed) source and drain resistors of a degenerate common-source FET stage are both $5.6\,\text{k}\Omega$. If the transconductance of the FET is $1.2\,\text{mA/V}$, by what percentage does the mid-band voltage gain from gate to drain differ from unity?

11.4. An n-channel junction FET has the following parameters:

$$I_{DSS(min)} = 2\,\text{mA} \quad I_{DSS(max)} = 5\,\text{mA} \quad V_{P(min)} = -3\,\text{V} \quad V_{P(max)} = -6\,\text{V}.$$

Design a small-signal CS stage so that the quiescent drain current falls between 0.8 and 1.2 mA. Assume a supply of 15 V and calculate the gate resistors R_1 and R_2 so that the input resistance is about $100\,\text{k}\Omega$.

11.5. A junction FET common-source amplifier stage with an 8.6 kΩ load resistor, is driven from a signal generator having an internal resistance of 10 kΩ. Neglecting the effects of biasing resistors, what is the high-frequency -3 dB point if the following parameters apply:

$$g_{fs} = 2 \text{ mA/V} \quad C_{iss} = 8 \text{ pF} \quad C_{oss} = 3 \text{ pF}.$$

11.6. A junction FET common-source stage uses an 18 V supply across which are connected biasing resistors R_1 and R_2, the common point of which is connected to the FET gate. A source resistor R_S and a drain load R_D are included.

For the FET used V_P lies between -2 V and -6 V, and I_{DSS} lies between 3 mA and 9 mA.

If the quiescent current is to be within the limits 1.0 mA to 1.6 mA, determine the value of R_S and choose R_1 and R_2 so that the input resistance to the stage is about 37 kΩ.

11.7 Design a CS junction FET amplifier stage using a single 12 V power supply and biased by connecting the gate to ground via a 2.2 MΩ resistor. Calculate the value of the source resistor R_S required to define a quiescent channel current of 1 mA and choose a drain resistor R_D to give a voltage gain of about -6.3.

If a sinusoidal signal generator having an internal resistance of 10 kΩ is applied to the input via a coupling capacitor, at what frequency f_H will the output signal be 3 dB below the mid-band level? Assume that $V_P = -6$ V, $I_{DSS} = 4$ mA, $C_{gs} = 8$ pF, and $C_{gd} = 5$ pF.

11.8. A degenerate CS stage has a drain resistor R_D and an unbypassed source resistor R_F. Its gate is connected to ground via a large resistor R_B. If the quiescent channel current is to be 1 mA, determine the value of R_F using a junction FET with $V_P = -6$ V and $I_{DSS} = 5$ mA.

What would be the value of R_D to provide a mid-band voltage gain of -3 from gate to drain?

11.9. For the bootstrap-follower of figure 11.17, calculate the values of R_{F1}, R_{F2}, and C_c if the quiescent channel current is to be 1 mA and the low-frequency -3 dB point is to be 1.5 Hz. Assume that the quiescent voltage at the output is 5 V, and that for the FET used, $V_P = -1.2$ V and $I_{DSS} = 6$ mA. Also assume that the gate resistor R_3 is 2 MΩ.

11.10. For the square-wave oscillator circuit of figure 11.23(a), calculate values of R_{F1} and R_{F2} so that the frequency of the square wave is given by:

$$f = 1/CR.$$

Assume that only 2 mA are available from the amplifier output for driving R_{F1} and R_{F2}, and that this output swings over ±14 V.

11.11. With the aid of a circuit diagram, describe how a junction FET can be used as an analog gate which will accept input signals of both polarities.

If such a FET has a V_P of -4 V and is to accept an input signal ranging from $+1$ V to -1 V, what gate-to-ground voltage is needed to hold it OFF under all conditions? What is the magnitude of the square pulse which will switch it ON under any conditions? (Assume that the load R_L is very large.) Sketch the waveform appearing across the load resistance during the ON pulse when the input voltage is at -1 V. Include the transients you would expect, and also the square drive waveform so that the transients and the waveform may be related.

11.12. A junction FET is used as a current-limiter to provide the long-tail current I_X for a bipolar difference stage. If the two quiescent collector currents are 1 mA each, calculate the value of the necessary source resistor if $V_P = -3$ V and $I_{DSS} = 4$ mA.

11.13. Below pinch-off the drain current of an IGFET is given by

$$I_D = -b\left(V_{DS}(V_{GS} - V_{GS(th)}) - \frac{V_{DS}^2}{2}\right).$$

Obtain r_{DS} from this expression and use it to calculate the percentage of E appearing at the input capacitor of the amplifier in figure 11.63 when the chopper IGFET is ON. The chopper drive square wave has a peak value of 10 V. What is the polarity with respect to ground?

The following values and parameters are applicable: $E = +10$ mV; $R_g = 1\,\mathrm{k\Omega}$; $R_1 = 2.7\,\mathrm{k\Omega}$; $V_{GS(th)} = -6$ V; and $b = +0.834 \times 10^{-3}$.

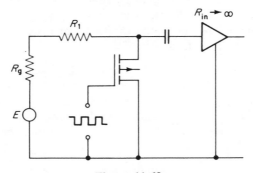

Figure 11.63.

11.14. For the integrator of problem 7.14, calculate the value of a switched capacitor which would replace R_G if the associated switching IGFETs had gate drive PRFs at 1 MHz.

11.15. A p-channel junction FET is connected as a CS amplifier stage with its gate taken to ground via a 10 MΩ resistor. The drain load resistor is 2.7 kΩ and the quiescent drain current is −2 mA. A signal generator with an internal resistance of 10 kΩ is applied to the gate via a coupling capacitor.

Draw the relevant circuit and determine the value of the (by-passed) source resistor R_S. Then calculate the mid-band voltage gain and the high-frequency −3 dB point f_H. The following parameters apply:

$$V_P = 4 \text{ V} \quad I_{DSS} = -10 \text{ mA} \quad C_{iss} = 10 \text{ pF} \quad C_{oss} = 4 \text{ pF} .$$

12

Triggers, Thyristors and Triacs

The semiconductor devices discussed so far have been capable of both continuous and switching operation but there also exists a class of component which exhibits inherently two-state operation. Some devices in this class have only one junction, whereas most have multiple junctions, but all depend on their capability of exhibiting *negative resistance slopes*. This attribute is actually a concomitant of positive feedback in circuits such as multivibrators using normal bipolar or field-effect transistors, but is inherent in the operation of the devices detailed below.

12.1 THE UNIJUNCTION TRANSISTOR (UJT)

The basic construction of the UJT is shown in idealized form in figure 12.1(a), where a bar of n-type silicon is shown to have a pair of ohmic (nonrectifying) contacts welded to each end. These are called "base 1" and "base 2," although their functions are not all analogous to the functions of a true base region in a normal bipolar transistor. The "emitter" junction is formed part-way along the bar, as shown.

When a voltage V_{BB} is applied between the two ohmic "bases" a current flows as shown in figure 12.1(b) and the bar acts as a simple silicon resistor. The fraction of V_{BB} appearing at the junction is termed the *intrinsic stand-off ratio* η, and when the emitter is made more positive than ηV_{BB} a current will flow to base 1. When this happens, a normal diode voltage V_D will appear, so that

$$V_E = \eta V_{BB} + V_D .\qquad (12.1)$$

The large number of holes injected into the n-region between the emitter

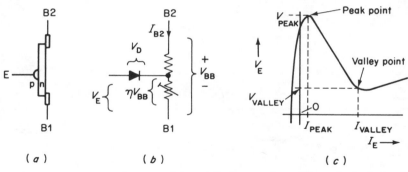

Figure 12.1. The unijunction transistor: (*a*) construction; (*b*) equivalent circuit; (*c*) characteristic.

and base 1 means that the adjacent part of the bar has a greatly reduced resistance so that V_E falls, leading to the negative resistance slope shown in figure 12.1(*c*).

Another way of looking at the device is to note that before the applied voltage reaches ηV_{BB}, the input current is that of a reverse-biased diode. When ηV_{BB} is reached the diode becomes forward-biased and the current rises rapidly, whilst the voltage decreases. Thus, the UJT is a voltage-operated device which presents a high input resistance until it breaks over, when it becomes a low-resistance device; that is, it is an OFF-to-ON voltage-controlled switch.

The peak point voltage shown in figure 12.1(*c*) can be designed to range from about 3–30 V, and the emitter current can usually accommodate a few tens of milliamps.

The basic circuit for almost all unijunction applications is that of a voltage sensor. Figure 12.2(*a*) shows a simple timing circuit using this principle. The capacitor C is charged via R, and when the peak point voltage is reached, C is discharged as the unijunction transistor switches ON. The value of R is made so large that the current then available is much less than the valley point current, and so the unijunction transistor switches OFF again. The values of R and R_{B1} can be calculated so that the cycle is repetitive and leads to the series of pulses shown.

Figure 12.2(*b*) includes the load-line representing R on the characteristic; and also the slope representing R_{B1}. For the switching cycle to be repetitive, the load-line R must cross the negative-resistance part of the characteristic so that its upper limit is bounded by the broken line representing R_{max}. This line crosses the characteristic just above the peak point, V_P, I_P. Thus, if the supply voltage is V_{CC}, then,

$$R_{max} < \frac{V_{CC} - V_P}{I_P}. \tag{12.2a}$$

If R_{max} is greater than this the UJT will never turn ON.

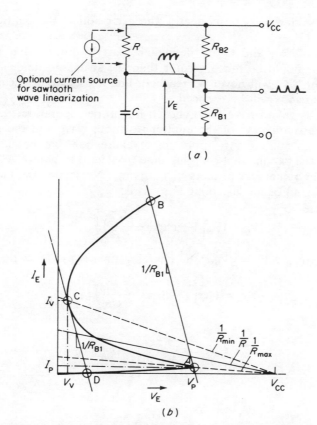

Figure 12.2. (*a*) Unijunction transistor circuit and (*b*) load-line diagram.

Similarly, R must not be less than the value shown by the broken line $R_{\min}$, which is given approximately by,

$$R_{\min} > \frac{V_{CC} - V_V}{I_V} . \tag{12.2b}$$

If $R_{\min}$ is less than this, the UJT will remain in the ON condition.

In general, R will be found to lie between $3\,\text{k}\Omega$ and $3\,\text{M}\Omega$ for contemporary devices.

In operation, the capacitor charges to the peak point "A", when the device switches ON and the working point moves rapidly to "B," which is defined by the slope of R_{B1}. As the capacitor discharges the working point moves to "C," then back to "D," which is again defined by the slope of R_{B1}. From here the cycle repeats.

The value of R_{B1} is normally very small ($<100\,\Omega$) so that its slope should be nearly vertical in the diagram (the shape of the characteristic has been exaggerated for clarity). Consequently, the excursion of the emitter waveform is nearly $(V_P - V_V)$.

The values of R_{B1} and C are defined in part by noting that the time constant $R_{B1}C$ should be much greater than the switching time of the UJT otherwise point "B" will never be reached, and if oscillation occurs at all, the signal excursions will be reduced.

If the various values have been chosen for proper operation, the period (and hence the frequency) of the emitter sawtooth wave and the B1 pulse train may be calculated. An approximate expression can be derived by assuming that the exponential charging time from zero to point "A" is much greater than the remainder of the switching time. Neglecting the fact that in operation charging begins at about V_V, this is:

$$V_E = V_{CC}\left[1 - \exp\left(-\frac{t}{CR}\right)\right].$$

At $V_E = V_P = \eta V_{BB} + V_D \simeq \eta V_{BB}$, the charging time is $t = T$, so that,

$$\eta V_{BB} = V_{CC}\left[1 - \exp\left(-\frac{T}{CR}\right)\right].$$

Hence,

$$\exp\left(-\frac{T}{CR}\right) = 1 - \frac{\eta V_{BB}}{V_{CC}}$$

or

$$-\frac{T}{CR} = \ln\left(1 - \frac{\eta V_{BB}}{V_{CC}}\right).$$

This gives,

$$T = CR \ln\left(\frac{1}{1 - (\eta V_{BB}/V_{CC})}\right). \tag{12.3a}$$

If R_{B1} and R_{B2} are sufficiently small that $V_{BB} \simeq V_{CC}$, this reduces to,

$$T = CR \ln\left(\frac{1}{1 - \eta}\right). \tag{12.3b}$$

The excursion of the pulse waveform developed across R_{B1} will be the difference between the peak point voltage, V_P, and the voltage defined by point "B" in figure 12.2(b).

The usefulness of R_{B2} is largely in temperature compensation. In equation (12.1) which gives V_E at the peak point, V_P, the diode drop, V_D, has a negative temperature coefficient, so that V_P falls as the temperature rises.

Conversely, the interbase bulk silicon resistance, R_{BB}, rises with temperature, so that if R_{B2} is correctly chosen, the actual voltage between the bases, V_{BB}, can be made to rise such that V_P remains essentially temperature independent. An expression for R_{B2} is usually provided by the manufacturer, and for the 2N2646 it is:

$$R_{B2} \simeq \frac{10\,000}{\eta V_{CC}} . \qquad (12.4)$$

Example

Using a 15 V supply and a 2N2646 UJT, design a pulse generator for a pulse rate frequency of 10 kHz. Assume that $\eta = 0.56$–0.75 and $R_{BB} = 4.7$–9.1 kΩ.

Using the circuit of figure 12.2(a), the value of R_{B2} may be determined using equation (12.4):

$$R_{B2} = \frac{10\,000}{\eta V_{CC}} .$$

That is,

$$R_{B2(max)} = \frac{10\,000}{0.65 \times 15} = 1190\ \Omega$$

$$R_{B2(min)} = \frac{10\,000}{0.75 \times 15} = 690\ \Omega .$$

Taking a preferred value of 1 kΩ, the true interbase voltage, V_{BB}, can be determined:

$$V_{BB} = \frac{V_{CC} R_{BB}}{R_{BB} + R_{B2}} .$$

That is,

$$V_{BB(max)} = \frac{15 \times 9.1}{9.1 + 1} = 13.5\ V$$

$$V_{BB(min)} = \frac{15 \times 4.7}{4.7 + 1} = 12.4\ V .$$

The value of CR may now be found from equation (12.3a) taking the widest possible limits:

$$CR = T \left[\ln \left(\frac{1}{1 - (\eta V_{BB}/V_{CC})} \right) \right]^{-1}$$

where $T = 0.1$ ms. That is,

$$CR_{(max)} = 0.1 \left[\ln \left(\frac{1}{1 - (0.56 \times 12.4/15)} \right) \right]^{-1} = 0.161\ \text{ms}$$

$$CR_{(min)} = 0.1\left[\ln\left(\frac{1}{1-(0.75\times13.5/15)}\right)\right]^{-1} = 0.089 \text{ ms}.$$

Taking a trial value of $C = 0.001\ \mu\text{F}$, then,

$$R_{(max)} = \frac{0.161}{0.001} = 161\text{ k}\Omega$$

$$R_{(min)} = \frac{0.089}{0.001} = 89\text{ k}\Omega.$$

These results suggest that R should be comprised of a 78 kΩ resistor in series with a 100 kΩ trimmer potentiometer.

Finally, the value of R_{B1} will be calculated with reference to the expected load.

If the voltage across C is applied to a high input impedance stage (e.g., a field-effect transistor) a sawtooth wave can be extracted. The exponential rise of this wave can be linearized by replacing R with an FET current limiter, or other constant current device (shown dotted in figure 12.2(a).

If R is made adjustable, an accurately calibrated pulse generator results; or, conversely, the first pulse can be used to switch on a relay which is thereafter held ON by a pair of its own contacts, while another pair deactivates the circuit. The latter device is, of course, a time delay network.

If a comparatively long delay is necessary, then the OFF-state leakage into the emitter is of importance (as is the leakage of the capacitor itself). This can range from the 30 μA of the 2N2646 at 30°C down to 0.02 μA for more sophisticated devices. Figure 12.3(a) shows how the leakage current can be isolated from the charging current using a low-leakage diode. It is sometimes cheaper to use this arrangement than to utilize a comparatively expensive low-leakage UJT.

The modification of figure 12.3(b) shows the circuit being fed from a diode pump (described in chapter 10). Here, a series of pulses—perhaps derived from another unijunction transistor—is allowed to charge up the

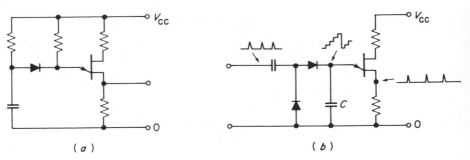

Figure 12.3. (a) A UJT leakage isolation circuit and (b) a pulse divider circuit.

tank capacitor C. If these pulses are of a uniform size, a fixed number will raise the potential of C to a little above the peak point voltage, and C will thereupon discharge. It is clear that such an arrangement represents a pulse counter or divider. Often, the circuit constants are defined so that the unijunction transistor fires after every tenth pulse, and a series of such circuits then forms a decade counter. Each unijunction transistor must, of course, feed some form of read-out device in such an application.

An alternative use for the circuit is as a staircase wave generator. Here, the voltage at the input of the unijunction transistor is applied to a high input impedance amplifier and the waveform shown is thereby amplified and rendered useful.

The unijunction transistor so far described has been the pn type, but it should be pointed out that the complementary, or np, version is also available. The circuitry is as for the pn type but with all polarities reversed. The construction is, however, different in that it consists of two transistor and two diffused resistor structures which are all fabricated as part of the monolithic planar process. This form of construction leads to greater parameter stability than is the case for the simple "bar" or "cube" structure of the pn type.

The actual switching action of the complementary unijunction transistor is similar to that of the four-layer diode, and this will be described in the next section. In the present context, however, it is relevant to mention that the full four-layer planar structure also lends itself to the programmable unijunction transistor (PUT) and this will be described at the end of the next section.

12.2 THE FOUR-LAYER OR SHOCKLEY DIODE

The four regions of the diode are represented in figure 12.4(a), from which it will be seen that the center junction is reverse-biased by the applied voltage, while the two outer junctions are forward-biased. Consequently, the device presents the high impedance typical of a reverse-biased junction,

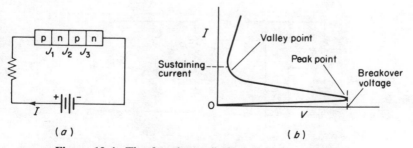

Figure 12.4. The four-layer diode and its characteristic.

and the very small current which does flow is composed of several components. Firstly, a normal leakage current will flow, and this is due to the intrinsically generated carriers at each side of the junction J_2. Secondly, some of the holes crossing J_1 from the adjacent p-region will succeed in reaching J_2, where they will be accelerated across by the high field existing at J_2. Finally, some of the electrons crossing J_3 from the adjacent n-region will reach J_2 and also be accelerated across.

As the applied voltage is increased, the field across J_2 becomes of progressively greater strength, and eventually the carriers are accelerated across it to velocities high enough to cause secondary ionization. This leads to avalanche breakdown, when the current rises rapidly and the voltage across J_2 falls as the junction becomes forward-biased.

This sequence of events results in the characteristic shown in figure 12.4(b), which is similar to that of the unijunction transistor. The pnpn diode is, however, rather faster in operation, and relaxation oscillators may easily be designed for frequencies up to about 0.5 MHz. Also, the breakdown voltage can be much higher, units having a 90–100 V peak point being readily available. The reverse characteristic is simply that of a normal diode since both J_1 and J_3 are reverse-biased under these circumstances.

Figure 12.5(a) shows a very simple time delay unit wherein a capacitor C is charged via a resistor R. When the breakover voltage of the four-layer diode is reached, C will discharge through the relay coil, so closing the contacts which then hold the relay ON. If desired, a further pair of contacts can be used to open-circuit the supply to R and C.

Figure 12.5(b) shows the same circuit used as a ramp or sawtooth generator. The repetition frequency can be controlled by making R variable, or linearity can be achieved by substituting a constant current generator for R.

A number of more sophisticated circuits have been designed around the four-layer diode, and it can be seen that it essentially complements the switching properties of the unijunction transistor by virtue of providing negative resistance slope switching action at the higher voltage levels.

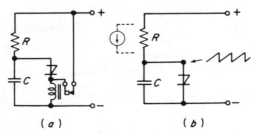

Figure 12.5. Typical four-layer diode circuits.

12.2.1 The Diac

If two four-layer diodes are connected in anti-parallel, a symmetrical break-over characteristic will result. A device based on this principle is the *diac*, and its symbol and characteristic are shown in figure 12.6.

The principal application of the diac is in the triggering of higher power four-layer switching devices, as will be described later. For the present, it will serve as an example of the way in which switching devices which have reverse-blocking properties may be connected back-to-back in order to produce symmetrical characteristics, so leading themselves to bilateral operation; that is, they become useful in current reversal circuits, including those involving AC mains. The break-over voltage for the diac is typically 28–36 V of either polarity.

12.2.2 The Programmable Unijunction Transistor (PUT)

A modification of the four-layer structure leads to the programmable unijunction transistor (PUT). Figure 12.7(a) indicates that a third connection is made to the sandwiched n-layer, and is taken to a potentiometer consisting of resistors R_1 and R_2. When the voltage at the emitter, V_E, exceeds the potentiometer voltage, ηV_{BB}, the pn diode begins to conduct, normal four-layer regenerative action occurs, and the PUT breaks over.

By comparing the voltages in figure 12.7(a) with those in figure 12.5(b), it will be clear that

$$\eta = \frac{R_1}{R_1 + R_2}$$

and the interbase resistance is $R_1 + R_2$.

Thus the device is a unijunction transistor with predetermined or programmable parameters. The resistors R_1 and R_2 normally form part of the circuit resistors, as shown in figure 12.7(b), where the PUT is shown operating as a simple pulse generator which develops its pulses across R_L.

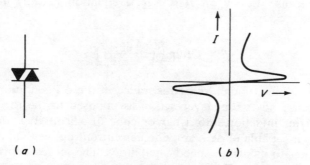

(a)   (b)

Figure 12.6. The diac symbol and characteristic.

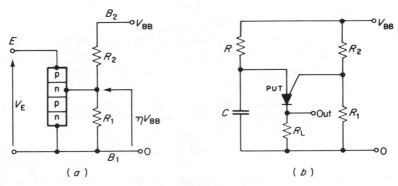

Figure 12.7. The programmable unijunction transistor (PUT).

As for the normal unijunction transistor, conduction commences when

$$V_E = V_P = \eta V_{BB} + V_D$$

or, in this case, when

$$V_E = \frac{R_1}{R_1 + R_2} V_{BB} + V_D \,.$$

The frequency of operation is again determined by assuming it to be essentially the time taken to charge the capacitor to V_E. That is:

$$V_P \simeq \left(\frac{R_1}{R_1 + R_2}\right) V_{BB} = V_{BB}\left[1 - \exp\left(-\frac{T}{CR}\right)\right]$$

giving

$$T = CR \ln\left(1 + \frac{R_1}{R_2}\right). \tag{12.5a}$$

Note that because $V_P \simeq V_{BB}R_1/(R_1 + R_2)$, equation (12.5a) may also be written:

$$T = CR \ln\left(\frac{V_{BB}}{V_{BB} - V_P}\right). \tag{12.5b}$$

The PUT exhibits a much lower resistance in the ON condition than does the UJT, because of its four-layer action as opposed to the UJT mechanism involving carrier injection into a bar or cube of silicon. It is this property which makes it capable of delivering higher current pulses than the UJT, and also contributes to its higher speed capabilities. The ON resistance of the PUT is typically $3\,\Omega$.

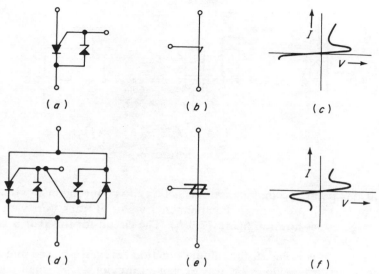

Figure 12.8. Operative equivalent circuits, symbols and characteristics of the sus (parts (a), (b), (c)) and the sbs (parts (d), (e), (f)).

12.2.3 Other Four-Layer Switching Devices

A variation of the PUT is the *silicon unilateral switch* (sus). This is essentially a PUT in which a Zener diode (or more correctly, an avalanche diode) structure is connected from the trigger electrode to the cathode, as shown in figure 12.8(a). When the voltage across the sus reaches the Zener break-over voltage, V_Z, a current is a injected into the trigger electrode and sus turns ON. Hence, it can be regarded as a PUT with a preprogrammed peak point voltage.

As in the case of the four-layer diode, the sus may be fabricated in the form of two structures in anti-parallel, as in figure 12.8(d). This forms the *silicon bilateral switch* (sbs) which will turn ON when the applied voltage reaches V_P in either direction.

Both the sus and sbs are designed so that the Zener diode allows breakdown at about 6–10 V, which means that insofar as circuit applications are concerned, they may be regarded as lower-voltage equivalents of the Shockley diode and the diac (most of which are designed for 28–36 V break-over).

12.3 THE THYRISTOR OR SILICON CONTROLLED RECTIFIER (SCR)

If a third electrode is connected to a four-layer diode as shown in figure 12.9(a), then it is possible to modify the break-over voltage by injecting a

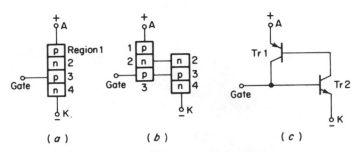

Figure 12.9. The thyristor or SCR.

current at this point. To understand this action, it is convenient to visualize the four layers as representing a complementary pair of transistors with two common regions as shown in figure 12.9(*b*). The circuit for this pair is given in figure 12.9(*c*).

When the gate is isolated, the middle junction is reverse-biased and only a small leakage current flows, as was explained previously. A small part of this current is due to a few holes from region 1 and a few electrons from region 4 arriving at the middle junction. If these currents are visualized as flowing in the circuit of diagram 12.9(*c*) it will be apparent that they would cause the two transistors to switch hard ON unless the two current gains $h_{FE(Tr1)}$ and $h_{FE(Tr2)}$ were very low indeed. This is because the collector current of Tr1 is the base current of Tr2, and the collector current of Tr2 is the base current of Tr1. This constitutes a positive feedback loop, and if the product $h_{FE(Tr1)}h_{FE(Tr2)}$ were greater than unity the loop current would build up rapidly.

It has previously been pointed out that h_{FE} is a function of the current through a transistor (figure 2.10) and in the present case the very low leakage current which flows results in an h_{FE} product which is much less than unity. However, when this current increases due to avalanche break-down or any other reason, the two current gains increase rapidly and enable the positive feedback mechanism to quickly saturate the transistors.

One way of initiating the process is to inject a current into the base of the npn section, that is, to energize the gate of the thyristor. This will increase the current in the Tr2 section and raise the value of $h_{FE(Tr2)}$, so initiating the switch-ON sequence.

The usual symbol for the thyristor, and the relevant current and voltage directions are shown in figure 12.10(*a*).

Figure 12.10(*b*) shows two thyristor characteristics, one for zero gate current, I_{t0}, and one for a gate current I_{t1}. If a load R_L is connected in series with the thyristor, and a voltage V_{CC} is applied, which is smaller than the break-over voltage for zero gate current, $V_{BR(I_{t0})}$, then a stable working point "A" will exist, with the thyristor OFF.

If a gate current I_{t1} (or more) is applied, the break-over voltage falls, and

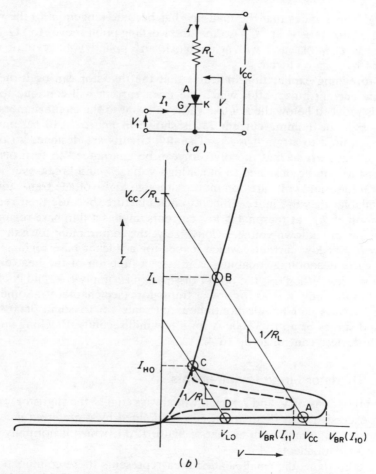

Figure 12.10. (*a*) The thyristor and (*b*) its characteristic at $I_t = 0$ (full curve) and I_{t1} (broken curve).

when the characteristic becomes tangential to the load-line, the working point will move rapidly to "B" as shown. This point is also stable, and the load current I_L will flow *even though the gate current may be reduced to zero*; that is, it is impossible to turn the thyristor OFF using the gate. (Note the similarity to the gas-filled *thyratron* tube, which requires a gate *voltage* to turn it ON and conduction continues even after removal of the gate voltage. It is by analogy with the thyratron that the electrodes of the thyristor have been termed the anode (A), cathode (K) and gate (G), as labelled in figures 12.9 and 12.10(*a*).)

If the gate current is reduced to zero then the thyristor will turn OFF *provided* that the applied voltage falls below V_{L0}, as shown in figure

12.10(b). This implies that the load-line has become tangential to the valley of the I_{t0} characteristic at "C," so that the working point moves to "D." The thyristor is now OFF and will again block an applied voltage until again triggered by a gate current.

The foregoing explanation indicates that the thyristor can be turned ON by a gate current pulse, after which the main current will continue to flow until it is reduced below the *holding* value relevant to the circumstances. For the I_{t0} case, the holding current I_{H0} is shown in figure 12.10(b), and the voltage required to attain it is V_{L0}. Usually, circuits are designed so that V falls to about zero so that the thyristor can be guaranteed to turn OFF.

Thyristors can be designed to block high voltages, and break-over values of several thousand volts are common. Although high-voltage transistors are also available, they are more difficult to manufacture, because their requirement of a high h_{FE} at reasonably low currents implies a thin base region and hence a low breakdown voltage. Conversely, the requirement for a thyristor is a low h_{FE} at low currents, which makes for a thicker base region.

The current handling capability is in part a function of the heat dissipation properties. When ON, the power dissipation is simply $I_L V$, and cooling is effected in the same way as for power transistors (see chapter 9). Some large modern devices can handle continuous currents up to some 3500 A and withstand surges of up to 32 000 A for a few milliseconds. Blocking voltages for such devices can range up to 4.5 kV.

12.3.1 Thyristor Firing Requirements

The combinations of V_t and I_t which will always trigger the thyristor, yet not lead to excessive heat dissipation, can be defined by a *preferred gate drive area* on an I_t, V_t graph such as that of figure 12.11, which is normally found in thyristor data sheets.

Near the origin, the small shaded area represents those combinations of V_t and I_t which may not trigger the thyristor at all, and so should be avoided.

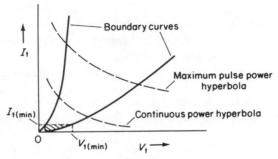

Figure 12.11. Thyristor triggering preferred areas.

The next area between the two boundary curves and the maximum continuous power hyperbola (typically at 0.5 W), contains all those V_t, I_t combinations which can be used without limiting the time of application; this is called the DC triggering area.

The upper hyperbola (often at about 5 W) limits the maximum instantaneous V_t, I_t combinations; béyond this, the gate junction may be damaged.

Between the two hyperbolas, trigger pulses can be used which, when integrated over a long period, must not exceed the energy in watt seconds over the same period at the DC triggering power.

12.3.2 DC Thyristor Circuits

Figure 12.12(*a*) shows a simple thyristor latching circuit, such that when a trigger pulse arrives (or a DC triggering level exists for not less than about 0.1 ms), the thyristor will fire, and will remain ON until the switch is opened, so cutting of the main current.

This example raises the basic question of main current interruption in thyristor circuits, for clearly the use of a switch is most inappropriate since it is the very device a thyristor is supposed to supplant.

Figure 12.12(*b*) shows one method of contactless current reduction. Here a second thyristor is connected as shown and triggering pulses are applied via capacitors. Suppose that Th1 is ON and Th2 is OFF. If a triggering pulse arrives, it will have no effect on Th1, but will turn Th2 ON. As Th2 saturates, the voltage at its anode, V_{A2}, falls rapidly from V_{CC} to the saturation voltage for Th2, which is very small. This negative-going pulse is transmitted to the anode of Th1 via the capacitor C_A, which takes V_{A1} below zero, so turning Th1 OFF.

The next pulse will reverse the procedure, turning Th1 ON, and hence Th2 OFF.

This capacitive turn-OFF method has many variations and some problems.

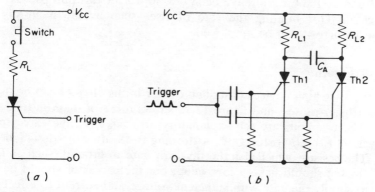

Figure 12.12. (*a*) A simple latching circuit and (*b*) a self-turn-OFF circuit.

The most common requirement is that it is desired to energize only a single load, so that Th2 is redundant from this point of view. To reduce costs, therefore, Th2 can be a much smaller device than Th1, and R_{L2} can have a high value. With such a system, it is important to establish that both thyristors cannot come ON together, and this is easily achieved by making R_{L2} sufficiently large that the current through Th2 is less than the holding current, I_{H0}. Hence, Th2 will only be ON transiently, while it is discharging C_A. This again makes possible a much lower power rating for Th2 than for Th1.

There are many methods utilized for thyristor turn-OFF in DC circuits, and all involve either capacitive or inductive methods of reducing the main current below I_{H0}. In particular, very high-power inverters can be designed using thyristors, and these complement the switching transistor circuits described in chapter 10.

12.3.3 The dv/dt and di/dt Problems

In high-power circuits particularly, the question of voltage and current risetimes becomes important. Like all semiconductor junction devices, the thyristor presents interelectrode capacitances. Hence, if a fast voltage pulse arrives at the anode, the resulting current (given by $i = C\, dv/dt$) may be sufficient to initiate the turn-ON sequence. For a given thyristor type, it is usual for the manufacturer to quote a *critical voltage rate of rise*, below which spurious triggering can be guaranteed not to occur. (For some modern high-speed thyristors, $dv/dt_{(crit)}$ can be as high as 2000 V/μs.)

The question of rate of rise of the main current, di/dt, is more serious. This is because, if the current rises too rapidly, it will have to pass through the small area of silicon at which conduction is first initiated, prior to the spread of conduction over the whole available area. The resulting high current density may irreversibly damage the thyristor. This situation is not uncommon in inverter circuits, for example, and thyristors with high di/dt ratings (such as 1000 A/μs) have been developed for this purpose. Alternatively, circuits for limiting the rate of rise—such as the inclusion of a saturable inductor—can be employed.

12.3.4 Triggering

Any of the switching devices described earlier in the chapter can be used to generate trigger pulses, and figure 12.13 shows three of them in time delay circuits. These circuits are self-explanatory; in each case, the capacitor, C, discharges into the thyristor gate, so turning it ON after the designed delay time T. The resistor which ties the thyristor gate to ground (such as R_{B1} in the UJT circuit) should have a low value; but the capacitor should be large enough to supply the gate with sufficient voltage and current to effect firing in spite of this.

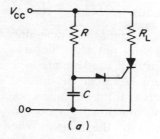

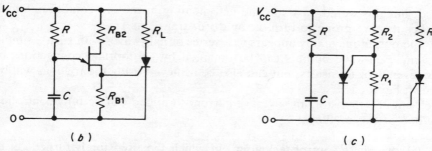

Figure 12.13. (*a*) Thyristor triggering by four-layer diode; (*b*) UJT; (*c*) PUT. (Thyristor turn-OFF method not shown.)

A complete flasher circuit using a UJT is shown in figure 12.14 and this demonstrates both the capacitor turn-OFF method, and starvation operation of one thyristor.

In figure 12.13, the means of thyristor turn-OFF is not shown, but may be any of the DC methods, or conversely an AC supply may be used. This latter method introduces the most important application of the thyristor—as an AC control element.

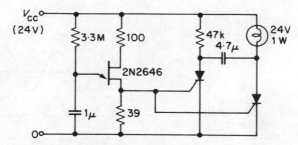

Figure 12.14. A filament lamp flasher circuit (for approximately 1 s flashes with values shown).

12.3.5 AC Thyristor Circuits

In AC circuits at moderate frequencies—including mains frequencies—the thyristor turn-OFF problem does not arise, because the supply falls to zero twice per cycle. If the thyristor is used simply as a solid state relay, circuits such as that shown in figure 12.15 can be utilized. Here, the resistor R is used to provide trigger current from the mains when the switch is OFF; and diode D1 prevents the gate junction from becoming reverse-biased during the negative-going half-cycle of the mains. When the switch is ON, the triggering current is shorted to neutral, and the gate junction is held OFF. (Note that the switch could also have been placed in series with the gate lead, but this would have left the gate floating in the OFF condition, which makes it more prone to turn-ON by dv/dt transients.)

The output inevitably appears as a series of half-waves with such a simple system. Full-wave outputs can be obtained by appropriate connections of diodes and/or thyristors, but can also be achieved using the triac, as will be shown later.

Using the thyristor, the average current to a load can be fully controlled by one of two methods:

(i) the phase-control technique, in which the thyristor is turned ON at some point in each half-cycle which is determined by the triggering circuit; or

(ii) the burst-control method, in which bursts of several complete cycles are allowed to reach the load. (This method is more appropriate for high power levels, where the gross distortion of waveshape in (i) cannot be tolerated. Integrated circuit triggering modules for burst-firing are normally used in modern systems.)

The basic circuit for phase control is shown in figure 12.16(*a*). Here, C is charged through R and the diode D1 during the positive half-cycle; and it discharges when the trigger element fires. If R is variable, then the time taken for C to reach the firing voltage may be predetermined, so that the

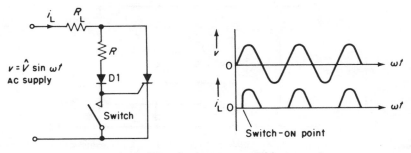

Figure 12.15. The thyristor as a solid state relay.

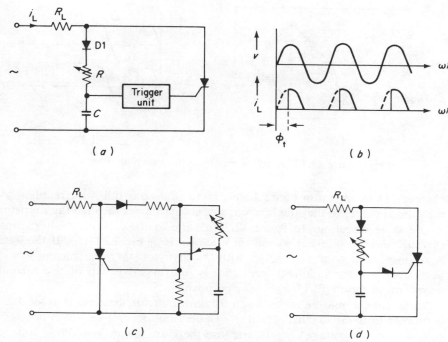

Figure 12.16. (*a*) The half-wave phase-control method and (*b*) the resultant current waveform. (*c*) The UJT and (*d*) the four-layer diode in half-wave thyristor control circuits.

point on the wave at which the thyristor is triggered and conduction commences is controllable. The resulting waveshape is shown in figure 12.16(*b*). Note that when the thyristor has fired, the voltage across the D1–*R*–*C* combination is very small, being only the thyristor saturation voltage. When the wave goes negative, the thyristor blocks, and diode D1 prevents the capacitor charging negatively.

If a DC voltage is required at the output, then smoothing can be applied (see chapter 14) and the DC level may be controlled by *R*. Conversely, R_L may accept AC, as is the case for a filament lamp dimmer circuit. (Here, however, great care in thyristor selection is necessary, because the cold resistance of a filament lamp is low, and a very large in-rush current appears initially, which can destroy the thyristor.)

Figures 12.16(*c*) and 12.16(*d*) show how a UJT and a four-layer diode may be used as the triggering elements in a half-wave phase control system. In such cases, it must be established that the triggering time (equation (12.3) for the UJT) is controllable over an appropriate portion of the half-period of the AC waveform.

Full-wave control may also be achieved using only one thyristor by employing four diodes in a full-wave bridge rectifier configuration (see

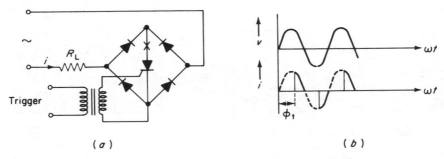

Figure 12.17. A full-wave thyristor control method.

chapter 14) as shown in figure 12.17. Here, the AC is full-wave rectified by the diode bridge, so that the load current always passes through the thyristor in the same direction. In figure 12.17(a), the load R_L receives alternating current, and the relevant waveform is shown in figure 12.17(b). If the load were inserted at point X in series with the thyristor, then the current would be fluctuating, but unidirectional; that is, the negative half of the current waveform in figure 12.17(b) would go positive.

Some care should be taken when using this circuit, because if inductance is present in the load, thyristor turn-OFF may not occur. Also, it is not very suitable for low-voltage circuits, because there are always two diode voltage drops plus one thyristor drop in series with the load.

Various multi-thyristor full-wave circuits exist, but the preferable alternative is use of the *triac*, which can be thought of in many ways as the equivalent of two thyristors connected in anti-parallel.

12.4 THE TRIAC

The triac is a device which will pass current in either direction, when triggered by either a positive or a negative pulse *irrespective of the polarity of the main voltage*. These properties make the triac ideal as an AC control element, and lead to considerable simplification of the triggering circuitry.

The (idealized) construction and the symbol for the triac are shown in figures 12.18(a) and (b), and its break-over characteristic in (c).

From figure 12.18(a), it will be seen that the device consists essentially of a pnpn structure in parallel with an npnp structure. Main terminal MT2 contacts both the upper p- and the upper n-regions; while main terminal MT1 contacts the lower p- and n-regions. This means that the main current can flow in either direction, and figures (d)–(g) show it doing so. If the direction of the main current is followed in each of the diagrams, it will be seen that it always flows along a p–n–p–n path, as in all four-layer devices. It is for this reason that the characteristic appears as in figure (c), which shows that the triac presents a typical pnpn breakdown characteristic in each direction.

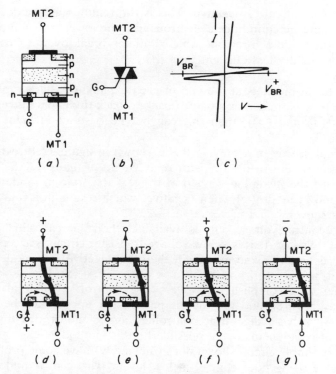

Figure 12.18. (a) The triac structure (idealized); (b) its symbol; (c) its break-over characteristic; (d), (e) current flows for positive triggering; (f), (g) current flows for negative triggering.

Now consider the triggering requirements. With respect to MT1, the figures (d)–(g) represent the following conditions:

(d) MT2 positive; G positive. This constitutes normal thyristor operation.

(e) MT2 negative; G positive. Here, the action is that electrons are injected into the lower p-region and some are collected by the middle n-region, which initiates the normal four-layer breakdown process. (This is an inefficient mode, and should be avoided if possible.)

(f) MT2 positive; G negative. In this mode, the lower p-region forms a forward-biased junction with the gate n-region, so that, from the top down, a conducting p–n–p–n$_{\text{gate}}$ four-layer diode appears. This implies that the main n–p junction has broken down, so that the main p–n–p–n path is also available for conduction. (This is known as the junction-gate thyristor principle, and it is less efficient than mode (d) but not as poor as mode (e).)

(*g*) MT2 negative; G negative. This is the remote-gate mode, in which the gate junction injects electrons into the lower p-region and hence to the middle n-region, which initiates break-over. (This mode is only slightly less efficient than (*d*).)

The notes in brackets at the end of each explanatory paragraph indicate that if the triac is treated as a pair of back-to-back thyristors (as represented by modes (*d*) and (*g*)), then this can be the basis of efficient full-wave operation.

The simplest way of achieving this is shown in figure 12.19, which is the full-wave equivalent of the thyristor relay circuit of figure 12.15. When the switch is ON, the incoming waveform triggers the triac in mode (*d*) when positive, and in mode (*g*) when negative, which results in a full-wave load current as shown.

If a DC control voltage only is available, then this can still be used to trigger the thyristor; that is, a low-voltage transistor circuit is acceptable, but because it is best to avoid mode (*e*), then the negative gate modes (*f*) and (*g*) should be chosen if possible.

In order to accomplish full-wave phase control, a bilateral triggering device may be used and the diac is particularly appropriate here. Figure 12.20 shows a basic diac–triac controller, and demonstrates the extreme economy of components made possible by this method.

The very simple circuit shown does (inevitably) have some problems, the most notable being poor control at low levels. This can be understood by referring to figure 12.21. Here, the voltage applied to the diac (which is the capacitor voltage, V_C) is shown; and V_{dBR}^{+} and V_{dBR}^{-} are the two break-over voltages of that diac. When saturated in the ON condition, the voltage across the diac and triac G–MT1 combination is V_{sat}^{+} or V_{sat}^{-}.

When V_C reaches V_{dBR}^{+}, the diac breaks over, and partially discharges the capacitor into the triac gate. The value of V_C is then V_{sat}^{+}, from which it follows the supply waveform through zero, then builds up negatively to V_{dBR}^{-} and fires the triac in the opposite direction; that is, the value of V_C has

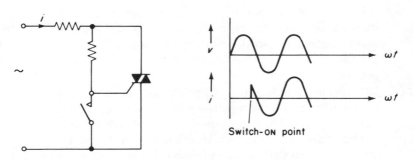

Figure 12.19. A triac solid state relay.

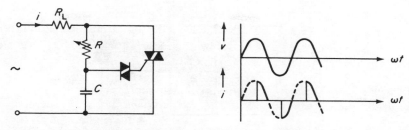

Figure 12.20. A full-wave diac–triac phase controller.

been required to build up (negatively) through a minimum voltage range of $(V_{sat}^{+} - V_{dBR}^{-})$.

Now suppose that R is increased so that on the next (positive) half-cycle, V_C fails to build up to V_{dBR}^{+}. This condition is shown as the "first firing failure," in figure 12.21, and after it, V_C would have to build up (negatively) to V_{dBR}^{-} in order to fire the triac on the next (negative) half-cycle. This voltage excursion, which approaches $(V_{dBR}^{+} - V_{dBR}^{-})$, is greater than $(V_{sat}^{+} - V_{dBR}^{-})$, which means that the triac will certainly not fire on that or any subsequent half-cycle.

If R is reduced, eventually a point is reached when V_C is able to build up through $(V_{dBR}^{+} - V_{dBR}^{-})$ and firing recommences. After a single firing, however, V_C is once more required to build up only through the lower voltage $(V_{sat}^{+} - V_{dBR}^{-})$ each half-period, so that firing will commence earlier in each half-period than was intended.

The practical result of this sequence of events is that if R has been increased so that firing has just stopped, then when it is again decreased, firing recommences at a smaller phase angle in each half-cycle; that is, the

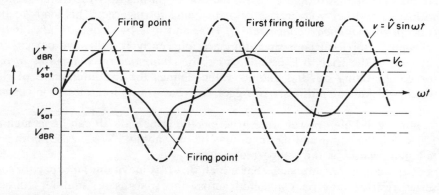

Figure 12.21. Capacitor voltage waveform for the diac–triac controller of figure 12.20.

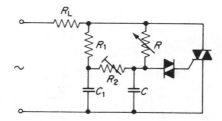

Figure 12.22. Diac–triac controller with improved low-current performance.

average load current snaps suddenly from zero to some intermediate value, rather than increasing gently.

Avoidance of snap-on can be achieved in two ways. Either the diac may be replaced by an asymmetric silicon bilateral switch (ASBS) or additional circuit components may be added. This latter technique usually takes the form illustrated in figure 12.22, where a second resistor and capacitor are included. After C has partially discharged, it is able to recharge from C_1 so that its voltage is raised towards V_{dBR} from V_{sat} after each break-over, and symmetry of operation is thereby improved. This makes for smoother control at low levels. In this circuit R_2 is adjusted for optimum performance at average load-current levels just above the minimum. (In the case of an incandescent lamp controller, this optimum condition is directly observable.)

12.4.1 Full-Wave dv/dt and di/dt Problems

If a thyristor is working in the half-wave mode, it has a complete half-period to recover its blocking capabilities each cycle. When it has fully recovered, its ability to withstand another voltage rise is defined by the *critical voltage rate of rise*, as has been mentioned. However, if it is used in a full-wave circuit such as that of figure 12.17, then before it has fully recovered it must withstand another voltage rise prior to being retriggered. This leads to the concept of *reapplied voltage rate of rise*, and it is by definition *always* the relevant factor for triacs, since they are inherently AC full-wave devices. In the case of the triac, it is usually called the *commutating voltage rate of rise* (or commutating dv/dt), and its value defines the maximum rate of rise which will fail to spuriously retrigger the triac via its interelectrode capacitances.

For a good high-power triac, whereas the critical dv/dt can be as high as 100 V/μs, the commutating dv/dt may be only about 5 V/μs. Also, both will fall significantly as the temperature rises.

If the conditions are such that a high dv/dt is inevitable, a capacitor and series resistor may be connected across the triac as a "snubber" circuit (figure 12.23) to reduce this rate of rise to an acceptable level.

The more serious problem of high in-rush current is similar in the case of

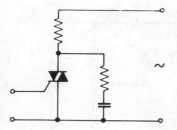

Figure 12.23. A "snubber" circuit for dv/dt reduction.

the triac as for the thyristor, and if necessary, a saturable inductor may be connected in series with the triac to reduce the initial rate of rise. After the initial transient, any subsequent high current will saturate the inductor so that it presents only its internal resistance to the circuit.

If a snubber circuit is included for dv/dt reduction, the resistor should be so chosen as to restrict the discharge of the capacitor through the triac, otherwise this transient may itself exceed the di/dt rating.

When inductive loads are involved, the fact that the voltage leads the current means that when the main triac current has fallen to zero, there will be a reverse voltage across it. Usually, the triac will still block under these circumstances, but for highly inductive circuits, particularly at high powers, more sophisticated triggering methods are needed.

12.4.2 Trigger Circuit Isolation

Both thyristors and triacs are three-terminal devices, which implies that one terminal must be common to both the main and the trigger circuits. For the thyristor, this is the cathode (K), and for the triac, main terminal one (MT1). There are many cases when it is desirable to isolate a low-voltage trigger network from a high-voltage main circuit, and two major techniques exist for this purpose (other than the obvious use of an electro-mechanical relay for slow applications). These use either a transformer or an optically coupled device.

The simplest use of a transformer is shown in figure 12.24(a), which is a modification of the thyristor relay circuit of figure 12.15. Here, a small, saturable-core transformer is used in such a way that when the signal winding, N_s, carries no current, the highly inductive primary, N_p, allows most of the triggering current through R_1 to enter the gate and fire the thyristor on each positive-going half-cycle. However, when N_s is energized using direct current the core saturates so that the (low) resistance of N_p shunts the gate–cathode junction and short-circuits the trigger current; that is, the saturable transformer acts in the same way as the switch in figure 12.19, but the control signal is isolated from the main supply.

A full-wave circuit is shown in figure 12.24(b). Here, a normal transformer is used in such a way that when the switch is closed only the

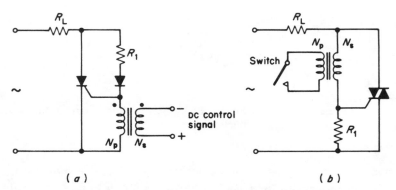

Figure 12.24. Use of a transformer for control circuit isolation.

resistance of the secondary winding, N_s, is in circuit, and the triac is triggered each half-cycle. When the switch is open the secondary presents a high reflected resistance so that insufficient current is available to trigger the triac. (The resistor R_1 shunts the transformer magnetizing current to ground.) The switch may be a small, low-voltage reed type, for example, if the winding ratio N_s/N_p is large.

Figure 12.25(a) shows a circuit which uses a pulse transformer for isolating a low-voltage pulse-generating trigger network from the main circuit. This pulse transformer must be so designed that it can accept the various waveforms which are commonly met in trigger circuits, and inject appropriate pulses into the thyristor or triac gate. For example, if a transistor circuit is designed to deliver a series of square pulses as in figure 12.25(b), the transformer could modify their shape as shown. This may be understood by referring back to figure 9.19, in which a transformer equivalent circuit is given. Here it is the leakage inductance L_1 which limits the risetime, so that for a good pulse transformer, L_1 must be made as small as possible.

If the total resistance in the gate circuit is R_{tg} (which includes the transformer winding resistance referred to as the primary), then the risetime constant will be L_1/R_{tg}.

Conversely, the magnetizing inductance, L_m, shunts the pulse, so that part of the available energy is lost to the gate. This implies that L_m should be large; if it is not, the square wave will droop excessively.

Figures 12.25(c) and (d) show how a pulse transformer can be allied with a UJT. The primary waveform is essentially the characteristic exponential decay of a discharging capacitor; and it will be modified by the pulse transformer as in figure 12.25(d). Here, it has been assumed that the capacitor has delivered sufficient energy to saturate the transformer at time τ_{sat}.

All transformers have self-capacitance, and in the present case this may

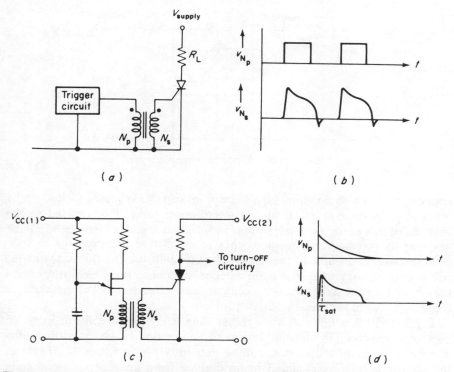

Figure 12.25. (*a*), (*b*) Use of the pulse transformer with a square-wave trigger. (*c*), (*d*) A UJT/pulse transformer trigger method.

lead to excessive peaking and damped oscillation (or "ringing") with the self-inductance. Hence, pulse transformers should also be designed for low self-capacitance.

The requirement of a high L_m suggests the use of a good soft magnetic material such as a ferrite (which must be in the form of a closed magnetic circuit, not a rod); and that of a low self-capacitance precludes bifilar winding methods.

The main alternative method of isolating the trigger circuit is shown in figure 12.26. Here, a light-emitting diode is optically, but not electrically, coupled to a phototransistor. When a pulse is applied to the diode, light (usually infrared) is emitted, which releases charge carriers in the transistor. These act as base current, which is amplified, and the output or collector current is used to trigger the thyristor.

This arrangement constitutes one version of the opto-coupler, which is described in chapter 13. Other versions replace the transistor by a photo-diode, a photo-Darlington pair, or a small photothyristor.

When a thyristor or triac is triggered part-way into a half-cycle, the sudden rise in current constitutes a transient (repetitive in phase control

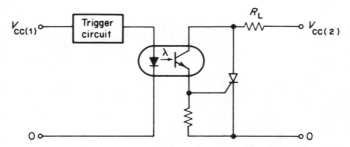

Figure 12.26. A photocoupled trigger.

applications), which contains high harmonics capable of being picked up by other apparatus as radio frequency interference (RFI). This has led to the concept of *zero crossover* triggering, in which the gate is energized at the moment the applied waveform reaches zero, so that the thyristor or triac conducts over the *whole* of the next half-cycle. Because the thyristor or triac will switch off *only* when the main current reduces to near zero, this means that *both* turn-ON and turn-OFF occur at zero crossover. (This method of triggering also minimizes the di/dt problem.)

If full control is desired, then rather than using phase control, in which part of each cycle is utilized, bursts of conduction, each consisting of a discrete number of half-cycles, can be generated, as in figure 12.27. Here, the control voltage is assumed to be derived from an appropriate sensing and signal conditioning network; and is applied to a zero crossover (ZCO) circuit which delays the development of a trigger voltage until the applied

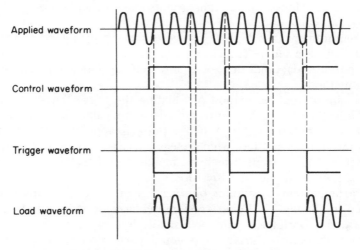

Figure 12.27. Burst control waveforms.

waveform crosses zero. (In figure 12.27, the use of a triac has been assumed, which is why the DC triggering voltage has been shown to be negative; this establishes that the triac works in the efficient modes (f) and (g) in figure 12.18.)

Numerous ingenious discrete-component circuits have been devised for achieving the ZCO function, but the use of monolithic microcircuits is now the optimum method in terms of component count and cost. Usually, these produce bursts of triggering pulses, rather than the DC of figure 12.27; and some are appropriate for use with thyristors rather than triacs.

Integrated ZCO are extremely versatile, and many offer facilities which make for extreme simplification in burst control circuits. For example, a thermistor may be connected directly to some ZCO integrated circuits so that if the thyristor or triac supplies a heater then a complete temperature controller results. Also, an override terminal is often provided, so that an incoming analog or digital signal can inhibit operation of the entire circuit for safety reasons. Such matters are dealt with comprehensively in manufacturers' literature.

12.5 GATE TURN-OFF (GTO) DEVICES

Normally, it is necessary to reduce the current through a four-layer device to less than the small holding current in order to turn if OFF. If this is not possible, it becomes necessary to employ either high-power switching transistors or IGFETs, or to specify the comparatively rare breed of *gate turn-off* (GTO) thyristors. The prime advantage of these devices over transistors is that they exhibit a lower saturation voltage at high currents; and that they can be triggered ON *and* OFF using pulses, rather than needing a continuous base or gate drive. The main disadvantage is that circuit provision must be made to produce the substantial negative pulse required to interrupt conduction.

In a normal thyristor having a pnpn structure, carriers of both polarities are injected at the two main terminals, and their high concentration in the intermediate "base" region maintains high conductivity throughout. When the main current is interrupted, no further carriers are injected, and the existing ones recombine, when the thyristor blocking properties re-emerge. This is why a finite time is needed for the device to recover after current interruption, and why reapplied (or commutating) dv/dt is less than the critical dv/dt.

In principle, a thyristor can be turned OFF by applying a negative voltage to the gate, so extracting enough of the main current so that the current gain product $h_{FE(1)}h_{FE(2)}$ in the two-transistor model falls below unity. In practice, this is only possible at very low currents, near the holding current level, I_H, because the carrier extraction is limited by the gate resistance and the maximum (reverse) voltage which it can tolerate.

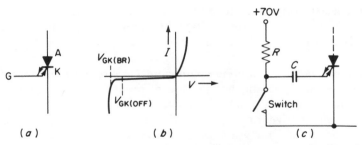

Figure 12.28. (*a*) The symbol, (*b*) gate characteristic, and (*c*) a turn-OFF circuit for the GTO.

For a GTO device, the gate geometry and doping levels are so designed that application of a reverse voltage—albeit a fairly high one, such as 70 V or so—will extract sufficient carriers to bring the remaining main current collected by the cathode to below I_H. This is not achieved without sacrifices in other parameters, however. In particular, the current gain product is inherently smaller than for a regular thyristor.

The high reverse breakdown voltage, $V_{GK(BR)}$, of the GTO is reflected in figure 12.28(*b*) where the gate-cathode characteristic is depicted. The (high) gate turn-OFF voltage, $V_{GK(OFF)}$, is also shown on this diagram, and the relevant circuit must be so designed that this is less than the reverse breakdown voltage. The provision of a high negative pulse can be an embarrassment in circuit design, so that a method such as that of figure 12.28(*c*) becomes appropriate. When the GTO is conducting and the switch is open, the right-hand plates of capacitor C are held near 0 V, whilst the left-hand plates charge to +70 V. When the switch is closed, these left-hand plates are suddenly shorted to 0 V, so that the right-hand plates pull the gate down to nearly −70 V. This extracts sufficient current to turn OFF the GTO. (In this circuit, if R is replaced by an inductor, L, then part of the necessary energy can be stored in the inductor as $\frac{1}{2}LI_G^2$, which allows the supply voltage to be reduced to about 35 V.)

Because of the large gate extraction current needed, plus the other limitations in performance of the GTO, it is not a popular alternative to the high-current switching transistor or IGFET, except where it is impossible to provide a continuous base or gate drive. However, very large GTOs will pass up to 2500 A with surge ratings of some 12 000 A, and will block up to 4 kV.

12.6 SUMMARY

The prime function of the four-layer devices introduced in this chapter is to control power for lamps and for motors and other electro-mechanical devices; that is, like power amplifiers, they form the last link in a signal-processing chain. However, because they exhibit complex switching behavior, much of the circuitry in which they appear is not subject to simple mathematical analysis. In fact, elementary consideration of phase control circuits leads to equations which are amenable only to graphical solution. An appropriate design procedure is therefore to use the manufacturers' data and graphs in the light of a proper understanding of the characteristics and limitations of the devices in question.

REFERENCES

1. Bird, B. M. and King, K. G., 1983, *An Introduction to Power Electronics* (New York: John Wiley).
2. Bradley, D. A., 1987, *Power Electronics* (New York: Van Nostrand-Reinhold).
3. Lander, C. W., 1987, *Power Electronics* (New York: McGraw-Hill).
4. Williams, B. W., 1987, *Power Electronics—Devices, Drivers and Applications* (New York: Macmillan).

QUESTIONS

1. In a unijunction transistor circuit, the timing components are R and C. What criteria determine the maximum and minimum possible values which R may take?

2. The resistor R which charges the capacitor C in a simple unijunction transistor circuit is replaced by a junction FET current limiter which produces a current I. Rewrite equation (12.3) accordingly.

3. What are the advantages of a PUT over a UJT?

4. What are the basic "turn-OFF" requirements of a thyristor?

5. Define the four possible operating conditions for a triac and quote those which require the least triggering energy.

6. What is the difference between phase control and burst control of a triac?

7. What is difference between "critical voltage rate of rise" and "reapplied voltage rate of rise" for four-layer devices?

8. What is the major disadvantage of the GTO?

PROBLEMS

12.1. In a UJT circuit like that of figure 12.2(a), a 24 V supply is used along with a timing capacitor of 0.1 μF. The UJT has an intrinsic stand-off ratio of 0.6, and the resistors R_{B1} and R_{B2} are small enough to neglect. What would be the PRF of the output pulse train if the timing resistor R were 82 kΩ? If an FET current-limiter were substituted for R, what would be its defined current for the same output PRF?

12.2. If a PUT were substituted for the UJT of problem 12.1, calculate the values of the potential divider resistors R_1 and R_2 if 6 mA were available to energize them (see figure 12.7(b)).

12.3. When a 48 V supply is applied to a relay circuit, it is required that the relay turn ON 2 s later. Design a circuit which will accomplish this using a thyristor and a 30 V Shockley diode.

12.4. Describe how the circuit of figure 12.29 functions, and calculate the frequencies of the output waveform when $RV1$ is at its upper and lower limits.

Assume that all components are correctly chosen and that Tr1 and Tr2 are identical structures. The intrinsic stand-off ratio for the Unijunction transistor is 0.63, its diode voltage drop is 0.6 V and V_{BB} may be taken as equal to the power supply voltage.

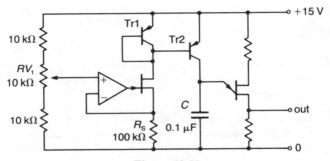

Figure 12.29.

12.5. The output from the circuit of figure 12.30 consists of a 1 kHz pulse train. Calculate the value of R_E. Assume that for the 2N2646, the intrinsic stand-off ratio is 0.56 and that its diode drop is 0.6 V. Assume also that the value of V_{BE} for transistor Tr1 is 0.6 V and that $kT/q = 26$ mV. The values of R_{B1} and R_{B2} are very small.

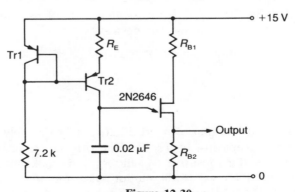

Figure 12.30.

12.6. A circuit is required which will sound an audible alarm for about 5 s after a 12 V supply is switched on. Design such a circuit using the following components, and explain how it works with reference to a neat diagram. Also, calculate the value of the timing resistor R.

Components available are: one 2N2646 unijunction transistor having an intrinsic stand-off ratio of 0.6, a V_D of 0.5 V and an R_{BB} which is much larger than R_{B1} or R_{B2}; one small thyristor having a holding current value below 5 mA; one small-power N-channel enhancement IGFET having a threshold voltage below 4 V; one 4.7 μF low-leakage capacitor; one audible alarm for 8–16 V, 10 mA operation. Resistors: 100 Ω; 1 kΩ; 2.2 kΩ; R (to be calculated).

12.7. A circuit is needed which will turn ON a 12 V 24 W filament lamp about 10 s after a switch is closed. Design such a circuit using a thyristor, a unijunction transistor, a 10 μF low-leakage capacitor, and any necessary resistors. Include a circuit diagram.

Assume that the rating of the thyristor is adequate, and that it needs a trigger voltage above 4 V. Also assume that the intrinsic stand-off ratio for the unijunction transistor is 0.56, that its diode drop is 0.6 V, and that its interbase resistance is much greater than R_{B1} or R_{B2} (neither of which need be calculated).

The power supply is a 12 V storage battery.

12.8. The filament lamp in the circuit of figure 12.31 is ON every alternate second. Calculate the value of R_E.

Assume that for the bipolar transistors $V_{BE} = 0.7$ V and $kT/q = 26$ mV; and for the unijunction transistor $\eta = 0.65$, $V_D = 0.6$ V, and R_{B1} and R_{B2} are negligibly small.

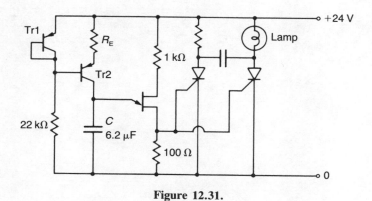

Figure 12.31.

12.9. Design a sawtooth waveform generator using a unijunction transistor and a junction FET current-limiter and operating from a 12 V supply. Choose components which will result in a frequency of about 5 kHz.

Assume that the intrinsic stand-off ratio for the UJT is 0.6, the interbase resistance is 7.5 kΩ and the diode volt drop is 0.7 V. The current-limiter is an n-channel 0.1 mA device.

Draw the circuit and also show how a normal JFET could be used in place of the CL, along with a preset variable resistor, to correct the frequency error resulting from component tolerances.

13

Semiconductor Transducers

The term "transducer" implies the conversion of one form of energy into another, and in the context of electronics it usually refers to those devices which convert physical or chemical quantities, which it is desired to measure, into electrical signals. Apart from the purely electrical applications so far presented, semiconductor devices can be used as transducers in several ways; for example, it has already been seen that diodes and transistors are very temperature sensitive, so that applications in temperature measurement are immediately suggested. Other applications include the conversion of radiation (including light) into electrical signals; the production of light (both visible and infrared) from electrical signals; and the detection and measurement of such widely divergent phenomena as pressure and the concentration of inflammable gases. Much—though not all—of the relevant technology is based on silicon, and modern microcircuit fabrication has resulted in devices as complex as a gas chromatograph on a single chip, diffusion column included[1,2]. This, however, is an unusual example, and it is more realistic to consider simpler and more common transducers such as those for temperature measurement.

13.1 TEMPERATURE TRANSDUCERS

As detailed in chapter 5, the collector currents of two identical transistors are approximated by:

$$I_{C1} \simeq I_{ES} \exp\left(\frac{qV_{BE1}}{kT}\right)$$

and

$$I_{C2} \simeq I_{ES} \exp\left(\frac{qV_{BE2}}{kT}\right)$$

which may be written:

$$\frac{q}{kT} V_{BE1} \simeq \ln \left(\frac{I_{C1}}{I_{ES}} \right)$$

and

$$\frac{q}{kT} V_{BE2} \simeq \ln \left(\frac{I_{C2}}{I_{ES}} \right)$$

so that

$$(V_{BE1} - V_{BE2}) = \frac{kT}{q} \ln \left(\frac{I_{C1}}{I_{C2}} \right). \tag{13.1}$$

That is, $\Delta V_{BE} \propto T$.

If the two collector currents are defined by current sources (such as described in chapter 7) then ΔV_{BE} may be used to measure temperature in a circuit like that of figure 13.1(a). Here, the two values of V_{BE} are applied to an instrumentation amplifier which performs the subtraction, so that its output voltage is directly proportional to the temperature. (Note that the collector currents must be kept sufficiently low to prevent significant self-heating).

Modern versions of such sensors convert ΔV_{BE} to a current rather than a voltage, so that two-terminal operation over wide supply voltage ranges becomes possible[3]. Further, instead of using identical transistor structures along with two different current sources, monolithic transistors having closely controlled emitter area ratios are fabricated, leading to the same result but with identical collector currents. Consider, for example, the double current-mirror circuit of figure 13.1(b), in which two transistors Tr1 and Tr2, having unequal emitter areas, are driven by equal currents derived from the mirror configuration Tr3/Tr4. The leakage currents of Tr1 and Tr2 will be proportional to their emitter areas and so will not cancel out as was the case for the identical structures to which equation (13.1) referred. If the ratio of areas and hence of leakage currents is r, then equation (13.1) becomes:

$$\Delta V_{BE} = (kT/q) \ln r = I_{E1} R_E.$$

But

$$I_{E1} = I_{E2} = \tfrac{1}{2} I$$

where I is the total current taken by the circuit, so that $I \propto T$ (if R_E is temperature-independent).

Several modern two-terminal temperature transducers are based on this

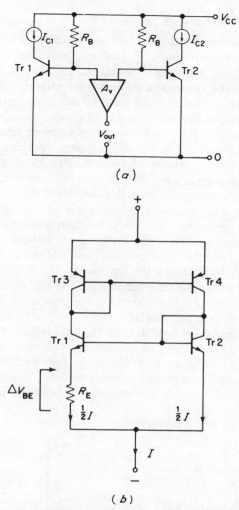

Figure 13.1. (*a*) Matched transistors used in a temperature-measuring circuit. (*b*) Double current-mirror temperature transducer.

principle, such as the Analog Devices AD590, and Timko[3] has shown how the basic circuit is extended to produce such practical devices. Basically, the temperature dependence of semiconductors arises because some valence electrons are given sufficient thermal energy to cross the band gap to the first conduction band. This mechanism is valid not only for single-crystal devices, but also for polycrystalline materials. The generic term *thermistor* encompasses the many variations of such temperature-sensitive devices, and is a contraction of "thermal resistor."

13.1.1 The Thermistor

The physical forms taken by thermistors are many, and depend entirely on the use for which they are intended[4]. Both negative (NTC) and positive (PTC) temperature coefficient types are available, the former being the most common. This NTC type consists essentially of a mixture of oxides, such as those of iron, cobalt, chromium and manganese. Some mixtures will, after sintering, exhibit temperature coefficients of about -3 to $-6\%/°C$. Figure 13.2(a) shows a family of characteristics for a series of NTC thermistors; whilst figure 13.2(b) gives a curve for a typical PTC unit, which employs different oxides, including those of barium, strontium and titanium. (It should be noted that PTC silicon resistors are also available, and have much smaller temperature coefficients than this—about $0.7\%/°C$ in fact.)

When the power dissipated in a thermistor is very small, so that there is no self-heating, its resistance will be a function of ambient temperature. This leads to applications such as the temperature stabilization of transistors (figure 5.24), or in fact, the actual measurement of temperature. The latter is best accomplished by inserting the thermistor into a bridge network as in figure 13.3(a). Here, the out of balance signal at terminals ab can be used either to deflect a suitably calibrated meter, or to operate a relay, when the arrangement will act as a thermostat.

If the current is great enough to heat the thermistor, then its resistance will change irrespective of ambient conditions. This situation is represented for an NTC thermistor in figure 13.4. Here, the log–log plot not only allows the very wide resistance range of a thermistor to be presented, but also means that power dissipation loci appear as straight lines rather than hyperbolae.

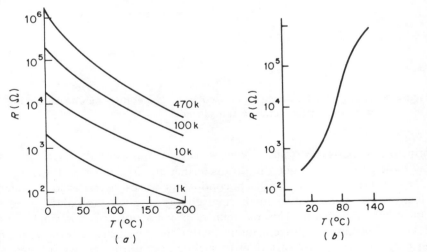

Figure 13.2. NTC and PTC thermistor characteristics: (a) resistance of some typical NTC bead thermistors as a function of temperature; (b) for a typical PTC unit.

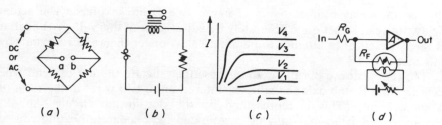

Figure 13.3. Some thermistor applications: (*a*) temperature measurement; (*b*) time delayed relay; (*c*) time delay characteristics; (*d*) gain control by indirectly heated thermistor.

The family of characteristics rises linearly at very low power levels, showing that the thermistor obeys Ohm's law, and has a resistance which depends only on the ambient temperature. When self-heating occurs, however, the thermistor resistance begins to decrease and eventually the curves coalesce as the self-heating process becomes the dominant factor.

If the thermistor is used in the self-heating mode, it lends itself well to sundry forms of timing circuit. Figure 13.3(*b*) shows an NTC thermistor connected in series with a relay. When the supply voltage is switched ON, the thermistor prevents the flow of sufficient current to operate the relay. However, as self-heating occurs, the resistance falls until the relay is able to close. The delay time must obviously be a function of the applied voltage;

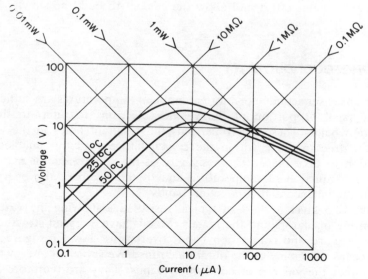

Figure 13.4. Typical NTC thermistor *V/I* characteristics (at three ambient temperatures).

that is, the delay characteristics of figure 13.3(c) are involved. The actual delay time can be set quite accurately by adjustment of this voltage though it is rarely practical to attempt to obtain a delay longer than a few seconds by this method.

The thermistor applications so far described dictate the use of bead-type devices; that is, the oxide mixture is sintered on to two wires in the form of a small bead. (This is represented by the dot in the circuit symbol.) However, this bead may also be surrounded by a separate heating coil so that the thermistor becomes an indirectly heated device. This implies a high degree of electrical isolation between the input (heater) and the output (bead), and this property has been put to use in many ways. For example, figure 13.3(d) shows how the gain of an operational amplifier may be controlled by an external direct current. Unfortunately, the response of this circuit is inevitably slow, and this, like most applications involving the use of the thermistor as a controlled resistor, has been superseded by the FET working below pinch-off. However, in cases where the long delay is not disadvantageous, the indirectly heated thermistor remains very useful. A case in point is the measurement of RMS current when the waveform is complex—the complex current is simply passed throught the heater, and the bead resistance is measured.

A larger mass (usually a rod) of the oxide mixture is often used to protect the filaments of lamps or vacuum tubes at the moment of switch-on. Pure metals have positive temperature coefficients, which means that when a voltage is applied to the cold filament, a large—and sometimes destructive—current can flow. An NTC thermistor connected in series with such filaments will counteract this effect and allow the current to build up slowly.

13.2 PHOTOCONDUCTIVITY

In an intrinsic semiconductor, normal ambient temperatures are sufficient to cause a small proportion of the valence electrons to jump to the first conduction band. These electrons, plus the resultant holes in the valence band, are free to move within the crystal lattice, and this enables some conduction to take place. The resistivity of the semiconductor material is therefore a function of temperature and of the width of the energy gap between the valence and conduction bands.

Figure 13.5 gives a chart of typical resistivities showing the vast range between the insulators and the metals. Metals have a crystal structure such that the valence and conduction bands overlap as shown in figure 2.1(c), which means that many of the outer electrons have sufficient energy to exist within either the valence or conduction bands. They are therefore free to move and exist as an "electron gas" within the crystal, which accounts for the good conducting properties of metals.

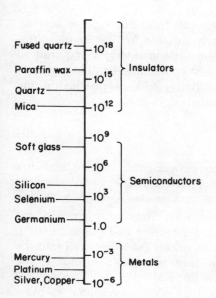

Figure 13.5. Resistivities of materials (in Ω cm).

Returning to the intrinsic semiconductor, it is reasonable to suppose that thermal agitation does not represent the only way in which energy can be supplied to the valence electrons. In fact, radiation in various forms can effect the necessary transitions. When this radiation is in the form of photons, as opposed to sub-atomic particles, the phenomenon is known as photoconductivity. The term implies that a shift of sufficient numbers of electrons into the conduction band will result in a change in the conductivity of the material. Devices utilizing this principle are often called photo-resistors.

Another photoelectric phenomenon is that of photoemission. Here, the incident photons are sufficiently energetic to remove electrons from the surface of a material, the process being represented by the well-known equation:

$$h\nu \geq W_f \tag{13.2a}$$

where $h\nu$ is the energy of a photon (Planck's constant × frequency) and W_f is the work function of the material in question.

This equation indicates that the incident photon must have a certain minimum energy in order that photoemission should take place—the actual photon density (i.e., radiation intensity) does not enter into the expression.

A great deal of work has been done on photoemissive devices such as photocells, photomultipliers and the various forms of image tube, and they will not be considered further here. It is, however, pertinent to point out that the high energy photons needed to effect the removal of electrons from the surface of materials of even very low work function, such as cesium, make the photoemissive devices particularly suitable for use in the ultra-violet and visible regions. In contrast, the photon energies required to transfer electrons from the valence to the first conduction band in semicon-ductors are much lower, leading to sensitivity peaks appearing in the infrared or visible regions. The relevant equation is now:

$$h\nu \geq W_g \qquad (13.2b)$$

where W_g is the width of the energy gap between the valence and first conduction bands. For silicon, W_g is about 1.1 eV whilst W_f for cesium it is about 1.9 eV. Photons having energies equal to these two figures would have wavelengths of 11 180 Å and 6450 Å, respectively. They both represent the longest wavelengths which can cause photoconduction and photoemission, respectively, and indicate that a transition in silicon can occur well into the infrared region, whilst electron emission from a cesium surface is not possible for wavelengths longer than the visible red.

Inevitably, the picture of photoconductivity so far presented is over-simplified, and several modifications and secondary processes are involved. Some of these will now be discussed.

Transitions do not cease abruptly when the wavelength of the incident light becomes longer than the threshold wavelength, which represents a photonic energy equal to the semiconductor energy gap. Instead, a "tail" appears as shown in figure 13.6. This appearance of photoconductivity at photon energies below W_g is due to thermal effects. There is a probability that a photon may give up its energy to an electron already possessing an amount of thermal energy such that the sum of the two is sufficient to raise that electron to the conduction level. Furthermore, W_g itself is not entirely constant, but will fluctuate somewhat depending on the magnitude of the thermal vibrations within the crystal lattice. This means that at some point, the random vibrations of the local atoms may result in an increase in density (with a lowering of W_g) at the instant when a photon of sub-threshold

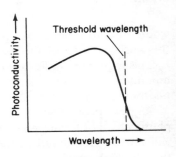

Figure 13.6. Photoconductivity as a function of wavelength.

energy arrives. Under these circumstances a valence electron acquiring this energy might be able to cross the lowered energy gap.

Apart from this, photoconductivity does not occur at wavelengths longer than the threshold, which means that photons of lower energies may pass through the material comparatively unimpeded; that is, the semiconductor will become transparent to the longer wavelengths if other effects do not become prominent. This situation does in fact occur, and optical components such as lenses and windows for infrared light are made from germanium and other semiconductors.

If both electrons and holes are present in a semiconductor, then recombination must occur, and the mean length of time before this takes place is known as the carrier lifetime. The probability of spontaneous recombination occurring is, however, very small, for momentum must be conserved in such a reaction, which implies closely matched energies unless a third body is provided by any one of the various imperfections or impurities which normally exist within a crystal, in particular at its surface.

An imperfection which results primarily in recombination is called a recombination center, and the concentration of such centers has a considerable effect upon the carrier lifetime in the photoconductor. There is also the trapping center, which is a form of imperfection capable of trapping a carrier within the lattice, but capable also of releasing it without recombination. Both forms of imperfection have a marked effect upon the speed with which the resistivity changes after a sudden alteration in the intensity of the incident light.

These mechanisms indicate that the spectral and electrical characteristics of an intrinsic semiconductor depend on many factors[5], some of which might reasonably be expected to be subject to external control during the fabrication process. To some extent this is true, and both the spectral response and speed of operation can be predetermined, within limits, by both physical control of the processes involved in manufacture and careful regulation of the amounts of added impurities. (Light doping with "activator" impurities, discussed more fully later, may be involved. For CdS cells, these include gallium, copper, and silver).

At present, the technology involved in the manufacture of photoresistors allows a very accurate peak spectral response to be predetermined for CdS and CdSe photoresistors, but it will in general be restricted to the range between 5300 Å and 7000 Å. (CdSe cells, though approximately an order of magnitude faster than CdS cells, are much more temperature sensitive.)

The electrode connections to the photoresist material are also of importance. For the CdS or CdSe type, gold is sometimes used since it is an inert metal, but this type of junction is nonohmic, that is, the voltage–current relationship is not linear, especially at very low voltages. Ohmic contacts can be made using indium or tin electrodes.

If extremely low energy photons are to be detected—that is, the cell is to be sensitive in the far infrared—then an intrinsic semiconductor may be heavily doped with an activator impurity. This is a material whose first conduction band exists at only a very slightly higher energy level than the valence band of the semiconductor. Low energy photons are thus enabled to raise electrons from this valence band to the conduction band of the impurity, where they are free to move. The holes generated in the semiconductor valence band are also free to move. Thus the activator impurity makes possible both hole and electron conduction, which leads to an ohmic characteristic.

Contemporary infrared detectors include germanium doped with gold (sensitive up to about 9 μm) mercury (up to 15 μm), or copper (up to 29 μm). There are also some germanium–silicon alloy cells activated with gold (up to 14 μm), or zinc (up to 16 μm).

The prime disadvantage of these detectors is that they must be cooled to very low temperatures, for normal thermal agitation energies are more than sufficient to raise electrons to the activator level. Such cells are cooled in double Dewar flasks using liquid gases, typically nitrogen in the outer chamber and helium in the inner. Cooled infrared windows of coated germanium or silicon or the less rugged cesium iodide or barium fluoride are also utilized.

The response of these cells is extremely fast, and time constants of the order of a microsecond or less are normal.

13.2.1 The Bolometer

For the detection of energy in the near infrared, it is often posible to use suitable forms of thermistor. A general survey of thermistors and their applications is given by Hyde[4], but here it is pertinent to consider briefly the variant known as the bolometer. This consists essentially of a mixture of semiconducting oxides, such as those of manganese and nickel, which are heated and partly reduced to form a flake of the required resistance. Though such devices are very sensitive (they will operate with an incident radiation down to about 10^{-9} W) their response is slow, for they depend, like all thermistors, on the heating effect of the radiation. This does, however, imply one great advantage—that the thermistor bolometer is essentially a "flat response" detector; that is, its response is independent of wavelength over a wide range. It is for this reason that thermistor bolometers are widely used as detectors when obtaining the spectral response curves of wavelength sensitive transducers; and when determining the spectral distribution patterns of light sources.

13.2.2 The Junction Photocell

The phenomena occurring at a pn junction have been described in chapter 2, and it has been shown how a potential difference occurs across this junction due to the migration of carriers across it.

If hole–electron pairs are generated continuously within or near such a junction, then it is reasonable to suppose that a current will pass around an external circuit. This is the basis of the *photovoltaic cell,* which is so constructed as to allow light to arrive at the junction and liberate carrier pairs. This form of operation is represented by the fourth quadrant of figure 13.7, which will be discussed more fully later.

If the junction is reverse-biased, then only the small leakage current due to intrinsically generated carriers will flow; that is, electrons generated in the p-region and holes generated in the n-region will be assisted across the junction. If light is allowed to reach the junction, the rate at which these intrinsic carriers are generated will increase; a situation which can be represented by the family of curves shown in the third quadrant of figure 13.7. This is the principle of the *photojunction cell.*

Operation in this quadrant is dependent on the existence of a reverse polarizing voltage, and this has led to an analogy being drawn between the photojunction and the photoconductive cell. This, however, is not a justified analogy, for whereas the resistance of the photoconductive cell is a function of light intensity, it remains an ohmic resistance; that is, the current passed is proportional to the voltage applied. Conversely, the current passing through a reverse-biased photojunction, though a function of light intensity, depends but little on the applied voltage. It is, in fact, a reverse-biased diode whose leakage current is due largely to the incident radiation.

The photojunction cell is very much faster in operation than is the photoconductive cell. This is because practically all of the applied voltage appears across the junction itself, leading to a depletion layer having a very steep potential gradient indeed. Consequently, the hole–electron pairs generated within this layer are immediately swept away to the electrodes, resulting in a very fast build-up of current. The risetime of such cells can be of the order of a microsecond and better.

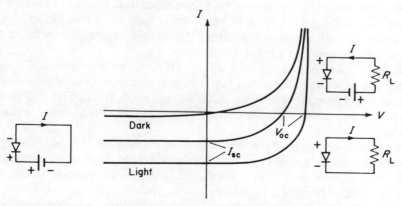

Figure 13.7. The effect of light in the three photodiode connection modes.

13.2.3   The Photovoltaic Effect

The representation of the photovoltaic effect shown in the fourth quadrant of figure 13.7 indicates that a voltage appears across a pn junction when that junction is illuminated. This phenomena has been known for several generations, but has only recently become fully understood as part of the general properties of the pn junction. However, photovoltaic cells have been available over much of the present century, the most successful ones (which are still readily obtainable) being the well-known selenium cells. These are very popular for use in light meters, their spectral response being similar to that of the human eye.

Some modern versions are made by depositing selenium—a p-type semiconductor even when highly purified—upon a metal base plate, typically aluminum or brass, then heating the unit to convert the selenium into its crystalline state. Cadmium is then allowed to diffuse into the selenium, forming the pn junction, while deposited cadmium oxide forms the n-type layer. Over this is deposited a very thin, transparent layer of gold to act as the alternate electrode. The process is completed by attaching lead-out wires to the electrodes and encapsulating the cell in a suitable resin.

The silicon photovoltaic cell is more efficient than the selenium cell as a transducer, but has a spectral response which peaks in the infrared. However, these two properties combine to make it an excellent device for the conversion of sunlight, hence its popularity as a solar battery. It is made in two forms, the p-on-n and the n-on-p. For many uses, including solar power conversion in artificial satellites, the latter is preferable owing to its greater resistance to degeneration by high energy radiation.

13.2.4   Thermoelectricity

For the detection of radiation in the near infrared, the thermopile is extremely useful. It may be considered to be the self-generating equivalent of the thermistor bolometer in that it utilizes the heating effect of the incident radiation rather than directly converting the photonic energy by production of carrier pairs. The term "thermopile" is taken to refer to two or more thermocouples connected in series.

The thermoelectric or Seebeck effect refers to the production of a small voltage across a circuit involving two different materials connected in series, one junction being hot whilst the other is kept cool. It was first observed in metals, but the most sensitive devices at present available utilize semiconductors.

The dominant thermoelectric mechanism involves the thermal generation of more carriers at the hot junction than at the cold, leading to a migration across the semiconductor. This implies that the semiconductor in question must be uniformly doped to a level where the expected temperatures are sufficient to raise electrons to the conduction band (if an n-type material is used) or to produce holes in the valence band (if a p-type is used).

The migration of carriers along the semiconductor represents a potential gradient and the resulting potential difference across the junctions is the useful thermoelectric voltage.

The thermopile must be used within the temperature range where the donors or acceptors are not "saturated," that is they have not all donated or accepted an electron. Were this not the case, a very small thermoelectric voltage would be produced owing only to the greater kinetic energies possessed by the carriers from the hot end over those from the cold end, leading to a slightly increased flow one way compared with the other.

13.2.5 Characterization of Photoelectric Devices

It is an unfortunate fact that there is no standard convention for characterizing photoelectric devices to suit a particular design purpose without too much confusion.

In general, a photosensitive device will have an output which depends on the level and wavelength of the incident radiation, applied voltage (if any), temperature, and the nature of the external circuit. The response itself will be in terms of voltage, current, resistance or power transferred, depending on the type of cell involved. Finally, the random noise produced by the cell—mainly a function of temperature—will also have to be taken into account in many applications.

The basic parameter describing a photocell is that of sensitivity; in other words the change in output signal as a function of the change in irradiation. Basically, irradiation is defined as the radiant power in watts per unit area falling upon the relevant surface, but when characterizing a particular cell, the area of the sensitive surface is often taken into account, and the total radiant power (in watts) falling upon the cell is considered. The units assigned to the numerical values given will differentiate between the two variants.

When sensitivity is being described, it is necessary to know the conditions under which the figure given is valid. This implies a knowledge of the nature of the incident radiation and the electrical conditions under which the cell is working. Consider the nature of the incident radiation.

If radiation within a narrow waveband is applied to a cell and its mid-wavelength is varied whilst maintaining constant the total radiant power reaching the cell, then the sensitivity of the cell as a function of wavelength can be plotted. This is called the *spectral sensitivity* of the cell, and the curve is usually normalized to represent a percentage of the maximum sensitivity, which will occur at one particular wavelength. This enables the spectral sensitivities of different cells to be compared directly as shown in figure 13.8. The actual sensitivity of each cell may be given in quite different units and under different circuit conditions, but providing that each sensitivity is expressed with reference to the irradiation per unit bandwidth, a direct spectral comparison can be made. Since semiconductor photocells are useful

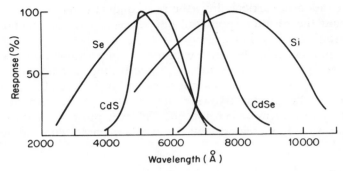

Figure 13.8. Comparison of spectral sensitivities of selenium, silicon (both photovoltaic), CdS and CdSe (both photoconductive) photocells (all normalized to 100% maximum). 1 Ångstrom unit (Å) = 10^{-10} m or 10^{-4} μm.

primarily in the visible and infrared regions, the wavelength of the incident light is most conveniently expressed in microns (micrometers), where 1 μm = 10^{-6} m. Using this unit, the visible band occurs between 0.4 μm and 0.7 μm. If desired, the millimicron (10^{-9} m) can be used to give a somewhat more practical figure.

As an example, the sensitivity of a photovoltaic cell might be given in microvolts per milliwatt per millimicron on open-circuit and at a given temperature.

Unfortunately, the plotting of a spectral sensitivity curve is quite difficult. In the first place a beam of heterogeneous radiation must be passed through a monochromator to obtain a narrow waveband at some known and controllable center wavelength, then the resulting light must be allowed to impinge upon the cell under test. Both the width of the waveband and the irradiation power it represents must be known. The former figure is difficult to determine though the latter can be measured using a transfer detector which is independent of wavelength, such as a bolometer or thermopile.

The difficulty involved in such measurements is one reason why manufacturers tend to express the sensitivity of their products in units which are more arbitrary, but simpler to determine. A common system is to express sensitivity as a function of irradiation from a filament lamp working at a given color temperature; and often the irradiation is given not in watts per unit area but in the units of illumination; that is, lumens per unit area. All these terms need defining at this juncture, but if the reader is already familiar with the terminology, the following digression may be omitted.

13.2.6 Some Notes on Irradiation and Illumination

All real radiant sources have a spectral energy distribution which is a function of wavelength. Incandescent sources exhibit a continuous spectrum, which can be plotted as a smooth curve, whilst discharge sources emit radiation in discrete lines or bands.

When a surface receives radiation from a source, it is said to be irradiated, and for uniform irradiation will receive E_e watts per unit area. If the irradiation is plotted as a function of wavelength, then for each unit increment of wavelength, the surface will receive $E_{e\lambda}$ watts per unit area per unit wavelength. Consequently,

$$E_e = \int_0^\infty E_{e\lambda} \, d\lambda \qquad (13.3)$$

where $E_{e\lambda}$ is the spectral irradiance and E_e is the total irradiance.

For incandescent sources, it is usual to compare the spectral distribution with that due to a *black body*, which is the most efficient possible radiator; that is, a body which radiates all the heat it receives without conduction or convection. It is possible to define the spectral distribution according to Planck's equation for any given temperature. A number of such curves are given in figure 13.9, from which it will be seen that—as might be expected—the hotter the black body, the more light at the blue end of the spectrum is emitted.

The tungsten filament lamp has a spectral distribution which is quite close to that of a black body, and this has led to the rather arbitrary concept of color temperature. This may be roughly defined as the real temperature of a black body which would emit light having the closest approximation to the color of the light produced by a filament lamp at some arbitrary working temperature; the two colors to be compared by projecting the light on to

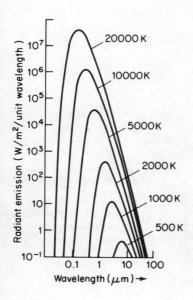

Figure 13.9. Spectral power radiated from a black body per unit area of that body. (Note that the radiant emission is the radiant power emitted per unit *source* area per unit wavelength. This should not be confused with spectral irradiance, $E_{e\lambda}$, which represents the spectral power *received* by a surface.)

adjacent parts of a white screen under similar geometrical conditions. By this means, a filament lamp working at a defined current can be calibrated in terms of color temperature even though the actual temperature of the filament is unknown.

In fact, some manufacturers define real filament temperature, which can only be measured using a radiation pyrometer.

As has been stated, the units of irradiation—ideally watts per square meter, in the SI system—often give way to the units of illumination. These arose as a result of many attempts to define a standard of intensity of visible light. This standard is now the candela, which is one-sixtieth the luminous intensity of one square centimeter of a black body working at the temperature of melting platinum. The luminous flux produced by a point source of one candela is 4π lumens.

The factor which relates the units of irradiation and those of illumination is the sensitivity of the human eye, the so called visibility or lamprosity curve. This is a curve which represents the sensitivity of the "average" human eye as a function of wavelength. It is normalized to unity, which means that if the ordinates of an irradiation curve (figure 13.10(a)) are multiplied by the ordinates of the visibility curve (figure 13.10(b)), the result is the illumination curve of figure 13.10(c).

Unfortunately, the units which would result from this transformation are youngs per unit area per unit wavelength, which are now not used.

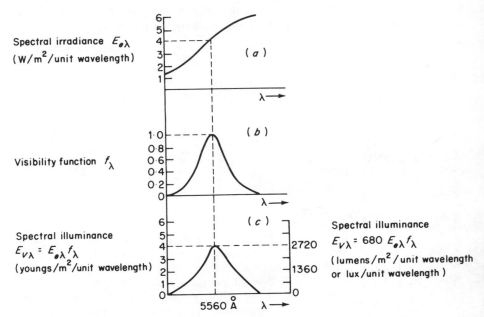

Figure 13.10. Relation between irradiance and illuminance units.

However, at the wavelength of peak sensitivity for the human eye (5560 Å), where the visibility function is unity,

$$1 \text{ young} = 1 \text{ W} = 680 \text{ lumens} .$$

Consequently, at any wavelength λ, the spectral illuminance is

$$E_{v\lambda} = 680 \, E_{e\lambda} f_\lambda$$

and the total illuminance is

$$E_v = \int_0^\infty E_{v\lambda} \, d\lambda = 680 \int_0^\infty E_{e\lambda} f_\lambda \, d\lambda . \tag{13.4}$$

If metric units are used,

$E_{e\lambda}$ = spectral irradiance in watts per square meter per unit wavelength
E_e = total irradiance in watts per square meter
$E_{v\lambda}$ = spectral illuminance in lumens per square meter (lux) per unit wavelength
E_v = total illuminance in lumens per square meter or lux.

Since a source of one candela emits 4π lumens, then the illumination received by the inner surface of a sphere of radius one meter, having a point source of N candelas at its center, is obviously N lux. This is the origin of the inverse square law:

$$E_v = I_v/d^2 \tag{13.5}$$

where I_v is the intensity in candelas.

This expression will be inaccurate for any system other than a point source and a sphere, but is approximately correct providing that the distance d to a real source is much greater than the maximum dimension of a flat illuminated area; and that the source itself is very small compared with d.

Unfortunately there are several units of illuminance other than the lux, including the ill-named foot-candle. Two of the most common have the following conversion factors:

$$1 \text{ foot-candle (1 lumen per square foot)} = 10.76 \text{ lux}$$

$$1 \text{ phot (1 lumen per square centimeter)} = 10\,000 \text{ lux}$$

13.2.7 Characterization of Photoconductive Cells

If a cell is intended for operation only in the visible region, it is reasonable to define its parameters with reference to the units of illumination. Ideally, the sensitivity at the most sensitive wavelength would be quoted, along with a graph of relative spectral sensitivity. From these data the result of using the cell with a discharge source having a line or band spectrum could be estimated. However, if the cell is expected to be used only in conjunction with a filament lamp, then the response to light of the relevant color temperature may be quoted.

When a cell is intended for use with infrared radiation only, then the sensitivity should be quoted with reference to units of irradiation, or conversely with reference to a color temperature which will be considerably lower than that required for its visible light counterpart.

Both irradiation and illumination units may be used to describe a cell covering a wide spectrum, and as an example the characteristics of the Philips 61SV will be considered. This is a lead sulphide cell having a useful operating waveband of from 0.3 μm to 3.5 μm, as shown in figure 13.11(a). When illuminated from a filament lamp at a color temperature of 2700°K and connected as shown in figure 13.11(b) the sensitivity is quoted as 3 mA (peak) per lumen at a level of 0.05 lumens falling upon the cell area. The peak current sensitivity refers to the fact that the incident light is chopped at 800 Hz, leading to an AC output.

The apparent difficulty in deciding how to produce a given color temperature is not serious providing that the color temperature quoted is approximately that which would be expected when a filament lamp is run under normal operating conditions. This is the case here, and the color tempera-

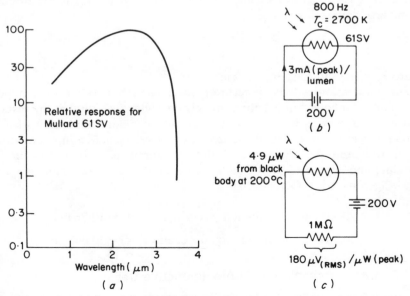

Figure 13.11. Sensitivity of the 61SV. (Reproduced by permission of Philips Components Ltd.)

ture of 2700°K is approximately that of a 75 W lamp running at normal voltage.

The infrared characterization is given at a real black body temperature of 200°C, which means that for practical measurements, a laboratory source would be needed. However, for design purposes a comprehensive set of Planckian radiation curves can be used.

For the 61SV under irradiation at a level of 4.9 μW from a black body at 200°C, the sensitivity is quoted as 180 mV (RMS) per μW (peak).

Here, the incident radiation is chopped at 800 Hz, and its peak value is considered. The output voltage, developed across a 1 MΩ load resistor as shown in figure 13.11(c), has an RMS value which is measured as a function of the peak irradiation.

This example reflects the subjugation of characterization methods to the expected application of a cell, and leads to four questions:

(1) How is the value of the load resistor chosen?
(2) How is the value of the applied voltage chosen?
(3) Why is the incident beam chopped and what determines the frequency?
(4) Why are RMS volts and peak irradiation power used?

The answers to these questions are in part logical and in part arbitrary. The load resistance for maximum power transfer should match the cell resistance, but whether maximum power transfer is wanted is another question, which depends for its answer on the nature of the amplifier input which the cell feeds. The applied voltage should approach the maximum rating of the cell, but the choice of an output RMS and an input peak measurement is arbitrary. The reason for chopping the incident beam is simply that for very sensitive detectors working at very low levels, a following AC amplifier may be used, which can result in an improved signal-to-noise ratio and an elimination of drift problems. The chopping frequency, however, must be well below the cut-off frequency for the device.

Some attempts have been made to rationalize the characterization of photoconductive cells, and one of the best known results is the definition of specific responsivity, S_1. This is numerically equal to the output in microvolts RMS across a matched load resistor when 1 V of bias is applied across the combination and the cell is irradiated at a level of 1 μW/cm^2.

Note that this definition does not specify the form which the irradiation takes; this information must be quoted along with the value of S_1. This makes possible the use of S_1 as a function of wavelength and some cells are in fact characterized by curves of S_1 plotted against wavelength.

The common usage of infrared cells at very low levels means that the noise voltage of the cell is of considerable importance. In this field the so-called noise equivalent power (NEP) is frequently used. This is the

irradiation which would produce a signal voltage equal to the noise voltage
or, alternatively, the irradiation which would lead to a signal-to-noise ratio
of unity.

Suppose that the output signal were V_s at an irradiation level of $E_e A$ W,
where A is the area of the cell. The power required to produce a signal
equal to the noise voltage V_N would be:

$$\text{NEP} = \frac{E_e A}{V_s} V_N . \tag{13.6}$$

The reciprocal of the NEP is called the detectivity[6] D, and this parameter
has become modified to take into account cell area in such a way that cells
of different areas might be directly compared. This is because the signal-to-
noise ratio varies inversely with the square root of the area[7]. If the noise
and signal voltages are measured over some bandwidth Δf at a center
frequency f, the specific detectivity D^* is

$$D^* = \frac{\sqrt{(\Delta f A)}}{\text{NEP}} . \tag{13.7}$$

This equation defines D^* as the signal-to-noise ratio which would be
obtained if the cell had unit area, and was irradiated with unit power, the
noise bandwidth also being unity. Note that the noise must be measured at a
defined chopping frequency f for it consists mainly of Johnson noise which is
a function of frequency (see chapter 6). If it is measured at a different
chopping frequency, then $\sqrt{f}$ should appear in the denominator[6].

In order to gain some idea of the quantities involved, consider the Philips
ORP 13 indium antimonide photoconductive cell, which is a liquid nitrogen
cooled device useful to 5.6 μm. From the data, NEP $= 2.3 \times 10^{-11}$ W at
$\Delta f = 1$ Hz and $A = 3$ mm^2 or 0.03 cm^2 giving

$$D^* = \frac{\sqrt{0.03}}{2.3 \times 10^{-11}} = 0.75 \times 10^{10} \text{ cm Hz}^{1/2} \text{ W}^{-1} .$$

When detectivity is quoted, the black body temperature of the source, the
chopping frequency, and the electrical bandwidth are sometimes given in
brackets. For the foregoing example this would be

$$D^*(500, 800, 1) = 0.75 \times 10^{10} \text{ cm Hz}^{1/2} \text{ W}^{-1} .$$

Note that D^* can also be a spectral function, in which case a curve of D_λ^*
against wavelength would be given and the units would become
cm Hz$^{1/2}$W^{-1} μm^{-1}.

The characterization of cells discussed so far has been concerned largely
with the more sophisticated types which are used for measurement purposes, particularly in the infrared. A more common form of photoconductive cell is the CdS or CdSe unit which can be used as a measuring

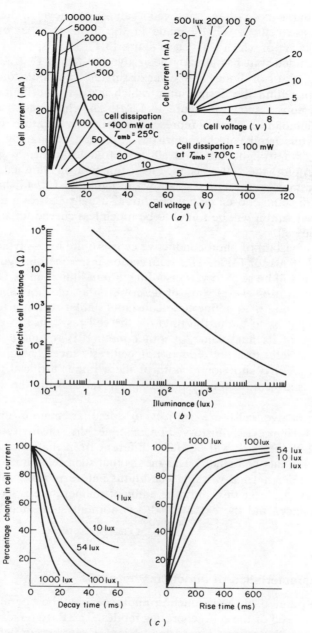

Figure 13.12. Curves for typical photoconductive cell with the lamp color temperature at 2700°K: (*a*) initial cell current plotted against cell voltage with illuminance as parameter; (*b*) effective initial cell resistance plotted against illuminance; (*c*) percentage changes in cell current plotted against decay time (left-hand diagram) and rise time (right-hand diagram) with illuminance as parameter. (Reproduced by permission of Philips Components Ltd.)

transducer, but is more often employed as an ON/OFF device. The characterization of these cells is usually made in the form of a set of graphs, a representative set of which is given in figure 13.12. (The spectral response is normally included, but has already been given in figure 13.8.)

Like a transistor or diode, the power dissipated by a photoconductive cell must be kept within the design limit, which leads to the limitation of the operating region as shown in figure 13.12(a). The characteristics at various levels of illuminance clearly represent the resistive nature of the cell. Sometimes, this family of characteristics is plotted on log–log axes, when each becomes a straight line, and all are parallel and at 45° to the axes. From such a plot, the cell resistance as a function of illumination is easily derived, and may be presented as shown in figure 13.12(b). This plot must be allied with some chosen circuit condition, usually that of a fixed voltage across the cell, and it will be found to be much less curved for the CdS than for the CdSe cell.

The time constant of photoconductive cells is quite long—typically 100 ms for CdS and 10 ms for CdSe. This information is graphed in figure 13.12(c), from which it will be seen to vary as a function of illuminance. Unfortunately, there are in use several ways of describing the relevant time constants. The most common is to define time constants similar to those for switching transistors; that is, to quote the time for the cell resistance to change from 10% to 90% of its final value. Another method is to quote similar time constants, but refer to a $1/e$ reduction in cell resistance. The graphical form of figure 13.12(c) is superior to both of these rather artificial "constants," and clearly shows how the response speed varies with the level of the illuminance.

Photoconductive cells are subject to some anomalous effects which depend on their previous history. For example, the sensitivity after having been stored in the dark will be quite different from that after exposure to light. (It will in fact be higher.) This means that such cells should be allowed to operate under normal working conditions before repeatability of results can be achieved. After this, only the ambient temperature is likely to affect the performance, and the extent of this is normally indicated by a further graph.

13.2.8 Characteristics of Junction Photocells

It has been pointed out that junction photocells may be operated in either the reverse-biased or the self-generating modes. The third quadrant of figure 13.7 represents the reverse-biased mode, and is redrawn in figure 13.13(b) in its more usual orientation, the photojunction cell itself being connected as in figure 13.13(a). The reverse current under dark conditions is simply the leakage current of the diode, and this increases markedly with irradiance. The curves of figure 13.13(b) indicate that this reverse current is slightly

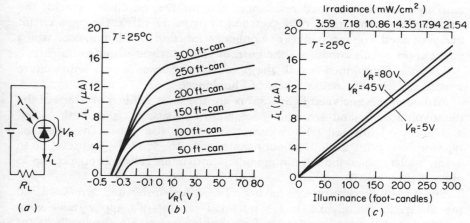

Figure 13.13. Characteristics of the 1N3734 photodiode: (*a*) photojunction circuit; (*b*) light current against reverse voltage; (*c*) light current against illuminance. Tungsten source operating at 2800°K, 1 foot-candle = 10.76 lux.

dependent on the applied voltage and that the incremental resistance presented is very high.

When a cell is operated in its self-generating or photovoltaic mode, its characteristics are represented by the fourth quadrant of figure 13.7. Here, a forward voltage and a reverse current appear, the relevant polarities being indicated in figure 13.14(*a*). The family of curves is usually reorientated as in figure 13.14(*b*). The generated voltage and current are clearly functions of

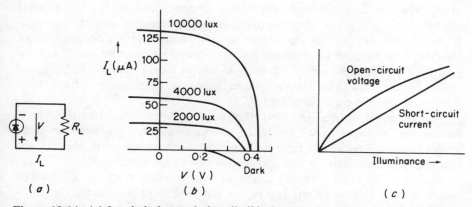

Figure 13.14. (*a*) Loaded photovoltaic cell; (*b*) characteristics of typical photovoltaic cell; (*c*) form of open-circuit voltage and short-circuit current curves for a photovoltaic cell.

both illuminance and load resistance, and for the extremes of load, the nature of the characteristics is sketched in figure 13.14(*c*). On open-circuit the generated voltage is a very nonlinear function of illuminance, whilst under short-circuit conditions the current characteristic is quite straight.

Like the photojunction cell, the photovoltaic sensor is very temperature sensitive, especially at low levels of illuminance.

Although the uncooled light sensors are the most familiar versions of the photovoltaic cell, and are mainly selenium and silicon types, both cooled and uncooled infrared photovoltaic transducers also exist. The uncooled types include both silicon and germanium cells, the latter being useful up to 2 μm, whilst cooled indium antimonide photovoltaic cells will operate up to about 5 μm.

Small photodiodes, though having a linear response when reverse-biased (as shown in figure 13.13), are intended for ON–OFF applications as in punched card and tape readers: in fact, multiple arrays are also available for this purpose. When it is necessary to measure illuminance, however, large-area cells become advantageous, and these can take several forms.

The simplest photojunction cell consists of either a p-on-n or an n-on-p planar diffused, oxide passivated structure. The most common is the n-on-p type, and this often incorporates a guard ring, as shown in figure 13.15(*a*). Here, the active part of the cell is the depletion region under the circular n^+-diffusion, and this is surrounded by a second, annular, n^+-diffusion which forms the guard ring.

The purpose of the guard ring is to extract that portion of the leakage current which is associated with the surface, as opposed to the bulk, of the material, and this is achieved as shown in the basic circuit of figure 13.15(*b*). Reduction of leakage, or dark current reduces the shot noise resulting from this current, and so makes the cell more sensitive. In other words the NEP is lowered. Surface leakage current is proportional to the root of the active area, whereas bulk leakage current is proportional to the area itself: consequently, the guard-ring technique is more successful with comparatively small-area devices (up to a few square centimeters) than with large. Further, it is possible to fabricate planar-diffused p-on-n photocells without guard rings, which have dark currents as low as n-on-p types utilizing guard rings. Hence, multiple arrays, where space is at premium, have been fabricated using p-on-n diffusions[8].

The speed of normal junction photocells may be limited to the microsecond region owing to the existence of comparatively long diffusion times in the field-free bulk material. If, however, a PIN structure is utilized, the rise and fall times can be reduced to the nanosecond region[9].

Basically, the PIN structure consists of near-intrinsic material sandwiched between very thin p and n layers, so that the field permeates the high-resistivity intrinsic material, and very little field-free bulk exists. Consequently, when hole–electron pairs are generated within the intrinsic

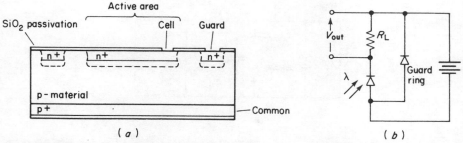

Figure 13.15. (*a*) An n-on-p planar-diffused guard-ring photojunction cell and (*b*) its basic circuit.

material, they are quickly swept apart by the field and collected at the electrodes.

The PIN structure is most conveniently fabricated in silicon by forming n-type material on one side of an intrinsic slice and backing it with a gold layer as one electrode; then oxidizing the opposite surface, etching a hole, and diffusing in an acceptor impurity to form a p-type island. The second electrode may then be an annular gold ring deposited at the periphery of the p-type island. The encapsulation is so designed that the incident light impinges upon the p-region, which, being very thin, allows most of the photons to pass through to the intrinsic layer to generate hole–electron pairs. This very thin p-region also contributes negligible bulk resistance, which also improves the high-frequency performance of the device.

A modification of the PIN diode employs a *Schottky barrier* in the place of the p-diffusion. Here, a thin, transparent gold film is deposited directly on to the surface of the n-substrate, so that when a negative potential is applied a depletion layer results. Light enters via the transparent gold film and produces charge carriers in the depletion layer and intrinsic region, both of which maintain voltage gradients, making for swift carrier collection[8].

Figure 13.16(*a*) is an equivalent circuit for a photodiode which incorporates a guard-ring structure. Here, the resistance between the guard-ring diffusion and the "active" diffusion is called the *channel* resistance, R_{ch}, and the bulk resistance of the semiconductor material and the contacts is called the *series resistance*, R_{sb} (this corresponds to $r_{bb'}$ in a transistor).

The complete list of constituent components is:

$$i_P = \text{photocurrent}$$
$$I_0 = \text{leakage, or dark current}$$
$$r_d = \text{diode incremental resistance}$$
$$C_j = \text{junction capacitance}$$
$$R_{sb} = \text{series resistance}$$
$$R_{ch} = \text{channel resistance.}$$

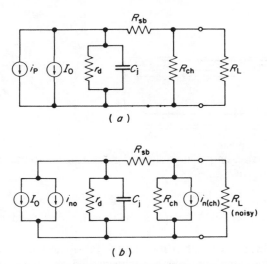

Figure 13.16. (*a*) An electrical equivalent circuit for the photodiode; and (*b*) its noise equivalent circuit for the dark condition (neglecting the thermal noise of R_{sb}).

(It has been assumed that the capacitances associated with the channel and series resistances are small.)

Note that in this equivalent circuit, the channel resistance can be combined with the load resistance, R_L, and along with C_j and R_{sb}, will define the cut-off frequency. If

$$R = R_{sb} + \frac{R_{ch}R_L}{R_{ch} + R_L}$$

then

$$f_H = \frac{1}{2\pi RC_j}. \tag{13.8}$$

For the reverse-biased photojunction, R_{sb} is small, but can be quite significant for unbiased photovoltaic operation; also, C_j is much larger for the unbiased case than for reverse-biased operation. Both of these phenomena lead to reverse-biased photojunction operation being much faster than photovoltaic operation.

There are two prime sources of noise in a photodiode; the shot noise associated with total current $I = (i_P + I_0)$, and the thermal noise arising in the series resistance R_{sb}. Also, if a guard-ring structure is involved, the channel resistance R_{ch} contributes another source of thermal noise.

The current in a diode is:

$$I = I_0 \left[\exp\left(\frac{qV}{kT}\right) - 1 \right].$$ (13.9)

For a reverse-biased photojunction, the exponential term is negligible, leaving the saturation current only. For the irradiated condition this includes the signal current, but under dark conditions it is simply I_0, the reverse leakage current.

The mean square shot noise associated with I_0 is, from equation (6.1),

$$\overline{i_{n0}^2} = 2qI_0\Delta f .$$ (13.10)

This is the dominant noise source for a reverse-biased photojunction under dark conditions. The thermal noise in the series resistance, R_{sb}, is usually small enough to be neglected; and the mean square noise in the channel resistance is (using equation (6.3)):

$$\overline{i_{n(ch)}^2} = \frac{\overline{v_{n(ch)}^2}}{R_{ch}^2} = \frac{4kTR_{ch}\,\Delta f}{R_{ch}^2} = \frac{4kT\,\Delta f}{R_{ch}} .$$ (13.11)

Here, the thermal noise voltage has been converted to a noise current so that it can be added to the (uncorrelated) shot noise:

$$\overline{i_{n0(total)}^2} = 2qI_0\,\Delta f + \frac{4kT\,\Delta f}{R_{ch}} .$$ (13.12)

A noise equivalent circuit is given in figure 13.16(b), from which it will be seen that the thermal noise of a load resistor can be included simply by observing that R_L and R_{ch} are in parallel, so that a single thermal noise for $R_L \| R_{ch}$ may be obtained.

For this reverse-biased photojunction mode, i_{n0} is dominant and the NEP and specific detectivity may be calculated if the dark current and spectral sensitivity are known. Recalling that the NEP is the irradiation which produces a signal equal to the noise under dark conditions, then using the spot-noise current, and noting the units involved,

$$\text{NEP} = \frac{\overline{i_{n0}}}{S} \frac{(A)(Hz)^{-1/2}}{(A)(W)^{-1}(Hz)^{-1/2}} = \frac{\overline{i_{n0}}}{S} \text{ W}$$ (13.13)

If the active area of the cell is known, then (if the cell area is in square centimeters)

$$D^* = \frac{(A\,\Delta f)^{1/2}}{\text{NEP}} \text{ cm Hz}^{1/2}\text{W}^{-1} .$$ (13.14)

As in all devices, there is a flicker noise contribution, and this can be taken into account by a rise in $\overline{i_{n0}^2}$ at approximately 6 dB/octave below about 30 Hz.

Example
A silicon photojunction cell has a dark current of 0.3 nA, an active area of 5.1 mm^2 and a responsivity of 0.45 A W^{-1} at 900 nm. What is the NEP and D^* at a light pulse frequency of 1 MHz?

$$\overline{i_{n0}} = (2qI_0)^{1/2} = (2 \times 1.6 \times 10^{-19} \times 0.3 \times 10^{-9})^{1/2}$$

$$= 0.98 \times 10^{-14} \text{ A Hz}^{-1/2}$$

$$\text{NEP}(900 \text{ nm}, 1 \text{ MHz}, 1 \text{ Hz}) = \frac{0.98 \times 10^{-14}}{0.45} = 2.2 \times 10^{-14} \text{ W Hz}^{-1/2}$$

$$D^*(900 \text{ nm}, 1 \text{ MHz}, 1 \text{ Hz}) = \frac{(A \Delta f)^{1/2}}{\text{NEP}} = \frac{(5.1 \times 10^{-2})^{1/2}}{2.2 \times 10^{-14}}$$

$$\simeq 10^{13} \text{ cm Hz}^{1/2} \text{W}^{-1} .$$

For photovoltaic operation into a very low load (such as the summing input of an operational amplifier) the voltage across the photodiode approaches zero, so that both terms of equation (13.9) have to be taken into account. Rewriting this equation for dark conditions[10],

$$I = I_0 \exp\left(\frac{qV}{kT}\right) - I_0 . \tag{13.15}$$

If each component is assumed to generate shot noise,

$$\overline{i_{n0}^2} = 2qI_0 \exp\left(\frac{qV}{kT}\right) \Delta f - 2qI_0 \Delta f . \tag{13.16}$$

Because $V \to 0$, this becomes

$$\overline{i_{n0}^2} = 4qI_0 \Delta f . \tag{13.17a}$$

In this equation, I_0 can be replaced by a term involving the dynamic or incremental diode resistance at the origin. From equation (13.15),

$$g_d = \frac{1}{r_d} = \frac{dI}{dV} = \frac{I_0 q}{kT} \exp\left(\frac{qV}{kT}\right)$$

and putting in $V = 0$ gives

$$g_0 = \frac{1}{r_0} = \frac{I_0 q}{kT}.$$

That is,

$$I_0 = \frac{kT}{qr_0}$$

so that equation (13.17a) becomes

$$\overline{i_{n0}^2} = \frac{4kT\,\Delta f}{r_0}. \qquad (13.17b)$$

Notice that this form of the equation is that which would result if the dynamic resistance were thought of as a thermal noise generator. From equation (6.3),

$$\overline{v_{n0}^2} = 4kTr_0\,\Delta f \qquad (13.17c)$$

which converts to the current noise form,

$$\overline{i_{n0}^2} = \frac{\overline{v_{n0}^2}}{r_0^2} = \frac{4kT\,\Delta f}{r_0}$$

which is equation (13.17b). (It should be noted that this is a grossly oversimplified explanation; a full treatment of noise in diodes is given by Buckingham and Faulkner[11].)

This noise is usually dominant in the photovoltaic mode, but the thermal noise of R_{sb} is at its maximum for the zero voltage condition. The equivalent circuit will be similar to that of figure 13.16(b) with the substitution of equation (13.17a) or (13.17b) for the shot noise current generator.

13.2.9 Applications of Semiconductor Photocells

It is neither reasonable nor possible to describe in detail even a few of the vast number of uses to which semiconductor photocells have been put. Instead, some general comments will be made which should enable the reader to embark upon his own design for any particular application.

Semiconductor photocell usage can be divided roughly into three fields: wide-band spectral and nonspectral measurement, narrow-band spectral and nonspectral measurement, and switching.

Wide-band measurement refers to the use of bolometers and thermopiles in the detection and measurement of radiation over a wide band of wavelengths. Their independence of wavelength makes them particularly useful in infrared spectrophotometry, where they logically take the place of the photomultiplier, which is basically an ultraviolet/visible device. They are also indispensable as transfer detectors for the characterization of wavelength-sensitive photocells and sources.

This independence of wavelength in the thermally sensitive detectors is bought at the price of response speed, for the time necessary for the active volume to heat up or cool down as the irradiance level changes restricts them to operation below 15–20 Hz. A detector of smaller volume would have a higher response speed but a lower sensitivity, and the better units are designed to represent an optimum compromise.

For very low-level operation, where stray pick-up signals are significant, it is necessary to shield the detector both electrostatically and magnetically. Usually this is done with concentric Mu-metal screening cans, and in some cases the cell is made in two adjacent sections, the wiring being so arranged that the wanted signals add whilst the pick-up signals cancel.

The wide-band measurement of irradiance is rarely performed spectrally by the wavelength-sensitive detectors; that is, E_e rather than $E_{e\lambda}$ is normally measured. However, in some spectrophotometers which rely upon the double-beam principle, a semiconductor detector may take over from a photomultiplier at the infrared end of the range. This is possible because the double-beam system renders the spectral sensitivity of the detector of minor importance providing that it works at all in the relevant region.

The narrow-band measurement of illuminance, E_v, has been the province of the selenium photovoltaic cell for many years. This is due to the early discovery of that cell, and to the good linearity of its current–illuminance characteristic at low load resistances. It also has the advantage of being able to operate a sensitive meter movement directly at moderate levels of illuminance (about 100 lux upwards) and this has led to its use in photographic exposure meters. Only in the last few years has the photoresistor been introduced into this field, and though a marked increase in sensitivity has resulted, the nonlinearity of the cell is a disadvantage.

If the load resistance is increased, the curvature of the characteristic changes, as is shown in figure 13.17, and it is possible to choose R_L so as to establish either of two unique conditions.

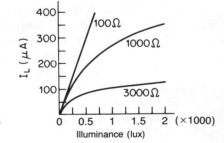

Figure 13.17. Characteristics at various loads for a typical photovoltaic cell.

Firstly, the characteristic may be arranged to approximate closely to a logarithmic law, which is useful where it is desired to measure optical density.

Secondly, R_L may be chosen to have a value equal to that of the internal resistance of the cell, in which case maximum power is transferred. The main application of this is for the generation of power, particularly where the weight of batteries militates against their use, as in artificial satellites. Silicon photovoltaic cells are normally used in such cases, for not only are they more efficient, but they have a response which extends further into the infrared than is the case for selenium. The correct load for this application can be found by drawing a hyperbola representing the power required and determining the illuminance required for a cell characteristic to touch it as shown in figure 13.18. The line from the origin to the point of contact represents the correct load.

The open-circuit voltage varies in a highly nonlinear manner with illuminance, and is therefore not a very useful quantity.

For the linear measurement of irradiance, the photodiode may be used in either the reverse-biased photojunction mode or the unbiased photovoltaic mode. A comparison between the characteristics of the two modes can be made by reference to the equivalent circuit of figure 13.16 and the photo-diode characteristics of figure 13.19 (which have been drawn in the correct quadrants) for a series of irradiance values from zero to E_{e3}.

Referring to this latter diagram, the slope of the E_{e3} characteristic at point A represents the internal resistance r_d for the self-generating mode corresponding to operation as a power source; and the load-line R_L for maximum power transfer should have the same value as this.

At point B, the slope of the characteristic has become very large, and if a low load is applied (as represented by the near-vertical load-line) the self-generated current is directly proportional to the irradiance; that is, the cell has become a linear radiant-power–current transducer. Such a low load can be conveniently provided by the summing point of an operational amplifier, as in figure 13.19(b).

If the photodiode is reverse-biased by a voltage V_D, the value of r_d becomes even larger, as at point C in figure 13.19(a); and the near-vertical

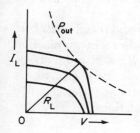

Figure 13.18. The photovoltaic cell as a power generator.

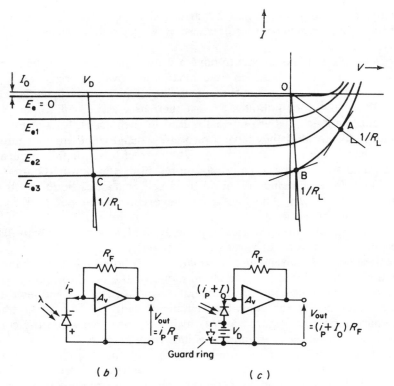

Figure 13.19. Photodiode characteristics and current-mode circuits.

load-line defined by the summing point in the circuit of figure 13.19(c) establishes linear operation in the photojunction mode.

If a guard-ring photodiode is used, the connection of the guard-ring diode will be as shown dotted in figure 13.19(c). For photodiodes of this form, the channel resistance, R_{ch}, appears in the equivalent circuit between the output terminals themselves, and this represents the resistance between the active part of the cell and the guard ring; that is, between the two n^+-diffusions in figure 13.15. However, it is effectively shunted by the very low load resistance at the summing point, and is therefore of little consequence in this form of operation.

Noise calculations for the circuits of figures 13.19(b) and (c) can be made as described in chapter 6 using values taken from data sheets in conjunction with the equivalent circuit of figure 13.16(b). Here, R_{sb} will normally be found to be small enough so that its noise contribution may be neglected. (A comprehensive treatment has been presented by Hamstra and Wendland[12] which includes a discussion of frequency response limitations.)

Other than its effect in subscribing to the limitation of high-frequency response, R_{sb} will slightly degrade the linearity of the cell at high values of i_p. This is because there will be a small voltage drop across it at the higher values of photocurrent: a situation which is similar to the degradation of h_{FE} by $r_{bb'}$ at high base currents in a transistor.

Both junction photocells and photoresistors are used in switching applications. Typically, the small photojunction cells having a fast response time will be used in such devices as bar-code readers, whilst the slower but more sensitive photoresistors appear in flame detectors, electro-optical production-line counters or burglar alarms.

Like some large-area photovoltaic cells, some photoresistors are designed to couple directly into loads without prior amplification. Since photoresistor characteristics are nonlinear, such loads rarely take the form of measuring instruments (except for some photographic light meters), but are often relays, as shown in figure 13.20(a).

If the resistance of the relay is R_L, it can be represented by a load-line on the characteristics as in figure 13.20(b). The currents at which the relay pulls in and drops out are seen to cross the cell characteristics at points which define the levels of illuminance for the ON and OFF states. The disparity between these current levels—the backlash—results in quite a wide range of illumination between these ON and OFF levels, and this can be improved by arranging a pair of relay contacts to bring in a resistor in parallel with the relay which has the effect of changing the position of the load-line.

If a more sensitive switching circuit is required, a photocell can be combined with a discrete or integrated transistor circuit. It has already been seen that a photodiode can be used with an operational amplifier to give an analog output; clearly a comparator (see chapter 10) can be used in place of the operational amplifier. Circuits which consist of photodiodes and comparators integrated on one chip are readily available, and can be obtained

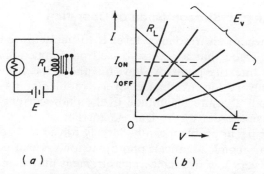

(a)   (b)

Figure 13.20. Direct operation of a relay using a photoresistor.

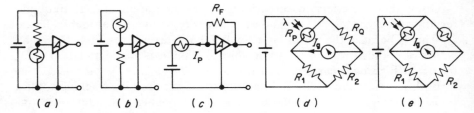

(a) (b) (c) (d) (e)

Figure 13.21. Some methods of loading a photoresistor.

not only as simple light level sensors, but as light–frequency converters.

There are numerous methods of loading photoresistors, and the following general remarks may be taken to apply to both (nonlinear) measuring and to switching circuits.

Firstly, an amplifier having a high input resistance may be used to amplify the voltage developed across either the photoresistor itself or across an associated load resistor, as shown in figures 13.21(a) and (b).

Secondly, the actual cell current could be fed into an operational amplifier as in figure 13.21(c), when the output voltage would be $I_P R_F$.

Finally, the resistance of the cell could be measured by the simple Wheatstone bridge network of figure 13.21(d). When the unbalance current I_g is brought to zero by adjustment of at least one of the bridge resistors, then $R_P = R_Q(R_1/R_2)$. In this type of circuit, one of the resistors, often R_Q, is made continuously variable, and is adjusted either by hand or via a servo-mechanism working from I_g. A simpler method is to measure R_P as a function of I_g.

Figure 13.21(e) shows one method of making the circuit less dependent on temperature by introducing a second photoresistor as a compensating element.

13.2.10   The Charge-Storage Mode of Operation

In applications where an irradiance level is to be measured, the foregoing material has indicated that a reverse-biased photojunction cell is the best choice from the linearity point of view, but the worst from the sensitivity aspect. Also, at very low irradiance levels, problems arise not only as a result of the small absolute magnitude of the photocurrent, but because it may be of the same order as the leakage current.

The leakage current problem can be largely solved by backing off against the leakage of a second, identical, photojunction, as will be shown below. The problem of very low irradiance measurement may then be approached by the charge-storage concept. In this method, a capacitor is first quickly charged to a predetermined voltage, and is then allowed to discharge via the junction photocurrent. Because the rate of discharge is linear under these conditions, either the time taken to discharge the capacitor to a second predetermined voltage, or conservely, the voltage to which the capacitor discharges in a fixed time interval, must be a function of the magnitude of the photocurrent. The relation is

$$\Delta V_C = \frac{I_P \Delta T}{C_0}$$

where ΔV_C is the voltage through which the capacitance C_0 is discharged by the photocurrent, I_P, in the time interval ΔT. That is,

$$I_P \propto \frac{1}{\Delta T} \quad \text{if } \Delta V_C \text{ is fixed}$$

or

$$I_P \propto \Delta V_C \quad \text{if } \Delta T \text{ is fixed.}$$

From the discussion above, it will be clear that I_P is the photocurrent *averaged* over the time interval ΔT, which is itself often a useful property of the technique. In other words, the amount of charge removed from the capacitor is proportional to the integral of the irradiation over the period ΔT, which is useful in that random noise in this incident irradiation will tend to cancel out if ΔT is large enough.

Weckler[13] initially detailed the charge-storage method of operating photojunctions, and Murphy and Kabell[14] extended the technique to the design of an integrating light meter. In this latter reference, they describe the operation of a light–frequency transducer in terms of the circuit in figure 13.22. Here, two identical junction photocells are connected in series, and

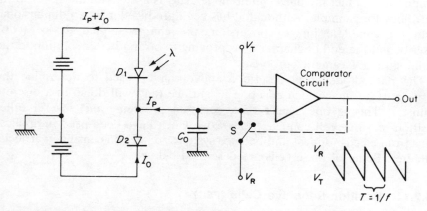

Figure 13.22. Principle of light–frequency converter (after Murphy and Kabell[14]).

one is occluded. This means that the leakage current (which is assumed to track with temperature if the cells are in close proximity) will flow through both cells and will not emerge at the common point of the two. However, the photocurrent due to the irradiated cell *will* emerge at the common point, and will be available to discharge linearly the capacitor, C_0. Here, C_0 is made up of the two cell capacitances which appear in parallel when measured from the common line of the power supply.

The comparator circuit is such that when the voltage on C_0 reaches the threshold voltage, V_T, the IGFET switch is closed long enough to recharge C_0 to the reference voltage, V_R. The resulting waveform is as shown in figure 13.22, and because C_0 charges very rapidly, and discharges comparatively slowly and linearly through D_1, the frequency of this waveform is given as follows:

$$\Delta V = \frac{\Delta Q}{C_0} \quad \text{or} \quad (V_R - V_T) = \frac{I_P T}{C_0}$$

giving

$$f = \frac{1}{T} = \frac{I_P}{C_0(V_R - V_T)}.$$

This means that the output frequency is directly proportional to the photo-current, and hence to the irradiation at the surface of the active photodiode.

Not only are monolithic light–frequency converters available, which use the "constant-voltage, variable-time" mode, but so are arrays of photo-diodes which operate in the "constant-time, variable-voltage" mode. Usually, these are linear arrays of 50 or more photodiodes which are charged sequentially from a shift register, and also interrogated sequentially. By this means a series of pulses can be extracted, of magnitudes representing the illumination levels across the length of the array.

This mode of operation is reminiscent of a single-line scan in a television camera, and in fact the linear photodiode array is used as an image sensor in metrology for example. Although it has been produced in a two-dimensional form for complete image conversion, the charge-coupled device (CCD) described at the end of chapter 11 is proving to offer a better solution to the "solid state TV camera" problem [15,16].

The photodiode array described above can be used to determine the position of an image focused upon its surface (and will do so in a discrete manner). This is one aspect of a common problem, and several other position-sensitive cell configurations exist which have been used in applications from electro-optical null-balance systems to horizon-sensing in artificial satellites. Some of these cells are described below.

13.2.11 Position-Sensitive Cells (PSC)

There are two basic analog approaches to PSC design. The first is to construct either a photodiode or photoconductive cell having a suitable geometry, and split symmetrically so that two separate but almost identical cells result. If a light spot is focused to fall symmetrically across the split, equal signals will appear from each half of the cell. If the light spot moves the signals will differ, and this difference signal will be a function of the displacement of the light spot from the null position.

This arrangement is illustrated in figures 13.23(a) and (b), whilst typical bridge circuits appear in figures 13.23(d) and (e). The potentiometer is included in these circuits to set the output at zero under conditions which are to be regarded as null.

An alternative arrangement is illustrated in figure 13.23(c). Here an extra, highly resistive stripe has been deposited upon the insulating substrate. This is connected to the lower conducting stripe via the photoconductive stripe only at the point where a bright image appears. The equivalent circuit, figure 13.23(f), is that of a potentiometer, which is again a position-sensitive device. This construction has the advantage that a position-defining signal will appear irrespective of the actual position of the image, whereas for the first two cells the output signal will be invariant when the image has left the slit completely and impinges only upon one half of the composite cell.

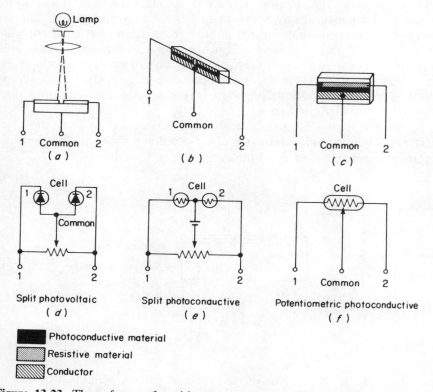

Figure 13.23. Three forms of position-sensitive photocell with associated circuits.

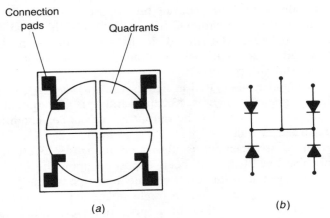

Figure 13.24. Typical plan-form of a four-quadrant photodiode (a) and its functional electrical equivalent (b).

Such devices may be made fully position-sensitive as illustrated in figure 13.24(a), which shows a typical arrangement for a four-quadrant silicon planar photodiode. Electrically, it may be thought of as a set of four individual diodes as in diagram (b), so that when a circular spot of light is exactly centered on the pattern, the photocurrents from each quadrant will be equal. Should the spot deviate from the center, the photocurrents will change in proportion to the illuminated areas. This obviously lends itself to either measurement via analog signal processing methods, or to servo control of the spot position, provided of course that the spot does not completely leave any quadrant.

The second approach, initially described by Wallmark[17] is typified by the UDT "Posicon," which consists of a Schottky-barrier structure having contacts at the bottom edges of the substrate, as shown in figure 13.25[18].

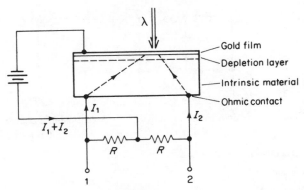

Figure 13.25. Mode of operation of the "Posicon" PSC (by permission of United Detector Technology Inc.).

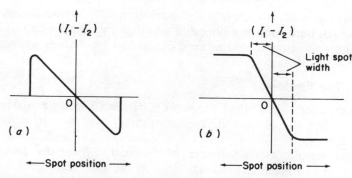

Figure 13.26. Difference current against light spot position for (*a*) the UDT "Posicon" and (*b*) a split photodiode.

When the light beam passes through the thin gold film and releases hole–electron pairs in the intrinsic region, the holes are collected by the gold film electrode and the electrons by the two bottom contacts. The electron current splits according to the resistances between the generation point and each contact; that is, according to the distances from the light spot to the edges.

If an instrumentation amplifier were connected so as to measure the current difference $(I_1 - I_2)$, the output characteristic would appear as in figure 13.26(*a*). Figure 13.26(*b*) shows the equivalent characteristic for a split photodiode, and clearly demonstrates how the signal is lost when the light spot leaves the split.

The "Posicon" structure may also be made fully position-sensitive by using four ohmic contacts instead of two.

13.2.12 The Phototransistor

The junction photodiode has already been discussed, and it has become apparent that an effective increase in sensitivity can be achieved by feeding the photodiode current into an amplifier. These two functions can reasonably be combined by using the base–collector junction of a transistor as the active photojunction. The result is the phototransistor, which has been made in both germanium and in silicon.

The base–collector junction is reverse-biased so that when the base is open-circuited, the leakage current I_{CBO} is amplified to I_{CEO} as was explained in chapter 2. In other words, I_{CBO} is amplified by a factor $(1 + h_{FE})$. When the junction is illuminated, I_{CBO} becomes a function of the illumination level, and this photocurrent is also amplified by $(1 + h_{FE})$. This means that the sensitivity is effectively greater than that of a comparable photodiode by the factor $(1 + h_{FE})$. (Actually, much of the photocurrent arises as a result of carriers created in the bulk of the base region for it is not possible for many photons to penetrate the junction itself.)

It is also possible to operate the phototransistor in the charge-storage mode, though here some considerable nonlinearity occurs as a result of the base–emitter junction governing read-out at low irradiation levels.

13.2.13   The Photo-FET

If the junction of a field-effect transistor is irradiated, charge carriers will be generated, and the normal leakage current I_G will be augmented by a photocurrent, I_P.

For the simple source-follower of figure 13.27(a), the leakage and photocurrent will produce a voltage drop in the gate resistor:

$$V_B = (I_G + I_P)R_g$$
$$\simeq I_P R_g \quad \text{if } I_P \gg I_G.$$

The effect of this voltage is to increase the channel current I_D, as is shown by the load-line shift of figure 13.27(b). (The slope of the load-line is, of course, $1/R_S$.) Hence, the photo-FET in this configuration can be regarded as a source-follower having irradiance as an input signal.

As a small-signal amplifier, it is, or course, very linear; and large-signal amplification and switching functions can also be realized in the same manner as for normal FETs.

The sensitivity of the photo-FET is given in terms of gate-current per unit irradiance; that is, the sensitivity is referred to the input. For example, in the case of the Siliconix P-102, the sensitivity is, typically,

$$S = 1.2 \ \mu\text{A mW}^{-1} \text{cm}^{-2} \quad \text{for radiation at } 0.9 \ \mu\text{m}$$

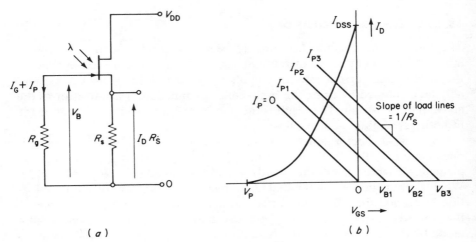

Figure 13.27. The photo-FET.

(i.e., S is quoted for the radiation emitted by a GaAs diode lamp in this case). In the simple circuit of figure 13.27(a), this sensitivity corresponds to an input voltage change of

$$\Delta V_B = SE_e R_g .$$

As the sensitivity is known, the change in I_D may be determined as described in chapter 11.

13.3 THE LIGHT-EMITTING DIODE (LED)

In addition to the transduction of irradiation to electrical energy, the semiconductor diode can produce radiation *from* electrical energy. This was discovered in the MIT Lincoln Laboratories in 1962, where it was found that a forward-biased gallium arsenide junction emitted infrared light. Gallium arsenide is a III-V compound (see table 2.1) and GaAs junctions are also used as solid state lasers for the production of coherent radiation[22]. This section will be concerned only with LEDs, however.

The light-producing mechanism is that of recombination. If an electron falls from the bottom of a conduction band to recombine with a hole at the top of a valence band, it must lose an amount of energy given by the width of the forbidden gap. This may be emitted as thermal energy (phonons) or as light quanta (photons).

If donor or acceptor impurities are present, electrons may be trapped temporarily before falling to the valence level. Hence, energy may be released in stages, each stage corresponding to an energy less than that of the forbidden gap.

For the direct band-gap transition, the wavelength of emitted light must be:

$$\lambda = \frac{hc}{E_g} = \frac{1.237}{E_g} \ \mu m \quad \text{if } E_g \text{ is in electron volts.}$$

For GaAs, $E_g = 1.37$ eV, so that

$$\lambda = \frac{1.237}{1.37} = 0.903 \ \mu m \quad \text{or} \quad 903 \ nm$$

which is in the infrared.

Conversely, for silicon carbide, $E_g = 2.1$ eV, so that

$$\lambda = \frac{1.237}{2.1} = 590 \ nm$$

which is in the green.

Unfortunately, silicon carbide is difficult to grow in single-crystal form, as are several other apparently appropriate materials. Gallium phosphide (GaP) is a material which can be grown consistently, and it exhibits an indirect transition involving the participation of a phonon, which though very inefficient, does release a photon in the green region. The inclusion of nitrogen to provide an alternative transition via impurity centers improves the quantum efficiency ("Quantum efficiency" is the average number of photons per electron: for a good LED, it is greater than 0.1%, taking into account absorption of photons in the various materials involved.) Further, GaP is transparent[19].

Most LEDs involve the epitaxial deposition of gallium arsenide phosphide (GaAsP) on to a GaAs substrate, because the P/As ratio used defines the color of the emitted light. This can range from infrared at about 900 nm, through visible red and yellow, to green at about 560 nm. The quantum efficiency at the red end is greater than that at the green, but the relative sensitivity of the eye at the green end makes up for this. The highest quantum efficiencies at the green end are obtained when nitrogen impurity is included in GaAsP having a P/As ratio over 49%; and when a (transparent) GaP substrate is used. The pn diode structure is made using zinc dopant into the n-type substrate, which may be tellurium- or sulphur-doped.

Figure 13.28 shows the spread of wavelengths available from a series of materials as a function of their band-gap energies.

LEDs are operated in the forward-biased mode at currents from 5 to 20 mA, when the drop is about 1.6 V. The current is most conveniently limited using a series resistor, but a constant-current drive transistor may be used if desired.

Apart from their usage as indicator lampes, LEDs have found a considerable field of application in opto-isolators (see later) and as generators of

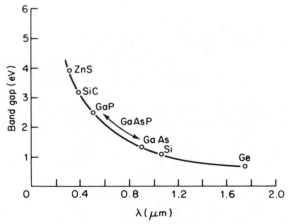

Figure 13.28. Wavelengths emitted by materials of various band-gap energies.

optical information. This is because the rise time is very fast, and nanosecond pulse operation is possible. Thus, optical information can be generated, and in fact can be transmitted using optical fibres in place of coaxial cables. The relevant receivers are the equally fast small-area reverse-biased photojunction devices[8].

Also, various arrays of linear-junction LEDs are obtainable for information display, the most common being seven-segment numerics for various forms of electronic instrument[19].

13.3.1 Characterization of Photoemitters[22]

It has been pointed out that the SI standard of luminous intensity is the candela. Therefore, the most logical way of describing the luminance of a surface is to define it in terms of candelas per square meter (or "nits," a regrettably disappearing term).

The luminance of a surface in a given direction is formally defined as the quotient of the luminous intensity in that direction and the orthogonal projection of the surface area:

$$L_{v\phi} = \lim_{A \to 0} \frac{I_\phi}{A_\phi} .$$

This form of the definition is important in that display devices, LEDs included, are strongly directional.

This does not represent the only way of characterizing sources of illumination, however, for it is equally valid to quote the so-called *luminous exitance* M_r in terms of emitted lumens per square meter. A geometrical derivation will show that the relationship between luminance and luminous exitance is as follows. If a surface emits M_r lumens per square meter, then,

$$L_r \text{ cd m}^{-2} = M_r/\pi \text{ lm m}^{-2} .$$

Note that some early units of luminance are still in use, such as the *foot-candle* (candelas per square foot) and the *stilb* (candelas per square centimeter). Also, the *lambert* (lumens per square centimeter) is still met as a unit of luminous exitance. Using the foregoing relationship, it is obvious that lamberts $= \pi$ stilbs.

Measurement of the total light output of a LED can be achieved by placing the diode at the bottom of a conical well, and a large-area calibrated photodiode at the (larger diameter) mouth of the well[20]. It is possible to obtain a photodiode having a computer-designed composite glass filter which makes the response of the combination either radiometric (flat, within limits), or photopic (corresponding to the spectral sensitivity of the light-adapted human eye)[21]. The latter can be used to compare the luminance of various LEDs directly.

13.3.2 The Photocoupler

A photocoupler is a light-tight encapsulation of a light source and a photocell and its primary characteristic is that it is *unilateral*; that is, a signal can pass from the light source to the photocell but not vice versa. Also, it can be designed so as to withstand a very high voltage difference between its input and output sides, so that it can be used as an *opto-isolator*. Examples of this usage include the measurement of current or transients on a high-voltage line and the isolation of a patient from mains-operated biomedical electonic equipment.

The simplest photocoupler consists of a filament lamp and one or more photoresistors, of which several commercial versions exist such as the Raytheon "Raysistor." The long thermal time constant implied by the filament lamp makes this device useful in feedback systems where one dominant lag is required, or when the delay is of little importance (as in the swell-pedal volume controls of some electronic organs for instance).

For fast operation, however, the use of a LED source is mandatory, and here, particularly good light-transfer efficiency is possible if a GaAs infrared LED is used along with a silicon photodiode which has its peak sensitivity in the infrared[19]. However, the photodiode is only one of the receptors which is used; others include phototransistors, photo-Darlingtons, photo-FETs and photothyristors. Such combinations exhibit improved *current transfer ratios* (CTR).

The CTR is simply defined as the ratio of output current to input current, expressed as a percentage. This definition is not a good one when the output current is determined in part by the external circuitry, as in the case of a photoresistor, but is adequate where current-source receivers (i.e., photo-diodes or phototransistors) are used. The CTR can range from a fraction of a per cent for a photodiode to upwards of 1000 per cent for a high-gain photo-Darlington pair.

Figures 13.29(a)–(f) shows a selection of symbols for a number of commercially available photocoupler types.

The photocoupler is inevitably a nonlinear device, and whereas this is of little importance insofar as the transmission of digital (or ON/OFF) signals is concerned, it does present some problems in analog circuits. For the more common devices utilizing either photodiode or phototransistor receivers, this nonlinearity can be depicted by plotting the incremental value of the CTR against the absolute value of the input current. In the case of a LED passing a current I_D irradiating a photodiode producing a photocurrent I_P, the incremental CTR would be:

$$\text{CTR} = dI_P/dI_D .$$

If this CTR were plotted against I_D, the resulting graph would appear as in figure 13.29(g)[8]. Here, it is apparent that there is a region where the CTR is

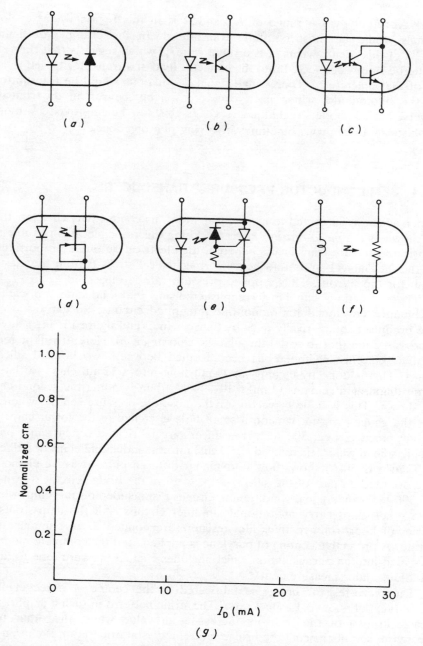

Figure 13.29. (a)–(f) Some photocoupler variants. (a) LED/photodiode; (b) LED/phototransistor; (c) LED/photo-Darlington; (d) LED/photo-FET; (e) LED/thyristor; (f) filament/photoresistor; (g) typical plot of (normalized) dI_P/dI_D for a LED-photodiode photocoupler.

fairly constant over a range of I_D, and it is in this region where analog signals would be most faithfully transmitted. In the example of figure 13.29(g) this occurs when I_D is around 20 mA, which suggests that the LED could be simply connected into the collector lead of a transistor biased to an I_Q of that magnitude. Then, small-signal fluctuations would be transmitted to the photodiode, which itself could reasonably be run in the (linear) reverse-biased mode. Techniques such as this can be improved by more sophisticated circuitry, sometimes involving photofeedback[23].

13.4 SEMICONDUCTOR PRESSURE TRANSDUCERS

When silicon is subjected to mechanical strain, its conductivity changes; this is known as the *piezo-resistive effect*[24]. Using this principle, it is possible to fabricate pressure transducers, wherein the electrical output is proportional to the mechanical strain and hence to mechanical stress (within the elastic limit for the silicon). Thus, applied pressure can be measured.

These functions can be accomplished using the solid-state fabrication techniques developed for monolithic integrated circuits. Such transducers are produced commercially (e.g. by Sensym Inc.) and figure 13.30(a) shows a cross-section (not to scale) through the chip involved. Here, it will be seen that a reference chamber has been formed between two bonded silicon chips. This is done by etching an 11 mil hole into a 12 mil chip, so that a 1 mil diaphragm remains (1 mil = 10^{-3} inch); then bonding the second chip as shown. This is done *in vacuo* for the absolute-pressure sensor, whereas for the gauge-pressure version a small hole is etched in the lower chip. In actual practice, over 300 such transducers are fabricated concurrently on a single silicon wafer, then diced (i.e., cut) into the individual chips described.

Figure 13.30(b) shows how four planar diffused piezo-resistive elements are oriented on part of the silicon chip, to form the bridge circuit shown in 13.30(c). Further signal-conditioning circuitry is included on this chip, while two other chips carry operational amplifier circuits. All the chips, plus a series of laser-trimmed thick-film resistors are bonded to an alumina substrate. A protective coating of parylene is applied, and the complete module is sealed into an encapsulation which includes a small pressure tube to take the fluid under measurement.

The piezo-resistor bridge is fabricated in the center of a rectangular diaphragm of about 3:1 aspect ratio. The strain patterns in such a diaphragm result in two of the resistors increasing in value whilst the other two decrease, so unbalancing the bridge, and the crystal orientation is chosen so as to optimize this effect.

If two opposing piezo-resistors increase their values to $R + \Delta R$, and the other two decrease to $R - \Delta R$ as a result of diaphragm strain under pressure, then the output voltage V_{out} is:

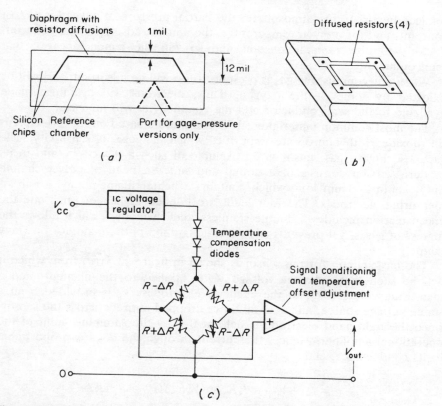

Figure 13.30. (a), (b) Part of oriented-crystal pressure transducer chip. (c) Basic sensor electronics (all integrated circuits).

$$V_{out} = \frac{V_B}{R} \Delta R$$

where V_B is the regulated bridge voltage.

ΔR and hence V_{out} can be shown to be proportional to the applied pressure.

Such transducer modules are available to cover ranges between 10 and 5000 psi full scale[25].

13.5 SEMICONDUCTOR GAS SENSORS

When a gas is adsorbed on to the surface of a semiconductor, an exchange of charges occurs, depending on the nature of the gas. In the case of oxygen adsorbed on to an n-type semiconductor, electrons move from the semiconductor to the adsorbed oxygen, which reduces the conductivity of that semiconductor. In the atmosphere, the partial pressure of oxygen is almost constant, and for a given temperature, the amount adsorbed by a semiconductor will also remain constant and so will the conductance of that semiconductor.

When any combustible gas is present, it too will be adsorbed, and will be oxidized by some of the oxygen. Thus, electrons will again be made available to the semiconductor and the conductivity will increase.

The most common material used for gas sensors based on this principle is tin dioxide, with appropriate impurities which increase its sensitivity. The Japanese Figaro gas sensor is an example of such a device[26], and in its modern version consists of a doped and sintered bead of polycrystalline SnO_2 about 2–3 mm long which contains a heater filament plus a pair of measuring electrodes. The four leads are bonded to header pins, and the encapsulation includes a double stainless-steel "window" which allows the ingress of gases, yet prevents any external ignition in the manner of a Davy lamp.

The basic circuit for the sensor is shown in figure 13.31(a) from which it will be seen that a heater voltage V_H is applied to the filament; and a measuring voltage, V_S, to the ancillary electrodes. V_S is provided from a stable voltage source, and is at all times dropped entirely across the sensor, since the right-hand electrode is held at $0\,\text{V}$ by the summing point of the operational amplifier. Hence, the current through the bead is proportional to its conductance, and

$$V_{\text{out}} = I_S R_F = (V_S R_F)G$$

where G is the bead conductivity.

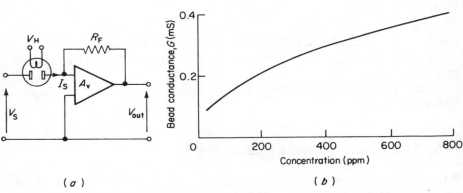

(a) $\qquad\qquad\qquad\qquad\qquad\qquad$ (b)

Figure 13.31. (a) Basic circuit for the Figaro gas sensor and (b) a typical sensitivity plot.

Figure 13.31(b) shows a typical plot of G against gas concentration.

Such devices can be made partially selective to light gases such as carbon monoxide or hydrogen (as opposed to hydrocarbons such as methane) by appropriate doping differences, and by running them at a lower temperature. The latter makes necessary an initial purging of adsorbed moisture and "stray" gases by running the sensor very hot for a short period prior to use. Hence, a practical circuit must incorporate a higher voltage supply and a means of temporarily applying it. (For modern devices, this is performed by a customised IC.)

Semiconductor combustible gas sensors demonstrate the well-known transducer compromise between sensitivity, linearity, and consistency in operation. For the measurement of gas concentrations, and consistent performance over long periods, the catalytic sensor is more appropriate. However, it is comparatively insensitive and two elements, one active and one passive, must be used in a bridge network. On the other hand, the semiconductor sensor is very sensitive indeed and is therefore appropriate for gas alarms where closely defined concentration measurement is not necessary.

In this context, the hysteresis comparator circuit of figure 10.27 is particularly useful, since it can operate from the same 5 V supply as a Figaro gas sensor. Figure 13.32 shows such an arrangement used in a gas alarm circuit. When an inflammable gas concentration is at a level such that the voltage drop V_{in} exceeds $V_{ref(hi)}$ (see figure 10.27), the audible alarm switches ON, while V_{ref} goes low. This ensures that the alarm remains ON irrespective of minor gas concentration changes until V_{in} falls below $V_{ref(lo)}$.

Semiconductor gas sensors are sensitive to humidity, ambient temperature, and air velocity, so that if they are used in situations involving closely defined gas concentration limits, their environmental working conditions should also be closely defined. Further, there is a long term drift of sensitivity which makes periodic recalibration desirable, again for critical situations only.

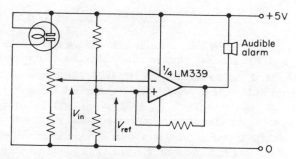

Figure 13.32. A gas alarm circuit (by permission of ENVIN Ltd.).

13.6 SUMMARY

In a single chapter, it has been possible to mention only a few of the more common solid-state transducers. Many others exist and are invaluable for the measurement of phenomena in fields as wide ranging as biomedical electronics[27] through automotive engineering[28]. In particular, it is now proving possible to integrate sensors and their associated electronics in both monolithic and hybrid form; and also exploit the chemical properties of known solid-state devices. A case in point to illustrate the latter application is the *ion-sensitive* FET[29], and various other allied devices[30,31]. In fact, the wide gap which has existed between the somewhat neglected technology of signal acquisition, and the possibly over-subscribed technology of signal and data processing, is now being slowly closed, in part by the development of reliable, accurate, and repeatable solid-state sensors.

REFERENCES

1. Middelhock, S, Angell, J. B. and Noorlag, D. J. W. 1980, Microprocessors get integrated sensors, *IEEE Spectrum* **17** 42–46 (February).

2. Barth, P. W., 1981, Silicon sensors meet integrated circuits, *IEEE Spectrum* **18** 33–39 (September).

3. Timko, M. P. 1976, A two-terminal IC temperature transducer, *IEEE J. Solid St. Circuits* **SC-11** 784–788.

4. Hyde, F. J., 1971, *Thermistors* (London: Iliffe).

5. Bube, R. H., 1960, *Photoconductivity of Solids* (New York: Wiley).

6. Clark Jones, R., 1953, Detectivity of Photoconductive Cells, *Rev. Sci. Instrum.* **24** 1035.

7. Zworykin, V. K., 1949, *Photoelectricity and its Applications* (New York: Wiley) p. 256.

8. Hewlett-Packard Staff, 1977, *Optoelectronics Applications Manual* (New York: McGraw-Hill).

9. Wendland, P., 1970, Silicon photodiodes revisited, *Electro-opt. Syst. Des.* 48–55 (August).

10. Robinson, F. N. H., 1969, Noise in junction diodes and bipolar transistors at moderately high frequencies, *Electron. Eng.* 218–220. (February).

11. Buckingham, M. J., and Faulkner, E. H., 1974, Noise in p–n junction diodes and bipolar transistors, *IERE Rad. Electron. Eng.* **44** 125–140.

12. Hamstra, R. H. and Wendland, P., 1972, Noise and frequency response of silicon photodiode/operational amplifier combination, *Appl. Opt.* **11** 1539–1547.

13. Weckler, G. P., 1967, Operation of p–n junction photodetectors in a photon flux integrating mode, *IEEE J. Solid St. Circuits* **SC-2** 65–73 (September).

14. Murphy, H. E. and Kabell, L. J., 1966, An integrating light meter, *IEEE J. Solid St. Circuits* **SC-1** 4–7 (September).

15. Chamberlain, S. G. and Kuhn, M. (eds.), 1978, *IEEE J. Solid St. Circuits* **SC-13** (February) (Special issue on optoelectronic devices and circuits).

16. Beynon, J. D. E. and Lamb, D. R., 1980, *Charge-coupled Devices and their Applications* (New York: McGraw-Hill).

17. Wallmark, T. J., 1957, A new semiconductor photocell using lateral photoeffect *Proc IRE* **45** 474–483.

18. Kelly, B. O., 1976, Lateral-effect photodiode, *Laser Focus* 38–40 (March).

19. Chappell, A. (ed.), 1976, *Optoelectronics—Theory and Practice* (Texas Instruments).

20. Grotti, D. R. and Major, L. D., 1970, Measuring light—the simple way, *Electro-Opt. Syst. Des.* 44–47 (July).

21. Wendland, P. H., Measuring LED outputs accurately, *Electro-Opt. Syst. Des.* 24–27 (November).

22. Wilson, J. and Hawkes, J. F. B., 1983, Optoelectronics: an Introduction (Englewood Cliffs, N.J: Prentice Hall).

23. Vettiger, P., 1977, Linear signal transmission with optocouplers, *IEEE J. Solid St. Circuits* **SC-12** 298–302.

24. Pfann, W. G. and Thurston, R. M., 1961, Semiconducting stress transducers utilizing the transverse shear piezo-resistance effects, *J. Appl. Phys.* (October).

25. Pressure Sensor Handbook, 1988 (Sensym Inc.).

26. Watson, J. 1984, The tin oxide gas sensor and its applications, *Sensors and Actuators* **5** 1, 29–42.

27. Cobbold, R. S. C., 1974, *Transducers for Biomedical Measurements: Principles and Applications* (New York: Interscience).

28. Allan, R., 1980, New applications open up for silicon sensors: a special report, *Electronics* 113–138 (November).

29. Moss, S. D., Johnson, C. C. and Janata, J., 1978, Hydrogen, calcium and potassium ion-sensitive FET transducers: a preliminary report, *IEE Trans. Biomed. Eng.* **BME-25** 49–54.

30. Wen, C. C., Chen, T. C. and Zemel, J. N., 1979, Gate-controlled diodes for ionic concentration measurement, *IEEE Trans. Electron. Dev.* **EC-26** 1945–1951.

31. Lanks, I. R. and Zemel, J. N., 1979, The Si_3N_4/Si ion-sensitive semiconductor electrode, *IEEE Trans. Electron. Dev.* **ED-26** 1959–1964.

32. Gillessen, K. and Schairer, W., 1987, Light Emitting Diodes—an Introduction (London: Prentice Hall).

QUESTIONS

1. Explain how a pair of matched bipolar transistors may be used in a temperature sensing circuit.

2. Why does a thermistor obey Ohm's Law only at low current levels?

3. What are the photodiode operating conditions relevant to (*a*) the photovoltaic and (*b*) the photojunction modes?

4. What is the fundamental difference between units of irradiation and illumination?

5. For a photosensitive transducer, what is meant by the "noise equivalent power" and the "specific detectivity"?

6. In a planar diffused photojunction cell structure, what is the purpose of the guard ring?

7. What is the dominant noise source in a reverse-biased photojunction under dark conditions?

8. Wavelength-independent irradiation sensors are normally very slow-acting devices. Why is this?

9. What is the "current transfer ratio" for a photocoupler?

10. Describe a simple way of measuring changes in conductivity using an operational amplifier.

PROBLEMS

13.1. An AD590 temperature transducer can be modeled as a current source which delivers $1\,\mu A/°K$. It is to be connected into an operational amplifier circuit which will measure temperatures between 17°C and 67°C and display this range on a meter having a full-scale-deflection of 10 V DC.

Draw the circuit, and calculate the values of the feedback resistor and a necessary zeroing resistor.

13.2. An E.G. & G. type SGD-100 photodiode has a quoted NEP of 9.6×10^{-14} W, (0.9 μm, 10^3, 1). Write down the full meaning of this abbreviation, and calculate D^* knowing that the active area of the device is $0.82\,mm^2$.

13.3. A photojunction diode of sensitivity 0.05 μA/lux, is used to measure luminance, and is loaded by a 10 kΩ series resistor R_D. Design a high-input-resistance operational amplifier circuit, the input of which is connected across R_D, and the output of which drives a digital voltmeter which can read 0 to +10 V DC. The maximum luminance expected is 1000 lux.

13.4. The circuit of figure 13.33 is a simple illuminance-to-frequency converter. The received light is predominantly at 5200 Å, at which wavelength the sensitivity of the photodiode is 5×10^{-9} A/lux, and the visibility function is 0.71. The reverse leakage current of the photodiode is negligibly small.

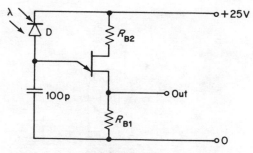

Figure 13.33.

The unijunction transistor is a 2N2646 having an intrinsic stand-off ratio of 0.75, a peak-point emitter current of 2 μA, and its forward emitter–base voltage drop can be taken as 0.5 V.

The values of R_{B1} and R_{B2} can be taken as much less than the interbase resistance R_{BB}; and it can be assumed that 1 W = 680 lumens at 5550 Å. Determine:

(a) the output PRF when the illuminance is 6.21 W/m^2
(b) the shape of the waveform at the emitter and
(c) the value of the lowest illuminance before oscillation ceases.

Also, sketch the unijunction I/V characteristic, and show the position of a typical load-line as defined by the photodiode. Is the PRF a linear function of the illuminance?

13.5. A very slowly changing signal from a pressure transducer varies between -20 mV and -80 V with respect to ground when the load on the transducer is not less than 1 kΩ. Design an operational amplifier circuit which, in conjunction with this transducer, will produce an output signal ranging from $+1$ V to $+4$ V with a worst-case drift error below 1% over the temperature range 10–50°C.

The maximal drift parameters for the operational amplifier

$$\frac{\Delta e_{0s}}{\Delta T} = 2 \,\mu V/°C \quad \frac{\Delta i_b^+}{\Delta T} = \frac{\Delta i_b^-}{\Delta T} = 4 \,nA/°C \quad \frac{\Delta i_d}{\Delta T} = 1 \,nA/°C \,.$$

13.6. A semiconductor gas sensor has a 5 V heater and two electrodes in accordance with the symbol of figure 13.34(a). The resistance between the electrodes is measured as a function of gas concentration with the result shown in figure 13.34(b).

Design a gas concentration measuring circuit using an operational amplifier with its associated components along with a voltmeter having a 0–10 V scale. Establish that the reading on the voltmeter is linearly related to gas concentration and that the full-scale deflection of 10 V corresponds to 1200 ppm of gas.

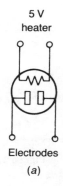

(a)

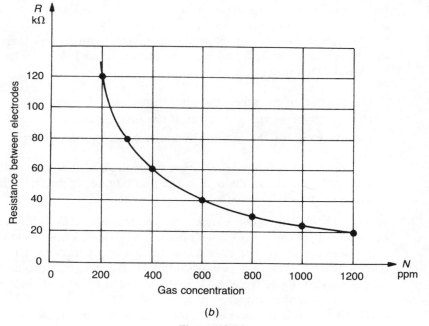

(b)

Figure 13.34.

Assume stabilized power supply voltages of ±15 V and a floating 5 V are available.

Plot the graph of output voltage versus gas concentration to confirm that it is a straight line.

13.7. A sensor for methane gas has a conductance which can be considered to vary in direct proportion to the concentration of this gas over a range zero to 0.2 millimhos (or millisiemens). It contains a filament which requires a 5 V supply, and this voltage is also to be used to energize the sensor electrodes.

Design an operational amplifier circuit which will allow a 0 to +10 V digital voltmeter to measure the relevant range of conductance, and hence gas concentration (in arbitrary units). Could the single 5 V supply also be used to power an appropriate operational amplifier?

13.8. The sensitivity of the photodiode shown in figure 13.35 is a 5 nA/lux. At what level of illumination will the lamp turn ON and at what level will it turn OFF?

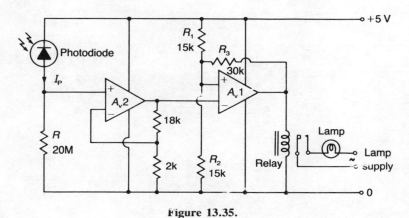

Figure 13.35.

14

Power Supplies

Almost all of the circuits presented in the preceding chapters have required DC power supplies, which in most cases have been assumed to be invariant, even though derived from an AC mains supply (which itself fluctuates) or from batteries, which may exhibit voltages that fall during discharge. These facts imply the necessity of providing equipment which will perform some or all of the following functions:

(a) When a mains supply is used, the voltage must be reduced to an appropriate level for transistor operation;

(b) The low-voltage AC must be rectified;

(c) The rectified current must be smoothed;

(d) The smooth DC must be stabilized against changes in mains voltage, load, and temperature.

When a battery supply is used, it may also be necessary to convert the voltage (down or up) to a level suitable for transistor operation, and then stabilize it as in item (d) above.

Modern mains-operated power supplies can be subdivided into two groups, as follows.

(i) Linear power supplies transform the mains voltage down to a suitable level, rectify and smooth the low-voltage AC, then stabilize the output using an appropriate electronic circuit. This stabilization circuit can also be obtained (typically in monolithic IC form) for inclusion on printed boards, so that critical parts of a circuit can be supplied with a closely-defined stabilized voltage. This has several advantages including the minimization of "regulation," or voltage drop along the thin printed board conductors.

(ii) Switching power supplies convert a DC source into square waves, which may then be transformed to an appropriate voltage level, rectified, and smoothed. The DC source itself may be some form of battery, or it may be the full mains voltage, rectified, and smoothed. This technique has several advantages which will be presented later; for the present, it should be noted that the transformed output square wave may be rectified, smoothed, and regulated as was the case for the linear power supply. However, if the output voltage level is fed back in such a way as to control the mark/space ratio of the switched square wave, thus effecting regulation in a manner similar to the Class D PWM amplifier described in chapter 10, the circuit becomes a *switch-mode power supply* or SMPS. It is also possible to obtain "on-board" SMPS units, a good example being where a 5 V logic supply is converted to ±15 V for an operational amplifier section.

Both types of supply are now described in more detail.

14.1 LINEAR-MODE POWER SUPPLIES

The actual design of small power transformers is outside the scope of this book, and it is a procedure rarely carried out by an electronics designer (except in special cases such as ferrite pot core work or special inverter transformers), because a very wide range of commercial mains transformers is readily available.

For rectification, the silicon diode is normally used, and devices having peak inverse voltages (PIV) up to several thousand are common. However, great care must be taken not to exceed the quoted PIV, otherwise the relevant diode may be destroyed, as it would also be by an excessive current. Hence, power supplies usually incorporate transient voltage and overcurrent protection circuits.

The simplest form of transformer/rectifier circuit is shown in figure 14.1(a). This is the half-wave circuit, so-called because the diode conducts

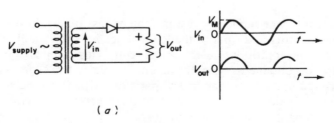

(*a*)

Figure 14.1. The half-wave rectifier.

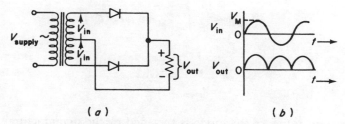

Figure 14.2. The full-wave rectifier.

only during alternate half-cycles, leading to the waveform shown. This circuit is not satisfactory for several reasons. Firstly, because only one half of each wave is utilized, the efficiency is low. Secondly, the output current requires a very comprehensive smoothing circuit owing to the lack of voltage for half of each cycle. Thirdly, the transformer core is magnetized only in one direction and is not taken through its complete B/H cycle. This means it is permanently magnetized to the remanence level.

An improvement is effected if the transformer is center-tapped as shown in figure 14.2(a). Here, both half-cycles are utilized, and the transformer core is taken through the entire B/H loop. However, only one side of the secondary is used at a time, which means that each side must develop the full voltage required and be able to carry the full load current. The output waveforms are satisfactory from the point of view of smoothing (figure 14.2(b), and the mean value of the unidirectional output voltage is clearly higher than was the case for the half-wave rectifier circuit.

The best combination of rectifier and transformer is the *bridge connection* shown in figure 14.3(a). Here, no center-tap is needed on the transformer secondary, and this single winding is in use at all times, thus effecting an economy in manufacture. It is of interest to note that at any instant, two of the diodes are in operation and that they are effectively in series with the load, and drop about 1.2–1.6 V.

The mean level of the output voltage may be obtained by integrating and dividing by the base-line over a half-period:

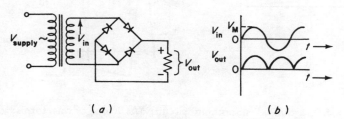

Figure 14.3. The bridge rectifier.

$$V_{\text{mean}} = \frac{1}{\pi} \int_0^\pi V_M \sin \omega t \, \mathrm{d}(\omega t)$$

giving

$$V_{\text{mean}} = \frac{2V_M}{\pi}. \tag{14.1}$$

The current rating of the diode is specified by the manufacturer, but may be implicit in the power rating. This is because the rectifier must be cooled in the same way as a transistor, and the better the cooling arrangements, the greater the allowable dissipation will be and the higher the value of maximum permissible forward current.

Associated with the dissipation figures are derating curves which show how the maximum permissible dissipation decreases as a function of ambient temperature. Such a curve has been given previously (figure 2.5(c)).

The most common method of smoothing the output wave from a rectifier assembly is to connect a large capacitor C_G across the load as shown in figure 14.4(a). The resistance R_G is comprised of both the internal resistance of the assembly and an external resistor which is often included to enhance the degree of smoothing. The internal resistance is due to the resistance of the whole of the secondary winding plus that presented by two diodes, since this example involves a bridge rectifier circuit. There will also be a series inductance due to the leakage reactance of the transformer. Ignoring this leakage inductance, which will be small for a good transformer, the time constant involved will be $(R_L + R_G)C_G$, and this should be much longer

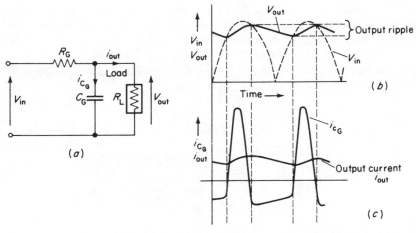

Figure 14.4. (a) Simple smoothing circuit; (b) voltage waveforms; (c) current waveforms.

than a half-period, leading to the output waveforms shown in figures 14.4(*b*) and (*c*).

It is clear that the output ripple voltage will decrease with increasing time constant, and this implies a high value for $(R_L + R_G)$, assuming that C_G is limited (as it will be owing to considerations of physical size). However, if R_G is made large, then the regulation will become poor; that is, the mean output voltage will fall rapidly with increasing load current. Conversely, if R_G is kept low, then the amount of ripple will increase with increasing load current. Moreover, if it is possible for R_L to become very small indeed, the rectifiers may pass currents which are large enough to cause damage. (Note that during the charging periods for C_G, shown in figure 14.4(*c*), the *mean* current is small compared with the peak charging current, and excess dissipation does not therefore arise.)

It is possible to calculate accurately the ripple and regulation for a power supply using capacitive smoothing and assuming a fixed load, but this is a tedious procedure owing to the complex charge–discharge law governing the behavior of i_{C_G}. A crude estimate may be made by assuming a fast capacitor charge followed by a linear discharge. If the ripple voltage is ΔV_{out}, then,

$$C_G \Delta V_{out} \simeq I_{out} \Delta t \, .$$

Here, I_{out} is the mean value of i_{out} and it is assumed that the capacitor is supplying almost all of this current during the discharge period Δt. Furthermore, if the charging time is indeed fast, then $\Delta t \rightarrow T/2$ or $1/2f$, where f is the supply frequency. Hence, given an acceptable ripple voltage level, an estimate of the smoothing capacitor can be made:

$$C_G \simeq \frac{I_{out}}{2f \, \Delta V_{out}} \, .$$

This unsurprising result shows that more smoothing is needed for a larger output current or a small ripple voltage.

An alternative smoothing circuit is shown in figure 14.5, where an inductance has been included in series with the output. If the value of L_G is very high, then the inductive reactance X_{L_G} is numerically much greater than R_L. Hence, the ripple voltage appears almost entirely across L_G, and very little appears across the load. Unfortunately, this means that the time

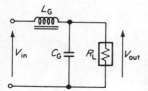

Figure 14.5. Inductive smoothing.

constant involved is now very large, and rapid changes in load current cannot be accommodated unless they can be supplied entirely by the capacitor.

The advantage of this *smoothing choke* is that its internal resistance can be quite small, which leads to a good regulation characteristic compared with the resistor–capacitor circuit. However, the use of the choke is declining, for the voltage stabilizing circuits now in general use permit wider regulation tolerances than would otherwise be admissible. To reduce ripple amplitude further, it is possible to add a second smoothing section as shown in figures 14.6(a) and (b).

When a diode is not conducting in any of the circuits so far discussed, it is subject to a reverse voltage. If a smoothing capacitor is present, then the simple half-wave rectifier circuit of figure 14.1 shows that during the nonconducting half-cycle, the reverse voltage across the diode can reach $2V_M$, the capacitor-maintained output voltage acting in series with V_{in}. This is also true for the full-wave rectifier, but any diode in the bridge rectifier can only be subjected to V_M. Thus the PIV for a diode to be used in a bridge need only be half that of a diode destined for either of the other circuits.

In addition to the normal reverse voltage applied to a rectifier diode, abnormal transients may occur which can exceed the value of V_M by many times. The most common transient encountered occurs when the mains supply is switched off (or a fuse blows). This involves cutting off the current in the primary of the transformer, and this leads to a damped oscillation occurring between this self-inductance and the associated leakage capacitance. The initial excursions of this transient can reach high amplitudes (see the section on power switching, chapter 10) which may damage a diode. Usually, it is possible to specify a PIV several times higher than V_M without incurring excessive cost, but otherwise a pair of Zener diodes can be connected back to back across the transformer secondary. Their breakdown voltages should be chosen to lie between the diode PIV and V_M, in which case they will remain nonconducting under conditions of normal operation but will shunt any transient currents when the associated voltage rises to their breakdown potential.

The protection afforded by Zener diodes can also be used at the output of the power supply, for under certain circumstances, such as those en-

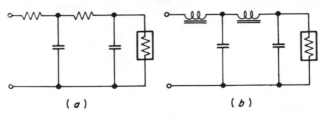

(a)   (b)

Figure 14.6. Two-section smoothing.

countered when an inductive load is switched off, a high voltage spike can return from the load side to the rectifier diodes.

For some purposes an unregulated power supply is entirely satisfactory, but where the following equipment is not only sensitive to its supply voltage, but also presents a fluctuating load, some form of voltage regulation is mandatory. Such regulators may take forms ranging from the simple two-element shunt stabilizer to multi-transistor feedback circuits. However, all power supply units may be characterized by parameters describing the change in output voltage with respect to load current, input voltage and temperature; that is,

$$V_{out} = f(I_L, V_{in}, T)$$

$$\Delta V_{out} = \frac{\partial V_{out}}{\partial I_L} \Delta I_L + \frac{\partial V_{out}}{\partial V_{in}} \Delta V_{in} + \frac{\partial V_{out}}{\partial T} \Delta T.$$

From this expression, three power supply parameters can be defined:

(a) *Load stability factor*

$$S_L = \left(\frac{\partial V_{out}}{\partial I_L} \right)_{(V_{in}, T)}. \tag{14.2}$$

This, of course, is simply the incremental output resistance of the unit.

(b) *Transfer stabilization factor*

$$S_V = \left(\frac{\partial V_{out}/V_{out}}{\partial V_{in}/V_{in}} \right)_{(I_L, T)}$$

$$= \left(\frac{\partial V_{out}}{\partial V_{in}} \frac{V_{in}}{V_{out}} \right)_{(I_L, T)}. \tag{14.3}$$

Note that this describes the fractional change in output voltage for a given fractional change in input voltage.

(c) *Temperature stability factor*

$$S_T = \left(\frac{\partial V_{out}}{\partial T} \right)_{(V_{in}, I_L)}. \tag{14.4}$$

It is also usual to quote the maximum or RMS values of ripple and noise to be expected at the output, and to state whether they vary with load or temperature.

All stabilized power units depend on the existence of an element which will provide a reasonably constant voltage, and this is often a Zener diode.

This can act either as a reference unit, with which the output may be compared, the difference voltage being used as a correcting signal; or it may itself act as a regulator. Both of these systems will be considered, but since the Zener diode is not a perfect voltage reference unit, its characteristics must first be described. (IC versions of the Zener diode and also V_{BE} and bandgap voltage references were treated in chapter 7.)

If a junction diode is operated with a reverse voltage (figure 2.4), this reverse voltage can be increased with little current flow until breakdown occurs. The current then rises rapidly and, unless limited externally, can generate sufficient heat to destroy the diode. The Zener diode is designed to work in this region, and over a wide range of currents will exhibit a voltage drop whose magnitude remains constant within very close limits. The device is so constructed that the breakdown current does not cause permanent damage unless a defined upper limit is exceeded.

Zener diodes can be divided roughly into two categories, whose characteristics are slightly different. The diodes exhibiting low voltage breakdown (3–5 V) utilize highly doped, low resistivity silicon. This material can produce a very narrow depletion layer at the junction of the p and n types, and when a reverse voltage is applied, breakdown occurs largely as a result of field emission. This is the true Zener effect, which is due to the quantum mechanical effect of "tunneling" at high field strengths (i.e., more than 10^5 V cm^{-1}). Tunneling refers to the capability of electrons of comparatively low energy to pass a forbidden energy gap which would be sufficiently wide to preclude their passage under the laws of classical mechanics.

Two important points arise when a Zener diode operates as a result of the true Zener mechanism. Firstly, the onset of breakdown is gradual, and the characteristic never exhibits a very steep slope; that is, the conductivity is never very high. This is clearly shown by figure 14.7, which gives the characteristics of a typical family of Zener diodes. Secondly, it would be expected that an increase in temperature would result in an increase in carrier concentration and a decrease in the width of the energy gap, leading to a reduction in breakdown voltage. This is indeed so, and the resultant negative temperature coefficient is plotted for a series of Zener diodes in figure 14.8.

For units having breakdown voltages above 6 V, the predominant mechanism is avalanche breakdown. This is again the result of a high field, but this time in less heavily doped, high resistivity material. (Actually, true Zener breakdown would also occur in such material, but at a higher field strength than that required for avalanche breakdown.) The process involves the acquisition of sufficient energy by electrons or holes crossing the depletion layer to ionize neutral atoms within that layer. Some 0.75 eV is needed to achieve this in silicon, and the combination of a sufficiently high accelerating field plus a comparatively long mean free path is needed to achieve this. Note that one electron or hole will release an electron–hole *pair*, each of which may achieve sufficient energy to ionize further atoms.

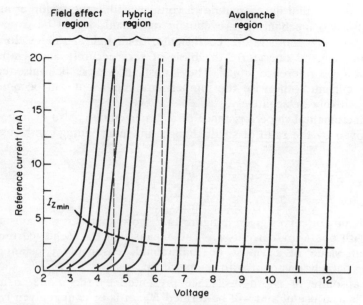

Figure 14.7. Characteristics for a typical family of Zener diodes.

This process may become cumulative, hence the term "avalanche breakdown."

Figure 14.7 shows the sharp cornered characteristics of those diodes operating in the avalanche mode, and also indicates that the resistance presented after breakdown is less than that of the true Zener diodes.

In avalanche diodes, a positive temperature coefficient is observed, for electron and hole mobility *decreases* with temperature, and a higher field is then required to maintain the same rate of ionization.

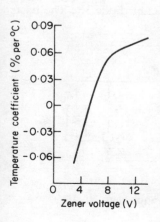

Figure 14.8. Typical temperature coefficient variation for a Zener diode family.

Those diodes that have breakdown voltages within the region around 5 V where the two mechanisms are equally important, would be expected to exhibit very low temperature coefficients. Such hybrid diodes do in fact exist, and selected ones form the basis of very accurate and temperature stable voltage reference units. They must, however, be operated at a constant current, otherwise the temperature coefficient will become finite either positively or negatively.

The incremental slope resistance of a Zener diode is also of importance, and is given by the ratio of small changes in Zener voltage with respect to current:

$$r_Z = \left(\frac{\Delta V_Z}{\Delta I_Z}\right). \tag{14.5}$$

The family of curves given in figure 14.7 shows that r_Z is not constant, but varies with operating current, I_Z. This means that the Zener current for a particular value of r_Z must be maintained reasonably constant. r_Z is, however, largely invariant with temperature.

The absolute or chord resistance presented by the diode to a direct current is also useful and will be termed R_Z. It is of course given by V_Z/I_Z and varies considerably with the operating point. As a first approximation, V_Z may be taken to be constant, in which case $R_Z \propto 1/I_Z$.

14.2 VOLTAGE REGULATOR CIRCUITS

The simplest method of regulating the output voltage from a power supply is by using a series dropping resistor and a shunt Zener diode as shown in figure 14.9(a). Here, the series resistor R is assumed to consist of both the internal resistance of the source, R_g, plus an external resistor R_R.

If the input voltage rises, then more current flows through the Zener diode and produces a greater volt drop across R. The output voltage V_{out} is, of course, always equal to the voltage across the Zener diode V_Z. The basic equation is

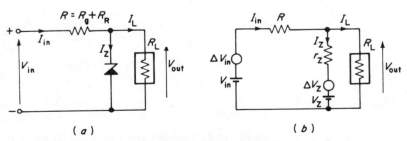

Figure 14.9. (a) Simple shunt regulator; (b) equivalent circuit.

$$I_{in} = I_Z + I_L .$$ (14.6)

The maximum power which can be dissipated within the Zener diode for this circuit can be found by assuming a condition of zero load current along with the maximum value of input voltage:

$$P_{diss(m)} = V_Z I_{Z(m)}$$ (14.7a)

where

$$I_{Z(m)} = \frac{V_{in(m)} - V_Z}{R}$$ (14.7b)

$P_{diss(m)}$ must not exceed the rated power dissipation capability of the diode, P_{tot}.

To find R, it is convenient to take the limiting case where V_{in} is minimal and the load current I_L is maximal. Here, R must be small enough to allow the minimum permissible Zener current $I_{Z(min)}$ to flow. ($I_{Z(min)}$ is shown by the broken line in figure 14.7.)

$$I_{L(m)} = \frac{V_Z}{R_{L(min)}}$$ (14.8)

so

$$R = \frac{V_{in(min)} - V_Z}{I_{Z(min)} + I_{L(m)}} .$$ (14.9)

Clearly, R must satisfy equations (14.7) and (14.9); if it does not do so, then another choice of Zener diode must be made.

The circuit is now fully defined except for the dissipation in the external part of R, that is R_R. This will not be greater than

$$P_{diss(R_R)} < I_{in(m)}^2 R_R$$

where

$$I_{in(m)} = \frac{V_{in(m)} - V_Z}{R} .$$

The degree of stabilization afforded by the circuit with respect to changes in supply voltage, load current, and temperature can be determined by using the equivalent circuit given in figure 14.9(b). Writing the nodal equation gives

$$I_{in} = I_Z + I_L \tag{14.6}$$

or

$$\frac{V_{in} - V_{out}}{R} = \frac{V_{out} - V_Z}{r_Z} + I_L$$

giving

$$V_{out} = V_{in}\left(\frac{r_Z}{r_Z + R}\right) + V_Z\left(\frac{R}{r_Z + R}\right) - I_L\left(\frac{r_Z R}{r_Z + R}\right).$$

This gives the load stability factor

$$S_L = \frac{\partial V_{out}}{\partial I_L} = -\frac{r_Z R}{r_Z + R} \simeq -r_Z \quad (\text{since } R \gg r_Z). \tag{14.10}$$

The negative sign shows that the output voltage decreases with load current. The transfer stabilization factor, S_V, is given by equation (14.2) as

$$S_V = \frac{\partial V_{out}}{\partial V_{in}} \frac{V_{in}}{V_{out}} = \frac{r_Z}{r_Z + R} \frac{V_{in}}{V_{out}}$$

$$\simeq \frac{r_Z}{R} \frac{V_{in}}{V_{out}}. \tag{14.11}$$

Finally the temperature stability factor is given by

$$S_T = \frac{\partial V_{out}}{\partial T} = \frac{\partial V_{out}}{\partial V_Z} \frac{\partial V_Z}{\partial T} = \frac{\partial V_{out}}{\partial V_Z} \theta_Z$$

where θ_Z is the temperature coefficient of the Zener diode (usually in millivolts per degree centigrade).

Thus,

$$S_T = \left(\frac{R}{r_Z + R}\right)\theta_Z \simeq \theta_Z. \tag{14.12}$$

An important advantage of this circuit is that the Zener diode tends to smooth out ripple in the input voltage; that is, if the ripple forms $\frac{1}{10}$ of the input voltage, then it only represents $\frac{1}{10}S_V$ at the output.

A disadvantage is that power is dissipated in the Zener diode and in the series resistor R_R. This factor, however, is often balanced by simplicity and low capital cost where power is not at a premium. It is useful in circuits of diverse sizes, for Zener diodes capable of dissipating from a few milliwatts to a hundred watts are readily obtainable. The slope resistance (and hence the value of S_L) is smaller for Zener diodes having high dissipation ratings than for low.

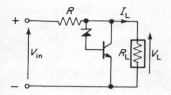

Figure 14.10. A shunt transistor regulator.

It is possible to use a small power Zener diode and still achieve a low output resistance by including a transistor as shown in figure 14.10. The actual stabilization itself depends on the constancy of V_Z in series with V_{BE}. This can be enhanced by choosing a Zener diode working in the avalanche mode and having a positive temperature coefficient which tends to compensate for the negative temperature coefficient of the base–emitter junction of the transistor.

Some highly-developed shunt regulators have been designed, such as the Texas Instruments TI4301, which is a three-terminal IC, available in either Silect or dual-in-line packages, and which will sink up to 100 mA at any preset voltage between 3 and 30 V. However, it is more usual to employ a voltage regulator IC which utilizes the series power controller (or series pass) method.

These are available in various forms which may be totally self-contained or may utilize an external series power controller. Figure 14.11 gives a block diagram of the general arrangment, and shows how a difference amplifier senses the error voltage (i.e., the difference between a reference voltage and a voltage derived from the output), and operates a series controller to complete a negative feedback loop and so minimize that error.

It is obvious that the circuit relating to figure 14.11 may be built up using an operational amplifier and some discrete components, and that appropriate overload and overvoltage circuitry may be included. However, for low power applications, a complete IC may be obtained to fulfil all these requirements, and it is this type which is used as an on-board component to stabilize a supply to localized parts of a large network.

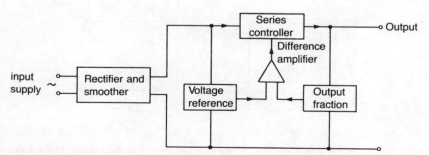

Figure 14.11. Basic elements of series-controlled stabilized power supply.

The regulator IC contains a reference element which may be a bandgap reference or a Zener diode (often the buried type), both of which are described in chapter 11[2,3]. If an external series controller is specified for high current work, a compound pair of bipolar transistors (as described in chapter 5) could be used so that the difference amplifier should not have to supply an excessively high drive current to the base; or alternatively, a power IGFET may be employed which, of course, needs only a voltage drive to its gate. Protection circuitry for the series controller depends upon the type of device chosen, but because (by definition) continuous current operation is mandatory, the continuous current limit on the safe operating area (SOA) diagram must not be exceeded. This is illustrated in figure 10.38, and appropriate protection circuitry for a bipolar power transistor appears in figure 10.39.

The most basic overcurrent protection[4] gives a characteristic like that shown in figure 14.12(a) where the regulated voltage V_{reg} is maintained up to a defined current I_{max}, after which a constant current takes over. The crossover point where the regulator changes its characteristic from constant voltage to constant current also defines the load resistance R_{CR} which dissipates maximum power, as shown. This technique may be extended to provide a feedback loop which can provide the *foldback* characteristic of diagram (b)[5]. Sensitive loads (such as 5-V logic) may be protected from overvoltages by use of the *electronic crowbar*, which often takes the form of a thyristor connected across the output. This is arranged to switch on when an overvoltage is detected by ancilliary circuitry, so reducing the output voltage to nearly zero, and activating the overcurrent protection circuit. Transient pulses may be limited using either a Zener diode or a transient suppressor, the latter being a zinc oxide voltage dependent resistor, or *varistor*, which behaves like a pair of back-to-back Zener diodes.

Voltage regulators are available for both positive and negative polarities, and for both fixed and variable outputs. Figure 14.13(a) represents an on-card regulator which may be used to provide a fixed voltage from a

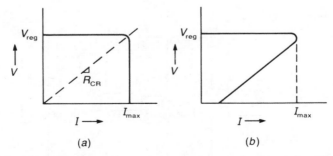

Figure 14.12. (a) A constant-voltage constant-current characteristic; (b) a foldback characteristic.

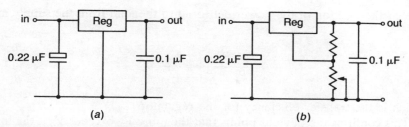

Figure 14.13. An IC fixed voltage regulator (*a*) and a pre-set regulator using the same device (*b*).

remotely located unregulated DC supply. This often makes an extra filter capacitor desirable to improve local smoothing, as shown. Also, a small (ceramic) capacitor at the output is useful for coping with the fast transients demanded by switching transistors, for example.

Many small regulators may be adjusted to provide nonstandard output voltages using very simple circuits like that of figure 14.13(*b*), and the equation giving the output in terms of the associated resistors is included in the manufacturers' data sheets.

The practical, and very simple, circuits of figure 14.13 reflect the ease of use provided by modern IC regulators, but when including them in a design, it is imperative to establish that their performances meet the stability factor values dictated by that design.

Unfortunately, the linear regulator is not efficient, because a significant portion of the power from the supply is dissipated by the series controller. This may be most easily seen by representing the regulator as a variable resistor, as shown in figure 14.14(*a*). From this, the power dissipated in the series controller and load respectively is:

$$P_R = (V_{in} - V_{reg})I_L \qquad (14.13)$$

and

$$P_L = V_{reg}I_L \qquad (14.14)$$

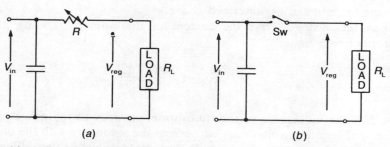

Figure 14.14. Basic equivalent circuit for a linear (*a*) and a switching regulator (*b*).

so that the fraction of the power delivered to the load from the supply must be:

$$\frac{P_{\text{L}}}{P_{\text{R}} + P_{\text{L}}} = \frac{V_{\text{reg}}}{V_{\text{in}}} \qquad (14.15)$$

which is the transfer efficiency for the regulator.

This confirms the obvious points that the more V_{in} exceeds V_{reg}, the lower the efficiency, and the more heat will be dissipated by the series controller. Unfortunately, V_{in} must always exceed V_{reg} by at least the voltage required to keep the series controller in an operating condition, and this dictates the lowest value of V_{in} that can be accommodated.

To obviate most of this dissipation, the series controller can be replaced by a solid-state switch, as shown in figure 14.14(*b*). This switch may be a bipolar or IGFET device operated with an ON-TO-OFF (or mark-space) ratio which controls the average voltage appearing at the output. This is actually the same concept as that which led to the class D or pulse-width-modulated amplifier already described in chapter 9 (see figure 10.9), though the aim is now to smooth this voltage and apply feedback to keep its value constant. The losses in the series switch will be very small, so that the transfer efficiency will always be very high.

This is the basis of the switch-mode power supply (SMPS)[6,7] and some approaches to its design now follow.

14.3 THE SWITCH-MODE POWER SUPPLY

The input voltage to an SMPS may be derived from the mains directly or via a transformer, and will usually follow a bridge rectifier and smoothing capacitor to provide an unregulated DC source. This DC is chopped by a solid-state switch, then processed and controlled, so that the unit may be referred to as a *DC-to-DC converter*.

Though many circuits have been designed to accomplish this, most are based on either the *flyback converter* or the *linear converter* techniques. Figure 14.15 shows the principle of the flyback converter, and assumes that V_{in} is the unregulated DC described above.

When the switch is ON, the current in the primary winding of the transformer rises linearly:

$$V_{\text{in}} = L \frac{\text{d}i}{\text{d}t} \quad \text{or} \quad \frac{\text{d}i}{\text{d}t} = \frac{V_{\text{in}}}{L}$$

where L is the inductance of the transformer referred to the primary.

During this time, a voltage appears across the secondary with the polarity shown by the dots, so that the diode D_1 prevents any current flow. When the

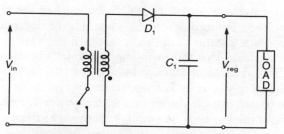

Figure 14.15. Principle of the flyback converter.

switch turns OFF, the energy stored in the core produces a secondary current because the secondary voltage is of the correct polarity to produce such a current flow in the absence of any other driving voltage. Hence, D_1 conducts and the current flows to the smoothing capacitor C_1 and the load. The mean charge on C_1 is, or course V_{reg}, and its value depends on the mark-space ratio of the switch timing, which implies that it is this which must be controlled by feedback from V_{reg}.

Figure 14.16 extends this concept in several ways. Firstly, a feedback path is shown which modulates the mark-space ratio of the switching device and hence stabilizes the value of V_{reg}. This may be achieved in several ways, including that depicted in figures 10.11 and 10.12; and the output-to-input isolation may utilize circuitry based on one of the opto-isolators of figure 13.29.

Figure 14.16 also shows a third winding and a diode D_2, the functions of which are to return stored energy to the supply in the event of the load

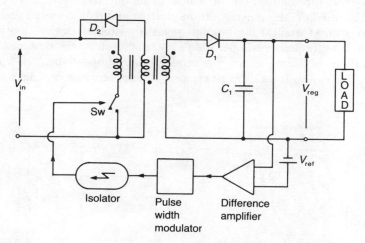

Figure 14.16. The flyback converter with isolated feedback control and energy return winding.

becoming an open circuit, or nearly so. If this part of the circuit were not present, the voltage induced when the switching device opened could be very high indeed and greater than the breakdown voltages of the associated semiconductor devices. This third winding is normally bifilar to ensure close coupling with little attendent leakage inductance (the term "bifilar" means that the windings for two coils are wound together).

The flyback circuit has the disadvantage that power is extracted from the supply only whilst the switch is ON (though more sophisticated versions, such as push-pull circuits, can obviate this), but it does call for very few components. In particular, if the mains voltage can be utilized directly, only one wound component—the transformer—is required. Also, if the PRF is high, not only is this transformer much reduced in size and weight, but the losses therein are minimized. In practice, flyback converters operate from about 20 kHz to 100 kHz, but it should be remembered that the higher the frequency, the greater the radio frequency interference (RFI) which is produced.

Figure 14.17 illustrates the second basic technique, the forward converter, and this circuit incorporates an inductor (or *choke*) L_1 in addition to a transformer. When the switch is ON, current flows from the secondary winding through diode D_1 and into the capacitor and load. However, when the switch turns OFF, the energy stored in L_1 causes current to continue flowing into the load and capacitor via D_2 (according to Lenz's Law). During this part of the sequence, energy stored in the transformer core is returned to the supply via the third winding and D_3. The associated control circuitry for the forward converter is similar to that for the flyback version, so the relevant diagram has been omitted.

Numerous manufacturers produce ICs which perform all the necessary functions for the realization of SMPSs including the voltage reference, difference amplifier, the pulse-width-modulator and its associated oscillator, the isolator and (usually) the bipolar or IGFET switching devices working in push-pull. The frequency of operation may be chosen by the designer within the limitations of the IC, as may the mode of operation—flyback or forward, single-ended or push-pull. Furthermore, these versatile ICs also usually

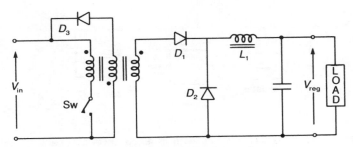

Figure 14.17. Principle of the forward converter.

contain overcurrent or foldback limitation facilities along with overvoltage and thermal protection[8].

Whereas complete power supplies may be obtained for most applications, occasionally special requirements call for custom designs, when either linear or switching control ICs may be specified along with their associated inductive components, the design of which should be regarded as part of electronics and not avoided by choice! In fact, the saturable reactor DC-to-AC inverters described in chapter 10 may also be used as very efficient converters by the addition of rectifying, smoothing and feedback control circuitry, and are particularly suited to high voltage power supplies for flash tube and similar work.

14.4 PROGRAMMABLE POWER SUPPLIES

In certain applications, notably test systems for various forms of electronic equipment, it is necessary to change supply voltages over wide ranges, and often quickly. This has led to the development of power supplies which can be controlled, or "programmed" as current jargon has it, to provide such changes[9]. The control signals may be in the form of voltages or currents, or variable resistances. They may also be digital signals provided by computerized sources.

Switching regulators do not lend themselves well to this application, but appropriately modified forms of linear regulators are ideal. This can be most easily understood by observing the similarity between a constant voltage power supply and an operational amplifier. Both have a very low output impedance, approximating to an ideal voltage source; and both are controlled by a signal at an input, this being the difference between a sample of the output voltage and a voltage reference in the case of the power supply. It is therefore reasonable to design a power supply as a high-power operational amplifier, so that it may be controlled by almost any input signal, appropriately conditioned.

The performance of the supply may now be couched in operational amplifier terms, where the concept of slew-rate is of particular relevance. A modern unit will slew at some 250 V/s—a figure which must be considered in the light of the several amps which the unit may be supplying at the time.

14.5 BATTERY POWER SOURCES

A power supply which produces a direct current (up to a limit) from a low-resistance source can be thought of as a direct voltage generator. In addition to the mains-derived power supplies already described, electrochemical cells can also approximate this function. Such a cell produces a low voltage—usually less than 3 V—but the individual cells may be connected in series to produce a higher voltage *battery*.

There are numerous situations in which it is necessary to utilize battery power supplies to operate electronic equipment. Cases in point include: land, sea, air, and space vehicles; environmental monitoring stations in areas remote from power; and in personal portable equipment such as radios, cameras, calculators, electronic watches, hearing aids and, of course, heart pacemakers.

Cells can be divided into two classes, *primary* and *secondary* (or *rechargeable*). In cells of both classes, an electrochemical process takes place, during which chemical energy is converted to electrical energy. However, in secondary cells, the process can be reversed by reversing the current; that is, the cell can be recharged, during which the chemical constituents are returned to their original states.

The most common types of cell in the two classes are listed below:

Primary	Secondary
carbon–zinc	lead–acid
alkaline (manganese dioxide)	nickel–cadmium
alkaline (mercuric oxide)	silver–zinc
zinc–air	
lithium	

It is usually evident whether a particular application calls for a primary or a secondary battery, not only in terms of the convenience or otherwise of charging, but also in that secondary cells can provide more power than primaries. However, the question of which battery to choose within the relevant category must be answered having regard to the electrical and environmental capabilities needed[10]. The most important of these characteristics are as follows:

Electrical	Environmental
voltage per cell	mounting position limitations
reduction in voltage over discharge period	temperature range (both working and storage)
current capability	pressure range to be encountered
internal resistance	acceleration to be withstood
capacity (ampere-hours)	
energy per unit weight	
energy per unit volume	
number of charges possible (for secondary cells)	

Brief descriptions of the most commonly used cells follow, along with notes on their prime advantages and disadvantages.

14.5.1 Primary Cells

The Leclanché cell was patented in 1866 and in its modern form (usually known as the carbon–zinc cell) has been predominant for several decades. This cell has a carbon rod cathode, an electrolyte consisting largely of an ammonium chloride solution in paste form, and a zinc anode, which forms the container. (Note that in electrochemical terminology, the positive terminal is called the "cathode" and the negative terminal is the "anode." This terminology is opposite to that employed by the engineer, who would not expect the positive carbon rod to be called a "cathode." It is, however, simply a reflection of the electrochemist's interest in the process going on inside the cell, rather than outside.)

The carbon–zinc cell also contains a depolarizer, which is a mixture of manganese dioxide and carbon, held around the carbon rod by the electrolyte paste.

The chief advantages of this cell are its cheapness and ubiquity; its disadvantages are many. During discharge, the norminal 1.5 V falls markedly, particularly when the cell is called upon to deliver high powers. In this context, "high power" would be more than about 0.5 W for a standard "D"-size cell. During discharge, the zinc is consumed, which is why leak-proof cells must employ a tube wrapped round the zinc container, with seals—usually plastic—at top and bottom. The reaction which takes place, leading to oxidation of the zinc, can be represented in the following simplified form:

$$Zn + 2MnO_2 \rightarrow ZnO.Mn_2O_3 .$$

A more complex form of the equation shows that the electrolyte takes part in the reaction, which is why its conductivity decreases during discharge.

Carbon–zinc cells are more suitable for intermittent rather than continuous operation. This is because of the formation of a layer of ionized hydrogen taking place more quickly than the depolarizer can remove it, which means that during a resting period, most of this layer will disperse, with a consequent partial voltage recovery.

The carbon–zinc cell will not operate at all below about −20°C and the capacity will be reduced by about 25% at freezing point. Also, although the capacity rises for higher-temperature operation, the cell will begin to deteriorate over about 45°C. For storage purposes, however, the shelf-life is improved at low temperatures, and diminished at high temperatures.

Although high-performance carbon–zinc cells are available, in which a low-resistance electrolyte-soaked paper replaces the paste, the various forms of alkaline cell provide better performance, in general. In the alkaline manganese dioxide cell, the electrolyte is potassium hydroxide (which is more conductive than ammonium chloride) whilst the case is of nickel plated steel, which is not consumed during discharge. The zinc anode takes the form of an internal cylinder composed of compressed, granulated zinc to provide a large surface area. The cell construction is therefore "inside-out" compared with the carbon–zinc cell, so that consumption of the zinc is immaterial, and the cell is inherently leak-proof. The polarity remains the same, however (positive stud, negative base), because it is constructed to provide a direct replacement for the carbon–zinc cell.

The outer can has an internal layer consisting of a mixture of manganese dioxide and carbon black to form the (positive) cathode, and is separated from the anode by a highly absorbent inert material.

The electrochemical equation is similar to that for the carbon–zinc cell, except that the electrolyte does not experience a chemical change. Hence, it maintains its conductivity, which in part accounts for the superior electrical performance of the cell.

The capacity of the alkaline manganese dioxide cell is slightly larger than that of a carbon–zinc cell of the same size at low current discharge levels, but becomes very much superior at high discharge rates. The nominal voltage is again 1.5 V, which also decreases rapidly with discharge. It will, however, work at lower temperatures than the carbon–zinc cell.

A marked improvement in voltage stability is obtained if a cathode of mercuric oxide and graphite is used, along with an electrolyte of potassium hydroxide with a zinc oxide inclusion. The anode is a pellet of compressed zinc powder, and the overall reaction is:

$$Zn + HgO \rightarrow ZnO + Hg \, .$$

The mercury released during this reaction is, of course, a good conductor, and so does not result in a fall in voltage during discharge. Hence, the mercury cell retains its normal voltage of 1.35 V (or 1.4 V if a small amount of manganese dioxide is added to the cathode) over most of its working life.

Mercury cells have a good energy/volume ratio and this, coupled with their good voltage retention, makes them particularly suitable in small ("button") sizes for watches, cameras, and hearing aids, or in stacked arrays for powering portable electronic test equipment. (The oscillator design of figure 8.18 works off a nominal voltage of 5.6 V, which implies that it will operate from a stack of four mercury cells.) The most remarkable use of mercury cells has been in some heart pacemakers, in which they are required to give utterly reliable service over some three years.

The mercury cell, in addition to being reliable and having good electrical characteristics, is also robust, and its double-steel-case form of construction enables it to withstand very marked pressure variations. It is, naturally, the most expensive of the three alternatives.

By far the best cell from the point of view of energy per unit weight or volume is the zinc–air cell[11]. This device depends on the controlled oxidation of a zinc anode to zinc oxide using atmospheric oxygen via a potassium hydroxide electrolyte. The cathode is a porous carbon structure which is not consumed during discharge.

Because atmospheric oxygen is used in this cell, it must be kept well ventilated during operation; but for storage, it should be kept in the hermetically sealed plastic bag in which it is normally supplied. In use, it can supply a very high current (up to 2 A for a D-size cell) without significant loss in capacity. Further, being dependent upon the presence of oxygen, it has the unique attribute of being usable also as an oxygen concentration transducer, the output voltage (usually 1.3–1.4 V) being a function of this concentration.

Another high-energy device in terms of both weight and volume is the *lithium cell*, in which lithium metal comprises the anode, carbon forms the cathode, and the electrolyte may be any of several organic or inorganic compounds. Lithium is very reactive, which means that a water-based electrolyte is not appropriate. Cells using sulfur dioxide dissolved in an organic solvent produce a nominal voltage of about 2.8 V; whereas some non-aqueous inorganic electrolytes result in nominal voltages of up to 3.4 V. Both systems lead to a cell which will operate over the remarkably wide temperature range of about $-55°C$ to $+75°C$.

In table 14.1 some of the properties of the primary cells previously mentioned are compared.

TABLE 14.1. A comparison of some primary cells (energy densities quoted for optimum working conditions).

	Carbon zinc	Alkaline (manganese dioxide)	Mercury	Zinc–air	Lithium (organic)	Lithium (inorganic)	Units
Nominal voltage	1.5	1.5	1.35 or 1.40	1.35	2.8	3.4	V
Energy density	55	80	100	450	330	500	$W\,h\,kg^{-1}$
Energy density	0.12	0.18	0.37	0.49	0.49	0.91	$W\,h\,cm^{-3}$
Working temperature range*	$\begin{cases} -10 \\ +50 \end{cases}$	-20 $+55$	0 $+55$	-10 $+50$	-55 $+75$	$\left.\begin{matrix} -55 \\ +75 \end{matrix}\right\}$	°C
Shelf-life*	4–12	18	24	48†	24	24	months

*Note: The working temperature range and shelf-life are very variable, and depend upon individual manufacturing techniques.
†Hermetically sealed.

14.5.2   Secondary Cells

The most common rechargeable battery is built up from the familiar lead–acid cell, which depends for its operation on the following overall reaction:

$$Pb + PbO_2 + 2H_2SO_4 \rightarrow 2PbSO_4 + 2H_2O .$$

During charging, the lead sulfate is converted back to lead and lead dioxide, and sulfuric acid is released. Unfortunately, though the (negative) lead anode and the (positive) lead dioxide cathode are made sufficiently massive to take little part in the reactions, they do disintegrate (largely because of positive plate corrosion) after a number of discharge–charge cycles, so limiting the lifetime of the cell.

The plate configuration and its additives, if any, vary considerably depending on the intended usage of the cell. Also, the electrolyte may take the form of liquid sulfuric acid, as in automotive batteries, or a thick paste, which allows the batteries to be used in any orientation.

Although the lead–acid cell is capable of providing extremely high currents for short periods, it will also exhibit a very stable terminal voltage (about 2 V) over long periods at low discharge rates. When the sulfuric acid becomes seriously depleted, however, the voltage falls rapidly from the "knee" of the curve. The specific gravity of the sulfuric acid is a good measure of the state of charge of the cells.

The lead–acid battery has an energy density of about $20\text{–}32\,W\,h\,kg^{-1}$, which is poor compared with primary cells, though the fact that it is rechargeable, and can provide high currents, outweighs this factor for many applications.

A somewhat better energy density of about $42\,W\,h\,kg^{-1}$ is provided by the nickel–cadmium secondary cell, which has become very common as a power source for portable electronic equipment.

This cell uses nickelic hydroxide as the positive (cathode) material, and cadmium as the negative. The reversible reaction is:

$$2NiO(OH) + Cd + 2H_2O \rightarrow 2Ni(OH)_2 + Cd(OH)_2 .$$

The potassium hydroxide electrolyte does not take part in the reaction and so is not depleted.

If the cell were over-charged, then oxygen and hydrogen would form at the two electrodes. In commercial cells, however, the positive plate is smaller than the negative plate, so that the normal reaction continues while oxygen alone is generated at the negative plate. This gas diffuses through the nylon or polypropylene separator grids, and reacts with the cadmium to form cadmium hydroxide. Thus the generation of gas is prevented during over-charge.

TABLE 14.2. A comparison of some secondary cells.

	Lead–acid*	Nickel–cadmium	Silver–zinc	Units
Nominal voltage	2.1	1.3	(a) 1.65	V
			(b) 1.5	
Energy density	32	42	132	W h kg^{-1}
Working temperature	−40	−20	−40	
range	+65	+80	+60	°C
Expected charging cycles	100–1000	250–10 000	10–30	cycles

*Sealed type, paste electrolyte.

The Ni–Cd cell exhibits a working voltage of about 1.3 V, which falls but little over a slow discharge period. Also, the cell can deliver high currents for short periods.

The energy density of the Ni–Cd cell is eclipsed by the silver–zinc cell at about 132 W h kg^{-1}. This cell uses silver oxide as the positive electrode, zinc as the negative, and potassium hydroxide as the electrolyte. The relevant equation is:

$$Ag_2O_2 + 2Zn + H_2O \rightarrow 2Ag + 2Zn(OH)_2 \, .$$

Unfortunately, both the silver oxide and the zinc dioxide are soluble in the potassium hydroxide electrolyte, so that a means of inhibiting this dissolving must be incorporated. This leads to some complexity in fabrication and the choice of separator material; and a limited cell lifetime.

During discharge, the cell voltage is 1.65 V while the divalent silver oxide is converted to the monovalent form; and this falls to about 1.5 V as the monovalent oxide converts to silver.

In table 14.2 three common secondary cells are compared.

14.5.3 Cell Capacity and Charging Rate

The energy content of a cell (either primary or secondary) is very much a function of discharge rate and ambient temperature. It also depends upon whether the discharging is accomplished continuously or intermittently: it has been explained how a carbon–zinc cell partially recovers during a resting period, for example.

Similarly, the *capacity* of a cell or battery is also a function of those parameters. The capacity C (in ampere-hours, A h) is the time integral of the current drawn:

$$C = \int i \, dt = It \quad \text{for a constant current} \, . \tag{14.27}$$

If a cell discharges, then the end of the discharge period will be defined when the voltage has fallen to a specified level (or by a given percentage of the nominal voltage).

With I assumed constant, the capacity is quoted by the manufacturer, who also offers curves showing how the cell will behave at various discharge rates. For example, if the $C/1$ rate were 4 A, then the current extracted for the $C/2$ rate would be 2 A. Conversely, at the $C/\frac{1}{2}$ rate, the current would be 8 A.

This concept is also useful when considering the charging of a secondary battery. For example, if a 4 A h cell were to be charged for 12 h, it would nominally require a current of $4/12 = \frac{1}{3}$ A. However, in practice, because of various losses, more energy must be injected than can be extracted, so that either a longer period or a higher current must be specified for charging. For a Ni–Cd cell, the excess is typically 20%.

The charging of a secondary cell is accomplished by passing a reverse current through it. To avoid excessive heat (and possibly gas) generation, the charger should incorporate a current-controlling circuit. Also, to avoid over-charging, the current should fall when the cell voltage has built up to an appropriate value; that is, the charger should have the *opposite* characteristic to a voltage-source power supply—it should have a constant current characteristic, followed by a constant voltage cut-off.

In the case of Ni–Cd cells, a low charging current (at a $C/10$ rate) may be maintained *ad infinitum* without damage to the cells. This makes them very useful as emergency power sources, where they can be continuously charged until the mains fails, when they can automatically be brought into operation.

14.6   SUMMARY

The foregoing material is necessarily only an overview of power-supply and battery methods, and some important topics have had to be omitted. For example, the speed of response of power supplies has hardly been mentioned, though this is of considerable importance, particularly in digital work and power electronics. Also, some of the newer, more exotic power sources have not been described, such as the superionic battery, which is essentially a capacitor of several *farads*, or the nuclear sources such as plutonium and polonium cells.

Keeping in mind that power supplies are nowadays usually bought rather than built, it is hoped that the chapter will enable the reader to evaluate the (sometimes over-optimistic) data issued by manufacturers; and to appreciate new developments in the light of existing power-source parameters. In particular, continuing progress in micropower electronics is making battery technology increasingly important, and is also bringing the solar cell into use not only as primary source of low power, but also as a means of charging secondary cells. A solar cell[12] is actually a large-area version of the photodiode described in chapter 13, and a great deal of work is currently in

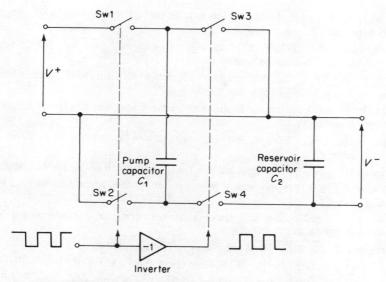

Figure 14.18. Principle of the Intersil "voltage mirror."

progress to produce cheap, large-area (and probably amorphous, as opposed to single-crystal) versions.

Also, with the rise of low-voltage, high-current, power supplies, Schottky-barrier (or metal-to-semiconductor) power rectifiers are becoming important[13], because they exhibit smaller forward voltage drops than do normal silicon diodes. Furthermore, germanium rectifiers and series-pass transistors should not be ignored in this context.

Finally, the use of IGFETs in switching applications (including the SMPS) is increasing, and sometimes taking quite novel forms, such as the Intersil "voltage mirror," the principle of which is depicted in figure 14.18. Here, a capacitor C_1 is charged from a positive supply V^+ via switching IGFETs represented by Sw1 and Sw2. These then open, and similar IGFETs Sw3 and Sw4 close, so that C_2 charges to the equivalent negative voltage V^-. This sequence is rapidly repeated, so that the positive input voltage is converted to a negative output voltage of almost the same magnitude. The ICL 7660 is a monolithic IC containing the switching and oscillator IGFET circuits, and when used with external pump and reservoir capacitors of 10 μF, is said to operate at 98% conversion efficiency, and present a 100 Ω output impedance.

REFERENCES

1. Schade, O. H., 1943, Analysis of rectifier operation, *Proc. IRE*, **31** (7).
2. Kesner, D., 1970, Monolithic voltage regulators, *IEEE Spectrum*, 24–32 (April).

3. Dobkin, R. C., 1979, Designers guide to ic voltage regulators, *EDN*, August 20 and September 5.
4. Annunziato, A., 1971, Prevent damage to loads and supplies, *Electron. Des.*, 42–97, April 29 and 64–67, May 13.
5. Hill, R. H., 1971, Greater regulator efficiency with foldback current limiting, *Electron. Eng.*, 67–69 (March).
6. Olla, R. S., 1973, Switching Regulators: the efficient way to power, *Electronics*, 91–95 (August 11).
7. Olla, R. S., 1980, Concepts in switching regulator design, *Electron. Prod. Des.*, 40–44 (July).
8. Dance, B., 1982, Switching regulator i.c.s, *New Electronics*, 36–42 (April).
9. Birman, P., 1975, Programmable power supplies, *IEE Electron. Power*, 239–240 (March).
10. Bell, T. E., 1988, Choosing the best battery for portable equipment. *IEEE Spectrum*, **25**(3) 30–35 (March).
11. Gregory, D. P., 1972, *Metal-air Batteries* (London: Mills and Boon).
12. Backus, C. E. (ed.), 1976, *Solar Cells* (New York, IEEE Press).
13. Fodor, G. T., 1980, Schottky power rectifiers, *Electron. Ind.*, 24–29, July.

General References
14. Hnatek, E. R., 1981, *Design of Solid-State Power Supplies* (New York: Van Nostrand, Reinhold).
15. *Voltage Regulator Handbook*, 1977 (Texas Instruments Inc.).
16. Mueller, G. A. (ed.), 1973, *The Gould Battery Handbook* (Gould Inc.).
17. Cahoon, N. C. and Heise, G. W., 1971 and 1976, *The Primary Battery*, Vols. 1 and 2 (New York: Wiley Interscience).
18. Gabano, J. P. (ed.), 1983, *Lithium Batteries* (New York: Academic Press).
19. Vincent, C. A., 1984, *Modern Batteries* (London: Edward Arnold).

QUESTIONS

1. What are the advantages offered by full-wave rectification over half-wave rectification?

2. Define the three stability factors for a power supply.

3. Why do Zener diodes which operate near 5–6 V exhibit better temperature stability than their equivalents which work at either lower or higher voltages?

4. With the aid of a block diagram, explain the principle of the series-regulated linear power supply.

5. What is an "electronic crowbar"?

6. Why is the switching regulator inherently more efficient than the linear series regulator?

7. What is a "flyback converter"?

8. What is a "forward converter" and what advantage does it have over the flyback converter?

PROBLEMS

14.1. A 12 V Zener diode is used as a regulator circuit like that of figure 14.9(a). If the maximum load current is to be 0.5 A and the minimum allowable Zener diode current is 20 mA, determine a suitable value for the series resistor R. Also, calculate the maximum power dissipated by both R and the Zener diode. Assume that the input voltage can vary between 16 V and 20 V.

Taking the nominal value of V_{in} to be 18 V, what is the value of the transfer stabilization factor S_V if the internal incremental resistance of the Zener diode is about 0.3 Ω?

14.2. Using the simple series regulator circuit of figure 14.19, calculate the value of R which will enable the smallest possible power transistor to be chosen, and determine $P_{diss(m)}$ for that transistor assuming that its value of V_{CE} must not fall below 2 V for satisfactory operation.

The input voltage can vary between 24 and 40 V, and the regulator is intended to apply 15 V across a 12 Ω load.

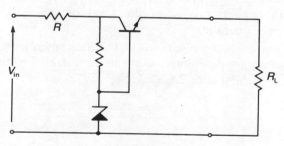

Figure 14.19.

14.3. A 5 V stand-by battery of 0.5 A h using Ni–Cd cells of nominal voltage 1.25 V each, is to be charged using the 24 V RMS winding of a mains transformer in series with a single rectifier diode and a resistor. If the maximum voltage per cell at full charge is 1.6 V, calculate the value of the resistor if the mean continuous charging current is to correspond to the 0.1 C level. (Take the diode volt drop as 0.7 V.)

14.4. The circuit shown in figure 14.20 is intended to regulate the current through the 12 V filament lamp to 1 A.

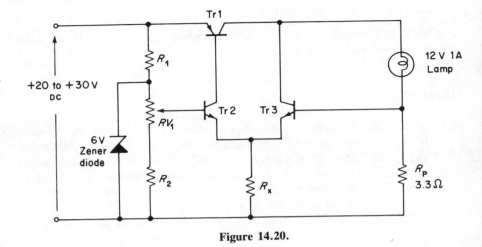

Figure 14.20.

(a) Explain how it works.

(b) For Tr1, determine minimum acceptable specifications for P_{tot}, V_{CEM} and I_{CM}.

(c) If $h_{\text{FE}} = 50$ for Tr1, determine a suitable value for R_x, assuming that V_{BE} for Tr2 and Tr3 is about 0.6 V. Is this an adequate value? If not, suggest possible alternatives.

(d) How much power does R_p dissipate? Does its change in resistance with temperature affect the level of the stabilized current in normal operation?

(e) When cold, the resistance of the filament is only 1/12 of that at operating temperature. Can this be a hazard to the circuit or any of its components?

15

CAD, ASICs and the Analog-Digital Interface

The foregoing chapters have introduced the most common semiconductor devices and structures, along with a series of basic discrete and integrated circuits which utilize them, a thorough knowledge and understanding of which is necessary to acquire competence in designing original and more complex circuits. When such a circuit design has been carried out, it becomes possible to employ analytical techniques to determine its performance, initially in terms of DC operating points, then followed by signal and noise analysis; but always with component tolerances catered for in "worst-case" design procedures.

Traditionally-designed circuits were constructed using "bread-boards," so-called because early designers would literally take a bread-slicing board from the kitchen, insert brass pins, and solder in the large components of the time! Today, several commercial equivalents of the bread-board exist, all of which will accept dual-in-line IC packages plus other components, and need no solder; and some of which are available complete with built-in power supplies. So, the tradition continues, and a "hands-on" designer will verify the network using sophisticated signal generators along with multi-range test instruments, cathode-ray oscilloscopes and spectrum analyzers, after which a printed circuit board will be designed, usually with computer assistance.

From this explanation, several disadvantages of bread-boarding will be apparent. Firstly, the procedure will become progressively more difficult as the size of the network increases, and its chances of working first time will become quite small! Secondly, trouble-shooting procedures will also become more exacting, and require very experienced designers indeed. Thirdly, when the circuit has been finalized, and all the trouble-shooting and component adjusting ("tweaking") has been carried out, it will not necessarily work well (or at all) with another set of components having parameters elsewhere in their tolerance spreads, which is particularly embarrassing at the printed-circuit board stage! However, it does make temperature characterization very convenient, given an appropriate test chamber.

These disadvantages can be overcome by the efforts of talented and experienced designers, but nowadays this approach—though still common— is following digital design into the computer era, and the inevitable frustrating period of gross user-unfriendliness is now largely past.

The fundamental problem facing computer-based analog network analysis is that the equations which define both the circuits and also the components or structures, are very complex, and need a considerable amount of computer power and memory, allied with a high degree of programming sophistication.

The basic computer approach to analog circuit analysis is as follows. The circuit in question is defined in terms of its nodes, and models representing the components between these nodes are inserted. If both power supplies and a signal source are also modeled, a circuit analysis program based on Kirchoff's current laws can then be run to determine at least the DC operating voltage levels and the signal performance.

This explanation conceals a great deal of sophistication, but many elementary programs have been written, some of which are quite small-scale, but very useful for the introduction of computer-aided analysis *per se*[1]. However, it was only in the latter half of the 1980s that powerful and user-friendly computer-aided analysis software packages became generally available; all the earlier ones necessitated the availability of major workstations such as the Sun or Apollo, driven by small main-frame host computers. However, useful work can now also be performed using desktop computers (usually IBM or IBM-compatible) and as always in electronics, prices are steadily reducing.

The primary problems in computer analysis of analog electronic circuits lie in the nature of the solid-state device mathematical models. In chapter 2, the Ebers–Moll equations were introduced, as was an associated equivalent circuit, this constituting a primitive model which nevertheless took account of both the quiescent *and* small-signal behavior of a bipolar transistor. However, to allow the designer to analyze his circuits comparatively easily, the normal procedure—which is the one followed in this book—is to design for quiescent operation first, then modify for small-signal operation based upon a perceived linearity which is valid only for incremental excursions about the operating points. Had it been reasonable to design using a single, all-encompassing model, then incremental equivalent circuits like the hybrid-π or h-parameter, would not have been necessary; nor would the "geometrical" approach which led to approximate large-signal design in the case of the power amplifiers of chapter 9. The more comprehensive versions of the Ebers–Moll model, however, are simply too large and contain too many parameters for convenient manual manipulation, but given enough computing power, it becomes both possible and reasonable to utilize them.

Currently, by far the most widely accepted models are those utilized in the SPICE (Simulation Program with Integrated Circuit Emphasis) program initially developed by Nagel[2] at U.C. Berkeley. In its early developmental forms, SPICE was very user-unfriendly, and required a deep syntactical knowledge for the formatting of circuit descriptions. However, it has now been massively developed by several companies, and their versions (some of which are listed in table 15.1) are offered within comprehensive analog design packages.

In figure 11.36, a small part of a SPICE-II equivalent circuit for an IGFET is given. A complete equivalent circuit is rarely seen, but a *netlist* of parameters for such a circuit would form part of a proprietary library. This netlist is actually a set of values of equivalent electrical elements which exist between the nodes of the circuit model for the relevant structure, component or IC, which means that it is in a form which a computer analysis program can accept. Thus, if a library model can be easily called-up, it becomes transparent to the user; that is, the designer need not know anything about either the model or its parameters, but need only know where in his circuit to place it.

As an example, a netlist for an IGFET in the Daisy Systems Corp. library contains 43 entries (as shown in table 15.2), the vast majority of which could not be obtained from the data sheets issued by the IGFET manufacturer. They include items such as the channel length modulation parameter λ, which appears in the simple analysis given in chapter 11 and is called LAMBDA in the netlist, plus many others drawn from the square-law Schichman–Hodges model and more sophisticated versions of Berkeley SPICE. The result is the netlist which exists as part of the proprietary DSPICE™ library. The *default* values shown are those which the computer will use if values specific to a given device or structure are not known, and may be changed by the user under the Daisy user-definable or UDM facility.

TABLE 15.1 A sample list of analog CAD vendors.

Firm	Software	Models
Analog Design Tools Inc.	Analog Workbench™ PC Workbench™	SPICE PLUS™
Daisy Systems Corp.	Analog Simulator	DSPICE™
Intergraph Corp.	Analog Circuit Simulator (ACS)	ACS Automodeller CSPICE
Mentor Corp.	Mentor Graphics™	MSPICE™ MSIMON™ (for IGFETs)
Valid Logic Systems Inc.	Analog Designer	ValidSPICE™
OrCAD Systems Corp.	Analog Simulation	MicroSim PSPICE™

TABLE 15.2 The Daisy Systems Corp. DSPICE™ netlist for IGFETs (reproduced by permission).

Name	Parameter	Units	Default
LEVEL	Model index	–	1
VTO	Zero-bias threshold voltage	V	0
KP	Transconductance parameter	A/V**2	2.0E − 5
GAMMA	Bulk threshold parameter	V**0.5	0
PHI	Surface potential	V	0.6
LAMBDA	Channel length modulation (MOS1, MOS2 only)	1/v	0
RD	Drain ohmic resistance	Ohm	0
RS	Source ohmic resistance	Ohm	0
CBD	Zero-bias B-D junction capacitance	F	0
CBS	Zero-bias B-S junction capacitance	F	0
IS	Bulk junction saturation current	A	1.0E − 14
PB	Bulk junction potential	V	0.8
CGSO	Gate-source overlap cap. per meter channel width	F/m	1E − 15F
CGDO	Gate-drain overlap cap. per meter channel width	F/m	1E − 15F
CGBO	Gate-bulk overlap cap. per meter channel length	F/m	0
RSH	Drain and source diffusion sheet resistance	Ohm/sq.	0
CJ	Zero-bias bulk junction bottom capacitance per square meter of junction area	F/m**2	0
MJ	Bulk junction bottom grading coefficient	–	0.5
CJSW	Zero-bias bulk junction sidewall capacitance per meter of junction perimeter	F/m	0
MJSW	Bulk junction sidewall grading coefficient	–	0.33
JS	Bulk junction saturation current per square meter of junction area	A/m**2	–
TOX	Oxide thickness–MOS1	meter	0
	–MOS2 & MOS3		1.0E − 7
NSUB	Substrate doping	1/cm**3	0
NSS	Surface state density	1/cm**3	0
NFS	Fast surface state density	1/cm**3	0
TPG	Type of gate material: +1 = opposite of substrate −1 = same as substrate 0 = Al gate	–	1
XJ	Metallurgical junction depth	meter	0
LD	Lateral diffusion	meter	0
UO	Surface mobility	cm**2/V − s	600
UCRIT	Critical field for mobility degradation (MOS2)	V/cm	1.0E4
UEXP	Critical field exponent in mobility degradation (MOS2)	–	0
VMAX	Maximum drift velocity of carriers–MOS2	m/s	0
	–MOS3		1E8
NEFF	Total channel charge (fixed & mobile) coef. (MOS2)	–	1
XQC	Coefficient of channel charge share attributed to drain	–	0
KF	Flicker noise coefficient	–	0
AF	Flicker noise exponent	–	1
FC	Coefficient for forward-bias depletion cap. formula	–	0
DELTA	Width effect on threshold voltage (MOS2 & MOS3)	–	0
THETA	Mobility modulation	1/V	0
ETA	Static feedback (MOS3)–UMAX defaulted	–	3.3
	–UMAX not defaulted		0
KAPPA	Saturation field factor (MOS3)	–	0.2
DL	Correction added to L on the device card	meter	0
DW	Correction added to W on the device card	meter	0

15.1 COMPUTER-AIDED DESIGN

The sequence of events followed by a user-friendly CAD procedure initially mirrors those followed by traditional design methods which are:

(i) A circuit is created by the designer which is intended to fulfil requirement specifications.

(ii) The circuit is analyzed, insofar as is possible, in terms of (a) quiescent conditions and (b) small-signals, large-signals, transients, frequency-response and noise where appropriate.

(iii) Worst-case limiting tolerance parameters are inserted where possible to determine their effects on the analyses in (ii).

(iv) A bread-board layout is constructed to verify the findings in (ii) using various test instruments and signal generators. Trouble-shooting procedures at this stage may lead to circuit modifications.

(v) A printed circuit layout is generated either manually or with computer assistance having regard to (a) the form (e.g., DIL-pac, surface mounting) and (b) the dimensions (e.g., pin or contact spacings) of the components.

These items may be taken in order and referred to the equivalent CAD procedure as follows:

(i) The circuit is created by the designer on the work-station using component symbols which appear in the relevant menu. These components may be moved, rotated or mirrored using simple mouse movements or keystrokes, and often appear in various colors for easy identification.

Note that the *designer* must be capable of doing this, and so must be familiar with contemporary components and ICs, and methods of incorporating them into circuits, just as in the "manual" design case. These is *no* substitute for this body of knowledge, and that offered in the present volume must be regarded as necessary but minimal! Insofar as the CAD facilities available are concerned, the designer should also know what the relevant library contains, and how to make parameter modifications or generate "custom" models where necessary.

(ii) In order to determine the quiescent conditions for the circuit, no input signal will be needed, but only power supply voltages. In computer terms, this implies specifying the appropriate power supply voltages existing at defined points in the circuit, and making sure that any signal voltage source is either a short circuit (or open circuit for a current source), or that the relevant value is set at zero.

Most programs use *icons* to depict test instruments. These are small line drawings of voltmeters, "scopes," signal generators etc., which can be moved to any designated "test point" defined by the user. Thereafter, the screen will display the relevant reading, or graph the waveform at that point. In the present case, voltmeter and ammeter icons would be used to determine the quiescent voltages and currents at appropriate points.

If the circuit is not too large, and the tolerances of the various components are known, the quiescent sensitivity of the circuit to parameter changes within these tolerance limits may be determined during this procedure, too. Otherwise a statistical approach based on Monte Carlo methods can be used to predict the effects of parameter tolerance variations, and appropriate software is also available as part of good CAD packages.

(iii) A signal generator icon may now be connected to the input, and the circuit response may be determined at the output (or anywhere else) in either the time or frequency domain. A wide selection of input signals is available, and in some packages, it is possible for the user to define an original input waveshape. For example, a swept sinusoidal input can result in a frequency response curve, or a square wave of high amplitude will illustrate the slewing capabilities of the circuit, which of course, are relevant to its switching performance. If the component model contains noise generators, the noise voltages at the output and at other points may also be determined at this stage.

(iv) The bread-board stage is avoided by the use of such CAD methods, and in fact is impossible where custom IC designs are concerned, for the very large expenditures involved imply that the test circuit *must* work properly the first time.

(v) Both IC metallization patterns and printed board layouts may also be designed using CAD, and all of the modern packages contain facilities for this purpose. Hence, color monitors are extremely useful where multi-layer printed circuit boards are involved, so that two or more layers are easily distinguishable. The art-work for producing the relevant masks are produced using associated high-definition printers. It is worthwhile noting here that analog printed board layouts are much less amenable to "full automation" than are their digital counterparts, because of their much higher external component count, and the fact that the components differ widely in geometry, size and pin location, and spacing.

The above CAD sequence does indeed mirror the "hands-on" design procedure. However, it also has some problems of its own, and in particular, the quite high cost of such a system, including a reasonably large library, must be justified in terms of cost-effectiveness to the purchasing organization. There is then the necessity of learning to use the system efficiently, and as in all computer work, it is easy to forget techniques without fairly constant practice. However, the circuit-symbol-and-icon approaches minimize this problem considerably. Also, for smaller organisations, attenuated versions exist which will run on personal computers, such as the PC Workbench™ of the Analog Design Tool Corp.

Inherent problems also exist in analog CAD methods. For example, when determining quiescent points, capacitance and inductance are neglected, but the models must contain the equations relating to the actual $V-I$ behavior of all the devices, such as exponential and near-square-law characterisation. When solving these equations, the computer will use an iterative numerical approach, such as the Newton–Raphson method. This involves "guessing" a solution, and using it, along with a first derivative, to linearize an equation and so generate an error which in turn may be used to generate a better guess. If the successive errors become smaller, this constitutes convergence to a solution. In principle, the system can oscillate around a solution, or it can in fact diverge. This led to problems in earlier CAD methods, though it has been largely obviated in modern packages such as Mentor's MSIMON (for IGFETs) which offers 100% convergence.

For small-signal analysis, the relevant equations are linearized about appropriate operating points to calculate the incremental performance. This, of course, is the computer equivalent of the function performed by the h-parameter or hybrid-π models, which again demonstrates how the CAD approach mirrors that of "manual" design methods.

The most common problem met by the designer is that, of the many thousands of available discrete and integrated components, only a small proportion can be contained in even an extensive library, though such libraries do cover a remarkably wide range, including (in the case of the Analog Workbench Corp.) models for nonlinear magnetic devices and cores along with a Power Design Module. It is for this reason that facilities for the user to insert new parameters are made available, these being acquired using appropriate instrumentation. For example, the Intergraph Corp. system interfaces its InterPro™ workstation directly to Hewlett-Packard test equipment to result in families of performance curves which are then used in ACS (Analog Simulation Software), so circumventing the SPICE type of model completely.

15.2 APPLICATION SPECIFIC INTEGRATED CIRCUITS

In addition to the analysis of networks comprising commercial discrete devices and integrated circuits, CAD methods may be used at structure level to analyze circuits destined to become ICs in their own right. This is the approach taken by manufacturers of general purpose ICs, and because the whole procedure, from the initial concept to final encapsulation, is very expensive, a careful estimation of minimum potential sales must be carried out before the design-to-production sequence is even begun. In the case of original equipment manufacturers (OEMs) who wish to incorporate one or more unique ASICs into their products, the IC manufacturer must enter the sequence at an early stage, and will eventually produce a *full custom* ASIC for the OEM. This procedure is also expensive, and so is found only where large sales are involved, or where copying by other manufacturers is to be discouraged! (Copying, or *reverse engineering* of ICs is difficult, but not impossible.)

A much less expensive method which is of major and increasing importance, involves the *uncommitted array*, or *semicustom* ASIC. Here, a manufacturer produces a chip containing an array of active devices plus resistors and capacitors, but omits the final metallization. The geometry of the array is then used by a customer to design an interconnection pattern which will result in a complete circuit to perform the desired function. This pattern is then sent to the chip vendor, who deposits the final metallization, then encapsulates samples of the ASIC and delivers them to the customer for verification and approval.

This procedure is analogous to that used for logic design, where the uncommitted gate array is common, but surprisingly, was actually preceded by the analog array, which first appeared around 1971 in the form of the user specific integrated circuit (USIC) of the EXAR Corporation. Modern forms of these are currently available in their Master-Chip^R series, and in addition to bipolar arrays, include ion-implanted BIFET, high-voltage and analog/digital arrays.

Early *mask-programmable* arrays were inefficient in terms of device utilization, which often fell below 50%, the prime reason being that they consisted of large numbers of structures in purely geometrical arrangements which made efficient interconnection design difficult. This problem has now been overcome by several manufacturers who group sets of active and passive structures into subunits often called *tiles* or *macrocells*. These may be more easily configured to form complete functional blocks including operational amplifiers and most other basic circuit elements. This means that the circuit designer becomes a system designer at the microchip level, but once again the point is made that familiarity with analog techniques is absolutely mandatory. A partial list of the of functional blocks which may be realized using these semicustom ASICs, and all of which have appeared earlier in the text, is as follows:

Operational amplifiers	Flip-flops
Transconductance amplifiers	Schmitt triggers
Comparators	Pulse generators
Analog gates	Pulse-width modulators
Current mirrors	Four-layer switch equivalents
Voltage references	Switched capacitor filters
Multipliers	Oscillators

TABLE 15.3. A sample list of semicustom array vendors.

Firm	Name of Array	Subunits
EXAR Corp.	FLEXAR™	Cells containing TWINSTOR PADSTOR and TWIN-BOOSTOR elements
Raytheon Corp.	Linear Macrocell Array	Macrocells
SIPEX Corp. (Datalinear Division)	High Performance Analog Array	Tiles
Plessey Semiconductors (Ferranti Interdesign Inc.)	Monochip™	Matrix of active and passive structures

Table 15.3 gives an example list of four manufacturers who adopt this form of approach with their own variations, and it is useful to consider these in further detail.

In order to realize block functions, the various forms of active structure must be augmented by resistors and capacitors, and the basic EXAR Flexible Linear Array*, or FLEXAR™ cell shown in figure 15.1(*a*), exemplifies this. Here, each of the three TWINSTORs consists of a dual-collector pnp bipolar transistor merged with a two commoned-collector npn structure. Various points on this composite structure are brought out to nine pads which may be interconnected by the final metallization to form any of the devices shown in figure 15.1(*b*). Notice that this listing implies that junctions may be used as either normal diodes or, given reverse biasing, as capacitances (before breakdown) or Zener diodes (after breakdown), Also, either npn and pnp transistors may be realized, or a combination of both may be connected to produce an npn/pnp switching equivalent like that of figure 12.9. Small resistances of about 500 Ω may also be realized by using two different contacts on either a p or an n region.

Returning to the cell of figure 15.1(*a*), arrays A and B contain various ion implant resistors, and also *crossunders*, which allow one connection to cross another. These crossunders are formed with a lower conductor layer which is used to partially connect the tiles, but is not otherwise accessible.

This family of BETA arrays contain various numbers of these cells (such as 60 in the BETA 180), plus PADSTORs and TWINBOOSTORs which offer different sizes and configurations of bipolar structure, the versatile PADSTOR acting also as a large-area connection pad.

When the designer wishes to use a given functional block (such as an operational amplifier), it is not often necessary to actually design the block using a knowledge of the individual cell contents. Instead, the so-called *soft cell* can usually be "called-up" by the associated software, which in the case

---

* All material relating to FLEXAR™ has been reproduced with the express authorization of EXAR Corporation.

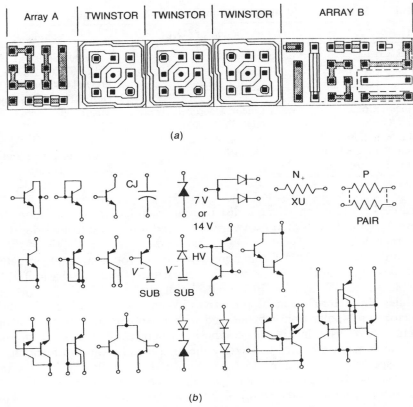

(a)

(b)

Figure 15.1. The EXAR TWINSTOR™ macrocell (a) and some of its connection schemes (b). Note that XU means a crossunder; and that the substrate (SUB) must go to the most negative voltage, V^-. (© EXAR Corporation, reproduced by permission.)

of FLEXAR™ systems, can be implemented on an IBM XT or AT personal computer. Both the schematic and its SPICE equivalent can also be called up, so that a complete circuit can be designed, then analyzed. Finally, a relevant netlist can be printed out and sent to the firm for fabrication. It is also worthy of note that for a commercially successful design, the firm can also translate the FLEXAR™ circuit into full custom form.

An alternative to—or perhaps as a final check of—the software design, is to manually make up a bread-board version, and this can be done using a kit of four pre-connected BETA arrays, the FLA series. These are ICs finished in DIL-pac forms ready for loading on to boards which may then be fully tested in the conventional manner.

A significant variation on the theme of macrocells is offered by Raytheon's RLA series, in which an array of *gain cells* is deposited along with extra transistors and blocks of silicon–chromium thin film resistors to provide appropriate circuit extensions and interfacing devices. The gain cell circuit is shown in figure 15.2(a) which also shows the set of contact pads. The general layout of this gain cell will appear familiar in view of the material presented in the earlier chapters: for example, the Tr14/Tr13

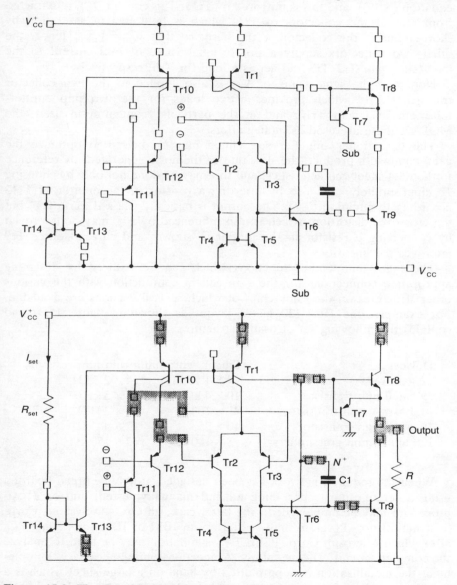

Figure 15.2. The Raytheon gain cell (a), connected as a MOPA2 operational amplifier (b) (reproduced by permission of Raytheon Co.).

combination is a current mirror pair and Tr2/Tr3 is a difference amplifier stage.

Figure 15.2(b) gives one example of how the gain cell metallization connections may be configured to produce an operational amplifier. Here, the bias current into Tr14 is defined by the resistor R_{set} (which is the R_4 of equation (5.46)); and this is mirrored by Tr13. This current, I_{set}, is extracted from one of the collectors of Tr1, which is held just in saturation by short-circuiting the collector to the base, so that $V_{CE} = V_{BE}$. Hence, the other two collectors supply a proportional value of bias current to the long-tailed pair Tr2/Tr3 and act as a load for Tr6, respectively.

Note that the same value of V_{BE} is also applied to the twin-collector transistor Tr10, which provides active loads for the two pnp emitter-followers Tr11 and Tr12; and in the particular configuration used, the MOPA2, these are used as input buffers.

The operational amplifier so defined may be biased to optimize the gain–bandwidth product, the slew rate or the noise generated, by reference to the data sheet characteristics, and this may be done not only by choosing I_{set} appropriately, but also by inserting a resistor in the emitter of Tr13 according to equation (7.2). The output is basically Class B (Tr8/Tr7) but the crossover distortion, which is also influenced by I_{set}, may be improved by connecting a resistor to the output as shown; and current-source Tr9 improves the linearity.

There are many other functions which can be realized by designing appropriate connections for the gain cell in conjunction with the various other structures on the same chip, of which several versions are available. For example, the RLA 120 has a layout as shown in figure 15.3, and contains the following set of usable structures:

Devices	Silicon–chromium thin film resistors	
12 gain cells	$24 \times 1.25\,\text{k}\Omega$	$24 \times 2.5\,\text{k}\Omega$
39 small npn transistors	$10 \times 4\,\text{k}\Omega$	$24 \times 5\,\text{k}\Omega$
4 100 mA npn transistors	$48 \times 10\,\text{k}\Omega$	$24 \times 20\,\text{k}\Omega$
16 small pnp transistors	$10 \times 25\,\text{k}\Omega$	$12 \times 40\,\text{k}\Omega$
148 aluminium crossunders	$8 \times 100\,\text{k}\Omega$	$12 \times 150\,\text{k}\Omega$

When the required network has been designed, it may be bread-boarded using a set of DIL-pac protyping ICs and manually tested; and/or a CAD procedure may be employed. In the latter case, an OrCAD Systems Corp. Schematic Entry Program may be run on an IBM or IBM-compatible PC, after which Microsim Corp. PSPICE^{TM} simulation may be used to analyze the complete circuit. The actual interconnection routing necessary to implement the metallization may be plotted by hand on a large sheet which is a complete version of the layout scheme exemplified by figure 15.3, and upon which all the contact pads and crossunders appear.

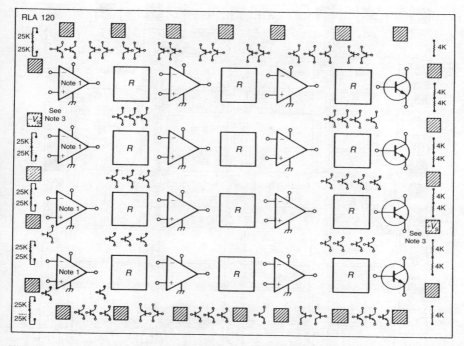

Figure 15.3. The Raytheon RLA 120 simplified schematic layout.

Notes: (1) The four macrocells include pnp input structures;
(2) resistor arrays "R" as listed in text; but right-hand column lacks 100 kΩ
value (reproduced by permission of Raytheon Co.).

Yet another variation on the theme of the tile or macrocell format is
offered by the DataLinear Division of the SIPEX Corporation. Here, a tile
of 12 npn and 12 pnp vertical bipolar transistors is fabricated using dielectric
isolation (DI as opposed to junction isolation or JI), which means that each
structure is diffused in a tub of SiO_2. Thus, not only is the necessity for
low-gain lateral or substrate pnp transistors removed, but any structure may
be connected anywhere without influencing other structures.

A typical uncommitted array chip such as the SP-1104 not only contains
16 tiles of small transistors (in slightly differing layouts termed "A" and "B"
tiles), but also larger and multi-emitter transistors, several dielectrically
isolated capacitors and buried Zener diodes. Furthermore, each tile is
flanked (top and bottom in figure 15.4) by uncommitted areas for user-
defined Nichrome resistor deposition. This has the advantage that the
designer may specify the actual resistor values needed without having to
compromise by employing array-defined values. They may also be configured to appear precisely where needed; but must be specified in terms of
area and geometry.

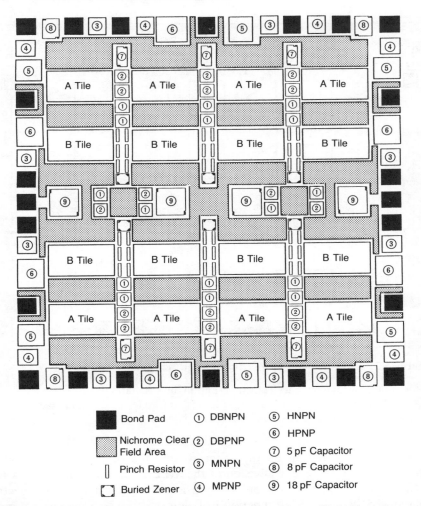

Figure 15.4. Organization and placement of the DataLinear SP-1104 analog array (reproduced by permission of SIPEX Ltd.).

Here again, a software library of fully characterized linear standard cells is available from the manufacturer, along with PSPICE™ and HSPICE™ simulations; also available is a series of bread-board ICs for practical evaluation.

Interdesign's Monochip™ is available in several different sizes and voltage ratings, and is essentially a matrix of a very large selection of active and passive structures including vertical low and high power npn transistors, lateral and substrate pnp transistors, numerous diffused and pinch resistors, capacitors and crossunders. Once again a set of kit parts is available for bread-boarding, and these consist of arrays of transistors and other components, and also a series of complete functional blocks. The metallization routing design may again be carried out manually using large-scale layout sheets, and to assist this, some three hundred functional block designs appear in the relevant instruction book. If so desired, software analyses may be implemented by the designer using the ASPEC program of the ISD Corp., and this may be accessed via CDC's CYBERNET data communication link, which operates worldwide.

Using semicustom methods such as those introduced above, it is possible to reduce the design-to-production costs compared with the equivalent process using standard ICs and discrete components, even when the production run is expected to be as low as a few thousand. However, care must still be taken to ensure that the desired functions are in fact possible using an available semicustom chip, especially where extremes in power, bandwidth or noise performance are required. So, once again, the designer *must* have sufficient background knowledge in electronics to make such a decision.

15.3 THE ANALOG-DIGITAL INTERFACE

The processing of analog signals, such as are provided by almost all transducers, may be carried out in three basic ways, as shown in figure 15.5. In diagram (*a*) the signal is amplified, filtered and otherwise processed using analog techniques throughout. There are, in fact, may cases where this approach is preferable, especially those where accurate and fast four-quadrant multiplication is needed. However, where the signal processing path calls for some form of digital input or control, such as where switched-capacitor filters are used, then it is necessary to include digital circuitry as in diagram (*b*). However, the most common contemporary systems first pre-process input signals using analog techniques, then convert these to digital data, process these data using digital methods, then reconvert back to a usable analog form. This system is represented by diagram (*c*) and a more comprehensive version has already been met in the Preface (figure P.1).

Earlier in the chapter, a few examples of semicustom ASICs were introduced, and since some of the uncommitted arrays described contained not only analog cells, but numerous individual transistors, it is obvious that they can support both analog and digital functions. Clearly, this is even more valid for full custom chips, and these are indeed areas of rapid growth in electronics. Perhaps the greatest problem lies in the complexity of the computational tools needed to analyze such networks, but the user-friendliness of these is unquestionably improving, as is evidenced by the Silvar-Lisco Corp. ANDI[TM] package, which performs quiescent and transient analyses on circuits containing both logic gates and basic analog functional blocks. It is therefore particularly useful where phase-locked loops, switched-capacitor filters and analog-to-digital and digital-to-analog converters are integral to the network.

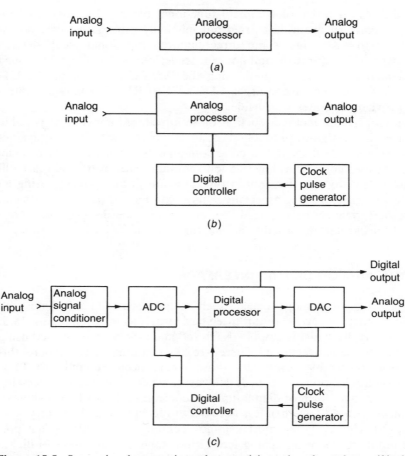

Figure 15.5. Some signal processing schemes: (*a*) analog throughout; (*b*) digitally-controlled analog processing; (c) conversion for digital processing.

With reference to figure P.1, not only does a digital signal processing facility exist but so do various digitally-controlled analog circuits, including the multiplexer (MUX), the sample-and-hold (SH) module, and the analog-to-digital converter (ADC). The MUX and the SH have been introduced earlier in the text, and it must now be assumed that the reader has also acquired a basic familiarity with digital electronics, including sampling theory, in order to treat the analog-to-digital process.

15.3.1 The Analog-to-Digital Converter[3]

Before considering any of the several common methods of A-D conversion, some overall basic factors must be taken into account relating to the performance of any conversion system irrespective of its *modus operandi*. These factors are best explained using a simple example, as shown in figure 15.6. In diagram (*a*), an analog voltage of 0–7 V is shown on the abscissa, and its digital equivalent is shown on the ordinate, expressed as a binary number of three bits. Here, there are eight possible levels, being given in general by 2^n where n is the number of bits utilized. Hence, in this case a change of state may occur at seven ($2^n - 1$) thresholds, or decision points, so

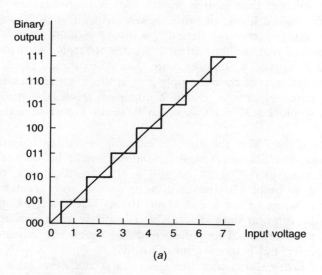

(*a*)

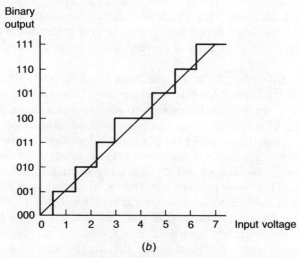

(*b*)

Figure 15.6. An ideal 3-bit ADC characteristic (*a*), and with linearity errors (*b*).

that the actual voltage cannot be accurately represented to better than one part in seven, which is 1 V; that is, in general, the *quantization error* is plus-or-minus half of the lowest significant bit, written $\pm\frac{1}{2}$ LSB, which is self-evident from figure 15.5(a).

This diagram is in fact ideal, because the changes in the binary number occur exactly half-way between each one volt increment. If they did not, there would be errors in *linearity*, as shown in diagram (b). However, even this case produces a binary representation which continues to increase with the analog voltage: that is, it is *monotonic*. A more serious shortcoming would have resulted if one or more downward transitions had occurred as the analog voltage increased, denoting a loss of monotonicity.

It is apparent that should a greater degree of *resolution* than ± 0.5 V be required, then an ADC which uses more than 3 bits should be chosen. For n bits, the resolution is of course simply 2^n, but this is sometimes expressed as an approximate percentage. Given 2^n *quantum levels*, the resolution of a 10-bit ADC would be $2^{10} = 1024$, which is about 1 part in a thousand, or 0.1%.

Note that if the 7 V of the previous example were digitized using a 10-bit ADC the *quantum size* (or simply *quantum*) would be only about 7 mV. Should the noise in the circuit be only a little less than this, there would obviously be no point whatsoever in using an ADC with greater resolution! Furthermore, when an ADC above about 10 bits is justified, its *offset* may be quoted in the relevant data sheet. This is entirely analogous to the offset voltage of an operational amplifier, and can usually be adjusted through zero empirically, but is temperature-sensitive.

The *scale factor* shows how the ADC is calibrated; that is, it defines the full-scale input voltage which gives rise to the maximum available binary number. This can be externally adjusted in most cases, a facility which can be very useful. Suppose, for example, that a 10-bit ADC is to be used to digitize 10 V. If the scale factor is adjusted to give a maximum output for an input of 10.24 V, then the LSB will correspond to exactly 10 mV rather than a little less than this.

Obviously, whatever the technique used to effect the A–D conversion, a finite time will be needed for this to take place. This is called the *conversion time*, and may vary from milliseconds to nanoseconds depending upon the particular type of converter used and its resolution. Returning to figure P.1, it is apparent that the SH must hold a sampled voltage for longer than the ADC conversion time. (Also it must not present another sample before the ADC has had time to recover, which is usually short.)

In Chapter 11, it was pointed out that a SH circuit also needs a certain time to acquire and stabilize a voltage level (see figure 11.48), but this is usually shorter than the conversion time for an ADC. This highlights the usefulness of the sampling technique as applied to single signals as well as the multiplexed-signal system of figure P.1. In the single-signal case, sampling theory[4] shows that the sampling rate must be at least twice the frequency of the highest-frequency component in the signal of interest, otherwise the original signal cannot be recovered. In practice, sampling rates of several times this value are normally employed, which indicates the importance of both the SH delays and the ADC conversion time. Figure 15.7(*a*) is an illustration of signal sampling and shows how each time the SH is pulsed, it holds the signal level for a period, which though short, must nevertheless be greater than the conversion time of the associated ADC. Figure 15.7(*b*) shows what would happen if the sampling frequency were too low—it would be possible to reconstitute a signal component of a frequency

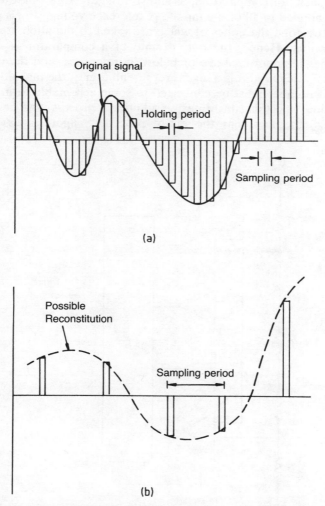

Figure 15.7. An adequately-sampled signal (*a*) and a possible effect of inadequate sampling (*b*).

much lower than that intended, leading to significant distortion of the original.

Finally, although the foregoing examples have used a binary number representation of the analog quantity, it is perfectly possible to design an ADC to produce any other code desired, such as two's complement or binary coded decimal (BCD), as required by the digital system which follows.

15.3.2 ADC Techniques

The most obvious way of converting an analog signal to a digital signal is to use a multiplicity of comparators, as shown in figure 15.8 where the analog signal v_{in} is applied to all of the inputs. A reference voltage V_{ref} is applied to a chain of resistors, the nodes of which are taken to the alternate inputs of the comparators. Hence, the output state of a comparator is dependent upon whether its input is above or below its reference, and the number of comparators which are ON is a digital representation of the magnitude of v_{in}. Clearly, this number must be converted to some acceptable form of binary representation, and this constitutes an additional process, and hence delay. Nevertheless, the fact that the entire procedure operates as a *parallel*

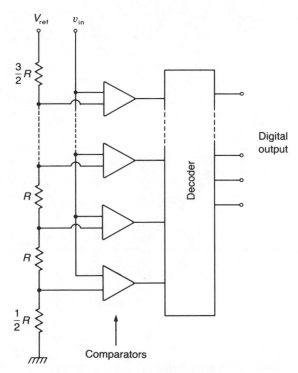

Figure 15.8. A parallel or flash ADC.

process, means that this form of converter can be very fast; and in fact, it is commonly known as a *flash converter* for this reason.

The lowest comparator must obviously change state when its input voltage rises to a level corresponding to $\frac{1}{2}$ LSB, which is why the lowest resistor has a value of $\frac{1}{2}R$, so that the succeeding resistances, of value R, lead to comparator state changes for each voltage rise corresponding to a full LSB.

If the output word is to be n bits long, then the number of comparators used must be $(2^n - 1)$. Thus, a four-bit ADC needs 15 comparators, a six-bit needs 63 and an eight-bit needs 255. Obviously, this is moving into the realms of VLSI, but fortunately, it is possible to cascade lower bit stages instead of allowing the comparator—and resistor—complexity to get out of hand. This technique, however, does also involve the use of digital-to-analog conversion.

Another technique, which is simpler but slower, requires the accurate generation of linear ramp voltages; this is done using operational amplifiers connected as integrators. Basically, this involves a determination of the time for such a ramp to reach the same amplitude as the input voltage, which can easily be detected using a single comparator. Of the various ways of implementing this concept, the *dual-slope* method is the most common, and the principle of its operation is illustrated in figure 15.9(a).

Here, the input voltage v_{in} is applied to an integrator, shown in diagram (b), for a fixed time T_1, which results in a linear ramp at the output. It is then replaced by a fixed reference voltage V_{ref} of the opposite polarity, which produces another linear ramp, but of opposite slope. This reaches zero (as detected by a zero-crossing comparator) in a time T_2 which is proportional to the voltage reached by the first ramp. Hence,

$$V_{ref}T_2 = v_{in}T_1 \quad \text{or} \quad v_{in} = V_{ref}\frac{T_2}{T_1}.$$

This expression shows that whereas the accuracy of the converter is dependent upon that of the reference voltage, it does not depend upon the stability of the clock which defines T_1 and T_2 provided that it does not drift significantly over the period $(T_1 + T_2)$. Similarly, the actual value of the integrating capacitor is unimportant provided it too is not subject to short-term drift, and is a low-leakage type. Therefore, provided care is taken to provide a stable V_{ref} (as from a band-gap reference or buried Zener diode) the dual-slope method can lead to an excellent ADC, albeit a slower one than the flash converter. This is the type of ADC that is used in most small digital panel meters.

The two examples given above are augmented by several other techniques, some of the most successful of which involve voltage-to-frequency (V–F) converters, and details of these will be found in most texts on digital electronics. However, to both conclude this section, and to introduce an

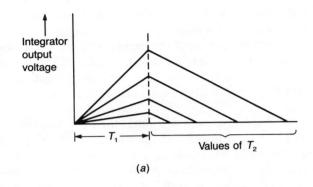

(a)

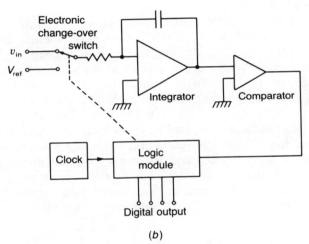

(b)

Figure 15.9. Principle (a) and block diagram (b) of the dual-slope ADC.

application of the digital-to-analog converter (DAC) it is appropriate to consider the *successive-approximation* ADC.

Figure 15.10(a) gives an example showing the principle of this form of DAC, whilst diagram (b) illustrates the basic layout. In the latter, a DAC, a *successive approximation register* (SAR) and a comparator will be seen to form a complete feedback loop. At the commencement of the conversion, the SAR injects a most significant bit (MSB) into the DAC, the output of which is compared with v_{in}. If this output is less than v_{in}, the MSB is retained and the next bit is injected, and so on until the comparator signals that the DAC output is equal to v_{in}. If the DAC output for an MSB input is greater than v_{in}, as in the example given in diagram (a), then the MSB is removed, and the next significant bit (which leads to half the voltage produced by the MSB) is injected, and so on, as shown.

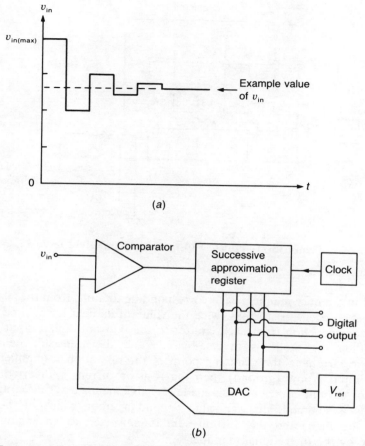

Figure 15.10. Principle (*a*) and block diagram (*b*) for succesive approximation ADC.

Such progressions will continue until the voltage between the comparator inputs is within a small range (as defined by its hysteresis performance) when the comparison is adjudged complete. Clearly, the sensitivity of the comparator must match the voltage defined by the LSB of the DAC and SAR.

15.3.3 The Digital-to-Analog-Converter

The conversion of digital to analog signals is actually easier than the converse, and the basic approach is to generate a current or a voltage which corresponds to each bit in the word to be converted. Figure 15.11 illustrates this using a set of identical transistors as current generators, all having the same base voltage V_{ref}, but with different emitter resistors. It is these emitter resistors which control the generated currents, and their values are

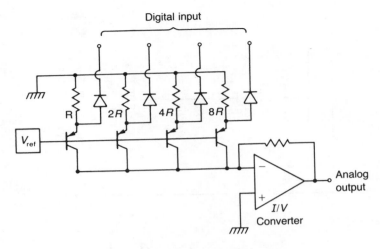

Figure 15.11. Principle of the weighted current DAC.

related in a binary manner, each corresponding to a bit, from the MSB to the LSB; that is, each resistor has twice the value of its next lower predecessor.

If the connection from each emitter to the relevant digital input point is high, then the diode is reverse-biased, and the transistor delivers its collector current to the summing point of the operational amplifier. If the digital input is low (ground) then the resistor current is diverted and no collector current contribution is made. By this means a set of contributory currents is summed, the net value depending upon which of the digital inputs are high; and this current sum is converted to an analog output voltage by the operational amplifier.

The reference voltage applied to the bases is chosen having regard to the digital voltage levels involved, and if these are 0 V and 5 V, the V_{ref} would be typically 1.2 V. This is not only an easy stabilized voltage to acquire, but is also of a convenient value having regard to standard tolerances in the logic levels.

A disadvantage of this form of DAC is that the resistors corresponding to the least significant bits can be very high, and hence difficult to fabricate in monolithic form, in addition to being very temperature-sensitive. This problem can be obviated by using blocks of current-generating transistors having similar sets of binary-weighted emitter resistors, and dividing down the final currents. However, an alternative solution is offered by another system, the R–$2R$ method, shown in figure 15.12.

Here, a reference voltage can drive a current into any of a set of resistors of value $2R$, which are connected to it via a corresponding set of electronic changeover switches, these being operated by the digital input lines. The bottom end of each $2R$ resistor can be connected to either ground or the

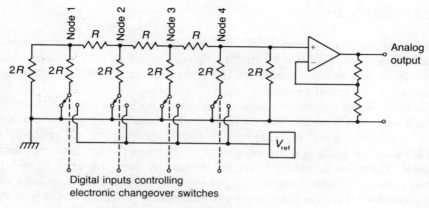

Figure 15.12. Principle of the $R-2R$ DAC.

reference voltage by its electronic change-over switch and the important point to note is that if V_{ref} is connected to any $2R$ resistor, then a current of $V_{ref}/3R$ will flow. This is because the current first flows up the $2R$ resistor, then divides at the upper node into two equal parts, because to the left or right of each node, the net resistance is always the same, and has the value R. This may be shown mathematically, but is also easy to see in the simple diagram given. Suppose that V_{ref} is applied only through the $2R$ resistor connected to node 2. Looking to the left of this node, a resistor R is in series with two $2R$ paralleled resistors, so that a net value of $2R$ appears. Looking to the right, a resistor R is in series with a combination whose value is obviously $2R$. Thus, the total resistance at node 2 is R, so that V_{ref} drives a current into a combined resistance of $3R$.

This means that if the bottom of any $2R$ resistor is connected to V_{ref}, then the top is at a voltage $V_{ref}/3$. If this appears at node 4, then the output voltage will be $V_{ref}/3$. However, if it appears at node 3, then the output will be only $V_{ref}/6$. Also, if it appears at node 2, then the output will be $V_{ref}/12$, and for node 1 it will be only $V_{ref}/24$. These voltages may be summed by superposition, and will result in a net output voltage which is binary-weighted, so resulting in a digital-to-analog conversion.

This concept is extendable to more bits, and in fact, by correctly tailoring the resistor values, codes other than pure binary may be directly generated. Further, because the conversion is dependent upon resistor ratios rather than absolute values, it is a very appropriate DAC for monolithic realisation.

Finally, it should be noted that any DAC which will accept a variable external reference voltage, becomes a *multiplying* DAC, because its output depends not only on the digital input, but also on the value of V_{ref}. This concept has a vast number of applications, including those which use V_{ref} as an input voltage, the amplitude of which may be varied (or "programmed") by the application of a second, digital input.

15.4 POSTCRIPT

The foregoing brief treatment of ADCs and DACs serves to show how analog and digital electronics are inextricably bound together, and how they both contribute to the performance of such tasks as the signal acquisition and processing system of figure P.1. Also, the earlier part of the chapter has indicated the directions in which design using both techniques is heading, which is towards realization in ASIC form, usually via CAD. However, semiconductors from single devices to highly complex ICs will continue to exist alongside ASICs, and irrespective of which route the designer chooses to follow, it remains necessary to recognize that the computer can only *aid* the design process, and not supplant it. The degree to which this assistance can be given will of course increase as CAD techniques progress, and it is in this area where knowledge-based computer systems are beginning to have a significant impact[5], but even here it must be remembered that the relevant knowledge has been derived from that possessed by competent designers.

REFERENCES

1. Snowden, C. M., 1988, *Interactive Circuit Analysis Package* (New York: Wiley).
2. Nagel, L., *SPICE 2: A Computer Program to Simulate Semiconductor Circuits*, Mem. #ERL-M250, Electronics Research Lab., University of California, Berkeley.
3. Seitzer, D., Pretzl, G. and Hamdy, N. A., 1983, *Electronic Analog-to-Digital Converters* (New York: Wiley).
4. Lynn, P. A., 1986, *Electronic Signals and Systems* (London: Macmillan Education).
5. Carley, L. R. and Rutenbar, R. A., 1988, *How to Automate Analog IC Designs*, *IEEE Spectrum*, 26–30 (August).

QUESTIONS

1. What are the disadvantages of "bread-boarding" in circuit design?

2. In a CAD netlist for a semiconductor device, what are "default values"?

3. In a user-friendly CAD system, what are test instrument icons?

4. What is a semicustom ASIC, and what are its main advantages over the full custom ASIC?

5. In the context of ADCs, explain the terms "quantization error," "monotonicity" and "resolution."

6. Describe three methods of analog-to-digital conversion.

7. Describe two methods of digital-to-analog conversion.

Appendix I

Decibels and Roll-offs

A.I.1 DECIBEL NOTATION

In chapter 5, the gain of amplifiers as a function of frequency was considered, and it was observed that these changes in gain were considerable. It is therefore useful to employ a convention which will express gain according to a logarithmic scale rather than a linear one, a further advantage being that the gains of a number of cascaded stages may, if expressed logarithmically, be added to give an overall gain. Decibel notation achieves these aims as follows.

Let the power input to an amplifier be P_{in} and the power output be P_{out}. The power gain is thus P_{out}/P_{in}. Expressed logarithmically,

$$\log_{10} (P_{out})/(P_{in}) = \text{power gain in bels}$$

and

$$10 \log_{10} (P_{out})/(P_{in}) = \text{power again in dB} .$$

If the input and output powers are dissipated in resistances of equal value, R, then

$$P_{in} = \frac{V_{in}^2}{R} \quad \text{and} \quad P_{out} = \frac{V_{out}^2}{R} .$$

Consequently,

$$\text{the power gain} = 10 \log_{10} \frac{V_{out}^2}{V_{in}^2} = 20 \log_{10} \frac{V_{out}}{V_{in}} . \tag{I.1}$$

In most cases, the condition of dissipation in equal resistances is not met, but the decibel system is still useful as a way of expressing voltage gains from input to output. However, it does become more credible when used to express the *ratio* of power or voltage gains at different points on the frequency spectrum for a given amplifier with a fixed load resistance.

Let the voltage gain at the mid-frequency be $V_{out(m)}/V_{in}$.

Let the voltage gain at another frequency be $V_{out(f)}/V_{in}$. Then the gain ratio is $V_{out(m)}/V_{out(f)}$ and in dB notation becomes

$$20 \log_{10} \frac{V_{out(m)}}{V_{out(f)}} . \tag{I.2}$$

For example, suppose $V_{out(m)} = 2V_{out(f)}$, then

$$20 \log_{10} 2 = 20 \times 0.3010 \simeq 6 \text{ dB} .$$

Here, the gain at f is said to be -6 dB, or 6 dB *down* compared with the gain at $f_{(m)}$. Table I.1 was constructed in the same way.

A particularly important figure is the so-called *cut-off* or half-power point. Suppose that $V_{out(f)}$ has fallen to 0.707 (that is, $1/\sqrt{2}$) of $V_{out(m)}$. Then:

$$20 \log_{10} \sqrt{2} \simeq 20 \times 0.15 = 3 \text{ dB} .$$

This voltage ratio has been treated in chapter 5 where the high- and low-frequency cut-off points were discussed, but its alternative term, the *half-power* point, arises when the ratio of output *powers* is considered:

TABLE I.1.

Voltage gain ratio	Approximate dB equivalent
2	6
4	12
8	18
16	24
10	20
20	26
40	32
80	38
10	20
100	40
10^4	80
10^8	160

TABLE I.2.

Ratio of power gains	Approximate dB equivalent
2	3
4	6
8	9
16	12
32	15

$$\text{ratio of output powers} = \frac{V^2_{\text{out(m)}}}{R} \frac{R}{V^2_{\text{out(f)}}}$$

$$= \frac{V^2_{\text{out(m)}}}{V^2_{\text{out(f)}}} = 2:1 .$$

Power output ratios are given by $10 \log_{10}(P_1/P_2)$ and table I.2 was constructed using this expression.

A.I.2 THE *RC* CIRCUIT

When a resistance is shunted by a capacitance (as in the case of the hybrid-π equivalent circuit, where $r_{b'e}$ is shunted by C_{in}) then the voltage drop across this combination falls as the frequency rises. If this parallel RC circuit is driven from a current source I, then the rate of roll-off can be determined as follows, using sinusoidal quantities:

$$V = IZ \quad \text{where} \quad Z = \frac{1}{\sqrt{[(1/R^2) + (1/X_C^2)]}}$$

so that

$$V = \frac{IR}{\sqrt{[1 + (R^2)/(X_C^2)]}} = \frac{IR}{\sqrt{(1 + 4\pi^2 f^2 R^2 C^2)}} .$$

At low frequencies where $1 \gg 4\pi^2 f^2 R^2 C^2$,

$$V = IR$$

whilst at high frequencies where $4\pi^2 f^2 R^2 C^2 \gg 1$,

$$V = \frac{I}{2\pi f C} .$$

That is, $V \propto 1/f$, which means that a doubling in frequency (i.e., an increase of *one octave*) will lead to a decrease in voltage by a factor of two (and vice versa). In other words, the *roll-off* is said to be 6 dB/octave.

When $|X_C| = R$, $V = IR/\sqrt{2}$, so that under these conditions, V is 3 dB down and the high-frequency cut-off point is given by

$$f_H = \frac{1}{2\pi RC}. \tag{I.3}$$

The phaseshift is given by

$$\tan \phi = \frac{R}{X_C}. \tag{I.4}$$

At high frequencies, where $X_C \ll R$, $\tan \phi \to \infty$, or $\phi \to 90°$, the voltage *lagging* the current.

A simple RC circuit of this form is called a *single lag*. Its dual, a series RC circuit driven from a voltage source, is shown in figure I.1(*a*). Here,

$$V_{\text{out}} = \frac{EX_C}{\sqrt{(R^2 + X_C^2)}} = \frac{E}{\sqrt{[(R^2/X_C^2) + 1]}}$$

$$= \frac{E}{\sqrt{(4\pi^2 f^2 R^2 C^2 + 1)}}.$$

Again, when $4\pi^2 f^2 R^2 C^2 \gg 1$, $V_{\text{out}} \propto 1/f$ and the slope is 6 dB/octave. The phaseshift will again be given by $\tan^{-1} R/X_C$ and the output voltage will lag behind the input voltage—a fact that is clearly seen from the phasor diagram in figure I.1(*b*).

If a single lag circuit were to be connected in cascade with a second, similar circuit, the roll-off slope would become 12 dB/octave and the phaseshift would approach 180°. For three similar lags, the slope would be 18 dB/octave, the phaseshift would approach 270° and so on.

The *single lead* circuit of figure I.2(*a*) will be recognized as being the type of circuit often used at the input of an AC amplifier. Here, the reactance C is

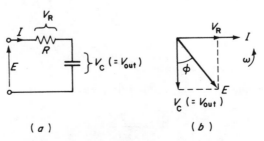

Figure I.1. A lag circuit and its phasor diagram.

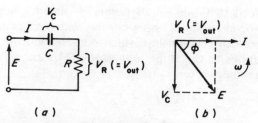

Figure I.2. A lead circuit and its phasor diagram.

predominant at *low* frequencies, and the voltage across R will rise 6 dB/octave, while V_R leads E.

The magnitudes and phaseshifts for up to three lags and leads are plotted in figure 5.20.

A.I.3 THE SINGLE-LAG, SINGLE-LEAD CHARACTERISTIC

An AC amplifier which is so designed that its frequency response is that of a single-lag single-lead network will have a gain–frequency characteristic which may be idealized as in figure I.3. Also shown in the figure are the relevant phaseshifts. This composite diagram is known as a Bode plot.

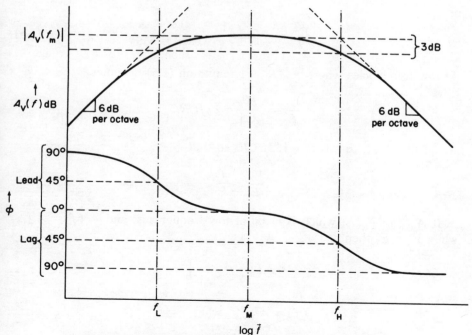

Figure I.3. Bode plot for single-lag, single-lead amplifier.

It will be noticed that the $-3\,\mathrm{dB}$ points, f_L and f_H, appear where the produced $-6\,\mathrm{dB/octave}$ roll-off slopes cut the mid-band gain level. This is a convenient geometrical relationship which comes about as a result of the straight line equations to the slope asymptotes. For example, when $f \gg f_\mathrm{H}$, equation (7.9) may be written

$$|A_\mathrm{v}(f)| = A_0 f_\mathrm{H}/f \,. \tag{I.5}$$

In equation (I.5) A_0 appears, since the equation was derived for a DC amplifier with only a single-lag roll-off. However, for an AC amplifier, it is necessary only to substitute the mid-band gain, $A_\mathrm{v(fm)}$, for A_0, that is,

$$|A_\mathrm{v}(f)| = A_\mathrm{v(fm)} f_\mathrm{H}/f$$

so that when $f = f_\mathrm{H}$, $|A_\mathrm{v}(f)| = A_\mathrm{v(fm)}$ and the phaseshift $(\tan^{-1} f/f_\mathrm{H})$ is $45°$ lagging.

The single-lead circuit which represents the coupling capacitor and input resistance of the amplifier is shown in figure I.2(a). This circuit gives rise to the low-frequency roll-off shown in figure I.3, and the $6\,\mathrm{dB/octave}$ asymptote may be shown to cut the $A_\mathrm{v(fm)}$ level as follows.

From figure I.2(a),

$$V_\mathrm{R} = \frac{ER}{\surd(R^2 + X_\mathrm{C}^2)}$$

or

$$\frac{V_\mathrm{R}}{E} = \frac{1}{\surd[1 + (X_\mathrm{C}^2/R^2)]} \,. \tag{I.6}$$

At low frequencies where $X_\mathrm{C}^2/R^2 \gg 1$, equation (I.6) becomes

$$\frac{V_\mathrm{R}}{E} = \frac{R}{|X_\mathrm{C}|} = 2\pi f C R \,.$$

At f_L, $|X_\mathrm{C}| = R$, that is $f_\mathrm{L} = 1/2\pi C R$, so that

$$\frac{V_\mathrm{R}}{E} = \frac{f}{f_\mathrm{L}} \,.$$

If $A_\mathrm{v}(f) \propto V_\mathrm{R}/E$, which is usual at low frequencies, and $A_\mathrm{v}(f) = A_\mathrm{v(fm)}$ when $V_\mathrm{R} = E$, then

$$A_\mathrm{v}(f) = A_\mathrm{v(fm)} \frac{f}{f_\mathrm{L}}$$

and

$$A_\mathrm{v}(f) = A_\mathrm{v(fm)} \quad \text{when } f = f_\mathrm{L} \,.$$

At f_L, ϕ is of course $45°$ leading.

Appendix II

Four-pole Parameters

Any linear, bilateral circuit having a pair of input and a pair of output terminals can be treated by general circuit theory, under which system it is called a "four-pole network," a "four-terminal network" or a "two-port network." The two input terminals are designated 1 and 1', and the two output terminals 2 and 2' as in figure II.1. The current convention is "positive-in, negative-out" as shown, and the relevant voltages comply with this.

If the currents and voltages are known, as are their relations, it is possible to describe the circuit in six different ways:

E_2 and I_2 in terms of E_1 and I_1	(output in terms of input)	(II.1)
E_1 and I_1 in terms of E_2 and I_2	(input in terms of output: a-parameters)	(II.2)
E_1 and E_2 in terms of I_1 and I_2	(the impedance set: z-parameters)	(II.3)
I_1 and I_2 in terms of E_1 and E_2	(the admittance set: y-parameters)	(II.4)

$$\left. \begin{array}{l} E_1 \text{ and } I_2 \text{ in terms of } E_2 \text{ and } I_1 \\ E_2 \text{ and } I_1 \text{ in terms of } E_1 \text{ and } I_2 \end{array} \right\} \text{ hybrid relationships} . \qquad \begin{array}{l} \text{(II.5)} \\ \text{(II.6)} \end{array}$$

A transistor, when correctly biased, and operating under small-signal conditions, approximates to a four-pole network, and chapter 3 introduced one of the sets of hybrid parameters. These were in the form of two equations:

$$v_{\text{in}} = h_i i_{\text{in}} + h_r v_{\text{out}}$$
$$i_{\text{out}} = h_f i_{\text{in}} + h_o v_{\text{out}} \qquad \text{(II.7)}$$

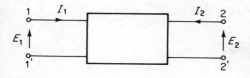

Figure II.1. The general four-pole network.

659

which will be rewritten in general form according to equation (II.5):

$$E_1 = h_{11}I_1 + h_{12}E_2$$
$$I_2 = h_{21}I_1 + h_{22}E_2 \,. \tag{II.8}$$

Although these particular hybrid parameters are common, and easily measurable, they are by no means the only ones in current use. For example, the y-parameters turn out to be very appropriate where high-frequency work is involved, and HF transistors are often characterized by these parameters expressed in complex form. The y-parameter equations, and their general circuit theory equivalents; that is, the admittance set, equation (II.4), are:

$$i_{\text{in}} = y_i v_{\text{in}} + y_r v_{\text{out}}$$
$$i_{\text{out}} = y_f v_{\text{in}} + y_o v_{\text{out}} \tag{II.9}$$

or

$$I_1 = y_{11}E_1 + y_{12}E_2$$
$$I_2 = y_{21}E_1 + y_{22}E_2 \,. \tag{II.10}$$

Since all of the six sets of equations describe the same four-terminal network, it must be possible to calculate any one set from any other. To do this, the sets of equations are written in matrix form and the interrelationships appear by applying the standard rules of matrices. For example, equations (II.7) and (II.9) become:

$$\begin{bmatrix} v_{\text{in}} \\ i_{\text{out}} \end{bmatrix} = \begin{bmatrix} h_i & h_r \\ h_f & h_o \end{bmatrix} \begin{bmatrix} i_{\text{in}} \\ v_{\text{out}} \end{bmatrix} \tag{II.11}$$

having a determinant $\triangle = h_i h_o - h_r h_f$ and

$$\begin{bmatrix} i_{\text{in}} \\ i_{\text{out}} \end{bmatrix} = \begin{bmatrix} y_i & y_r \\ y_f & y_o \end{bmatrix} \begin{bmatrix} v_{\text{in}} \\ v_{\text{out}} \end{bmatrix} \tag{II.2}$$

having a determinant

$$\triangle = y_i y_o - y_r y_f$$

and the relationships are found to be:

$$h_i = \frac{1}{y_i} \qquad y_i = \frac{1}{h_i}$$

$$h_r = \frac{-y_r}{y_i} \qquad y_r = \frac{-h_r}{h_i}$$

$$h_f = \frac{y_f}{y_i} \qquad y_f = \frac{h_f}{h_i}$$

$$h_o = \frac{\triangle}{y_i} \qquad y_o = \frac{\triangle}{h_i} \, .$$

It would seem pointless to give complete tables of such relationships, for data on transistors are easily available and are normally quoted in terms of the parameters most suited to the transistor in question. It does, however, happen that multi-stage circuits can be analyzed using matrices throughout, and this process can be greatly simplified by choosing the correct set of parameters. For example, if a cascade connection is made—and this is the most common connection—then it transpires that a *composite* matrix of parameters can be obtained by simply multiplying together the parameter matrices for the individual stages *providing* the parameters involved are the *a*-parameters relating to equation (II.2). If such an analysis is to be made, then it is clearly worth while converting to *a*-parameters. However, the simpler transistor circuits rarely justify this approach and for this reason it will not be treated here. Unfortunately, the *h*-parameter matrices can only be simply multiplied in this way when the inputs of four-terminal networks are all in series and the outputs are all in parallel.

Appendix III

Noise Bandwidth and Noise Factor Minimization

A.III.1 NOISE BANDWIDTH

If the total noise voltage at the output of an amplifier is applied to a load resistor, R_L, then the power dissipated will be

$$\frac{\overline{v_{no}^2}}{R_L}$$

where $\overline{v_{no}^2}$ is the mean square of this total noise voltage.

The value of the mean square noise at the output of the amplifier may be obtained if the frequency functions of both the noise referred to the input and also the voltage gain are known. Then,

$$\overline{v_{no}^2} = \int_0^\infty \overline{v_{ni}^2}(f)|A_v(f)|^2 \, df \, . \tag{III.1}$$

In the special case where v_{ni} is independent of frequency, that is, when the signal source is purely resistive, and the $1/f$ and HF ranges are excluded, this integral may be rewritten using the spot value of $\overline{v_{ni}^2}$:

$$\overline{v_{no}^2} = \overline{v_{ni(spot)}^2} \int_0^\infty |A_v(f)|^2 \, df \, . \tag{III.2}$$

The factor under the integral sign may be redefined as follows:

$$\int_0^\infty |A_v(f)|^2 \, df = A_v^2 \Delta f \tag{III.3}$$

where A_v is the maximum voltage gain and Δf is the *noise bandwidth*.

The mean square noise voltage at the output under these circumstances may now be expressed by combining equations (III.2) and (III.3):

$$\overline{v_{no}^2} = A_v^2 \overline{v_{ni(spot)}^2} \Delta f = A_v^2 \overline{v_{ni}^2} \tag{III.4}$$

and the RMS noise at the output will be

$$(\overline{v_{no}^2})^{1/2} = A_v (\overline{v_{ni}^2})^{1/2}$$

which is equation (6.10).

The noise bandwidth of a single-lag, single-lead amplifier quoted in equation (6.2) is derived from equation (III.3) as follows:

$$\Delta f = \frac{1}{A_v^2} \int_0^\infty |A_v(f)|^2 \, df. \tag{III.5}$$

In the case of a simple single-lag, single-lead amplifier,

$$A_v(f) = A_v \frac{1}{1 + j(f_L/f)} \frac{1}{1 + j(f/f_H)}$$

where f_L and f_H are the -3 dB cut-off frequencies.
From this,

$$|A_v(f)|^2 = A_v^2 \frac{1}{1 + (f_L/f)^2} \frac{1}{1 + (f/f_H)^2}$$

$$= A_v^2 \frac{(f/f_L)^2}{1 + (f/f_L)^2} \frac{1}{1 + (f/f_H)^2}.$$

Putting this into equation (III.5) and performing the integration gives, according to a table provided by Cherry and Hooper[1]:

$$\Delta f = \frac{\pi}{2} \frac{f_H^2}{f_L + f_H}.$$

This may be converted to a useful approximation as follows:

$$\frac{\pi}{2} \frac{f_H^2}{f_L + f_H} \frac{f_H - f_L}{f_H - f_L} = \frac{\pi}{2} \frac{f_H^2(f_H - f_L)}{f_H^2 - f_L^2}.$$

For a bandwidth of a decade or more, $f_H^2 \gg f_L^2$, so that this becomes

$$\Delta f \simeq \tfrac{1}{2}\pi(f_H - f_L)$$

which is equation (6.2).

A.III.2   NOISE FACTOR MINIMIZATION

For a bipolar transistor, the noise factor is given by equation (6.22) as:

$$\mathrm{NF} \simeq 1 + \frac{kT}{2qI_C R_G} + \frac{qI_C R_G}{2kTh_{FE}} .$$

To determine $\mathrm{NF}_{(min)}$, this may be differentiated with respect to I_C, keeping in mind that h_{FE} is a function of I_C, and equating to zero:

$$\frac{d(\mathrm{NF})}{dI_C} \simeq \frac{-kT}{2qR_G I_{C(opt)}^2} + qR_G\left[2kTh_{FE}\left(1 - \frac{I_{C(opt)}}{h_{FE}}\frac{dh_{FE}}{dI_C}\right)\right]^{-1} = 0$$

giving

$$I_{C(opt)} \simeq \frac{kT}{qR_G}\left[h_{FE}\left(1 - \frac{I_{C(opt)}}{h_{FE}}\frac{dh_{FE}}{dI_C}\right)^{-1}\right]^{1/2} . \qquad (\mathrm{III}.6)$$

The contribution of the term in parentheses in equation (III.6) depends on the rate of change of h_{FE} with I_C, which can be large in the case of some transistors, such as alloy-junction types. Thornton et al[2] derive this slope from the following expressions:

$$I_C = I_{ES} \exp\left(qV_{BE}|kT\right)$$

and

$$I_B = I_{BS} \exp\left(qV_{BE}|nkT\right) .$$

Here, n takes account of effects which become significant at low levels such as the base current due to recombination at localized traps, particularly near the surface.

$$\frac{I_C}{I_B} = h_{FE} = \frac{I_{ES}}{I_{BS}}\exp\left[\frac{qV_{BE}}{kT}\left(1 - \frac{1}{n}\right)\right] = \frac{I_{ES}^{1/n}}{I_{BS}}I_C^{1-(1/n)}$$

so that

$$\frac{dh_{FE}}{dI_C} = \frac{I_{ES}^{1/n}}{I_{BS}}\left(1 - \frac{1}{n}\right)I_C^{1/n} = \frac{n-1}{nI_B} .$$

Hence,

$$\frac{I_C}{h_{FE}}\frac{dh_{FE}}{dI_C} = \frac{I_C}{h_{FE}}\frac{n-1}{n} = \frac{n-1}{n} . \qquad (\mathrm{III}.7)$$

From plots such as those of figure 6.6, an approximate numerical value for this appears to be about 0.15, so that equation (III.6) becomes:

$$I_{C(\text{opt})} \simeq \frac{kT}{qR_G} \sqrt{\left(\frac{h_{FE}}{0.85}\right)}$$

$$\simeq \frac{28}{R_G} \sqrt{h_{FE}} \text{ mA} \quad \text{at 25°C if } R_G \text{ is in ohms}$$

which is equation (6.24*b*).

Answers to Problems

CHAPTER 1

1.1. 16 V
1.2. 40
1.3. npn; p-channel
1.4. (b) and (c) are impossible
1.5. -224; -5 V
1.6. $+0.179$ mA, -1 V

CHAPTER 2

2.1. 0.57 V; 2.05 W
2.2. 0.026 Ω; 260 Ω
2.3. 100
2.4. When $I_B = 0.1$ mA
2.5. 10.1 μA
2.6. 1.54 A
2.7. $h_{FB} = -0.98$, $h_{FE} = +49$, $h_{FC} = -50$

CHAPTER 3

3.1. $h_{ie} = 150$; $h_{oe} = 30$ μs; $h_{ie} = 850$ Ω; $h_{re} = 1.5 \times 10^{-4} = \triangle_e = 3 \times 10^{-3}$
3.2. -0.993
3.3. 195 mA/V; 770 Ω
3.4. 365 kHz
3.5. 250 Ω
3.6. 90 V

CHAPTER 4

4.2. 2.53 V
4.3. $R_C = 6$ kΩ; $R_B = 540$ kΩ
4.4. -140
4.5. 1.79 kΩ
4.6. $R_{in} = 103.4$ kΩ; $A_v = 0.987$
4.7. $R_{out} \simeq 67$ kΩ
4.8. $K = 0.048$
4.9. $A_V = 0.9912$ (or 226/228)

CHAPTER 5

5.1. $R_{in} \simeq 3.6$ kΩ
5.2. C_c dominates at 103 Hz
5.3. $A_v \simeq -351$; $f_H = 123$ kHz
5.4. (a) -4.45; (b) -21.6; (c) 318.3 Ω
5.5. $0.5R_E = 2.15$ kΩ; $R_C = 5.7$ kΩ; $A_v = -2.65$; $f_L = 40.2$ Hz
5.6. (i) 1 mA; (ii) 2 mA; (iii) 7.8 V; (iv) 4.6 kΩ; (v) 702 kΩ; (vi) -140; (vii) 101.4 Hz defined by C_E
5.7. (i) 2 mA; (ii) 9.6 kΩ; $v_{out} = 2.67$ V
5.8. (i) 0.112 V; (ii) 0.648 mV
5.9. (i) 5 kΩ; (ii) Not strictly necessary except for balance; (iii) 1 mA; (iv) Zero; (v) -4.7; (vi) No; (vii) -344
5.10. $+40$; 5.3 mA
5.11. R for current mirror $= 29.3$ kΩ; $R_{C_1} = R_{C_2} = 15$ kΩ; $V_{out\ (DC)} = 7.5$ V
5.12. 16.06 mV
5.13. $A_V = -210.6$; $f_L = 68.3$ Hz

CHAPTER 6

6.1. NF $= 1.71$; No, $R_{G(opt)} = 8$ kΩ
6.2. NF $= 2.5$ dB; NF$_{(min)} = 2.06$ dB; $R_{G(opt)} = 5$ kΩ; Less by 1.9 dB
6.3. NF $= 3.52$; 14.2%
6.4. NF $= 1.13$; 0.17 μV RMS
6.5. NF $= 0.34$ dB
6.6. Pink noise $= 0.127$ μV RMS; White noise $= 0.25$ μV RMS; Total noise $= 0.28$ μV RMS; Johnson noise $= 0.23$ μV RMS

6.7. NF $= 3.5$; NF$_{(min)} = 3.47$
6.8. 63.36 μV
6.9. 72 dB

CHAPTER 7

7.1. Simple current mirror 3.85%; Wilson current mirror 0.077%
7.2. 104.6 Ω
7.3. 20
7.4. $R_2 = 600 \, \Omega$; $R_1 = 1.2 \, k\Omega$
7.5. 20.4 kHz
7.6. 75 μA
7.7. 16.1 mV
7.8. R_d is necessary, when $\Delta V_{out} \leq 7.52$ mV
7.9. Up to 10 mV
7.10. Noninverting, $R_{F_1} = 9 \, k\Omega$, $R_{F_2} = 1 \, k\Omega$ and $R_d = 19.1 \, \Omega$
7.11. Error up to 1.06% so R_d not needed.
7.12. $\approx 0.1\%$
7.13. $R_F = 20 \, k\Omega$; Summing resistor to $+15$ V line $= 250 \, k\Omega$
7.14. 90° leading; 1.6 V peak
7.15. 0.68 s
7.16. 47 $k\Omega$
7.17. (a) 20 mV; (b) 4 V; (c) 2.04 V; CM component is 2% of the total output
7.18. 3.83%
7.19. Yes. (3.15% to 0.894%)

CHAPTER 8

8.1. If $C = 0.01 \, \mu$F, then $R = 15.9 \, k\Omega$
8.2. 2.6 V

CHAPTER 9

9.1. $V_{CC} = 24$ V; $I_{CM} \geq 0.5$ A; $P_{tot} = 3$ W
9.2. $P_L = 7.5$ W in theory, about 4.5 W in practice; $\theta_{sa} = 1.33$°C/W
9.3. $P_L = 3.7$ W in theory, about 2.3 W in practice; $P_{tot} = 750$ mW
9.4. 18.77 W reduction

9.5. 42 W; $(1.74 + 1.74):1$; $V_{CEM} = 60$ V
9.6. $1:1.31$; $P_{tot} \geq 2.5$ W; $V_{CEM} = 48$ V; $I_{CM} \simeq 1$ A
9.7. 44.7 V $(\simeq 50$ V$)$
9.8. $P_{diss(m)} = 2.38$ W; Yes, $T_j = 62.5°C$ max
9.9. $P_{tot} \geq 5.7$ W; $V_{CEM} \geq 30$ V; $I_{CM} \geq 3.75$ A; 28.2 W
9.10. $V_{CC}^+ = V_{CC}^- = 70$ V each; $P_{tot} \geq 20.24$ W; $I_{CM} \geq 2.83$ A; $R_1:R_2 = 1:2$
9.11. Type "B" must be chosen
9.12. Type "C" must be chosen with a ± 40 V supply

CHAPTER 10

10.1. (i) 10 W; (ii) 0.53 W
10.2. 1 μS; 5 μS
10.3. $R_c = 2.4$ kΩ; $R_B = 120$ kΩ; $C = 0.06$ μF
10.4. 94 kΩ
10.5. If $C_t = 0.1$ μF then $R_A = R_B = 15$ kΩ
10.6. (i) Sawtooth 50 ms long and 10 V high; (ii) Insert emitter resistor $R_{E_2} = 360$ Ω; (iii) Negative
10.7. $N_P = 139$ turns; $I_{C(mag)} = 0.26$ A
$V_{CEM} \geqslant 37.5$ V; $I_{CM} \geq 0.6$ A

CHAPTER 11

11.1. (a) 16 kΩ; (b) 1.76 kΩ; (c) -7.46; (d) 30.38 pF; (e) 167 kHz; (f) 16.5 Hz
11.2. $I_Q = 0.54$ mA; $V_B = 3$ V; $f_H \simeq 500$ kHz; R_1 may be 2.2 MΩ whence $R_2 = 330$ kΩ
11.3. 13% low
11.4. $R_S = 4.9$ kΩ; $V_B = 2.32$ V; $R_1 = 532$ kΩ; $R_2 = 123$ kΩ
11.5. $f_H \simeq 267$ kHz
11.6. $R_S \simeq 4.4$ kΩ; $V_B = 3.53$ V; If $R_1 = 190$ kΩ then $R_2 = 46$ kΩ
11.7. $R_S = 3$ kΩ; $R_D = 9.5$ kΩ; $f_H \simeq 360$ kHz
11.8. $R_F = 3.3$ kΩ; $R_D = 13.9$ kΩ
11.9. $R_{F_1} \simeq 710$ Ω; $R_{F_2} \simeq 4.3$ kΩ; $C_c \simeq 0.01$ μF
11.10. $R_{F_1} \simeq 5.3$ kΩ; $R_{F_2} \simeq 1.8$ kΩ
11.11. $V_S^- = -5$ V; $V_S^+ = 1$ V
11.12. $R_S = 1.1$ kΩ
11.13. $r_{DS} = 300$ Ω; 75%; negative

11.14. 0.001 μF
11.15. $R_S = 1.1\,\text{k}\Omega$; $A_v = -6.1$; $f_H = 463\,\text{kHz}$

CHAPTER 12

12.1. 133.1 Hz; 0.19 mA
12.2. $R_1 = 2.4\,\text{k}\Omega$; $R_2 = 3.6\,\text{k}\Omega$
12.3. Let $C = 4.7\,\mu\text{F}$ whence $R = 430\,\text{k}\Omega$
12.4. 100 Hz to 50 Hz
12.5. $R_E \simeq 348\,\Omega$
12.6. $R = 1\,\text{M}\Omega$
12.7. $R = 1\,\text{M}\Omega$
12.8. $R_E = 614\,\Omega$
12.9. $C = 0.0025\,\mu\text{F}$ plus typical UJT load resistors

CHAPTER 13

13.1. $R_F = 200\,\text{k}\Omega$; $R \simeq 52\,\text{k}\Omega$
13.2. 9.4×10^{11} ($0.9\,\mu\text{m}$, 10^3, 1) cm $\sqrt{\text{Hz}}/\text{W}$
13.3. Noninverting op. amp. with $R_{F_1} = 72\,\text{k}\Omega$ and $R_{F_2} = 5\,\text{k}\Omega$ potentiometer
13.4. (a) 6753 Hz; Linear sawtooth 0.83 W/m^2
13.5. Inverting op. amp. with $R_G = 1\,\text{k}\Omega$, $R_F = 50\,\text{k}\Omega$ and $R_d = 980\,\Omega$
13.6. Inverting op. amp. with $R_F = 40\,\text{k}\Omega$
13.7. Inverting op. amp. with $R_F = 10\,\text{k}\Omega$; No, ± supplies are needed
13.8. 2 lux and 3 lux

CHAPTER 14

14.1. $R = 7.7\,\Omega$; $P_R = 4.16\,\text{W}$; $P_Z = 12.5\,\text{W}$; $S_V = 0.056$
14.2. $R = 5.6\,\Omega$; $P_{\text{diss(m)}} = 22.5\,\text{W}$
14.3. $R = 74.1\,\Omega$
14.4. (b) $P_{\text{tot}} \geq 15\,\text{W}$; $V_{\text{CEM}} > 30\,\text{V}$; $I_{\text{CM}} > 1.02\,\text{A}$; (c) $R_x = 67.5\,\Omega$; (d) $P_{\text{RP}} = 3.3\,\text{W}$; Yes; (e) No

Index

Printed in Singapore by Chong Moh Offset Printing Pte Ltd